W0235235

Open Problems in Mathematics
and Computational Science

Çetin Kaya Koç

Editor

Open Problems
in Mathematics
and Computational Science

 Springer

Editor
Çetin Kaya Koç
Department of Computer Science
University of California, Santa Barbara
Santa Barbara, CA
USA

ISBN 978-3-319-10682-3   ISBN 978-3-319-10683-0 (eBook)
DOI 10.1007/978-3-319-10683-0
Springer Cham Heidelberg New York Dordrecht London

Library of Congress Control Number: 2014957413

Printed on acid-free paper

Springer is part of Springer Science+Business Media (www.springer.com)

Preface

A selected group of invited speakers and more than 150 students and researchers attended a special conference on September 18–20, 2013, in "Said Halim Pasha Palace" in Istanbul. There had never been a conference of this kind in Turkey, where "open" or "unsolved" problems are discussed, and even in the world there have only been a few examples.

In principle, mathematicians, scientists, and engineers attend conferences to speak about problems they have solved and to "impress" and inform the academic community about their methods and the final solution. It is not generally expected that a researcher would take the stand in a conference to talk about a problem she or he could not (yet) solve. However, all scientific processes start with hypotheses whose ramifications we do not know or problems whose solutions are not clear yet. Either for personal reasons or in accordance with the expectations of scientific conferences and their attendees, researchers tend to push the open/unsolved problems to the back burner and talk about what they have solved, understood, or proved. Still, once in a while (perhaps every 5–10 years), some researchers come together to discuss problems they have not solved yet or problems whose solutions seem rather challenging. Since the 1970s, there have been 7 such conferences.

Therefore, I am very happy that we were able to organize this *Open Problems in Mathematical and Computational Sciences Conference* with support from the Scientific and Technological Research Council (TÜBİTAK) of Turkey.

A large number of young researchers, MSc, and PhD candidates from Turkey, as well as several from neighboring countries, attended the conference. The invited scientists of the conference are among the most prolific mathematical and computational scientists in the world. They come from various countries, demonstrating that science and engineering are culturally very diverse now. The list of countries and number of scientists from each country were a good reminder of this fact: Belgium (2), Brazil (1), Canada (2), China (2), France (3), Germany (2), Japan (1), Norway (1), Romania (1), Turkey (3), and the USA (2).

The Open Problems Conference was held in Said Halim Pasha Palace, one of the most beautiful seaside palaces in Istanbul, whose history goes back at least 150 years and as far as Egypt!

Said Halim Pasha was the son of Mehmet Abdülhalim Pasha who was one of the four sons of Mehmet Ali Pasha from Kavala, the second largest city in Northern Greece. Mehmet Ali Pasha (Muhammad Ali of Egypt) was an Ottoman commander of Albanian origin and is regarded as the founder of modern Egypt because of the dramatic reforms in the military, economic, and cultural spheres he instituted. Said Halim Pasha was born in Cairo in the year 1863 and completed his education in private lessons in Cairo, where he learned Arabic, Persian, English, and French. He studied politics for 5 years in Switzerland. The palace had become the property of Prince Abdülhalim Pasha in the year 1876 and was reconstructed to its current appearance by the travelling architect, Petraki Adamandidis of the Dardanelles. The property was inherited by the nine children of the Abdülhalim Pasha after his death in 1890. After going through several owners, the Said Halim Pasha Palace was restored following a fire in 1995 under the name "Prime Ministry Official Guest House."

Several peoples' names need to be mentioned with gratitude, they made both the Open Problems Conference and the Open Problems Book possible.

First of all, I sincerely thank Ronan Nugent for his valuable advice and the Editorial Office of Springer for their help in getting the book published.

On behalf of the invited speakers, I am also sincerely grateful to TÜBİTAK for agreeing with us about the vision of the Open Problems Conference and their subsequent work that produced this book and for providing the financial support. I would also like to thank to Şükran Külekci, İsa Sertkaya, Birnur Ocaklı, Mehmet Sabır Kiraz, and Osmanbey Uzunkol for working around the clock several days before, during, and after the conference.

Santa Barbara, CA, USA   Çetin Kaya Koç

Contents

About Open Problems

Çetin Kaya Koç

Abstract A small group of computer scientists and mathematicians from industry and academia convened in a historical home ("Said Halim Pasha Palace") overlooking the Bosphorus Straits to discuss several difficult problems they and others in similar fields are tackling. The motivation of the *Open Problems in Mathematical and Computational Sciences Conference* was to enable and encourage the academic community, particularly young researchers and Ph.D. candidates, to hear about unsolved, open problems in mathematical and computation sciences, directly from the scientists who are rigorously investigating them.

1 The Conference

In general, scientists go to conferences to present discoveries that are already made, to explain results or to expose and excite the community about connections within various theories or structures, and to share their insights and proofs. Conferences are places where we get to see and hear about solutions, ask questions about them, and hope to understand them better in this process. Rarely is there an opportunity to talk about problems that have not been solved yet or solutions which are not yet satisfactory, except during the lunches, coffee breaks, or at other quiet times.

In many instances, scientists working on problems whose solutions are difficult to obtain will state that asking the right question is the real challenge. It is imperative to stop and think once in a while in order to understand the background of the tools and the mechanisms needed for tackling the problems we are working on. Conferences that deal with open problems are rare, but they are useful avenues for such objectives. Almost all conferences are for presenting the solutions to certain classes of problems whose origins we may not have any idea about.

Ç.K. Koç (✉)
University of California Santa Barbara, Santa Barbara, CA 93106, USA

Mathematical and Computational Sciences Labs, TÜBİTAK BİLGEM, Gebze, Kocaeli, Turkey
e-mail: koc@cs.ucsb.edu

© Springer International Publishing Switzerland 2014
Ç.K. Koç (ed.), *Open Problems in Mathematics and Computational Science*,
DOI 10.1007/978-3-319-10683-0_1

In a world replete with information, what matters most is sometimes not the answers but rather the context, the origin and the body of questions for which answers are sought or obtained.

This conference was planned with these ideas in mind. One purpose of the *Open Problems in Mathematical and Computational Sciences Conference* is to encourage, motivate, and excite the mathematical and computational sciences community to discuss open problems. We would like to hear them formulate the questions and present processes which will be helpful in the quest for answers.

Of course, we all know about certain open problems or conjectures in mathematics such as the Goldbach conjecture or the twin primes conjecture or the Riemann hypothesis. Some well-known problems have been resolved during the last 20 years, three excellent examples being Fermat's last theorem by Andrew Weil in 1995, the Poincaré conjecture by Grigori Perelman in 2003, and the prime gap problem by Yitang Zhang (and later by the Polymath Project participants) in 2013. The list of difficult problems in mathematics is pretty long, and solutions come in decades or even centuries. And when they come, they are deservedly celebrated, and the international media and thus the public pay attention; stories are made and impressions are created. Furthermore, mathematics institutes around the world, for example, the Clay Institute, publish problem lists and offer prizes which further publicize the phenomena.

However, we are limiting our attention to computational problems in this conference; there is also a long list of unsolved problems in computer science, such as:

- $P = NP$ problem
- Existence of one-way functions
- Is the graph isomorphism problem in P?
- Is factoring in P?
- Is primality testing in P?
- What is the fastest algorithm for the multiplication of integers?
- What is the fastest algorithm for matrix multiplication?

The list is not complete, and our intention is not to complete the list, but to bring the best minds to describe, elucidate, and explain some of these open problems in the mathematical and computational sciences, particularly the problems they themselves are interested in or working on or for which they have formulated partial or near-complete solutions. We want them to tell us how they approach such problems and what are the mechanisms and tools they are using and share with us and excite us with the creative energy they are applying to such problems.

A perfect example from the above list was the question "Is Primality Testing in P?" This was affirmatively answered by Manindra Agrawal, Neeraj Kayal, and Nitin Saxena of the Indian Institute of Technology Kanpur, by giving the first deterministic polynomial time algorithm for primality testing. The implications of this development are indeed great for cryptography, coding, and finite fields, where primality plays a central role.

To summarize, one of the underlying purposes of our 3-day conference was to encourage young researchers, particularly Ph.D. candidates, to learn about exciting, interesting, and important (yet) unsolved problems in mathematical and computation sciences, directly from the researchers who are thinking about them. I believe the informal atmosphere of the conference allowed them to listen to the seminars, ask questions, interact, and discuss possible answers or pose new questions to the invited speakers.

We believe such a close interactive environment served as a catalyzing event and hopefully will synchronize local research communities with the best, most challenging, and perhaps most useful problems the world's best minds are working on. Hopefully, in several years, perhaps even as early as the next Open Problems Conference, a few of these challenging problems will find their partial or complete solutions.

2 The Participants

The following people attended the conference as invited speakers:

- Paulo Barreto, Universidade de Sao Paulo
- Claude Carlet, Université Paris 8
- Guanrong Chen, City University of Hong Kong
- Ömer Eğecioğlu, University of California, Santa Barbara
- Gerhard Frey, Göttingen Academy of Sciences
- Tor Helleseth, University of Bergen
- Antoine Joux, Université de Versailles Saint-Quentin-en-Yvelines
- Andrew Klapper, University of Kentucky
- Alfred Menezes, University of Waterloo
- David Naccache, Université Paris II
- Koji Nakano, Hiroshima University
- Ferruh Özbudak, Middle East Technical University
- Daniel Panario, Carleton University
- Bart Preneel, KU Leuven
- Gheorghe Păun, Romanian Academy
- Jean-Jacques Quisquater, Université catholique de Louvain
- Henning Stichtenoth, Sabancı University
- Murat Tekalp, Koç University
- Han Vinck, University of Duisburg-Essen

We thank our speakers for taking time to come to Istanbul to talk about problems that excite them and to share them with us. There were more than 150 participants, most of whom were from Turkey, as expected; however, about 10 % of the participants were from other European countries, including Bulgaria, Denmark, France, and Romania.

3   The Book

As we were planning the conference, we also developed a plan to publish a book arising from the presentations.

This book contains *selected and revised* papers from the conference. We gave a window of about 6 months to the speakers to create the chapters in this book, revising and expanding their work by adding an introduction section and an annotated bibliography. The introduction section of each chapter is intended to provide the background of the topic of the chapter, assuming the reader is a first-year graduate student who has the general knowledge of electrical engineering, computer science, programming, and computational mathematics via his/her undergraduate education and has just started reading books and papers in the area of the chapter. Therefore, the chapters attempt to give all basic definitions, introduce the context, and summarize algorithms, theorems, and proofs. On the other hand, the bibliography aims to introduce the most important references to follow up, giving a short description of these papers and books, and their importance to the field. I hope you will find these chapters to your liking.

The Past, Evolving Present, and Future of the Discrete Logarithm

Antoine Joux, Andrew Odlyzko, and Cécile Pierrot

Abstract The first practical public key cryptosystem ever published, the Diffie–Hellman key exchange algorithm, relies for its security on the assumption that discrete logarithms are hard to compute. This intractability hypothesis is also the foundation for the security of a large variety of other public key systems and protocols.

Since the introduction of the Diffie–Hellman key exchange more than three decades ago, there have been substantial algorithmic advances in the computation of discrete logarithms. However, in general the discrete logarithm problem is still considered to be hard. In particular, this is the case for the multiplicative groups of finite fields with medium to large characteristic and for the additive group of a general elliptic curve.

This chapter presents a survey of the state of the art concerning discrete logarithms and their computation.

1 Introduction

1.1 The Discrete Logarithm Problem

Many popular public key cryptosystems are based on discrete exponentiation. If G is a multiplicative group, such as the group of invertible elements in a finite

A. Joux
CryptoExperts, Paris, France

Chaire de Cryptologie de la Fondation de l'UPMC, Paris, France

Sorbonne Universités, LIP6, UMR 7606, UPMC Univ Paris 06, Paris, France
e-mail: Antoine.Joux@m4x.org

A. Odlyzko
School of Mathematics, University of Minnesota, Minneapolis, MN 55455, USA
e-mail: odlyzko@umn.edu

C. Pierrot (✉)
DGA/CNRS, Sorbonne Universités, LIP6, UMR 7606, UPMC Univ Paris 06, Paris, France
e-mail: Cecile.Pierrot@lip6.fr

© Springer International Publishing Switzerland 2014
Ç.K. Koç (ed.), *Open Problems in Mathematics and Computational Science*,
DOI 10.1007/978-3-319-10683-0_2

field or the group of points on an elliptic curve, and g is an element of G, then g^x is the discrete exponentiation of base g to the power x. This operation shares basic properties with ordinary exponentiation, for example, $g^{x+y} = g^x \cdot g^y$. The inverse operation is, given h in G, to determine a value of x, if it exists, such that $h = g^x$. Such a number x is called a discrete logarithm of h to the base g, since it shares many properties with the ordinary logarithm. If, in addition, we require some normalization of x to limit the possible answers to single valid value, we can then speak of *the* discrete logarithm of h. Indeed, without such a normalization, x is not unique and is only determined modulo the order of the element g.

Assume for simplicity that G is a cyclic group generated by g and that the notation $\log_g(h)$ denotes a value such that $h = g^{\log_g(h)}$. Then, as with ordinary logarithms, there is a link between multiplication of elements and addition of logarithms. More precisely, we have:

$$\log_g(h \cdot j) \equiv \log_g(h) + \log_g(j) \mod |G|.$$

We say that we solve the discrete logarithm problem (DLP) in G if given any element g^x in G, we are able to recover x. To normalize the result, we usually ask for x to be taken in the range $0 \leqslant x < |G|$. In many applications, in particular in cryptography, it is sufficient to be able to solve this problem in a substantial fraction of cases. (The usual theoretical standard is that this fraction should be at least the inverse of a polynomial in the logarithm of the size of the group.)

The main interest of discrete logarithm for cryptography is that, in general, this problem is considered to be hard. The aim of this chapter is to provide state-of-the-art information about the DLP in groups that are used for cryptographic purposes. It gives pointers to the latest results and presents observations about the current status and likely future of the DLP.

1.2   *Applications of Discrete Logarithms*

In some sense, the discrete logarithm has a long history in number theory. It is just an explicit way to state that an arbitrary cyclic group containing N elements is isomorphic to $(\mathbb{Z}_N, +)$. Still, before the invention of the Diffie–Hellman protocol, the problem of efficiently computing discrete logarithms attracted little attention. Perhaps the most common application was in the form of Zech's logarithm, as a way to precompute tables allowing faster execution of arithmetic in small finite fields.

The role of the DLP in cryptography predates Diffie–Hellman. Indeed, the security of secret-key cryptosystem involving linear feedback shift registers (LFSR) is closely related to the computation of discrete logarithms in finite fields of characteristic two. More precisely, locating the position where a given subsequence appears in the output of an LFSR is, in fact, a DLP in the finite field defined by the feedback polynomial.[1]

The main impetus to intensive study of discrete logarithms came from the invention of the Diffie–Hellman method in 1976 [DH76]. Much later, the introduction of pairing in cryptography in 2000 (journal versions [Jou04, BF03]) increased the level of attention on some atypical finite fields, with composite extension degrees and/or medium-sized characteristic.

1.2.1 Diffie–Hellman Key Exchange

Let us recall the first practical public key technique to be published, which is still widely used, the Diffie–Hellman key exchange algorithm. The basic approach is as follows. If Alice and Bob wish to create a common secret key, they first agree, on a cyclic group G and a generator g of this group.[2] Then, Alice chooses a random integer a, computes g^a, and sends it to Bob over a public channel, while Bob chooses a random integer b and sends g^b to Alice. Now Alice and Bob can both compute a common value, which then serves as their shared secret:

$$(g^b)^a = g^{a \cdot b} = (g^a)^b.$$

The security of this system depends on the assumption that an eavesdropper who overhears the exchange, and thus knows g, g^a, and g^b, will not be able to compute the shared secret. In particular, this hypothesis assumes that the eavesdropper is unable to solve the DLP in G. Indeed, if the DLP for this group is solvable, he can compute either a or b and recover the shared secret $g^{a \cdot b}$. However, it is not known whether the problem of computing g^{ab} given g, g^a, and g^b, which is known as the computational Diffie–Hellman problem (CDH), is equivalent to the computation of discrete logarithms. Moreover, to prove the security of many cryptographic protocols, it is often necessary to consider the associated decision problem: given g, g^a, g^b, and h, decide whether h is the correct value of g^{ab} or not. This latest problem is called the decision Diffie–Hellman problem (DDH).

There are also many generalized computational and decision problems somehow related to the DLP that have been introduced as possible foundations for various cryptosystems. Since it is not easy to compare all these assumptions, in an attempt to simplify the situation, Boneh et al. [BBG05] have proposed the *uber-assumption* which subsumes all these variations and can be proven secure in the generic group model (see Sect. 2.5).

However, the DLP itself remains fundamental. Indeed from a mathematical viewpoint, it is a much more natural question than the other related problems, and

---

[1]Assuming that it is irreducible, which is usually the case.

[2]The group G and generator g can be the same for many users and can be part of a public standard. However, that can lead to a reduction in security of the system.

in practice, none of these other problems has ever been broken independently of the DLP. Since the introduction of the Diffie–Hellman key exchange, this concern has motivated a constant flow of research on the computation of discrete logarithms.

Another extremely important assumption in the above description is that the eavesdropper is passive and only listens to the traffic between Alice and Bob. If the attacker becomes active, then the security may be totally lost, for example, if he can mount a *man-in-the-middle* attack where he impersonates Bob when speaking to Alice and conversely. This allows him to listen to the decrypted traffic. To avoid detection, the attacker forwards all messages to their intended recipient after reencrypting with the key that this recipient has shared with him during the initial phase. One essential issue when devising cryptosystems based on discrete logarithms is to include safety measures preventing such active attacks.

1.2.2 Other Protocols

After the invention of the RSA cryptosystems, it was discovered by El Gamal [Gam85] that the DLP can be used not only for the Diffie–Hellman key exchange, but also for encryption and signature. Later Schnorr [Sch89] gave an identification protocol based on a zero-knowledge proof of knowledge of a discrete logarithm, which can be turned into Schnorr's signature scheme using the Fiat–Shamir transform [FS86].

There are many more cryptosystems based on the DLP which will not be covered here. However, let us mention the Paillier encryption [Pai99]. This system works in the group $\mathbb{Z}_{N^2}^*$, where $N = pq$ is an RSA number of unknown factorization. In particular, this is an example of a discrete logarithm-based cryptosystem that works within a group of unknown order. This system possesses an interesting property, in that it is additively homomorphic; the product of the Paillier encryption of two messages is an encryption of their sum.

Another very interesting feature of discrete logarithms is the ability to construct key exchange protocols with additional properties, such as authenticated key exchange, which embed the verification of the other party identity within the key exchange protocol. Perfect forward secrecy, in which disclosure of long-term secrets does not allow for decryption of earlier exchanges, is also easy to provide with schemes based on discrete logarithms. For example, in the Diffie–Hellman key exchange, Alice's secret a and Bob's secret b are ephemeral, and so is the shared secret they used to create, and (if proper key management is used) are discarded after the interaction is completed. Thus, an intruder who manages to penetrate either Alice's or Bob's computer would still be unable to obtain those keys and decrypt their earlier communications. It is also possible to mix long-term secrets, i.e., private keys, and ephemeral secrets in order to simultaneously provide perfect forward secrecy and identity verification.

1.2.3 A Powerful Extension: Pairing-Based Cryptography

Besides the Diffie–Hellman key exchange, a natural question to ask is whether there exists a three-party one-round key agreement protocol that is secure against eavesdroppers. This question remained open until 2000 when Joux [Jou04] devised a simple protocol that settles this question using bilinear pairings. Until then, building a common key between more than two users required two rounds of interaction. A typical solution for an arbitrary number of users is the Burmester–Desmedt protocol [BD94].

The one-round protocol based on pairing works as follows. If Alice, Bob, and Charlie wish to create a common secret key, they first agree on $G_1 = \langle P \rangle$ an additive group with identity $\mathcal{O}$, a multiplicative group G_2 of the same order with identity 1, and a bilinear pairing from G_1 to G_2. Let us recall the definition

Definition 1.1 A symmetric bilinear pairing[3] on (G_1, G_2) is a map

$$e : G_1 \times G_1 \to G_2$$

satisfying the following conditions:

1. e is bilinear: $\forall\, R, S, T \in G_1$, $e(R + S, T) = e(R, T) \cdot e(S, T)$,
   and $e(R, S + T) = e(R, S) \cdot e(R, T)$.
2. e is non-degenerate: If $\forall\, R \in G_1$, $e(R, S) = 1$, then $S = \mathcal{O}$.

Alice randomly selects a secret integer a modulo the order of G_1 and broadcasts the value aP to the other parties. Similarly and simultaneously, Bob and Charlie select their one secret integer b and c and broadcast bP and cP. Alice (and Bob and Charlie, respectively) can now compute the shared secret key

$$K = e(bP, cP)^a = e(P, P)^{abc}$$

We know that the security of DH-based protocols often relies on the hardness of the CDH and DDH problems. Likewise, the security of pairing-based protocols depends on the problem of computing $e(P, P)^{abc}$ given P, aP, bP, and cP, which is known as the computational bilinear Diffie–Hellman problem (CBDH or simply BDH). This problem also exists in its decision form (DBDH). However, little is known about the exact intractability of the BDH, and the problem is generally assumed to be as hard as the DLP in the easier of the groups G_1 and G_2. Indeed, if the DLP in G_1 can be efficiently solved, then an eavesdropper who wishes to compute K can recover a from aP and then compute $e(bP, cP)^a$. Similarly, if the DLP in G_2 can be efficiently

---

[3]In general, asymmetric pairings are also considered. For simplicity of presentation, we only describe the symmetric case.

solved, he could recover bc from $e(bP, cP) = e(P, P)^{bc}$, then compute bcP, and finally obtain K as $e(aP, bcP)$.

One consequence of the bilinearity property is that the DLP in G_1 can be efficiently reduced to the DLP in G_2. More precisely, assume that Q is an element of G_1 such that $Q = xP$, then we see that $e(P, Q) = e(P, xP) = e(P, P)^x$. Thus, computing the logarithm of $e(P, Q)$ in G_2 (to the base $e(P, P)$) yields x. This reduction was first described by Menezes et al. [MOV93] to show that supersingular elliptic curves are much weaker than random elliptic curves, since the DLP can be transferred from a supersingular curve to a relatively small finite field using pairings.

After the publication of the Menezes, Okamoto, and Vanstone result, cryptographers started investigating further applications of pairings. The next two important applications were the identity-based encryption scheme of Boneh and Franklin [BF03] and the short signature scheme of Boneh et al. [BLS04]. Since then, there has been a tremendous activity in the design, implementation, and analysis of cryptographic protocols using bilinear pairings on elliptic curves and also on more general abelian varieties, for example, on hyperelliptic curves.

1.3   Advantages of Discrete Logarithms

A large fraction of the protocols that public key cryptography provides, such as digital signatures and key exchange, can be accomplished with RSA and its variants. Pairing-based cryptosystems are a notable exception to this general rule. However, even for classical protocols, using discrete logarithms instead of RSA as the underlying primitive offers some notable benefits.

1.3.1   Technical Advantages

Smaller Key Sizes The main advantage of discrete logarithms comes from the fact that the complexity of solving the elliptic curve discrete logarithm problem (ECDLP) on a general elliptic curve is, as far as we know, much higher than factoring an integer of comparable size. As a direct consequence, elliptic curve cryptosystems currently offer the option of using much smaller key sizes than would be required by RSA or discrete logarithms on finite fields to obtain a comparable security level.

In truth, the key size reduction is so important that it more than offsets the additional complexity level of elliptic curve arithmetic. Thus, for the same overall security level, elliptic curve systems currently outperform more classical systems.

Perfect Forward Secrecy When using RSA to set up a key exchange, the usual approach is for one side to generate a random secret key and send it to the other encrypted with his RSA public key. This grants, to an adversary that records all the traffic, the ability to decrypt every past communications, if he ever gets hold of the corresponding private key.

By contrast, as we have already mentioned in the introduction, a correctly designed key exchange protocol based on the DLP can avoid this pitfall and achieve perfect forward secrecy, thus preventing an adversary to decrypt past communications [DOW92].

1.3.2 Algorithmic Diversity

Cryptographers have learned from history that it is unwise to base security on a single assumption, as its violation can lead to simultaneous breakdown of all systems. For this reason it is important to have a diversity of cryptosystems and have candidate replacement systems. Schemes based on discrete logarithms provide an alternative to those derived from RSA and other algorithms whose security depends on difficulty of integer factorization.

However, we should note that both integer factorizations and discrete logarithms would be easy to obtain from quantum computers. Hence it is important to investigate even more exotic cryptosystems, such as those based on error-correcting codes and lattices.

The chapter is organized as follows. Section 2 deals with generic algorithms, i.e., those that assume no special knowledge about the underlying group and consider group operations as black boxes. By contrast, Sect. 3 presents the index calculus method, a very useful framework to obtain a family of algorithms that make extensive use of specific knowledge of the group. Section 4 presents concrete algorithms to solve the DLP in finite fields, and Sect. 5 describes the state of the art about the DLP on algebraic curves.

2 Generic Results

This section discusses some general results for discrete logarithm that assume little knowledge of the group. In the most general case, we only ask for a group whose elements can be represented in a compact way and whose law is explicitly given by an efficient algorithm. We also consider the case where the order of the group and possibly its factorization are also given. This case is interesting because for many groups that are considered in practice, this information is easily obtained. Typically, for an elliptic curve, the group order is efficiently found using point counting algorithms. Moreover, system designers usually choose curves whose order is a small multiple of a prime, in which case, factoring the order becomes easy.

Throughout the section, we use the same notations as in Sect. 1.1. First, we describe some general complexity results that relate the hardness of the DLP to classical complexity theoretical classes. Second, assuming that the factorization of the order of G is given, we show that computing discrete logarithms in G is no harder than computing discrete logarithms in all subgroups of G of prime order. This is the outcome of the Pohlig–Hellman algorithm [PH78], which is a constructive method to compute discrete logarithms in the whole group from a small number of computations of discrete logarithms in the subgroups of prime order. We also describe Pollard's rho algorithm [Pol78] that allows the computation of discrete logarithms in a group G in $O(\sqrt{|G|})$ operations. Combining Pohlig–Hellman with Pollard's rho essentially[4] permits the computation of discrete logarithms in time $O(\sqrt{p})$, where p is the largest prime factor of the order of a generic group.

Finally, we discuss the issue of computing many independent discrete logarithms in the same generic group, amortizing part of the computation cost; we also briefly present the generic group model as proposed by Shoup in [Sho97b] and the lower bound on the complexity that is related to it.

2.1 Complexity Classes

In order to describe the exact level of hardness of a computational problem, the main approach is to describe the complexity classes the problem belongs to. To this end, the traditional approach is to work with decision problems, i.e., problems with a yes/no answer. Since the DLP itself is not a decision problem, the first step is to introduce a related decision problem whose hardness is essentially equivalent to computing discrete logarithms. This can be done in many ways, for example, let us consider the following problem.

Problem 2.1 (Log Range Decision) **Given a cyclic group G and a triple (g, h, B):**

- Output **YES** if there exists $x \in [0 \cdots B]$ such that $h = g^x$.
- Otherwise output **NO**.

An algorithm or oracle that solves this problem can be used to compute discrete logarithms using a binary search. This requires a logarithmic number[5] of calls to *Log Range Decision*. As a consequence, the hardness of *Log Range Decision* is essentially the same as the hardness of the DLP itself.

---

[4]If $|G|$ is a product of many small primes, possibly with multiplicity, this claim does not hold. However, this is not an interesting case for cryptographic purposes.

[5]In other words, the number of oracle calls is a polynomial in the bitsize of the answer.

2.1.1 Log Range Decision is in NP ∩ co-NP

To show that the problem is in NP, assume that there exists $x \in [0 \cdots B]$ such that $h = g^x$, then x itself is a witness to this fact which is easily tested in polynomial time.

When g is a generator of G and $|G|$ is known, giving a possible discrete logarithm of h to the base g is also a satisfying witness to prove that the answer is NO. Thus, in this simple case, the problem belongs to co-NP. However, in general, the situation is more complex: we need a generator[6] g_0 of G together with $|G|$ and its factorization to prove this fact. With g_0 in hand, the discrete logarithms of both g and h to the base g_0 suffice to determine whether h belongs to the subgroup generated by g and, if needed, to prove that none of the discrete logarithms of h to the base g belong to $[0 \cdots B]$. As a consequence, even in the general case, Log Range Decision is in co-NP.

2.1.2 Log Range Decision is in BQP

Another very important complexity theoretic result about the computation of discrete logarithms is that there exists an efficient *quantum* algorithm invented by Shor [Sho97a]. This algorithm works for arbitrary groups, assuming that the group operation can be computed efficiently. It is based on the quantum Fourier transform and belongs to the complexity class **BQP** (bounded-error quantum polynomial time) that corresponds to polynomial time computation on a quantum computer with a bounded-error probability.[7]

2.1.3 Computing $|G|$ Using a Discrete Logarithm Computation

When considering the DLP, we often assume that the group order $|G|$ is known. One justification is that it is often the case with the groups that are used in cryptography. Here, we point out another reason. When the DLP becomes easy in a group, it is possible to compute $|G|$ using a discrete logarithm computation. Assume that we are only given the bitsize of $|G|$, i.e., that we know that $|G| \in [2^{n-1}, 2^n - 1]$. In this context, given g a generator of G, we see that

$$|G| = 2^n - \log_g(g^{2^n}).$$

---

[6]Since there are many distinct generators of G, in fact $\varphi(G)$, g_0 is easy to find by testing random candidates.

[7]Typically, an error probability of $1/3$ can be used in the formal definition of BQP.

2.1.4 Average Case Hardness and Random Self-reducibility

When using hard problems to build cryptosystems, one important issue is to be sure that randomly generated instances of the problem are practically hard. This deviates from the standard definition of hardness in complexity. In cryptography, a problem that admits hard instances is not enough; we need the problem to be hard not only in its worst case, but also in its average case and also usually in most cases.

Concerning the DLP, we have a very nice property, *random self-reducibility* introduced in [AFK89]. This property shows that any instance of the DLP can be rerandomized into a purely random instance. As a consequence, if the DLP is easy in the average case, it is also easy in the worst case. Conversely, if there exists hard instances of the DLP in some group G, then the DLP is hard for random instances in G.

The reduction works as follows: assume that we are given an oracle that solves the DLP in G for random instances and some fixed instance of the problem, $h = g^x$. Choose an integer r modulo $|G|$ uniformly at random and define $z = hg^r$, then z follows a uniform random distribution in G. If the given oracle can compute $\log_g(z)$, we recover x from the relation $x \equiv \log_g(z) - r \pmod{|G|}$.

2.2 Pohlig–Hellman

Let G be a group of order n, g a generator, and h the element for which we want to compute the discrete logarithm x. We suppose further that we know the factorization of n:

$$n = \prod_{p_i \mid n} p_i^{e_i}.$$

The Pohlig–Hellman algorithm permits us to reduce the DLP in G to DLPs in cyclic groups of prime order p_i. We proceed in two phases:

1. First we reduce the DLP in G to DLPs in groups with orders a power of the primes p_i involved in the factorization of n. For each p_i, we set

$$n_i = \frac{n}{p_i^{e_i}}, \qquad g_i = g^{n_i} \quad \text{and} \quad h_i = h^{n_i}.$$

So $g_i^{p_i^{e_i}} = g^n = 1$ and the order of g_i is exactly $p_i^{e_i}$. Moreover $g_i^x = g^{n_i x} = h^{n_i} = h_i$. Thus, h_i belongs to the subgroup of order $p_i^{e_i}$ generated by g_i. More precisely, h_i can be considered as the projection of the element we want the discrete logarithm on the subgroup generated by g_i. Let us call x_i the discrete logarithm of h_i to the base g_i. We then have

$$x \equiv x_i \mod p_i^{e_i}.$$

Since the $p_i^{e_i}$ are pairwise coprime, if we know all the x_i, a simple application of the Chinese remainder theorem permits us to recover x.

2. A further simple reduction shows that solving the DLP in a group of prime order allows to solve the DLP in groups with orders that are powers of that prime.

To conclude, what has to be kept in mind is that computing discrete logarithms in G is no harder than computing discrete logarithms in all subgroups of prime order in G.

2.3 Discrete Logarithms in G in $O\left(\sqrt{|G|}\right)$

There are several methods for computing discrete logarithms in a group G in about $\sqrt{|G|}$ operations. The first and best known of these is the Shanks baby step/giant step technique.

2.3.1 Baby Step/Giant Step

Let n be the order of G, or even an upper bound of $|\langle g \rangle|$, and h be the element for which we want to compute the discrete logarithm x. Let m be equal to $\lceil \sqrt{n} \rceil$. If we let q and r be such that $x = qm + r$ with $0 \leqslant r, q < m$, which is possible, thanks to the size of m compared to n, then it is clear that finding x is exactly the same as recovering q and r. First we remark that we have

$$(g^m)^q = (g^{mq}g^r)g^{-r} = g^{mq+r}g^{-r} = hg^{-r}. \tag{1}$$

We create the first list:

$$\text{Baby} = \{(hg^{-r}, r) | 0 \leqslant r < m\}.$$

We call it the baby list, because we multiply each step by the inverse of g (which is considered to be small). If, by good luck, there exists a couple $(1, r')$ in this set, we have obtained $hg^{-r'} = 1$ and thus $x = r'$. If not, we create another list:

$$\text{Giant} = \{((g^m)^q, q) | 0 \leqslant q < m\}.$$

We call it the giant list because we multiply in this case each step by g^m. We sort the two lists to find a collision on the two first elements of each pair. When we obtain $(hg^{-r}, r) \in \text{Baby}$ such that $(g^m)^q = hg^{-r}$, thanks to (1), we have found q and r and thus x.

Checking for equality in two sorted lists of m entries each can be done in linear time (assuming that the representations of elements are compact enough). Hence the running time of the algorithm is dominated by the arithmetic required to compute the two lists and the time to sort them. This algorithm is deterministic and solves the DLP in $\tilde{O}\left(\sqrt{n}\right)$ operations.[8]

2.3.2 Pollard's Rho Algorithm

This algorithm runs in time comparable to the Shanks method, $O\left(\sqrt{n}\right)$ operations, but has the advantage that it is practically memoryless. Unlike the Shanks algorithm, though, it is probabilistic, not deterministic. It was proposed by Pollard in 1978 [Pol78] and works as follows.

Let us imagine that we have a partition of G into three subsets of roughly equal size A_1, A_2, and A_3. We define the map f by:

$$f(b) = \begin{cases} gb & \text{if } b \in A_1 \\ b^2 & \text{if } b \in A_2 \\ hb & \text{if } b \in A_3. \end{cases}$$

We take now a random integer x_0 in $\{1, \cdots, n\}$ and we compute $b_0 = g^{x_0}$. We consider the sequence $b_{i+1} = f(b_i)$. The algorithm relies on two facts. First, for each i we can rewrite b_i as

$$b_i = g^{x_i} h^{y_i} \tag{2}$$

where $(x_i)_i$ and $(y_i)_i$ are given by the initial choice of x_0, $y_0 = 0$ and:

$$x_{i+1} = \begin{cases} x_i + 1 \mod n & \text{if } b_i \in A_1 \\ 2x_i \mod n & \text{if } b_i \in A_2 \\ x_i & \text{if } b_i \in A_3. \end{cases} \quad \text{and} \quad y_{i+1} = \begin{cases} y_i & \text{if } b_i \in A_1 \\ 2y_i \mod n & \text{if } b_i \in A_2 \\ y_i + 1 \mod n & \text{if } b_i \in A_3 \end{cases}$$

Second, since we are computing a sequence in a finite group, there exist two integers $i \geq 0$ et $k \geq 1$ such that we have a collision $b_i = b_{i+k}$ (in practice we search collision of the form $b_i = b_{2i}$). Thanks to Eq. (2) we have

$$g^{x_i} h^{y_i} = g^{x_{i+k}} h^{y_{i+k}}$$

which yields a linear equation for $\log_g(h)$:

$$x_i - x_{i+k} \equiv \log_g(h)(y_{i+k} - y_i) \mod n.$$

If we can invert $y_{i+k} - y_i$ modulo n, then we can recover the discrete logarithm of h. If $y_{i+k} - y_i$ is not invertible, we need to remember that Pollard rho is usually

---

[8]As usual, the $\tilde{O}$ notation $\tilde{O}(n)$ is a shorthand for $O(n \log^\alpha n)$ for an arbitrary value of α.

used as a subroutine of Pollig–Hellman, which means that n is usually prime. As a consequence, the only option is to restart a different instance of computation, for example, using another choice for x_0.

The Pollard rho algorithm can be implemented so that it requires only $O(1)$ elements in memory and $O(\sqrt{n})$ operations. Some practical improvements of this algorithm are presented in [Tes00, BLS11, CHK12].

In practice, computations of discrete logarithms using generic algorithms use a combination of the Pohlig–Hellman and Pollard rho algorithms. Depending on the computer architecture used for the computations, there exist alternatives to Pollard's rho that are sometimes more appropriate (see the next section). However, the overall complexity using these algorithms remains $O(\sqrt{p})$ where p is the largest prime dividing the order of the group. In fact, Sect. 2.5 shows that generic group algorithms cannot outperform this complexity.

2.4 Scalability of Generic Discrete Logarithm Algorithms

From a purely theoretical viewpoint, a $O(\sqrt{|G|})$ algorithm that only uses a constant amount of memory is a very fine solution. However, for practical purposes, it is very useful to know whether such a computation can be distributed on a parallel computer or a network of independent computers. Indeed, this scalability issue often decides whether a computation is feasible or not.

In this setting, it is very useful to replace cycle finding algorithms by algorithms based on the distinguished point technique. According to [Den82, p. 100] the idea of the distinguished point technique was proposed by Rivest. Quisquater and Delescaille [QD89] used the technique to find collisions in the DES algorithm. The in-depth study made by van Oorschot and Wiener [vOW99] shows how the technique can be used in order to efficiently take advantage of parallelism for collision search.

Basically, the main idea of the distinguished point technique is to build chains of computations, starting from a random value and iterating a fixed function f to compute a chain of successors. Denoting the starting point x_0, we iteratively compute $x_{i+1} = f(x_i)$. We abort the computation when encountering a point x_N that satisfies some distinguished point property. Typically, this property is taken to be that the representation of x_N starts with a specified number of '0' bits. We then store the triple (x_0, x_N, N). Recall that as in Pollard rho, we wish to find a collision of f in order to compute the desired discrete logarithm. With the distinguished point technique, any collision between two distinct chains ensures that the two chains terminate at the same distinguished point. Conversely, given two chains ending at the same distinguished point, recomputing the two chains from their respective starting points, accounting for the length difference, usually leads to an explicit collision. Since the initial computations of chains are independent from each other, it is extremely easy to distribute them over a large number of distinct computers.

Recently, using a slight variation of this distinguished point technique, it was shown in [FJM13] that given L independent discrete logarithms to compute in the same group $|G|$, the computation can be achieved in time $O(\sqrt{L|G|})$ rather than $O(L\sqrt{|G|})$. Similar results were already known under the condition $L \leq O(|G|^{1/4})$ [KS01].

2.5   The Generic Group Model

In 1997, Shoup [Sho97b] introduces a theoretical framework to study the complexity of generic algorithms: the generic group model. In this model, he shows that any generic algorithm must perform $\Omega(\sqrt{p})$ group operations, where p is the largest prime dividing the order of the group. Since this lower bound essentially[9] matches the known upper bound, the generic group model emphasizes the fact that currently known generic algorithms for computing discrete logarithms are optimal.

In a nutshell, in the generic group model, group elements are identified by unique but arbitrary encodings. As a consequence, it is not possible to exploit any special properties of the encodings, and group elements can only be operated on using an oracle that provides access to the group operations.

One frequently encountered criticism of the generic group model is that it suffers from the same weaknesses as the random oracle model, which is considered with suspicion by many cryptographers. Namely, in these models, there exists secure protocols that cannot be securely instantiated [Den02, CGH00].

3   Index Calculus Method

The results from the generic group model no longer apply when extra information about the group structure is known. Indeed, this extra information can then be used to obtain faster algorithm. The most important example is the index calculus method which uses this additional knowledge to provide subexponential algorithms.

Though the index calculus method works both for factoring and for discrete logarithm, here we only consider its application to discrete logarithm computations.

3.1   General Description

The basic idea of index calculus algorithms relies on three main steps: the sieving phase (also called the relation collection phase), the linear algebra phase, and

---

[9]Up to logarithmic factors.

the individual logarithm phase. Basically, the first phase creates relations between the logarithms of elements belonging to a small subset of the considered group, the second one recovers those logarithms, and the last one permits to obtain the logarithm of any arbitrary element by relating it to the logarithms obtained during the first two phases. Those three steps work as follows:

1. **Sieving Phase or Relation Collection Phase.** For simplicity, assume that G is a cyclic group generated by g. We want to create a large number of multiplicative relations between elements belonging to a subset of the group G. This subset is usually constructed by selecting elements which can be considered to be *small*, in some sense that depends on the context. This subset of G is usually called the smoothness basis or the factor basis. Let $\{g_i, i \in I\}$ denote this smoothness basis and consider a relation of the form

$$\prod_{i \in I} g_i^{m_i} = \prod_{i \in I} g_i^{n_i}. \tag{3}$$

Then, taking the discrete logarithms of the two sides, we deduce

$$\sum_{i \in I} m_i \log_g g_i \equiv \sum_{i \in I} n_i \log_g g_i \mod |G|.$$

This becomes a linear equation between the logarithms of the g_i viewed as formal unknowns. We stop the sieving phase once we have collected enough such linear equations to obtain a system of codimension 1.

2. **Linear Algebra Phase.** The aim of the linear algebra step is to solve the previous system of linear equations. Thus, we get at the end of this phase all the discrete logarithms of the smoothness basis.[10]

   A very important observation that naturally applies in most index calculus algorithms is that the equations produced during the relation collection phase are very sparse. This is extremely important, because sparse system can be solved using special algorithms which are much faster than general linear system algorithms. This is detailed in Sect. 3.4.

3. **Individual Logarithm Phase.** To really solve the DLP in G, we should be able to compute the logarithm of any arbitrary element z of G. Roughly, the goal of this last phase is to decompose z into products of other elements, which can in some sense be considered smaller than z and iterate until z is finally expressed as a product of elements belonging to the smoothness basis. Plugging the values of the discrete logarithms obtained during the first two phases in this expression yields the logarithm of z.

---

[10]Or at least, a large fraction of these logarithms. Indeed, depending on the exact properties of the relation collection phase, a few elements of the smoothness basis might possibly be missing.

3.2   Collection of Relations

In order to design index calculus algorithms, we thus need to construct multiplicative relations as in (3).

The simplest approach for discrete logarithms modulo a prime p is to take a random integer a, compute $u \equiv g^a \mod p$ for u an integer such that $1 \leqslant u \leqslant p-1$, and check whether

$$u = \prod q_i$$

where the q_i are primes satisfying $q_i < B$ for some bound B. When the above congruence holds, we say that u is B-smooth and we call B the smoothness bound. For most values of a, u will not be smooth, and so will be discarded. However, even with this primitive approach, one can obtain running time bounds of the form[11] $L_p(1/2, c)$. Moreover, this approach provides a provable although probabilistic algorithm for solving the DLP in many finite fields.

Though this remains an interesting algorithm because it is at once simple and rigorous, it is possible to devise better algorithms with other strategies. One key idea is to represent the group G in which we want to compute discrete logarithms in two different but compatible ways. In other words, we want to be able to draw a commutative diagram like the one presented in Fig. 1. With this representation in hand, for all x in E, we can get two elements in G related in an algebraic way. Thanks to commutativity, we have an equality in the group G:

$$\varphi_1(\psi_1(x)) = \varphi_2(\psi_2(x)).$$

However, for this to be useful, we need to have a way to select some special relations among those created. To this end, we choose a small set in each intermediate set E_1 and E_2 of the diagram. Once these are chosen, in the sieving phase we keep only relations that involve elements of these two small sets and no other. We call the smoothness base (or factor base) the subset of G consisting of elements that

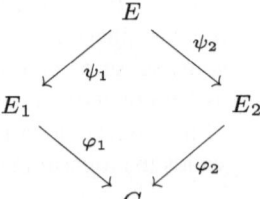

Fig. 1 Commutative
diagram for the sieving phase

---

[11]See Sect. 3.3 to understand the origin of this L notation. For the moment, just read $L_q(\alpha, c)$ as a shorthand for $\exp\left((c + o(1))(\log q)^\alpha (\log\log q)^{1-\alpha}\right)$.

can be obtained through these two small subsets[12] of E_1 and E_2. The number field sieve [Sch00, Gor93, JLSV06] and the function field sieve (FFS) [AH99, JL06] that have complexity of the form $L_{p^n}(1/3, c)$ both follow this general strategy. They are heuristic algorithms in that their analyses depend on plausible assumptions, but, ones that have not been proved rigorously. Despite the fact that they share a common algorithmic structure, there is a major difference between the NFS and the FFS. The former algorithm is based on multiplicative relations between algebraic integers in number fields while the latter works in function fields. At the bottom level, this means that one algorithm needs to factor integers while the other factors polynomials. This is a major difference since polynomials are much easier to factor than integers and also have more systematic properties which have been used in the recent algorithms reported in the next paragraph.

A small change in just the sieving phase can lead to a substantial improvement in the complexity of an algorithm. In fact, recent progress in the index calculus method for the DLP has come from better collections of relations. However, the notion of sieving tends to disappear since the new algorithms proposed to solve the DLP in finite fields with small characteristic rely on a new trick that directly creates those relations. Those new methods, developed in Sect. 4.2.2, have recently yielded complexities in $L_{p^n}(1/4, c)$ [Jou13b] for finite fields with a small characteristic. With an additional improvement made this time in the individual logarithm phase, this has led to a heuristic quasi-polynomial algorithm [BGJT13], again for large fields of small characteristic.

3.3 Smoothness

Index calculus algorithms depend on a multiplicative splitting of elements (integers, ideals, or polynomials) into elements drawn from a smaller set, typically consisting of elements that are in some sense considered to be *small*. Elements that do split this way are called smooth, and a fundamental problem in the analysis of index calculus algorithms is to estimate how the relation generation process produces those smooth elements. In most cases, the heuristic assumption is made that the elements that arise during the process essentially behave like random elements of the same size. This assumption was introduced to simplify the analysis of the algorithm, and it successfully led to many algorithmic improvements and to a large number of integer factorization and discrete logarithm records.

However, depending on such a heuristic is uncomfortable, and many researchers would like to come up with rigorous algorithms. Unfortunately, at the present time, the existing rigorous algorithms are much less efficient than their heuristic siblings. Quite surprisingly, the most recent advances are based on the fact that, in cases

---

[12]Note that those subsets are sometimes called the smoothness bases by some authors too.

where the classical heuristic assumption become, false, it is possible to use this failure to our advantage and produce more efficient algorithms.

To be more precise, let us give classic definitions and major theorems used in order to estimate this probability:

Definition 3.1 An integer is y-smooth if all its prime factors are lower than y.

Definition 3.2 A polynomial over a finite field is m-smooth if all its irreducible factors have degree lower than m.

Canfield et al. [CEP83] gave in 1983 the probability of smoothness of integers. More than a decade later, Panario et al. [PGF98] generalized this estimation to the probability of smoothness of polynomials in finite fields. A less general result in this direction was obtained earlier in [Odl85]. These two main results that are surprisingly close can be summarized in the following estimate:

Estimate 3.1 *The probability for an arbitrary integer lower than x to be y-smooth (respectively for a random polynomial of degree less than n to be m-smooth) is*

$$u^{-u+o(1)}$$

where $u = \dfrac{\log x}{\log y}$ *(respectively* $u = \dfrac{n}{m}$*).*

The very first analyses of the asymptotic running time of index calculus algorithms appeared in the 1970s and were of the form $\exp\left((c + o(1))(\log p)^{1/2}(\log\log p)^{1/2}\right)$. In fact, index calculus algorithms not only have in common their structure in three phases but also the expressions of their asymptotic complexities. To simplify these expressions, we usually write them with the help of the following notation:

$$L_q(\alpha, c) = \exp\left((c + o(1))(\log q)^{\alpha}(\log\log q)^{1-\alpha}\right)$$

where α and c are constants such that $0 < \alpha < 1$ and $c > 0$. This is linked to the smoothness probability of elements since it directly comes from Estimate 3.1. The simple notation $L_q(\alpha)$ is often used when c is not specified, and the expression $L_q(\alpha, c + o(1))$ is abbreviated in $L_q(\alpha, c)$ where $o(1)$ is for $q \to \infty$. The most important parameter is the first one, since it governs the transition from an exponential time algorithm to a polynomial time one. In fact, if α tends to 1, $L_q(\alpha)$ becomes exponential[13] in $\log q$, and on the other hand, if α tends to 0, $L_q(\alpha)$ becomes polynomial in $\log q$.

This notation permits not only to write the complexities in a simple and compact form but also to give an indication concerning the different ranges of application of algorithms for finite fields.

---

[13]Note that since $\log q$ is the number of bits necessary to encode elements of the group we are considering, it is the natural parameter to consider when expressing the complexity of algorithms.

3.4   Sparse Linear Systems over Finite Fields

Index calculus algorithms use linear algebra to recover the logarithms of the elements of the smoothness basis. Since these logarithms are determined modulo the order of the considered group, we need to solve a large system of linear equations over a residue ring $\mathbb{Z}/m\mathbb{Z}$. For a long time in the 1970s and early 1980s, this step was regarded as a major bottleneck, affecting the asymptotic running time estimates of algorithms. This was due to the cubic complexity of solving linear systems with classical methods such as Gaussian elimination.

Even today, the linear algebra step remains difficult and it is a more serious problem for discrete logarithm than for factoring. The main difference is that for factoring, we need solutions modulo 2, while for discrete logarithm we require solutions modulo large numbers. This is one of the reasons of the persistent gap between factorization and discrete logarithm records in $\mathbb{F}_p$, with p, a prime. Fortunately, the linear systems of equations produced by index calculus algorithms are sparse, often to a very large extent.

A sparse matrix is a matrix that contains a relatively small number of non-zero entries. Very frequently, it takes the form of a matrix in which each line (or each column) only contains a small number of non-zero entries, compared to the dimension of the matrix. With sparse matrices, it is possible to represent in computer memory matrices with much larger dimension, describing each line (resp. column) as the list of positions containing a non-zero coefficient, together with the value of the corresponding coefficient. When dealing with a sparse linear system of equations, using plain Gaussian elimination is a bad idea. Indeed, each pivoting step increases the number of entries in the matrix and after a relatively small number of steps, the matrix can no longer be considered as sparse. As a consequence, if the dimension of the initial matrix is large, Gaussian elimination quickly overflows the available memory. In order to deal with sparse systems, a different approach is required.

Three main families of algorithms have been devised to deal with linear algebra in the case of sparse matrices. These methods behave better than general purpose linear algebra algorithms.

The first family, structured Gaussian elimination, initially proposed in [Odl85] and implemented in [LO90] contains variants of the Gaussian elimination algorithm that perform pivot selection in a way that minimizes the fill-in of the matrix throughout the algorithm. These methods are used to reduce the dimension of the original system and produce a reduced-size system which remains reasonably sparse. This reduced system is then solved using an algorithm from one of the other two families.

A common property of the two other families is that they use a matrix involved in the linear algebra in a very restrictive way. In fact, it only appears in matrix–vector products, where some variable vectors are multiplied either by the considered matrix or its transpose. The first of these two families contains Krylov subspace methods which have been adapted from numerical analysis and construct sequences of mutually orthogonal vectors. In particular, this family contains the Lanczos and conjugate gradient algorithms, already described for the discrete logarithm context in [COS86]. The second family contains the Wiedemann algorithm [Wie86] and its generalization for parallel processing, Block Wiedemann. To put it in a nutshell, the algorithms in this family find a solution of a linear system by computing the minimal polynomial[14] of the considered matrix.

Both the Krylov subspace and Wiedemann families of algorithms cost a number of matrix–vector multiplications equal to a small multiple of the matrix dimension. Thus, for an $N \times N$ matrix containing λ entries per line on average, the global cost is $O(\lambda N^2)$.

4 Discrete Logarithm in Finite Fields

4.1 A Short History

The earliest methods introduced to compute discrete logarithms are generic. Of course, the fact that discrete logarithms can be computed using exhaustive search is self-evident. However, the algorithmic techniques to outperform this simple approach are more recent. The first method to achieve this is the baby step/giant step, initially introduced in 1971 by Shanks [Sha71] for the computation of class numbers in quadratic fields. The next technique, proposed in 1978, is the Pollard rho method [Pol78], a variation on Pollard rho factoring algorithm [Pol75] from 1975.

Interestingly, the link that Pollard's Rho algorithm shows between factorization of integers and the computation of discrete logarithms modulo prime is much more general, and most of the algorithms known to solve one of the problems admit variants that apply to the other. There are some exceptions. For example, it is not known how to obtain a variation of the elliptic curve factoring method (ECM) of Lenstra to compute discrete logarithms. However, a variation of ECM can be used [MW96, JN03] to provide a relationship between the hardness of CDH and DLP. Until recently, it was believed that this relationship between the hardness of integer factorization and discrete logarithm computations could be extended to arbitrary finite fields. However, due to the recent results on the computation of discrete logarithms in small characteristic, this is no longer clear.

In 1976, the invention of Diffie–Hellman key exchange kindled renewed interest on the DLP in finite fields. In 1977, the discovery of RSA also renewed the interest in the integer factoring problem. At that time, the state of the art in factoring was not far in advance of what was described in a book published in 1922 by Kraitchik [Kra22], which shows how the use of quadratic forms can speed up factorization. The same book also provides methods for the computation of discrete logarithms; however, the

---

[14]More precisely, this is the goal of Wiedemann algorithm. The block version computes something somewhat different but quite similar.

terminology of Kraitchik used the French word *indice* instead of discrete logarithm. This terminology spawned the name *index calculus* for these algorithms. The early index calculus algorithms where proposed first for prime fields $\mathbb{F}_p$ [Adl79]. They achieve complexity in $L_p(1/2, c)$ for a certain constant c. The main advantage of these initial algorithms is that they can be turned into provable version as shown in [Pom87]. Moreover, they were generalized to finite fields of the form $\mathbb{F}_{p^k}$ with fixed p by Hellman and Reyneri in [HR82].

Nonetheless, the original value of c was too high for practical application. The situation was largely improved by the Gaussian integer method introduced in 1986 by Coppersmith et al. [COS86] which lowered the value of c to 1. At that time, it was also discovered that a variation of this algorithm obtained by replacing numbers by polynomials could be used to compute discrete logarithms in small characteristic finite fields.

A drastic change occurred in 1984 when Coppersmith proposed, in the case of characteristic 2, a heuristic algorithm with complexity $L_{2^n}(1/3)$ [Cop84]. This initial progress quickly led to the introduction of several other heuristic algorithms with $L(1/3)$ complexity both for factoring and discrete logarithms computations. A survey on the early effective implementations of these algorithms for discrete logarithms appeared in 1996 [SWD96]. For a long time, $L(1/3)$ algorithms focused on field with small characteristic, prime fields, and occasionally fields of the form $\mathbb{F}_{p^k}$ for small values of k [Sch00]. The view changed in 2006, with two articles that showed that taken together, the number field sieve [JLSV06] and the FFS [AH99, JL06] are enough to cover the whole range of finite fields with heuristic $L(1/3)$ algorithms. Essentially, the result was to split the finite field in three groups, small characteristic with complexity $L(1/3, (32/9)^{1/3})$, medium characteristic with complexity $L(1/3, (128/9)^{1/3})$ and large characteristic with complexity $L(1/3, (64/9)^{1/3})$.

In 2013 and 2014, several algorithmic improvements on the complexity of discrete logarithm algorithms have appeared: two variants of the number field sieve have been designed for finite fields with medium to high characteristic [JP13, BP14], and a breathtaking step forward [Jou13a, GGMZ13, Jou13b, BGJT13, GKZ14] has been made for finite fields with small characteristic. We discuss this in Sect. 4.2. This history of discrete logarithms is summarized in Fig. 2 and the history of records by Table 1.

4.2 Current Discrete Logarithms

Current discrete logarithms algorithms for finite fields vary with the relative sizes of the characteristic and the extension degree. In order to choose the one that is well suited for a given field $\mathbb{F}_{p^n}$, we write $p = L_{p^n}(l_p, c_p)$, with $0 \leq l_p \leq 1$ and c_p as a value or reasonable size (i.e., close to 1). As for complexity, the first parameter is the most important one in this notation. In fact, for a fixed size of finite field, when the characteristic is very small, l_p is close to 0. Conversely, when the finite field is

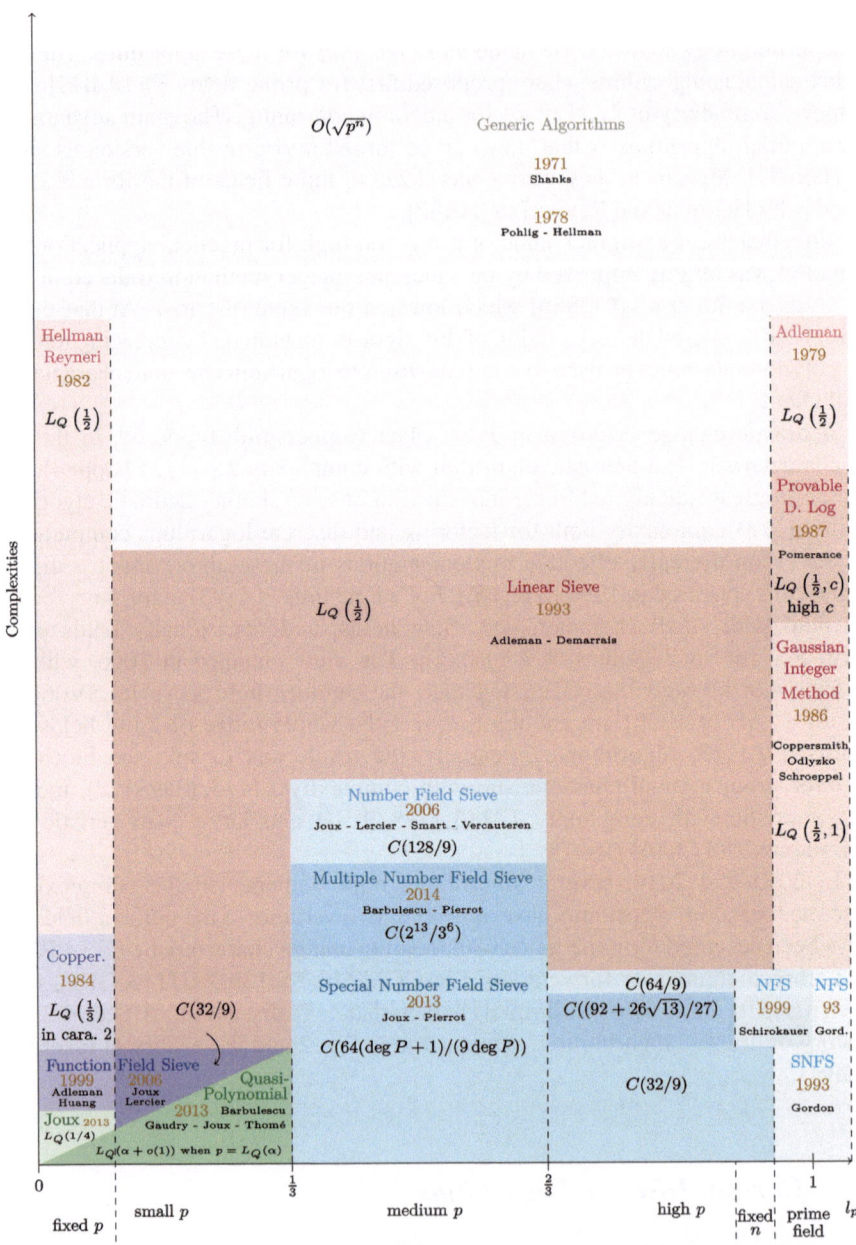

Fig. 2 Bird's-eye view of algorithms for discrete logarithm in finite fields and their complexities from 1970 to 2014. We warn the lector that his drawing is not to scale since a difference in the first or in the second parameter in the L_Q notation does not have the same effect on the complexity at all. Yet, the main *color* of each algorithm illustrates the variation of the first parameter: we depict $L_Q(1/2)$ algorithms in *red*, $L_Q(1/3)$ in *blue*, and $L_Q(1/4)$ and quasi-polynomial in *green*. For a fixed *color*, the darker an algorithm is drawn, the more recent it is. Furthermore, we introduce in this drawing the notation $C(x) = L_Q(1/3, (x)^{1/3})$

Table 1 History of discrete logarithm records

Date	Field	Bitsize	Cost (CPU.hours)	Algorithm	Authors
1992	2^{401}	401	114,000	[COS86]	Gordon, McCurley
1996	p	281	?	[COS86]	Weber, Denny, Zayer
1998/02	Special p	427	12,500	[Gor93]	Weber
1998/05	p	298	2,900	[COS86]	Joux, Lercier
2001/01	p	364	290	[JL03]	Joux, Lercier
2001/04	p	397	960	[JL03]	Joux, Lercier
2001/09	2^{521}	521	2,000	[JL02]	Joux, Lercier
2002	2^{607}	607	>200,000	[Cop84]	Thomé
2005/06	p	431	350	[JL03]	Joux, Lercier
2005/09	2^{613}	613	26,000	[JL02]	Joux, Lercier
2005/10	65537^{25}	400	50	[JL06]	Joux, Lercier
2005/11	370801^{30}	556 ★	200	[JL06]	Joux, Lercier
2007	p	530	29,000	[JL03]	Kleinjung
2012/06	$3^{6\cdot97}$	923	895,000	[JL06]	Hayashi, Shimoyama, Shinohara, Takagi
2012/12	p^{47}	1,175 ★	32,000	[Jou13a]	Joux
2013/01	p^{57}	1,425 ★	32,000	[Jou13a]	Joux
2013/02	2^{1778}	1,778 ★	220	[Jou13b]	Joux
2013/02	2^{1991}	1,991 ★	2,200	[GGMZ13]	Gologlu, Granger, McGuire, Zumbragel
2013/03	2^{4080}	4,080 ★	14,100	[Jou13b]	Joux
2013/04	2^{809}	809	19,300	[AH99, JL06]	The Caramel Group
2013/04	2^{6120}	6,120 ★	750	[GGMZ13, Jou13b]	Gologlu, Granger, McGuire, Zumbragel
2013/05	2^{6168}	6,168 ★	550	[Jou13b]	Joux
2014/01	$3^{6\cdot137}$	1,303	920	[Jou13b]	Adj, Menezes, Oliveira, Rodriguez-Henriquez
2014/01	2^{9234}	9,234 ★	398,000	[Jou13b]	Granger, Kleinjung, Zumbragel
2014/01	2^{4404}	698-bit subgroup	52,000	[Jou13b]	Granger, Kleinjung, Zumbragel

The ★ in the bitsize column indicates (possibly twisted) Kummer extensions

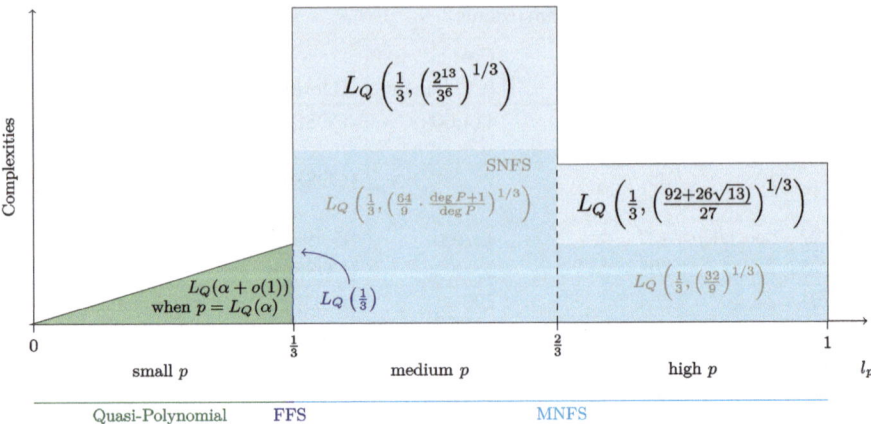

Fig. 3 Current algorithms for discrete logarithms in finite fields and their complexities. For a finite field of fixed size $Q = p^n$, this figure shows how asymptotic final complexities vary with the relative sizes of the characteristic p and n. We recall that we write $p = L_{p^n}(l_p)$

a prime field (the extension degree is thus equals to 1), the natural choice is to set $l_p = 1$. More precisely, finite fields split in three groups:

- Finite fields with high characteristic, when $l_p \geq 2/3$.
- Finite fields with medium characteristic, when $1/3 \leq l_p \leq 2/3$.
- Finite fields with small characteristic, when $1/3 \leq l_p$.

Each case is related to one algorithm which is examined in details in the sequel. The two boundary cases when l_p equals $1/3$ or $2/3$ are a little bit more intricate since several algorithms are available in those cases. They will not be treated here, but let us simply recall that the FFS is still the best option for some fields in the first boundary case.

We give in Figs. 3 and 4 two different viewpoints of the current situation. The first figure summarizes which algorithm has to be chosen for a given finite field, whereas the second one shows which sizes of field are weak compared with a given complexity. The axes of Fig. 4 may seem surprising to some innocent reader. However, since $\log Q = n \log p$, it is natural to compare the size of n with the size of $\log p$ (and not with the size of p).

4.2.1 Medium and High Characteristic

For a finite field with medium or high characteristic, Joux, Lercier, Smart, and Vercauteren presented in 2006 an adaptation of the number field sieve (NFS) that has a complexity in $L_{p^n}(1/3)$. For finite fields with high characteristic, it extended the variant of Shirokauer that had the same complexity, namely, $L_{p^n}(1/3, (64/9)^{1/3})$, but was available only for finite fields with fixed extension degree. The NFS as

Fig. 4 Current domains for discrete logarithm algorithms in $\mathbb{F}_{p^n}$. For each fixed size $Q = p^n$ correspond several relative sizes of p and n (see the *red line*) that lead to choose one algorithm or another. The *blue line* is an *iso-complexity line*: for a given complexity c, the DLP in each finite field that is represented in the *blue* part of the drawing (resp. exactly on the *blue line*) can be solved with an algorithm that has a complexity lower than c (resp. that equals exactly c)

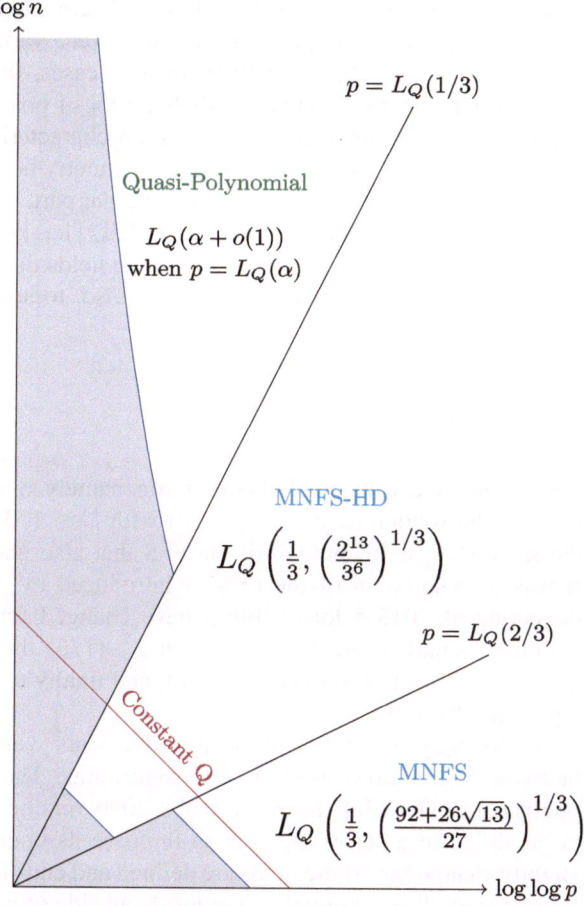

proposed in [JLSV06] is an index calculus algorithm that takes advantage of two representations of the finite field that rely on number fields. In a nutshell, the sieving process deals with linear polynomials and the smoothness basis consists in elements in the number fields that have norms lower than a certain predefined smoothness bound. A tricky post-process permits to associate each element of the smoothness basis to an element of the finite field.

For finite fields with medium characteristic, [JLSV06] proposed a variant of the classical number field sieve that leads to a final complexity in $L_{p^n}(1/3, (128/9)^{1/3})$. The polynomial selection used to represent the finite field is easier, however, the sieving can no longer be done on linear polynomials. Since high degree polynomials are used in the sieving phase, this variant of the number field sieve is often called the NFS-HD. This is also the reason why the complexity of the algorithm is higher in this case than in the high characteristic case.

The currently best known algorithm for discrete logarithms in medium and high characteristic is the multiple number field sieve, a variant of NFS proposed in 2014 by Barbulescu and Pierrot [BP14]. In both cases, the main idea of MNFS is to consider not only two number fields but a lot of possible paths in the diagram. A specific benefit is obtained in the medium characteristic case since each number field plays the same role. This notion of symmetry no longer exists in the high case where one of the number fields has a particular part.

Note that a special number field sieve [JP13] has been designed for both medium and high characteristic. It concerns all finite fields that have a sparse representation of their characteristic, and can be applied, so, to some finite fields coming from pairing-based constructions.

4.2.2 Small Characteristic

For a finite field with small characteristic, namely, a field $\mathbb{F}_{p^n}$ where the characteristic can be written as $p = L_{p^n}(l, c)$ with $l \leq 1/3$, Joux and Lercier presented the same year an adaptation of the FFS that also had a complexity in $L_{p^n}(1/3)$. It was an adaptation of the FFS as introduced by Adleman in 1993. Since the beginning of 2013 a lot of things have changed for those fields with small (or extremely small) characteristics. From $L_{p^n}(1/3)$ the complexity of the DLP has dropped to $L_{p^n}(1/4 + o(1))$ [Jou13b], and finally to a heuristic quasi-polynomial algorithm [BGJT13].

Surprisingly, several of these improvements work by falsifying the standard heuristic assumptions used in older algorithms. The first of these improvements published in [Jou13a] showed that the 2006 version of the FFS from [JL06] can be modified in a surprising way to improve its complexity. The basic idea is to slightly change how finite fields are defined and ends in a situation where the search for one smooth polynomial on the left-hand side of a relation can be amortized by constructing many possible right-hand sides from a single initial polynomial on the left. In the specific case of Kummer extensions, this can be improved further. For the first time, this new method takes advantage of the fact that the independence assumption between polynomials for the smoothness property does not hold in this context. This improvement especially focused on fields with characteristic close to $L(1/3)$ or, more generally, fields containing a subfield of size $L(1/3)$.

The next step concerns small characteristic fields, where it is possible to go well beyond the initial improvements. The basic idea can be viewed in two different ways: one can either consider a family of polynomials whose splitting probability is much higher than for random polynomials of the same degree as proposed in [GGMZ13] or start from a polynomial that splits and use a generalized version of the change of variable from [Jou13a] to construct many polynomials from this starting point. This latter approach is described in [Jou13b] and combined with a new method for computing individual logarithms; it yields a heuristic $L(1/4)$ algorithm. From an asymptotic point of view, this can be improved to a heuristic quasi-polynomial algorithm using another strategy for computing individual logarithms [BGJT13].

Note that these recent algorithms remain heuristic. However, it requires a new form of heuristic which is similar to but differs from the old one. Namely, whenever a polynomial occurs, we consider its probability of smoothness to be close to that of a random polynomial of the same degree *unless there is an explicit reason that falsifies this assumption*. Of course, whenever an explicit reason appears, by design, it largely increases the splitting probability. One of the main lines of research in small characteristic is now to try to build a heuristic-free algorithm. A first step has been done in this direction in [GKZ14], removing the smoothness heuristic of the descent phase.

5 Elliptic Curve Discrete Logarithm

Subexponential index calculus algorithms have been developed for a variety of DLPs. The one notable exception, where in general we still do not have algorithms better than those for the generic problem, is for elliptic curve discrete logarithms.

Most of the recent progress in discrete logarithm algorithms has come from developments in the index calculus method through exploitation of algebraic properties of finite fields. Unfortunately, this approach is in general not applicable to elliptic curve discrete logarithms. For elliptic curves, there exist some direct discrete logarithms algorithms that work for specific classes of curves and some indirect approaches that transfer the problem to finite fields [MOV93, FR94] or to higher genus curves [GHS02].

In general, the best known discrete logarithm algorithms for elliptic curves have exponential time complexities. However, this is not the case for higher genus curves, for which there exists index calculus algorithm. Moreover, some specific families of elliptic curves are also vulnerable to index calculus.

5.1 High Genus Curves

A very important result concerning curves of genus at least 3 introduced in [GTTD07] is that there exists an index calculus algorithm that applies to hyperelliptic curves of genus $g \geq 3$ defined over $\mathbb{F}_q$ and computes discrete logarithms in time $\tilde{O}(q^{2-2/g})$. This outperforms generic algorithms whose complexity in this case is $\tilde{O}(q^{g/2})$.

Note that there are similar results concerning non-hyperelliptic curves; for example, see [EGT11, DK13].

5.2   Elliptic Curves over Extension Fields

Where elliptic curves over extension fields are concerned, there are two main approaches: cover (or Weil descent) attacks and decomposition attacks. In addition, for some good configurations, it is possible to combine the two approaches into an even more efficient algorithm [JV12].

5.2.1   Weil Descent

This approach introduced in [GHS02] aims at transporting the DLP from an elliptic curve defined over an extension field to a higher genus curve defined over a smaller field. If the genus of the target curve is not too large, this can lead to an efficient discrete logarithm algorithm.

5.2.2   Decomposition

The basic idea of the decomposition method [Sem04] is to find relations between the smoothness basis elements by using the nth Semaev's summation polynomial to model the fact that n points on the curve sum to zero. Due to the symmetry of this polynomial, it is possible to reduce its degree by expressing everything in terms of the elementary symmetric polynomials in the abscissa of the solution points. Over an extension field of degree close to n above the base field, choosing the smoothness basis to be made of points with abscissa in the base field, it is possible to rewrite Semaev's polynomial as a polynomial system over the base field.

When the extension degree is much larger and cannot be decomposed into a favorable tower of extension, the situation is less clear. The typical case considered in [FPPR12, PQ12] is to take an elliptic curve over $\mathbb{F}_{2^p}$, where p is prime. The main difficulty is that contrary to the previous setting, none of the natural choices of smoothness basis are preserved when considering the symmetric polynomials in the abscissa. As a direct consequence, it is no longer possible to easily reduce the degree of Semaev's polynomial, which makes the asymptotic behavior of the method much harder to predict.

6   The Future

To date, the status of the DLP is quickly evolving. As a consequence, trying to predict future changes is extremely difficult. For this reason, we only sketch out the main open problems and give a small list of possible progress.

The most dangerous and notable risk for the DLP in general, which also applies to integer factorization, is the possibility of large-scale general purpose quantum computers. If such machines were to become available, the Shor algorithm would completely break these two hard problems. However, at the present time, it is unclear whether such machines will become available in the foreseeable future.

Concerning discrete logarithms in finite fields, several avenues for progress are open. First, in small characteristic, the present quasi-polynomial algorithm could be improved in many directions, with the removal of heuristic hypotheses, the improvement of the exponent in the polynomial part of the complexity, and the search for a polynomial time algorithm. In larger characteristic, the methods that have been recently discovered cannot be applied directly. Moreover, these methods deeply rely on specific properties of polynomials which do not seem readily adaptable to numbers. Yet, the $L(1/3)$ complexity no longer seems to be a natural bound, and one could possibly expect progress for the NFS in this range, stemming from totally new ideas. There is also a possibility for such eventual progress to improve the complexity of factoring.

The most difficult challenge for discrete logarithms is probably the search for a subexponential algorithm that would apply to general elliptic curves defined over large characteristic fields. However, even finding new index calculus algorithms to cover additional special cases of curves is already a very challenging and fascinating problem in this field of research.

References

[Adl79] L.M. Adleman, A subexponential algorithm for the discrete logarithm problem with applications to cryptography (abstract), in *FOCS* (1979), pp. 55–60

[AFK89] M. Abadi, J. Feigenbaum, J. Kilian, On hiding information from an oracle. J. Comput. Syst. Sci. **39**(1), 21–50 (1989)

[AH99] L.M. Adleman, M.-D.A. Huang, Function field sieve method for discrete logarithms over finite fields. Inf. Comput. **151**(1–2), 5–16 (1999)

[BBG05] D. Boneh, X. Boyen, E.-J. Goh, Hierarchical identity based encryption with constant size ciphertext, in *EUROCRYPT* (2005), pp. 440–456

[BD94] M. Burmester, Y. Desmedt, A secure and efficient conference key distribution system (extended abstract), in *EUROCRYPT* (1994), pp. 275–286

[BF03] D. Boneh, M.K. Franklin, Identity-based encryption from the Weil pairing. SIAM J. Comput. **32**(3), 586–615 (2003)

[BGJT13] R. Barbulescu, P. Gaudry, A. Joux, E. Thomé, A quasi-polynomial algorithm for discrete logarithm in finite fields of small characteristic. CoRR (2013). abs/1306.4244

[BLS04] D. Boneh, B. Lynn, H. Shacham, Short signatures from the Weil pairing. J. Cryptol. **17**(4), 297–319 (2004)

[BLS11] D.J. Bernstein, T. Lange, P. Schwabe, On the correct use of the negation map in the Pollard Rho method, in *Public Key Cryptography* (2011), pp. 128–146

[BP14] R. Barbulescu, C. Pierrot, The multiple number field sieve for medium and high characteristic finite fields. IACR Cryptol. ePrint Arch. **2014**, 147 (2014)

[CEP83] E.R. Canfield, P. Erdös, C. Pomerance, On a problem of Oppenheim concerning factorisatio numerorum. J. Number Theory **17**, 1–28 (1983)

[CGH00] R. Canetti, O. Goldreich, S. Halevi, The random oracle methodology, revisited. CoRR (2000). cs.CR/0010019

[CHK12]   J.H. Cheon, J. Hong, M. Kim, Accelerating Pollard's Rho algorithm on finite fields. J. Cryptol. **25**(2), 195–242 (2012)

[Cop84]   D. Coppersmith, Fast evaluation of logarithms in fields of characteristic two. IEEE Trans. Inf. Theory **30**(4), 587–593 (1984)

[COS86]   D. Coppersmith, A.M. Odlyzko, R. Schroeppel, Discrete logarithms in GF(p). Algorithmica **1**(1), 1–15 (1986)

[Den82]   D.E. Denning, *Cryptography and Data Security* (Addison-Wesley, Reading, 1982)

[Den02]   A.W. Dent, Adapting the weaknesses of the random oracle model to the generic group model, in *ASIACRYPT* (2002), pp. 100–109

[DH76]   W. Diffie, M.E. Hellman, New directions in cryptography. IEEE Trans. Inf. Theory **22**(6), 644–654 (1976)

[DK13]   C. Diem, S. Kochinke, Computing discrete logarithms with special linear systems. Preprint (2013)

[DOW92]   W. Diffie, P.C. Oorschot, M.J. Wiener, Authentication and authenticated key exchanges. Des. Codes Cryptogr. **2**(2), 107–125 (1992)

[EGT11]   A. Enge, P. Gaudry, E. Thomé, An $L(1/3)$ discrete logarithm algorithm for low degree curves. J. Cryptol. **24**(1), 24–41 (2011)

[FJM13]   P.-A. Fouque, A. Joux, C. Mavromati, Multi-user collisions: applications to discrete logs, Even-Mansour and prince. IACR Cryptol. ePrint Arch. **2013**, 761 (2013)

[FPPR12]   J.-C. Faugère, L. Perret, C. Petit, G. Renault, Improving the complexity of index calculus algorithms in elliptic curves over binary fields, in *EUROCRYPT* (2012), pp. 27–44

[FR94]   G. Frey, H. Georg Rück, A remark concerning m-divisibility and the discrete logarithm in the divisor class group of curves. Math. Comput. **62**, 865–874 (1994)

[FS86]   A. Fiat, A. Shamir, How to prove yourself: practical solutions to identification and signature problems, in *CRYPTO* (1986), pp. 186–194

[Gam85]   T. El Gamal, A public key cryptosystem and a signature scheme based on discrete logarithms. IEEE Trans. Inf. Theory **31**(4), 469–472 (1985)

[GGMZ13]   F. Göloglu, R. Granger, G. McGuire, J. Zumbrägel, On the function field sieve and the impact of higher splitting probabilities—application to discrete logarithms in and, in *CRYPTO (2)* (2013), pp. 109–128

[GHS02]   P. Gaudry, F. Hess, N.P. Smart, Constructive and destructive facets of Weil descent on elliptic curves. J. Cryptol. **15**(1), 19–46 (2002)

[GKZ14]   R. Granger, T. Kleinjung, J. Zumbrägel, On the powers of 2. Cryptology ePrint Archive, Report 2014/300 (2014)

[Gor93]   D.M. Gordon, Discrete logarithms in GF(p) using the number field sieve. SIAM J. Discrete Math. **6**(1), 124–138 (1993)

[GTTD07]   P. Gaudry, E. Thomé, N. Thériault, C. Diem, A double large prime variation for small genus hyperelliptic index calculus. Math. Comput. **76**(257), 475–492 (2007)

[HR82]   M.E. Hellman, J.M. Reyneri, Fast computation of discrete logarithms in GF(q), in *CRYPTO* (1982), pp. 3–13

[JL02]   A. Joux, R. Lercier, The function field sieve is quite special, in *ANTS* (2002), pp. 431–445

[JL03]   A. Joux, R. Lercier, Improvements to the general number field sieve for discrete logarithms in prime fields. A comparison with the gaussian integer method. Math. Comput. **72**(242), 953–967 (2003)

[JL06]   A. Joux, R. Lercier, The function field sieve in the medium prime case, in *EUROCRYPT* (2006), pp. 254–270

[JLSV06]   A. Joux, R. Lercier, N.P. Smart, F. Vercauteren, The number field sieve in the medium prime case, in *CRYPTO* (2006), pp. 326–344

[JN03]   A. Joux, K. Nguyen, Separating decision Diffie-Hellman from computational Diffie-Hellman in cryptographic groups. J. Cryptol. **16**(4), 239–247 (2003)

[Jou04]   A. Joux, A one round protocol for tripartite Diffie-Hellman. J. Cryptol. **17**(4), 263–276 (2004)

[Jou13a]   A. Joux, Faster index calculus for the medium prime case application to 1175-bit and 1425-bit finite fields, in *EUROCRYPT* (2013), pp. 177–193

[Jou13b]   A. Joux, A new index calculus algorithm with complexity $L(1/4+o(1))$ in very small characteristic. IACR Cryptol. ePrint Arch. **2013**, 95 (2013)

[JP13]   A. Joux, C. Pierrot, The special number field sieve in finite fields - application to pairing-friendly constructions, in *Pairing* (2013), pp. 45–61

[JV12]   A. Joux, V. Vitse, Cover and decomposition index calculus on elliptic curves made practical—application to a previously unreachable curve over $\mathbb{F}_{p^6}$, in *EUROCRYPT* (2012), pp. 9–26

[Kra22]   M. Kraïtchik, *Théorie des nombres* (Gauthier-Villars, Paris, 1922)

[KS01]   F. Kuhn, R. Struik, Random walks revisited: extensions of Pollard's Rho algorithm for computing multiple discrete logarithms, in *Selected Areas in Cryptography* (2001), pp. 212–229

[LO90]   B.A. LaMacchia, A.M. Odlyzko, Solving large sparse linear systems over finite fields, in *CRYPTO* (1990), pp. 109–133

[MOV93]   A. Menezes, T. Okamoto, S.A. Vanstone, Reducing elliptic curve logarithms to logarithms in a finite field. IEEE Trans. Inf. Theory **39**(5), 1639–1646 (1993)

[MW96]   U.M. Maurer, S. Wolf, Diffie-Hellman oracles, in *CRYPTO* (1996), pp. 268–282

[Odl85]   A.M. Odlyzko, Discrete logarithms in finite fields and their cryptographic significance. Adv. Cryptol. **209**, 224–314 (1985)

[Pai99]   P. Paillier, Public-key cryptosystems based on composite degree residuosity classes, in *EUROCRYPT* (1999), pp. 223–238

[PGF98]   D. Panario, X. Gourdon, P. Flajolet, An analytic approach to smooth polynomials over finite fields, in *ANTS* (1998), pp. 226–236

[PH78]   S.C. Pohlig, M.E. Hellman, An improved algorithm for computing logarithms over gf(p) and its cryptographic significance (corresp.). IEEE Trans. Inf. Theory **24**(1), 106–110 (1978)

[Pol75]   J. Pollard, A Monte Carlo method for factorization. BIT Numer. Math., **15**, 331–334 (1975)

[Pol78]   J. Pollard, Monte Carlo methods for index computations mod p. Math. Comput., **32**(143), 918–924 (1978)

[Pom87]   C. Pomerance, Discrete Algorithms and Complexity: Proceedings of the Japan-US Joint Seminar, June 4-6, 1986, Kyoto, Japan, D. S. Johnson, T. Nishizeki, A. Nozaki and H. S. Wilf (Editors), Academic Press, New York, (1987)

[PQ12]   C. Petit, J.-J. Quisquater, On polynomial systems arising from a Weil descent, in *ASIACRYPT* (2012), pp. 451–466

[QD89]   J.-J. Quisquater, J.-P. Delescaille, How easy is collision search. New results and applications to DES, in *CRYPTO* (1989), pp. 408–413

[Sch89]   C.-P. Schnorr, Efficient identification and signatures for smart cards, in *CRYPTO* (1989), pp. 239–252

[Sch00]   O. Schirokauer, Using number fields to compute logarithms in finite fields. Math. Comput. **69**(231), 1267–1283 (2000)

[Sem04]   I. Semaev, Summation polynomials and the discrete logarithm problem on elliptic curves. IACR Cryptol. ePrint Arch. **2004**, 31 (2004)

[Sha71]   D. Shanks, Class number, a theory of factorization and genera, in *Proceedings of the Symposium on Pure Mathematics* (1971), pp. 415–440

[Sho97a]   P.W. Shor, Polynomial-time algorithms for prime factorization and discrete logarithms on a quantum computer. SIAM J. Comput. **26**(5), 1484–1509 (1997)

[Sho97b]   V. Shoup, Lower bounds for discrete logarithms and related problems, in *EUROCRYPT* (1997), pp. 256–266

[SWD96]   O. Schirokauer, D. Weber, T.F. Denny, Discrete logarithms: the effectiveness of the index calculus method, in *ANTS* (1996), pp. 337–361

[Tes00]   E. Teske, On random walks for Pollard's Rho method. Math. Comput. **70**, 809–825 (2000)

[vOW99]   P.C. van Oorschot, M.J. Wiener, Parallel collision search with cryptanalytic applications. J. Cryptol. **12**(1), 1–28 (1999)
[Wie86]   D.H. Wiedemann, Solving sparse linear equations over finite fields. IEEE Trans. Inf. Theory **32**(1), 54–62 (1986)

Isogenies in Theory and Praxis

Gerhard Frey

Abstract We want to give an overview on arithmetical aspects of abelian varieties and their torsion structures, isogenies, and resulting Galois representations. This is a wide and deep territory with a huge amount of research activity and exciting results ranging from the highlights of pure mathematics like the proof of Fermat's last theorem to stunning applications to public-key cryptography. Necessarily we have to be rather superficial, and thus specialists in the different aspects of the topics may be disappointed. But I hope that for many, and in particular for young researchers, the chapter may serve as an appetizer and will raise interest for a fascinating area of mathematics with many open problems (some are very hard and worth a Fields Medal but others are rather accessible).

The first section of the chapter gives basic notions, definitions, and properties of abelian varieties. Disguised as examples one will find their theory over the complex numbers $\mathbb{C}$ and the special case of elliptic curves. The second section discusses the situation over finite fields, in particular the role of the Frobenius endomorphism, and over number fields where the most interesting results and challenging conjectures occur. Finally we discuss algorithmic aspects of isogenies, mostly of elliptic curves, and relations to cryptography.

1 General Theory

We begin by explaining the background of the subjects we shall discuss in the chapter. Instead of citing a large number of original papers, we mostly refer to the handbook [ACF] where the reader can find all relevant items mentioned below

This chapter is based on a lecture presented at the conference "Open Problems in Mathematical and Computational Sciences" in Istanbul. I would like to thank the organizers for the opportunity to participate in this very interesting and inspiring conference and for their warm hospitality in the most beautiful environment.

G. Frey (✉)
Institute for Experimental Mathematics, University of Duisburg-Essen, Essen, Germany
e-mail: frey@exp-math.uni-essen.de

© Springer International Publishing Switzerland 2014
Ç.K. Koç (ed.), *Open Problems in Mathematics and Computational Science*,
DOI 10.1007/978-3-319-10683-0_3

37

discussed on different levels of abstraction and with an extensive bibliography
helping to go deeper to details in his/her favorite subjects. The second standard
reference will be [M1] where the background for abelian varieties is explained.

1.1 Abelian Varieties

1.1.1 Notations and Definitions

In the whole chapter K denotes a field with $\mathrm{char}(K) = p \geq 0$, and overfields
containing K are denoted by L.

K_s is a fixed separable closure of K.

The *absolute Galois group* $G_K = \mathrm{Aut}_K(K_s)$ is the group of field automorphisms
of K_s that leave elements of K fixed.

G_K has a natural topology as *profinite group* in which subgroups of finite index
form a system of neighborhoods of the unit element. It is important that G_K is
compact with respect to this topology.

Affine Varieties Affine varieties $V_a \subset \mathbb{A}^n$ are zero sets of ideals I_{V_a} with
coordinate ring $K[X_1, \ldots, X_n]/I_{V_a}$ and, if I_{V_a} is a prime ideal, with function field
$F_{V_a} = Quot(K[X_1, \ldots, X_n]/I_{V_a})$. In this case V is *irreducible* in the Zariski
topology and the dimension $\dim(V_a)$ of V_a is the transcendental degree of F_{V_a}
over K.

Example 1 1. $\mathbb{A}^n$ is the affine space defined by the zero ideal in $K[X_1, \ldots, X_n]$.
2. Take $n = 2$ and $I_a = < f(X_1, X_2) > \neq \{0\}$. Then V_a is the *plane affine curve*
   defined by the equation

$$f(X_1, X_2) = 0.$$

Its coordinate ring is $K[X_1, X_2]/ < f(X_1, X_2) >$.

It is *irreducible* iff $f(X_1, X_2)$ is an irreducible polynomial.

In this case $F(V_a)$ is an algebraic extension of $K(X_i)$ iff $f(X_1, X_2)$ is not
constant as function of X_i.

For overfields L of K define

$$V_a(L) = \{x = (x_1, \ldots, x_n) \in L^n; \ f(x) = 0 \ \forall f \in I_{V_a}\}.$$

So $\mathbb{A}^n(L) = L^n$.

A *morphism* ϕ is a *polynomial* map from V_a to an affine variety W_a.

It induces a map ϕ^* of the coordinate ring of W_a to the coordinate ring of V_a, which extends to an inclusion

$$F_{W_a} \hookrightarrow F_{V_a}$$

if the ideals defining W_a and V_a are prime ideals.

Example 2 Let V_a be an irreducible plane affine curve. Take $W_a = \mathbb{A}^1$, $\phi(X_1) = X_1, \phi(X_2) = 0$.

Then ϕ is the projection of V_a to the line $X_2 = 0$.

Assume that ϕ is not the constant map. Then ϕ^* induces the natural injection $K(X_1) \subset K(X_1, X_2)$.

Projective Varieties The next important step is to define *projective* varieties. Recall that a polynomial $F(Y_0, \ldots, Y_n)$ is homogenous of degree d iff every monomial occurring in F with coefficient $\neq 0$ has degree d.

An ideal $I \subsetneq K[Y_0, \ldots, Y_n]$ is homogenous iff it is generated by homogenous polynomials.

For elements y, y' in $L^{n+1} \setminus \{(0, 0, \ldots 0)\}$, define

$$y \sim y' \text{ iff there is } \lambda \in L^* \text{ with } y = \lambda \cdot y'.$$

A *projective variety* V defined over K is the zero set $\mod \sim$ of a homogenous ideal $I_V \subset K[Y_0, \ldots, Y_n]$ for appropriate n. The L-rational points of V are

$$\{y = (y_0, \ldots, y_n) \in L^{n+1}; f(y) = 0 \forall f \in I_V\}/ \sim .$$

Example 3 1. The projective space $\mathbb{P}^n/K$ is the projective variety defined by the zero ideal in $K[Y_0, \ldots, Y_n]$. Its L-rational points are $\mathbb{P}^n(L) = L^{n+1}/\sim .$
2. Take $n = 2$ and $I = < F(X, Y, Z) >$ where F is a homogenous polynomial of degree d. Then V is the *plane projective curve* defined by the equation

$$F(X, Y, Z) = 0$$

It is *irreducible* iff F is an irreducible polynomial.

Affine Covers of Projective Varieties We recall the easy observation that every homogenous polynomial $F(Y_0, \ldots, Y_n)$ can be transformed into $n + 1$ polynomials $f_j(X)$ ($j = 0, \ldots, n$) in n variables by the transformation

$$t_j : Y_i \mapsto X_i := Y_i/Y_j.$$

We remark that t_j can be interpreted as rational map from $\mathbb{P}^n$ to $\mathbb{A}^n$ which is defined and bijective when restricted to U_j consisting of points with Y_j coordinates $\neq 0$. By the inverse transform, we embed $\mathbb{A}^n$ into $\mathbb{P}^n$ and so U_j is isomorphic to $\mathbb{A}^n$ as affine variety. Inside of $\mathbb{P}^n$ it is an open subset in the Zariski topology.

As result we get a finite open covering of $\mathbb{P}^n$ by $n + 1$ affine subspaces.

Remark 1 There are many possibilities to find such covers. But having chosen homogenous coordinates $(Y_0, \ldots, Y_n)$, the above cover is rather usual, and one occasionally calls the projective variety $U_0 : Y_0 = 0$ "infinite hyperplane."

Having an affine cover U_j of $\mathbb{P}^n$, one can intersect it with projective varieties V and get

$$V = \bigcup_j V_{j,a} \text{ with } V_{j,a} := V \cap U_j$$

as union of affine varieties.

Converse process: Given a polynomial $f(X_1, \ldots, X_n)$ of degree d, we get a homogenous polynomial $f^h(Y_0, \ldots, Y_n)$ of degree d by the transformation

$$X_i \mapsto Y_i / Y_0 \text{ for } i = 1, \ldots, n$$

and then clearing denominators.

Assume that V_a is an affine variety with ideal $I_a \subset K[X_1, \ldots, X_n]$. By applying the homogenization explained above to all polynomials in I_a, we get a homogenous ideal $I_a^h \subset K[Y_0, \ldots, Y_n]$ and a projective variety V with ideal I_a^h containing V_a in a natural way.

V is called a projective closure of V_a.

A bit misleading one calls $V \cap U_0 = V \setminus V_a$ "infinite points" of V_a.

Example 4 Take

$$f(X_1, X_2) = X_2^2 + a_1 X_1 X_2 + a_3 X_2 - X_1^3 - a_2 X_1^2 - a_4 X_1 - a_6 \text{ with } a_i \in K$$

and denote by E_a the corresponding affine plane curve.

Introducing the variable Y_0, we define the homogenized polynomial

$$F(Y_0, Y_1, Y_2) = Y_0 Y_2^2 + a_1 Y_0 Y_1 Y_2 + a_3 Y_0^2 Y_2 - Y_1^3 - a_2 Y_0 Y_1^2 - a_4 Y_0^2 Y_1 - a_6 Y_0^3.$$

The corresponding plane projective curve is denoted by E.

Then $E \setminus E_a$ consists of exactly one point P_∞ that is the projective class of $(0, 0, 1)$.

Remark 2 Example 4 introduces an important object. If E_a has no singular points,[1] then E_a is an *elliptic curve given by a Weierstrass equation* (see Definition 3).

---

[1] That is the tangent space of every point of E_a has dimension 1; see [ACF], Sect. 4.4.1.

A morphism between projective varieties V and W is a map from V to W that is, restricted to any affine piece of V, an affine morphism (i.e., a polynomial map) to an affine piece of W.

If V is a projective variety whose ideal I_V is a prime ideal, then the function field F_V of V is the function field of a non-empty affine Zariski-open part V_a of V. (This is independent of the choice of V_a.)

In this case the dimension of V is the transcendental degree of F_V over K.

Group Schemes For more details and proofs concerning the following notions and results, we refer to [ACF] or [M1], Chap. III, 11.

Definition 1 A group scheme is an affine or projective variety G with a morphism

$$\oplus : G \times G \to G,$$

the *addition law*, a morphism

$$\iota : G \to G,$$

the *inversion morphism* and a unit element

$$e \in G(K),$$

in a more highbrow language, the *zero section*, satisfying the axioms of composition in groups interpreted in the language of morphisms.

1. Associativity expressed as identity between maps from $G \times G \times G$ to G:

$$\oplus \circ (\oplus \times id_G) = \oplus \circ (id_G \times \oplus).$$

2. Existence of a neutral element:

$$\oplus_{|\{e\} \times G} = pr_2(\{e\} \times G)$$

where pr_2 is the projection to the second factor of the Cartesian product.

3. Existence of inverse elements:

$$\oplus \circ (id_G \times \iota)$$

is the constant map with image point e.

If the addition law is commutative, i.e., it is compatible with interchanging the components in $G \times G$, then G is a commutative group scheme.

We remark that for all overfields L of K, we get that $G(L)$ is a group; the addition law in $G(L)$ is given by rational functions with coefficients in K, and so for all fields $K \subset L \subset K_s$, the Galois group G_L acts on $G(K_s)$ with $G(L) = G(K_s)^{G_L}$.

Example 5 Define μ_n as affine variety with ideal generated by

$$X_1^n - 1$$

or homogeneously by

$$Y_1^n - Y_0^n.$$

Define

$$\oplus : \mu_n \times \mu_n \to \mu_n$$

by

$$(X_1, X_2) \mapsto Z_1 = X_1 \cdot X_2.$$

e is the point $X_1 = 1$ and $\iota(X_1) := X_1^{n-1}$.

The resulting group scheme is the scheme of the nth roots of unity.

For overfields L of K, one gets that $G(L)$ is the group of elements ζ in L with $\zeta^n = 1$.

Here comes the key subject for the chapter:

Definition 2 An abelian variety A is an **absolutely**[2] **irreducible projective group scheme**.

Because of the importance for theory and practice, the case $d = 1$ deserves an extra definition.

Definition 3 An abelian variety of dimension 1 is called **elliptic curve** E.

Theorem 1 *Let A be an abelian variety. Then A is a commutative group scheme, and hence, $A(L)$ is an abelian group.*

A proof of this result can be found in [M1], Chap. 2.4.

Example 6 (Abelian varieties over $\mathbb{C}$) We shall sketch the "classical" case: $K = \mathbb{C}$. For details we refer to [ACF], Section 5.1 or [M1], Chapter I.

Projective varieties are compact analytic varieties.

Let A be an abelian variety over $\mathbb{C}$ and denote by $A_{\mathbb{C}}$ the associated analytic variety. From the classification of compact commutative Lie groups it follows that

$$A_{\mathbb{C}} \cong \mathbb{C}^d / \Lambda \text{ with } d = \dim(A)$$

---

[2]That is, irreducible as variety over K_s.

and

$$\Lambda = \mathbb{Z}^d \oplus \Omega \mathbb{Z}^d$$

with symmetric *period matrix* Ω whose imaginary part $Im(\Omega)$ is positive definite. Hence, Ω is an element in the *Siegel upper half plane* $\mathbb{H}_d$. Ω is determined up to transformations with elements in $Sp(d, \mathbb{Z})$, the group of symplectic matrices with determinant 1 and integral entries.

The equivalence classes of elements of $\mathbb{H}_d$ modulo $Sp(d, \mathbb{Z})$ form a **moduli space** for abelian varieties of dimension d defined over C.

It is worthwhile to look at the special case $d = 1$, i.e., A is an elliptic curve E. Ω is a 1×1 matrix with entry

$$\tau \in \mathbb{H} := \{z \in \mathbb{C}; Im(z) > 0\}.$$

τ is unique up to Möbius transformations

$$z \mapsto \frac{az + b}{cz + d}$$

with elements $\begin{pmatrix} a & b \\ c & d \end{pmatrix} \in Sl(2, \mathbb{Z})$. To emphasize this connection we sometimes denote E by E_τ.

To find an equation for the curve E, one uses the j-function and so defines a one-to-one cover map from $\mathbb{H}/Sl(2/\mathbb{Z})$ to the affine line.

This very explicit theory provokes the question:

Can one find algebraic versions of period matrixes to define explicit moduli spaces for abelian varieties?

For $d = 1$ we have the very satisfying algebraic theory of elliptic curves that will be discussed below.

Much more difficult is the situation for $d > 1$.

The first groundbreaking step was done in a series of three celebrated papers of Mumford [M2] where he "translated" the classical theory of theta functions into an algebraic frame and introduced theta groups and used theta null points to define points corresponding to abelian varieties (with level structure) on the moduli space.

From the computational point of view, this representation is not optimal since the degree of the defining equations and the number of variables is large. An enormous step forward is done by recent work of Lubicz, Robert, Faugère, Gaudry and others and can be found in the beautiful paper [LR].

It opens a wide area for computational research, and so we encourage to go deeper to the (partly solved)

Open Problem 1 *Find fast algorithms to compute moduli points for given[3] abelian varieties over finite or $\mathfrak{p}$-adic fields, and conversely, attach to moduli points the corresponding abelian varieties with addition law as explicit and efficient as possible.*

1.2   Homomorphisms of Group Schemes

Let G_1, G_2 be group schemes defined over K.

Definition 4   1. A morphism

$$\phi : G_1 \rightarrow G_2$$

is a homomorphism iff it is compatible with the addition laws in G_i, i.e.,

$$\oplus_{G_2} \circ (\phi \times \phi) = \phi \circ \oplus_{G_1}.$$

In particular, ϕ induces a group homomorphism from $G_1(L)$ to $G_2(L)$ that is given by rational functions defined over K and hence compatible with the action of G_K on points over K_s.

The set of homomorphism from G_1 to G_2 defined over K is denoted by $\mathrm{Hom}_K(G_1, G_2)$.

2. The kernel $\ker \phi$ is the scheme-theoretical inverse image of the zero section of G_2 under ϕ.

It is a subgroup scheme of G_1.

Its K_s-rational points are the K_s-rational points of G_1 mapped under ϕ to e_{G_2}.

3. $\phi \in \mathrm{Hom}_K(G_1, G_2)$ is an isogeny iff:

   (a) $\ker(\phi)$ is a finite group scheme.
   (b) The image under ϕ of the connected component of the unit element of G_1^0 of G_1 in the Zariski topology has the same dimension as the connected component of the unit element of G_2. For instance, if G_1 and G_2 are irreducible, then $\dim(G_1) = \dim(G_2)$.

1.2.1   Isogenies of Abelian Varieties

Let A, B be abelian varieties.

First we note a remarkable "rigidity property" of abelian varieties.

---

[3]For example, by homogenous equations.

Theorem 2 *A morphism*

$$\phi : A \to B$$

is a homomorphism iff $\phi(0_A) = 0_B$.

The proof can be found in [M1], Chapter II, Corollary 1.

Now assume that $\phi \in \mathrm{Hom}_K(A, B)$ is an isogeny.

By definition $\ker(\phi)$ is a finite group scheme and $\dim A = \dim B$. So ϕ induces an embedding ϕ^* of finite index of the function field F_B into F_A.

The degree of ϕ is $[F_A : \phi^*(F_B)]$, its separable degree is $[F_A : \phi^*(F_B)]_s ep$.

ϕ is separable if its degree is equal to its separable degree, and this is so iff $\ker(\phi)$ is an étale group scheme.

In this case

$$| \ker(\phi)(K_s)| = \deg(\phi)$$

and $\ker(\phi)(K_s)$ is a G_K-module that determines ϕ uniquely.

Example 7 Take $n \in \mathbb{N}$ and $A = B$. Define the map $[n]$ as $n - 1$-fold composition of $\oplus_A$.

Then $[n]$ is an isogeny that maps A to A and hence is an isogeny in $\mathrm{End}_K(A) := \mathrm{Hom}_K(A, A)$.

The kernel of [n] is denoted by $A[n]$ and its points are called *n*-torsion points.

$[n]$ is separable iff n is prime to $\mathrm{char}(K) = p$.

The separable degree of $[p]$ is p^k with $0 \leq k \leq \dim_K(A)$. k is the p-rank of A and A is ordinary iff $k = \dim_K(A)$.

Scalar Multiplication We assume that A is an abelian variety of positive dimension.

For negative integers z, define $[z] = \iota_A[-z]$ and denote by $[0]$ the constant map with image e_A. One checks very easily that these definitions yield an injection of $\mathbb{Z}$ into $End_K(A)$. We mention without proof that one knows more: For "generic" abelian varieties we get that $\mathrm{End}_K(A) = \mathbb{Z}$, and abelian varieties for which this equality does not hold have usually interesting properties (see Example 8 below for elliptic curves).

The induced operation of $\mathbb{Z}$ on A is called scalar multiplication and is very important both for theoretical and practical applications. Hence, there is much work invested in order to develop fast algorithms to evaluate $[n]$.

A prominent example is to expand n dyadically and then use addition and doubling (i.e., evaluation of [2]) to get an algorithm of complexity polynomially in $\log(n)$. But there are many more refined ways applicable in generic or specific situations (e.g., using fast inversion, "dividing" by 2, using [3], and using the *Montgomery ladder*). Though a lot of work is done and there is a vast literature (see, for instance [ACF], Chapter 9), there is still room for faster algorithms in special situations. This is an interesting research area and motivates to formulate an

Open Problem 2 *Try to find optimal algorithms for scalar multiplication in interesting instances.*

Remark 3 Isogenies of abelian varieties are "quasi-isomorphisms": to $\phi : A \to B$ there exists an isogeny $\Psi : B \to A$ such that $\Psi \circ \phi = [\deg(\phi)]$. Hence, to be isogenous defines an equivalence relation defining *isogeny classes* of abelian varieties.

Example 8 We continue the discussion given in Example 6 and assume that A, B are abelian varieties over $\mathbb{C}$ of dimension d_A and d_B with lattices Λ_A and Λ_B.

Homomorphisms from A to B correspond to homomorphisms of the attached compact Lie algebras and hence are given by linear maps:

$$\alpha : \mathbb{C}^{d_A} \to \mathbb{C}^{d_B}$$

with the additional property that

$$\alpha(\Lambda_A) \subset \Lambda_B.$$

As a consequence we get that, up to isomorphisms, the isogenies from an abelian variety A correspond to sublattices Λ_B of rank d_A of Λ_A, and the degrees of the isogenies are equal to the indices of the sublattices in Λ_A.

In particular, the degree of $[n]$ is $n^{2 \dim(A)}$.

As application we determine the endomorphisms of elliptic curves E_τ given by the lattice $\mathbb{Z} \oplus \tau \mathbb{Z}$ with $\tau \in \mathbb{H}$. We look for isogenies η attached to $\alpha \in \mathbb{C}$ such that $\alpha = \mu_1 + \mu_2 \tau$ and $\alpha \cdot \tau = \lambda_1 + \lambda_2 \tau$ with $\mu_i, \lambda_i \in \mathbb{Z}$. Hence,

$$\mu_2 \tau^2 + (\mu_1 - \lambda_2)\tau - \lambda_1 = 0$$

and so we get that either $\mu_2 = 0$ and so $\eta = [\mu_1]$ or τ satisfies a quadratic polynomial over $\mathbb{Q}$ and all isogenies of E_τ are given by elements α in the imaginary quadratic field $\mathbb{Q}(\tau)$.

A closer look (see [De]) using more properties of elliptic curves shows that τ is an algebraic integer and that the isogenies of E_τ form an order[4] O_τ in $\mathbb{Q}(\tau)$.

It follows

Theorem 3 *The ring of endomorphism of elliptic curves E over fields of characteristic 0 is either equal to $\mathbb{Z}$ (generic case) or equal to an order in an imaginary quadratic field. In the second case, the period τ of E (interpreted in an obvious way over $\mathbb{C}$) is an integer in an imaginary quadratic field, and E has* **complex multiplication** *(or is a CM curve).*

---

[4]See Definition 2.81 in [ACF] or any textbook on algebraic number theory.

In particular, the ring of endomorphisms of an elliptic curves defined over a field of characteristic 0 is commutative.

Isogenies of Elliptic Curves and Modular Curves Let E be an elliptic curve defined over K.

A separable isogeny of E can be composed by a cyclic isogeny η of E of degree n (i.e., $\ker(\eta)(K_s)$ is a G_K- invariant cyclic subgroup of order n in $E[n](K_s)$) followed by a scalar multiplication.

Turning things round we look, for n prime to p, for the functor that associates to overfields L of K all pairs

$$\{(E, C_n)/L; \ C_n \ E \text{ elliptic curve over } L, \ C_n \subset E(K_s) \text{ cyclic of order } n, \ G_L\text{-invariant}\}/\sim$$

where $\sim$ denotes equivalence modulo isomorphisms of pairs.

This functor defines a *moduli problem* (over K) that has for $K = \mathbb{C}$ a geometric presentation. That means that there is a curve over $\mathbb{C}$ such that its points parameterize the above-described pairs for $K = \mathbb{C}$. The necessary ingredients for the construction of this curve are contained in Examples 6 and 8.

To be explicit, define

$$Y_0(N)_\mathbb{C} = \mathbb{H}/\Gamma_0(n)$$

with

$$\Gamma_0(n) = \left\{ \begin{pmatrix} a & b \\ c & d \end{pmatrix} \in Sl(2, \mathbb{Z}) \ c \equiv 0 \mod n \right\}.$$

This is an affine curve with a natural cover map to the affine line $\mathbb{H}/Sl(2, \mathbb{Z})$ parameterizing isomorphy classes of elliptic curves over C. Since isogenies of degree n of elliptic curves correspond to inclusions of lattices with index n, it follows that the points on $X_0(N)(\mathbb{C})$ parameterize isomorphy classes of pairs (E, η) of elliptic curves E with cyclic isogenies η of degree N over $\mathbb{C}$.

By general principles this yields the existence of the ***modular curve***

$$Y_0(n) \text{ defined over } \mathbb{Z}[1/n]$$

with $Y_0(N)$ isomorphic to $Y_0(N)_\mathbb{C}$ over $\mathbb{C}$ with the property that elements in $Y_0(n)(K)$ correspond to elliptic curves with cyclic isogenies of degree n.[5]

$Y_0(n)$ and its projective completion $X_0(n)$ (obtained by adding "cusps") is explicitly known and very well understood. It has a rich structure (keywords: Hecke operators and modular forms) that is responsible for deep connections with number theory, and we shall see below how the determination of rational points on modular

---

[5]Caution for specialists: because of the existence of twists, Y_0 is only a coarse moduli space.

curves leads to very interesting diophantine results and conjectures and hence to (deep and difficult) *open problems*.

We go back to the general situation and assume that A, B are abelian varieties over K. In the context of isogenies, natural questions arise, which we formulate as **Tasks:**

1. Decide whether A and B are isogenous,
2. If A is isogenous to B, find an isogeny (of low degree).
3. Compute explicitly the image B of a given isogeny of A when its kernel is known.
4. Compute explicitly the isogeny map from A to B if the kernel of the isogeny is known.

For elliptic curves a lot is known to solve these tasks (see [Le]). Nevertheless algorithmic problems are still open and challenging. We shall come back to this below.

The situation is much more difficult and unclear for higher dimensional abelian varieties. Here a big step forward (in particular for task 3) is made in [LR] and [FLR]. But many questions remain widely open if one asks the questions in this generality. For special cases the situation may be much better. As example see [S] or [FK2]. So it is a challenging

Open Problem 3 *Find interesting instances for which the tasks formulated above can be solved at least partly.*

1.2.2 ℓ-Adic and Galois Representations

The main reference for this subsection is [M1], Chapter IV. The facts with examples but mostly without proof can be found in [ACF].

Let as usual A be an abelian variety of dimension d and take $n \in \mathbb{N}$. In the whole subsection, we assume that n is prime to char(K).

We shall study $A[n]$ and derived objects.

For $K = \mathbb{C}$, it follows from Example 6 that as abelian groups

$$A[n] \cong (\mathbb{Z}/n)^{2d}.$$

By general arguments like Lefschetz principle and Hensel's lemma, we get that this is true in general:

$$A[n](K_s) \cong (\mathbb{Z}/n)^{2d}.$$

G_K acts on $A[n]$ and so yields a representation

$$\rho_{A,n} : G_K \to \text{Aut}((\mathbb{Z}/n)^{2d})$$

or, after a choice of a base in $A[n]$,

$$\rho_{A,n} : G_K \to Gl(2d, \mathbb{Z}/n).$$

Take a prime $\ell \neq p$ and $n = \ell^k$ and use the natural maps

$$[\ell] : A[l^{k+1}] \to A[\ell^k]$$

to define the projective limit

$$T_\ell(A) := \varprojlim_k A[\ell^k],$$

the ℓ-adic Tate module of A.

It follows that $T_\ell(A) \cong (\mathbb{Z}_\ell)^{2d}$ and that $V_\ell(A) := T_\ell(A) \otimes \mathbb{Q}_\ell$ is a $\mathbb{Q}_\ell$-vector space of dimension $2d$.[6]

G_K operates on $T_\ell(A)$. This action induces a $\mathbb{Z}_\ell$-adic representation attached to A given by the projective limit

$$\varprojlim_k \rho_{A,\ell^k}.$$

By tensorizing with $\mathbb{Q}_\ell$, we get the ℓ-adic representation

$$\tilde{\rho}_{A,\ell},$$

a representation of dimension $2d$ of G_K over the ℓ-adic numbers $\mathbb{Q}_\ell$ with representation space $V_\ell(A)$.

A quite similar construction can be made with homomorphisms

$$\phi : A \to B :$$

By restricting ϕ to $A[l^k]$, we get homomorphisms

$$\phi_{\ell^k} : A[\ell^k] \to B[\ell^k]$$

and so as projective limit an T_ℓ- homomorphism

$$\widetilde{\phi}_\ell : T_\ell(A) \to T_\ell(B),$$

which has a finite co-kernel if ϕ is an isogeny, and by tensorizing with $\mathbb{Q}_\ell$, we get an homomorphism between $V_\ell(A)$ and $V_\ell(B)$, also denoted by $\widetilde{\phi}_\ell$. It is easily seen that for isogenies ϕ, the map $\widetilde{\phi}_\ell$ restricted to $T_\ell(A)$ is injective, and it is an isomorphism between $V_\ell(A)$ and $V_\ell(B)$.

---

[6]$\mathbb{Z}_\ell$ is the ring of l-adic integers and $\mathbb{Q}_\ell$ the field of ℓ-adic numbers (see [ACF]).

We have a natural homomorphism from $\mathrm{Hom}_K(A, B)$ into $\mathrm{Hom}_{G_K}(T_\ell(A), T_\ell(B))$.

Taking $A = B$, we get an injective representation from $\mathrm{End}_K(A)$ into $\mathrm{End}_{G_K}(T_\ell(A))$, the group of endomorphisms of the $\mathbb{Z}_\ell$-module $T_\ell(A)$ that commute with the action of G_K. This representation is called the ℓ-adic representation of endomorphisms of A.

Remark 4 The Tate modules (and their p-adic counterpart, the Dieudonné module, which we do not discuss here) and the embedding of $\mathrm{Hom}_K(A, B)$ into $\mathrm{Hom}_{G_K}(T_\ell(A), T_\ell(B))$ play a key role for the study of abelian varieties, and they give a lot of information about the absolute Galois group of K (see [T] and [Fa]). They are the counterparts in the *étale cohomology* of the lattices in the complex theory.

Application: Endomorphisms of Elliptic Curves Every endomorphism $\eta \neq 0$ of E is an isogeny, and so $\mathrm{End}_K(E) \bigotimes \mathbb{Q}$ is a skewfield.

The action of $\mathrm{End}_K(E)$ on the ℓ-adic Tate module of E induces an injection of $\mathrm{End}_K(E)$ into $Gl(2, \mathbb{Z}_\ell)$.

From algebra it follows that $\mathrm{End}_K(E) \bigotimes \mathbb{Q}$ is equal to $\mathbb{Q}$, a quadratic field or a quaternion field. This information and some more ingredients from the theory of elliptic curves allow us to characterize $\mathrm{End}_K(E)$.

Case in which E cannot be defined over an absolute algebraic field (i.e., its absolute invariant j_E (see Example 4) is transcendental over its prime field): we get that $\mathrm{End}_K(E) = \mathbb{Z}$.

Case of number fields: We have seen already that over fields K of characteristic 0, the ring $\mathrm{End}_K(E)$ is commutative, and so quaternion fields are excluded.

Generically it is equal to $\mathbb{Z}$; in special cases we have complex multiplication (CM) and $\mathrm{End}_K(E)$ is an order in an imaginary quadratic field (see Example 8).

Case of finite fields: Over finite fields the generic case is the CM-case. In this case the elliptic curve E is ordinary, i.e., $E[p](K_s) \cong \mathbb{Z}/p$ (see 1.2.1).

If $[p]$ is purely inseparable, then $\mathrm{End}_K(E)$ is an order in a well-determined quaternion algebra and E is called *supersingular*. Supersingular elliptic curves are (up to twists) defined over $\mathbb{F}_{p^2}$ and isogenous to each other.

1.3 Jacobian Varieties

Till now abelian varieties occurred in a rather abstract way, and in spite of the work of Mumford and Lubicz–Robert, it is difficult and often too complicated to find explicit equations and addition laws.

The situation is much better for an important subclass of abelian varieties, which historically came first (already in the nineteenth century) and which motivated A. Weil to define abelian varieties: *Jacobian varieties attached to curves.*

Let C be a projective non-singular curve [7] of genus g over K (see [ACF], Definition 4.107) with *divisor group*

$$\mathscr{D}(K_s) := \{D = \sum_{P \in C(K_s)} z_P \cdot P; z_P \in \mathbb{Z} \text{ and almost all } z_P = 0\}.$$

The subgroup of divisors of degree 0 is

$$\mathscr{D}(K_s)^0 := \{D; \sum z_P = 0\}.$$

The Galois group G_K acts by linear extension in a natural way on $\mathscr{D}(K_s)$. For $K \subset L \subset K_s$, define

$$\mathscr{D}(L)^0 = (\mathscr{D}(K_s)^0)^{G_L}.$$

Examples for divisors of degree 0 are *principal divisors*: $0 \neq f \in F_C \cdot K_s$ has the principal divisor

$$(f) = \sum z_P P \text{ where } z_P \text{ is the order of vanishing of } f \text{ in } P.[8]$$

Obviously the set of principal divisors form a subgroup $\mathscr{P}$ of $\mathscr{D}(K_s)^0$. Define

$$\mathscr{P}(L) := \mathscr{P} \cap \mathscr{D}(L)^0$$

and

$$\mathrm{Pic}_C^0(L) := \mathscr{D}(L)^0 / \mathscr{P}(L),$$

the L-rational divisor class group of degree 0 of C.

Theorem 4 (Abel–Jacobi) *The functor*

$$L \mapsto \mathrm{Pic}_C^0(L)$$

*is representable by an abelian variety of dimension g, the **Jacobian variety** J_C, i.e., in a functorial way we have*

$$J_C(L) = \mathrm{Pic}_C^0(L).$$

---

[7] The tangent space of every point of C has dimension 1, see [ACF], Sect. 4.4.1

[8] Poles give rise to negative "order of vanishing".

The theorem of Riemann–Roch (*[ACF], Theorem 4.106*) *yields the following:*

$$J_C \text{ is birationally equivalent to } C^g / S_g$$

where S_g is the symmetric group of g letters acting on the g-fold Cartesian product of C by permuting the factors.

Hence, the addition on Jacobian varieties is reduced to the addition of divisor classes of curves, and the theorem of Riemann–Roch tells that there are distinguished representatives, namely, positive divisors of degree $\leq g$. It follows that addition of classes is possible if one can find for divisors of degree $\geq g + 1$ positive divisors in the same class but of degree $\leq g$.

Example 9 (Elliptic Curves as Jacobians) Assume that C is a projective regular curve of genus 1 **with a K-rational** point P_∞.

By the *theorem of Riemann–Roch* one gets the following: every L-rational divisor class c of degree 0 of E contains exactly one point $P \in C(L)$ with

$$P - P_\infty \in c.$$

The map

$$J_C(L) \to C$$

$$c \mapsto P$$

is an explicit isomorphism from $J_C(L)$ to $C(L)$.

Hence, C is an elliptic curve and $C(L)$ is an abelian group.

Weierstrass Equation The theorem of Riemann–Roch yields the following: we find a **Weierstrass equation** for E in the projective plane (see Example 4), and if $p \neq 2, 3$,[9] we can normalize to get

$$E : Y^2 Z = X^3 + aXZ^2 + bZ^3$$

with

$$\Delta_E = -16(4a^3 + 27b^2) \neq 0.$$

We refind the j-invariant that was classically defined as meromorphic function on $\mathbb{H}$: For $a = 0$, set $j_E = 0$; for $b = 0$ set $j_E = 12^3$; and for $ab \neq 0$, define

$$j_E = 12^3 \frac{-4a^3}{\Delta_E}.$$

---

[9]For $p|6$, see [ACF] 13.1.1 and 13.3.

We remark that j_E determines E up to twists and that to every $j \in K$ we find E with $j_E = j$ (see [ACF], 18.1.1). E has exactly one point with $Z = 0$. Choosing this point as $P_\infty = (0, 1, 0)$, we can describe the addition in coordinates and get the well-known *addition formulas*.

There is a vast literature in this area (see, for instance, [ACF] and many publications, e.g., by D. Bernstein and T. Lange), but nevertheless it is till nowadays not impossible to do even better, and so we formulate a (minor)

Open Problem 4 *Find optimal equations and algorithms for scalar multiplication for elliptic curves over given fields* $\mathbb{F}_q$ *(depending on the structure of* $\mathbb{F}_q$ *and the architecture of the used computer maybe).*

2 Abelian Varieties over Special Fields

2.1 $K = \mathbb{F}_q$

In this subsection we take $K = \mathbb{F}_q$, the field with $q = p^d$ elements, and denote by $\mathbb{F}_{p,\infty}$ its algebraic closure.

The Frobenius automorphism π_p of $\mathbb{F}_{p,\infty}$ is defined by

$$x \mapsto \pi_p(x) := x^p.$$

$\pi_q = \pi_p^d$ is a topological generator of the absolute Galois group $G_{\mathbb{F}_q}$ of $\mathbb{F}_q$.

2.1.1 The Frobenius Isogenie

We attach to the Galois element π_q a **geometric object** by extending its operation to points in $\mathbb{P}^n(\mathbb{F}_{p,\infty})$.

This yields a *homogenous polynomial map*

$$(X_0, \ldots, X_n) \mapsto (X_0^q, \ldots X_n^q)$$

and so the *Galois* element induces *morphisms* of varieties V over $\mathbb{F}_q$ which, by abuse of notation, we also denote by π_q.

We assume that V is irreducible. Going to affine pieces and choosing affine coordinates $X_1, \ldots, X_n$, one easily see that

$$\pi_q^*(V)$$

is the subfield of F_V generated by $X_1^q, \ldots, X_n^q$ and so $F_V/\pi_q^*(V)$ is purely inseparable of degree $q^{\dim(V)}$.

The Frobenius morphism π_q is compatible with polynomials with coefficients in K and so with the addition on abelian varieties A over $\mathbb{F}_q$. Hence, π_q is a purely inseparable isogeny of degree $q^{\dim(A)}$ called Frobenius endomorphism.

Since

$$\pi_q \in \mathrm{End}_{\mathbb{F}_q}(A) \setminus \mathbb{Z} \cdot id_A \neq \emptyset,$$

we get that $\mathrm{End}_{\mathbb{F}_q}(A)$ has elements different from scalar multiplications.

The Characteristic Polynomial of the Frobenius Endomorphism Since G_{Fq} is topologically generated by π_q, it follows that the representations $\rho_{A,n}$, respectively $\tilde{\rho}_{A,\ell}$ of abelian varieties A, are determined by $\rho_{A,n}(\pi_q)$ respectively $\tilde{\rho}_{A,\ell}(\pi_q)$.

A fundamental result of Tate [T] is that $\tilde{\rho}_{A,\ell}$ is a semi-simple representation, i.e., it is determined by its characteristic polynomial

$$\chi(T)(\tilde{\rho}_{A,\ell}(\pi_q)).$$

We vary the primes ℓ (always $\neq p$) and get a *globalization* that is due to A. Weil:

Theorem 5 $\chi(T)(\tilde{\rho}_{A,\ell}(\pi_q)) \in \mathbb{Z}[T]$ *is a monic polynomial* $\chi_{A,q}(T)$ *of degree* $2\dim(A)$ *independent of* ℓ, *and for all* $n \in \mathbb{N}$

$$\chi_{A,q}(T) \equiv \chi(T)(\rho_{A,n}(\pi_q)) \mod n.$$

It follows that $\chi_{A,q}(\pi_q)(A) = \{0_A\}$.

This theorem justifies the statement that $\chi_{A,q}(T)$ is *the characteristic polynomial* on A of π_q.

Point Counting Here comes one of the most important applications of the Frobenius endomorphism.

Since $A(\mathbb{F}_{p,\infty})^{G_{\mathbb{F}q}} = A(\mathbb{F}_q)$ and since $\pi_q - id_A$ is a separable isogeny, it follows from

$$A(\mathbb{F}_q) = \ker(\pi_q - id_A)$$

Theorem 6

$$|A(\mathbb{F}_q)| = \chi_{A,q}(1).$$

Hence a strategy to determine $|A(\mathbb{F}_q)|$ is to compute $\chi_{A,q}(T)$.

The deep basic result for these computations is due to Hasse ($d = 1$) and Weil ("Riemann hypothesis for curves"):

Theorem 7 *The eigenvalues of* π_q *are complex integers with absolute value equal to* $q^{1/2}$.

Hence, $| A(\mathbb{F}_q)| = q^{\dim(A)} + \mathcal{O}(q^{\dim(A)-1/2})$.

An immediate consequence is that the ith coefficient of $\chi_{A,q}(T)$ is an integer with absolute value bounded by $\binom{2\dim(A)}{i} q^{(2\dim(A)-i)/2}$ ([ACF], Corollary 5.8.2). Hence, to determine $\chi_{A,q}(T)$ is enough to compute an approximation of sufficient precision.

Example 10 For elliptic curves E defined over $\mathbb{F}_q$, we have

$$|| (E(F_q) | +1 - q | \le 2 \cdot q^{1/2}.$$

2.1.2 The Isogeny Theorem over Finite Fields

Finally we stress the importance of the Frobenius isogenies by the following result of Tate [T]:

Theorem 8 *Let A, B be abelian varieties defined over $\mathbb{F}_q$ with Tate modules $T_\ell(A)$ and $T_\ell(B)$.*

(i) *A is isogenous to B iff for one $\ell \ne p$, the Galois module $T_\ell(A) \otimes \mathbb{Q}$ is isomorphic to $T_\ell(B) \otimes \mathbb{Q}$.*

(ii) *A is isogenous to B iff the characteristic polynomials of the Frobenius endomorphisms on A and B are equal.*

We remark that this result "reduces" Task 1 in Sect. 1.2.1 to the computation of the characteristic polynomial of abelian varieties. We shall see in Sect. 3.2 how one can attack this task. Because of its importance, we formulate it already here as one major

Open Problem 5 *Find fast algorithms to compute for abelian varieties A defined over $\mathbb{F}_q$ the characteristic polynomial of the Frobenius endomorphism.*

2.2 Abelian Varieties over Number Fields

We look at the mathematically most interesting case: the field K is a number field, i.e., a finite algebraic overfield of $\mathbb{Q}$. The exciting task is to relate arithmetical properties of these fields with diophantine properties of geometric objects, and it turned out that abelian varieties are a very useful tool for this.

We begin with a by now classical result of Serre [Se1].

Theorem 9 *Assume that the elliptic curve E over K has no complex multiplication. There is a number n_E such that for all primes $\ell > n_E$, we have*

$$\rho_{E,\ell}(G_K) = Gl(2, \mathbb{Z}/\ell).$$

In particular E has only finitely many K-rational cyclic isogenies.

How can one determine n_E for given E?

   What are the exceptions?

Open Problem 6 (Conjecture Due to J.P. Serre) *Can one find n_0 depending only on K such that for all E (outside a finite exceptional set) $n_E = n_0$?*

Remark For $K = \mathbb{Q}$ and elliptic curves one knows more: Mazur has determined a list of all isogenies of all elliptic curve and exceptional small images of $\rho_{E,n}$ are understood (up to the non-split Cartan case).

   For general number fields K, the order of rational torsion points of elliptic curves over E can be bounded by an estimate depending on the degree of K over $\mathbb{Q}$ only (theorem of Merel and Parent).

Open Problem 7 *Can one generalize Theorem 9 to abelian varieties of dimension ≥ 2?, For example, is it true for abelian varieties with $\operatorname{End}_K(A) = \mathbb{Z}$ that for almost all rational primes ℓ, the image of $\rho_{A,\ell}$ contains $GSp(2\dim(A), \mathbb{Z}/\ell)$, the symplectic group of dimension $2\dim(A)$ over $\mathbb{Z}/\ell$?*

All results obtained in this direction rely on work of Serre [Se2]. Interesting progress is made by Hall in [Ha].

2.2.1 Local-Global Methods

How can one prove results like Theorem 9? Besides the specific properties of the investigated objects, one looks at the arithmetical structure of number fields given by a system of valuations with well-known completions.

   To be concrete take $K = \mathbb{Q}$.

   First, we have the absolute value $|\ |$ (an archimedean valuation) with completion $\mathbb{R}$ and algebraic closure $\mathbb{C}$.

   Next we have the ring of integers $\mathbb{Z}$ with prime ideals $p \cdot \mathbb{Z}$ which give rise to non-archimedean p-adic valuations w_p with

$$w_p(x) = \text{maximal power of } p \text{ dividing } x,$$

completion $\mathbb{Q}_p$, its algebraic closure $\mathbb{Q}_{p,s}$ with absolute Galois groups G_p, and residue field $\mathbb{F}_p$. It is crucial that G_p can be identified (uniquely up to conjugation) with a subgroup of $G_\mathbb{Q}$, the decomposition group of an extension of w_p to $\mathbb{Q}_s$.

   For general K, replace $|\ |$ by metrics induced by embeddings of K in $\mathbb{C}$, $\mathbb{Z}$ by its integral closure O_K in K and w_p by valuations attached to prime ideals $\mathfrak{p}$ of O_K containing p.

   Diophantine objects over K can be interpreted over the completions (*localization*) or modulo $\mathfrak{p}$ (*reduction*).

   This relates diophantine problems over finite fields, $\mathbb{C}$, p-adic fields, and number fields.

The aim is to get local-global information (going in both direction). Here is a first prominent example.

2.2.2 CM Theory

We use an embedding of K in $\mathbb{C}$ and look at elliptic curves E over K as

$$E = \mathbb{C}/(\mathbb{Z} + \tau\mathbb{Z}).$$

We recall that E has complex multiplication if τ is an algebraic integer generating an imaginary quadratic field $K_E := \mathbb{Q}(\tau)$ and then $\mathrm{End}_C(E)$ is an order $O_E \in O_{K_E}$.
   Class field theory tells more:
   The $\mathbb{C}$-isomorphy classes of elliptic curves E' isogenous to E correspond one-to-one to the ideal classes of orders O_E in O_{K_E}, the absolute invariant of E' generates the ring class fields H_E of O_E, and $\rho_{E,n}(G_{H_E})$ is an abelian group and so not containing $Sl(2, \mathbb{Z}/n)$.
   From number theory we know that for given n, there are only finitely many orders in imaginary quadratic fields with class number $\leq n$, and so there are, up to twists, only finitely many elliptic curves with CM defined over K(hence, only finitely many twist classes of elliptic curves are excluded in Theorem 9).
   The relation of elliptic curves with CM over number fields to elliptic curves over finite fields is given by a central result, **Deuring's lifting theorem**.

Theorem 10 *Let E be an ordinary elliptic curve over $\mathbb{F}_q$. There is an elliptic curve $\tilde{E}$ defined over a number field K and a prime ideal $\mathfrak{p}$ of O_K such that $\tilde{E}$ mod $\mathfrak{p} = E$ and $\mathrm{End}(\tilde{E}) = \mathrm{End}(E)$.*

Hence $\mathrm{End}(E)$ is an order in an imaginary quadratic field $K_{\tilde{E}}$ and the Frobenius endomorphism π_q corresponds to an imaginary quadratic algebraic integer with norm q. The discriminant of its characteristic polynomial $\chi_{E,q}(T) = (T - \lambda_1)(T - \lambda_2)$ is negative and so $\lambda_1\lambda_2 = q$ and $\mathrm{trace}(\phi_q)^2 - 4q < 0$. But then $(|E(\mathbb{F}_q)| - q - 1))^2 - 4q = \mathrm{trace}(\phi_q)^2 - 4q < 0$.
   So we get a proof (due to Deuring–Hasse) of the "Riemann hypothesis for elliptic curves" (Theorem 7):

$$||E(\mathbb{F}_q)| - q - 1)| < 2\sqrt{q}.$$

Due to **Shimura–Taniyama** there is a beautiful generalization of CM theory to abelian varieties of higher dimension replacing imaginary quadratic fields by CM-fields of larger degree. For abelian varieties of dimension 2 and 3 this is explained in [ACF], Chapter 18.

Open Problem 8 *Generalize the algorithmic aspects of CM from elliptic curves to Jacobians of curves of small genus.*

Remark 5 For curves of genus 2 and 3, part of the work is done in the theses of A. Spallek and A. Weng.

2.2.3 Local-Global Principles for Galois Representations

We go deeper into the arithmetic of number fields K.

Let $\mathfrak{p}$ be a prime of K, L a Galois extension of K and $\tilde{\mathfrak{p}}$ a prime in O_L that contains $\mathfrak{p}$ with residue field $\mathbb{F}_q$. Assume that $\mathfrak{p}$ is unramified in L/K.[10]

A **Frobenius automorphism** $\sigma_{\mathfrak{p}}$ is an element in $G(L/K)$ that is continuous with respect to the $\tilde{\mathfrak{p}}$-adic metric and which acts modulo $\tilde{\mathfrak{p}}$ like π_q.

We remark that $\sigma_{\mathfrak{p}}$ is determined by $\mathfrak{p}$ (only) up to conjugation.

Let V be a finite dimensional vector space over $\mathbb{C}$ or over a finite field $\mathbb{F}_q$ or over an ℓ-adic field. We endow V with either the discrete topology ($K = \mathbb{C}$ or $K = \mathbb{F}_q$) or the ℓ-adic topology. Let

$$\rho : G_K \to \mathrm{Aut}(V)$$

be a continuous representation, which is *semi-simple*, i.e., ρ is determined by the characteristic polynomials of the images under ρ. We assume in addition that $K_s^{\ker(\rho)}/K$ is unramified outside of a finite set S of primes.

Theorem 11 (Density Theorem of Čebotarev) ρ *is uniquely determined by*

$$(\chi(\rho(\sigma_{\mathfrak{p}}))(T))_{\mathfrak{p} \notin S} \text{ prime of } O_K.$$

This theorem is the reason for the deep relations between Galois theory and arithmetic.

Remark 6 There is a constructive version of Theorem 11: given two representations

$$\rho_i : G_K \to \mathrm{Aut}(V); i = 1, 2$$

with $K_s^{\ker(\rho_1)} = K_s^{\ker(\rho_2)}$ there is a number n depending on arithmetical invariants of ρ_i like the discriminant of $K_s^{\ker(\rho_i)}$ such that

$$\rho_1 \cong \rho_2,$$

iff

$$\chi(\rho_1(\sigma_{\mathfrak{p}}))(T) = \chi(\rho_2(\sigma_{\mathfrak{p}}))(T) \text{ for all } \mathfrak{p} \text{ with } \mathrm{Norm}(\mathfrak{p}) \le n.$$

---

[10]That is, the normalized valuation attached to $\tilde{\mathfrak{p}}$ is a continuation of the one attached to $\mathfrak{p}$.

This result makes identification of Galois representation effective. Unfortunately, the bound n tends to be very large (even under the assumption of the generalized Riemann hypothesis GHR [Oe]), and so the result can only very rarely be used for computational investigations. But there are situations where one can do better, for instance, if one knows that the representations are related to modular forms [R].

Open Problem 9 *Find (or conjecture) effective versions of Theorem 11 in special but interesting instances.*

2.2.4 The Theorem of Faltings

Let A be an abelian variety defined over a number field.

Theorem 12 $\tilde{\rho}_{A,\ell}$ *is semi-simple.*

This is an extremely deep theorem obtained by Faltings in the celebrated paper [Fa]. Among others, it implies Mordell conjecture:

Curves of genus > 1 have only finitely many K-rational points
On the way to his result Faltings proved

Theorem 13 (Isogeny Theorem) *Abelian varieties A and B are isogenous iff for one prime ℓ*

$$\tilde{\rho}_{A,\ell} \cong \tilde{\rho}_{B,\ell}.$$

In fact Faltings proved that for given A, B there is a number $n(A, B)$ such that A is isogenous to B iff for one $n > n(A, B)$

$$\rho_{A,n} \cong \rho_{B,n}.$$

Warning: The following problem is difficult and is closely related to Open Problem 9.

Open Problem 10 *Give reasonable estimates for $n(A, B)$ in terms of the conductors of A, B. Hint:Look at the work of Masser–Wüstholz.*

2.2.5 Conjectures for Elliptic Curves

To show how deeply Galois representations and diophantine problem are related, we go to elliptic curves over number fields and formulate really challenging **OPEN PROBLEMS**, which, because of their importance and difficulty, are called

CONJECTURES.

They can be found in [FK1]. They express that, up to some exceptions, only isogenous elliptic curves should have groups of torsion points that are isomorphic as Galois modules.

Conjecture 1 (Darmon) *There is a number $n_0(K)$ such that for all elliptic curves E, E' over K and all $n \geq n_0(K)$ we get*

$$\textit{If } \rho_{E,n} \cong \rho_{E',n} \textit{ then } E \textit{ is isogenous to } E'.$$

A variant of this conjecture is

Conjecture 2 (Kani) *There is a number n_0 (independent of K) such that for $n \geq n_0$ there are, up to twist pairs, only finitely many pairs (E, E') of elliptic curves defined over K which are not isogenous and with $\rho_{E,n} \cong \rho_{E',n}$.*
For prime numbers n, we can choose $n_0 = 23$.

Much easier but also not proved is

Conjecture 3 (Frey) *We fix an elliptic curve E_0/K.*
There is a number $n_0(E_0, K)$ such that for all elliptic curves E over K and all $n \geq n_0(E_0, K)$ we get

$$\textit{If } \rho_{E,n} \cong \rho_{E_0,n} \textit{ then } E \textit{ is isogenous to } E_0.$$

We remark that this conjecture can be formulated in a much more general way ([Fr1], Conjecture 5), which is proved if we replace number fields by function fields in one variable.
We mention amazing consequences of this conjecture:
It implies the (in-)famous

ABC-conjecture

and the

asymptotic Fermat conjecture

and has implications to the theory of modular forms. These conjectures can also be found in [Fr1] (Conjecture 1 and Conjecture 2).
To give the flavor of these conjectures, we formulate a version of the ABC-conjecture over $\mathbb{Q}$ that is due to Masser and Oesterlé:

Conjecture 4 *For all $\epsilon \in \mathbb{R}_{>0}$ there is a number $c_\epsilon \in \mathbb{R}$ such that for integers A, B with $A \cdot B \neq 0$ and $\gcd(A, B) = 1$, we get*

$$|A| \leq c_\epsilon \cdot \left(\prod_{p | A \cdot B \cdot (A-B)} p \right)^{1+\epsilon}.$$

3 Algorithmic Aspects and Applications

In this section the focus lies on computational aspects of abelian varieties over finite fields $\mathbb{F}_q$. Many of the results are motivated and initiated by problems from public-key cryptography. A more detailed discussion of this fruitful interaction between algorithmic algebraic geometry and data security can be found in [Fr2] and [Fr3].

3.1 Addition on Jacobian Varieties over Finite Fields

Jacobian varieties are accessible to computations via curve arithmetic and enjoy the rich structure of abelian varieties. As first example we look at the addition on Jacobian varieties. We use the general theory of Jacobian varieties (Sect. 1.3) and recall that for the addition on them, one needs a reduction algorithm among divisors in the same class. This problem was solved by Heß [He] and by Diem and leads to an outstanding result inside of the rapidly progressing algorithmic algebraic geometry.

Theorem 14 (Diem, Heß) *Let C be a curve of genus g over $\mathbb{F}_q$.*
The arithmetic in the degree 0 class group of C can be performed in an expected time which is polynomially bounded in g and $\log(q)$.

In practice it is still challenging to find algorithms that are fast enough for applications. A lot of work is done (even for curves of genus 1) to find equations for C for which the addition is optimal, and till now there are many publications that give special fast addition algorithms for special instances of curves and fields. So we find an

Open Problem 11 *Implement the addition algorithm efficiently for Jacobian varieties of curves of low genus (e.g., $g \leq 4$) and find optimal equations (maybe depending on the field $\mathbb{F}_q$).*

3.2 Point Counting

A major task is the computation of the Frobenius endomorphism π_q.

This is motivated by the outstanding role this endomorphism plays in theory (Theorem 8) and practice (point counting).

Special (but nevertheless sufficiently "random") instances are found by using the CM-theory and hence to *begin* with the **ring of endomorphisms** of Jacobians over $\mathbb{C}$.

To compute the characteristic polynomial of π_q for large q and for "random" abelian varieties, one uses its action on an accessible vector space (usually a cohomology group) and an approximation algorithm. This becomes effective because of the Hasse-Weil estimates of the coefficients (Theorem 7).

To proceed one uses the whole arsenal of arithmetic geometry, namely:

- *étale cohomology* that leads to algorithms first introduced for elliptic curves by R. Schoof, which become practical for elliptic curves because of using isogenies instead of points (Atkin–Elkies), and so usually one calls them SEA-algorithms
- *p-adic cohomology* (work of Kedlaya, Vercauteren, Gerkmann, and many others)
- *p*-adic *lifting* by effective *p*-adic versions of Deuring's lifting theorem (Theorem 10) for elliptic curves and versions for higher dimension (keyword *canonical lifts*) given by *p*-adic theta functions, cf. Open Problem 1 (work of Satoh, Lubicz, Carls, Mestre, and many others)
- *deformation* theory (geometric-algebraic or differential-geometric) (Lauder, M. Li)

An extensive discussion of these methods can be found in [ACF], Chapter 17.

Result: In cryptographic relevant ranges we get:

- We can count points on random elliptic curves.
- We can count points on Jacobians of random curves over fields of small (and even medium) characteristic.
- We have still problems with random curves of genus 2 (but see work of Gaudry and Schost [GS] and [CL]), and we have many *special* families of curves whose members are accessible for point counting (e.g., by CM-methods) ([ACF], Chapter 18).

Open Problem 12 *1. Count points on Jacobians of genus 2 (without CM) and of genus 3 (with or without CM).*
2. There is a lifting theorem for ordinary *abelian varieties analogous to Deuring's lifting theorem for elliptic curves.*
Study algorithmic aspects of the lifting theorems.

3.3 Computation of Isogenies

We come back to the tasks formulated in Sect. 1.2.1 but now restricted to the case that $K = \mathbb{F}_q$. One of the question was: Can one, for given A, compute explicitly isogenies η as concrete functions?

An optimistic answer would be: yes, with complexity polynomial in $\log(q), \dim(A), \deg(\eta)$.

In fact, this is true for elliptic curves and relies on the computation of equations for the modular curve $Y_0(n)$. The basic work was done (after Deuring) by Vélu [V], and accelerations that make the algorithm efficient are due to Couveignes, Lercier, Elkies, and many others. These algorithms are responsible for the efficiency of point counting on elliptic curves by SEA-algorithms. It turns out that the cost for the computation of an isogeny of degree ℓ is

$$\mathcal{O}(\ell^2 + \ell \log(\ell) \log(q)).$$

There are hopeful beginnings of a similar theory for genus 2 curves [CL,FLR,GS] that promise to become a fascinating area of mathematical research.

So we state it as an

Open Problem 13 *Find effective formulas for isogenies between abelian varieties or Jacobian varieties of genus 2 and 3.*

The big disadvantage of the formulas for isogenies is that they are polynomial in the degree of the isogenies.

So they are only usable for isogenies of small degree. To repair this one uses more number theory and assumes in addition that the abelian variety is of CM-type with endomorphism ring O that is an order in a CM-field K. (For elliptic curves E this is equivalent with the condition that E is ordinary.)

We sketch the strategy.[11]

An isogenous variety A' has also CM with a ring of endomorphism $O' \subset K$. First, assume that $O \subset O'$. By definition O and O' are lattices of dimension $d = \dim(A)$ and so correspond to abelian varieties $\tilde{A} = \mathbb{C}^d/O$ and $\tilde{B} = \mathbb{C}^d/O'$ (Example 6). The inclusion of O' in O induces an isogeny from $\tilde{A}$ to $\tilde{B}$. If $[O' : O]$ is small, one can hope to describe the corresponding isogeny. (One has a good chance that in practical cases this will be so.)

The next step is to assume that $O = O'$ (or that at least the degree of the isogeny η one wants to compute is a prime not dividing $[O' : O]$).

For simplicity assume that $B \cong \tilde{B}$. Isogenies of degree ℓ to B correspond to ideals $\mathfrak{L}$ in O with norm ℓ. But one has more freedom. Changing by isomorphisms means to change $\mathfrak{L}$ by a principal ideal, and one of the main results of CM theory is that the isomorphism classes of abelian varieties with endomorphism ring O correspond to *ideal classes* of O. This gives an idea how to treat isogenies of large prime degree between abelian varieties with endomorphism ring O: one has to find prime ideals $\mathfrak{p}_1, \ldots, \mathfrak{p}_k$ in O with small norm and k "not large" such that $\prod_i \mathfrak{p}_i$ is in the same ideal class as $\mathfrak{L}$, and then compute the chain of isogenies with kernel $\mathfrak{p}_i$. There are theorems in algebraic number theory (Minkowski's theorem and smoothness results known from algorithms to factor numbers) and *heuristics* (like GRH) that predict that with a high probability, this search will be successful.

In the next paragraph we shall write down the results for isogenies of elliptic curves relying on these principles. We formulate already here the

Open Problem 14 *Assume that C_1, C_2 are curves of genus 2 over $\mathbb{F}_q$ with Jacobian varieties of CM-type that are isogenous.*

Use CM theory to compute isogenies.

---

[11]In the following we simplify by looking at abelian varieties with *principal polarization* (e.g., Jacobian varieties) and then neglect some more subtle points concerning these polarizations.

Finding Isogenies of Elliptic Curves over $\mathbb{F}_q$ A good part of the following results rely on the groundbreaking paper [K] of Kohel. We apply the considerations from above to ordinary elliptic curves E, E' defined over $\mathbb{F}_q$ with endomorphism ring O_E. It is evident that the class number h_E of O_E and so the discriminant Δ_{O_E} of O_E will play an important role. For random E we have to expect that h_E is of size $\mathcal{O}(q^{1/2})$ and so that the algorithms to find isogenies are exponential in $\log(q)$. The beautiful result of Galbraith and Stolbunov in [GSt] is

Theorem 15 *The cost for finding an isogeny between elliptic curves whose endomorphism ring is O_{K_E} is*

$$\mathcal{O}(q^{1/4+o(1)}\log^2(q)\log\log(q)).$$

This result hints that for large q and randomly chosen E, it is hard to find isogenies, and in fact there are cryptographic schemes that propose to use this problem as crypto primitive (for one version of such schemes, see cf. 3.3.1 below).

In the discussion above, we have remarked that there are similarities with algorithms factoring numbers. In fact, an approach due to Jao and Soukharev shows (under "reasonable" heuristics like GRH) the following.

Theorem 16 ([JS], Theorem 4.1) *Assume that E is an ordinary elliptic curve given in Weierstrass form with given Frobenius endomorphism π_q (i.e., $|E(\mathbb{F}_q)|$ is known) and endomorphism ring O_E.*

Take $n \in \mathbb{N}$ and assume that $[O_E : \mathbb{Z}[\pi_q]]$ is prime to $|E(\mathbb{F}_{q^n})|$ and let $\mathfrak{L}$ be an ideal of O_E whose norm is a prime number ℓ.

Take $P \in E(\mathbb{F}_{q^n})$.

Then there is an algorithm that computes an elliptic curve E' and an isogeny

$$\eta : E \to E'$$

*with kernel $\mathfrak{L}$ and the X-coordinate of $\eta(P)$ in running time that is polynomial in $\log(\ell)$, $\log(q)$, n and **subexponential** in $\log(\Delta_{O_E})$ (for the explicit estimate, see [JS]).*

3.3.1   Two Applications

Equivalence of Discrete Logarithms in Isogeny Classes   A very important crypto primitive for public-key cryptography is the discrete logarithm (DL) in the group of rational points $E(\mathbb{F}_q)$ of elliptic curves E over finite fields. The (till now justified) hope is that the complexity of DL is exponential in the order of the largest prime dividing $|E(\mathbb{F}_q)|$. But it is well known that one has to be careful since some elliptic curves (e.g., supersingular curves) can be attacked by algorithms with subexponential complexity. Very often, this is done by a transfer, i.e., by a *subexponentially computable* map into another group in which the DL is vulnerable (see [ACF], Chapter 22).

An obvious question is whether one can use isogenies as transfer maps.

The answer is no because of a very nice result that uses, besides the above discussed methods to compute isogenies, the equivalence of the isogeny graph of elliptic curves with the same ring of endomorphism over $\mathbb{F}_q$ with a graph of ideals in this endomorphism ring (again Deuring's lifting theorem is crucial). With properties of this graph induced by classical analytic number theory of imaginary quadratic number fields, one gets

Theorem 17 (Jao et al. [JMV]) *Discrete logarithms in isogenous elliptic curves over $\mathbb{F}_q$ are subexponentially equivalent.*

Open Problem 15 *Prove the same result for Jacobian varieties of CM-type attached to curves C of genus 2.*

The Couveignes–Stolbunov Crypto System This system is a cryptosystem based on a *principally homogeneous space*.

We continue to assume that E is ordinary. We denote by S_E the set of isomorphy classes (over $\mathbb{F}_{p,\infty}$) of elliptic curves $E'/\mathbb{F}_q$ with

$$\text{End}_{\mathbb{F}_{p,\infty}}(E') = \text{End}_{\mathbb{F}_{p,\infty}}(E) = O \subset \mathbb{Q}(\sqrt{-d}).$$

Again we use the one-to-one correspondence between S_E and the ideal class group $Cl(O)$ of O.

In fact, S_E is a *principal homogenous* space with translation group $Cl(O)$ with the following action:

Lift E to $\tilde{E}$ (Deuring's lifting theorem). Without loss of generality assume that the lattice defining $\tilde{E}$ over $\mathbb{C}$ is O. Take an ideal $\mathfrak{a} \subset O$ with divisor class c.

Then $c \cdot [E]$ is the isomorphy class of the elliptic curves E' whose Deuring lift is over $\mathbb{C}$ defined by the lattice $\mathfrak{a}$.

This can be used for a crypto system going back to Couveignes and implemented by Stolnikov.

As private key, take c, and as public key, the j-invariant of E'.

To make this computable, one has to find in each ideal class of O an ideal that is the product of prime ideals with small norm. Hence, one has to use the same techniques as in Sect. 3.3.

Remark 7 • The system is slow for one cannot use a square and multiply algorithm.

• It can be shown that the crypto primitive is NOT the DL in $Cl(O)$, and so a direct application of Shor's algorithm for quantum computers does not work.

• Nevertheless there is an algorithm using quantum computer that breaks the system in subexponential time.

3.4 Constructions of Isogenies by Correspondences

We end by describing a general construction of isogenies between abelian subvarieties of Jacobian varieties. This construction can be done over arbitrary ground fields K. It is important in our context because of its immediate applications to DL systems attached to divisor classes of curves over finite fields.

Correspondences of curves C, D are induced by morphisms

$$f_1 : H \to C \text{ and } f_2 : H \to D$$

(hence H is a common cover of C and D) and application of conorm, respectively norm maps, on divisor class groups:

$$\text{Pic}^0(C) \xrightarrow{f_{2,*} \circ f_1^*} \text{Pic}^0(D).$$

Under mild conditions one can assure that

$$\eta : J_C \to J_D$$

has finite kernel.

If the degrees of f_i are not too large, one can compute the maps on divisor classes explicitly.

Very often one uses curves C with a cover

$$f : C \to \mathbb{P}^1,$$

and takes for H the Galois closure of this cover and for D the fixed curve under a subgroup of the Galois group ("monodromy group") of f. By this, one has natural connections with *Hurwitz spaces* and their very rich theory ([FK1] and [FK2]).

One example for this method is Weil descent if $\mathbb{F}_q \neq \mathbb{F}_p$ that may transfer a seemingly hard DL problem to an easier one.

Another example was worked out in [FK2] explaining B. Smith's isogeny of degree 8 mapping hyperelliptic curves of genus 3 to *non*-hyperelliptic curves of genus 3 (and so weakening the DL [Di]). The result of Smith is

Theorem 18 (Smith) *There are $\mathcal{O}(q^5)$ isomorphism classes of hyperelliptic curves of genus 3 defined over $\mathbb{F}_q$ for which the discrete logarithm in the divisor class group of degree 0 has complexity $\mathcal{O}(q)$, up to log-factors.*

Since $|\text{Pic}^0(C)| = \mathcal{O}(q^3)$, the DL system of these hyperelliptic curves of genus 3 is weak.

To get this result Smith has to use certain heuristics.

The advantage of the approach by Hurwitz spaces is, besides delivering a structural background, that these spaces are often accessible for explicit description. For instance, in the case discussed here, one can determine the four-dimensional subspace in the moduli space of hyperelliptic curves of genus 3 consisting of curves that are in the image of Smith's isogeny, and so justify his heuristics [FK3].

Open Problem 16 *Find interesting correspondences of low degree between Jacobian varieties induced by correspondences between curves and (possibly) attached to Hurwitz spaces.*

References

[ACF] H. Cohen, G. Frey (eds.), *Handbook of Elliptic and Hyperelliptic Curve Cryptography* (CRC, Providence, 2005)

[CL] R. Carls, D. Lubicz, A p-adic quasi-quadratic time point counting algorithm. Int. Math. Res. Not. **4**, 698–735 (2009)

[De] M. Deuring, Die Typen der Multiplikatorenringe elliptischer Funktionenkörper. Abh. Math. Sem. Hamb. **14**, 197–272 (1941)

[Di] C. Diem, An index calculus algorithm for plane curves of small degree, in *Proceedings of ANTS VII*, ed. by F. Heß, S. Pauli, M. Pohst. Lecture Notes in Computer Science, vol. 4076 (Springer, Berlin, 2006), pp. 543–557

[Fa] G. Faltings, Endlichkeitssätze für abelsche Varietäten über Zahlkörpern. Invent. Math. **73**, 349–366 (1983)

[FLR] J.-Ch. Faugère, D. Lubicz, D. Robert, Computing modular correspondences for abelian varieties. J. Algebra **343**, 248–277 (2011)

[FK1] G. Frey, E. Kani, Curves of genus 2 with elliptic differentials and associated Hurwitz spaces. Cont. Math. **487**, 33–82 (2009)

[FK2] G. Frey, E. Kani, Correspondences on hyperelliptic curves and applications to the discrete logarithm, in *Proceedings of SIIS, Warsaw 2011*, ed. by P. Bouvry, M. Klopotek, F. Leprévost, M. Marciniak, A. Mykowiecka, H. Rybiński. Lecture Notes in Computer Science, vol. 7053 (Springer, Berlin, 2012), pp. 1–19

[FK3] G. Frey, E. Kani, Normal Forms of Hyperelliptic Curves of Genus 3, preprint

[Fr1] G. Frey, On ternary equations of Fermat type and relations with elliptic curves, in *Modular Forms and Fermat's Last Theorem*, ed. by G. Cornell, J.H. Silverman, G. Stevens (Springer, New York, 1997), pp. 527–548

[Fr2] G. Frey, Applications of arithmetical geometry to cryptographic constructions, in *Proceedings of Finite Fields and Application* (2001), pp. 128–161

[Fr3] G. Frey, Relations between arithmetic geometry and public key cryptography. Adv. Math. Commun. **4**, 281–305 (2010)

[GSt] St. Galbraith, A. Stolbunov, Improved algorithm for the isogeny problem for ordinary elliptic curves. Appl. Algebra Eng. Commun. Comput. **24**, 107–131 (2013)

[GS] P. Gaudry, E. Schost, Hyperelliptic point counting record: 254 bit jacobian, June 2008. http://webloria.loria.fr/~gaudry/record127

[Ha] C. Hall, An open-image theorem for a general class of abelian varieties. Bull. Lond. Math. Soc. **43**, 703–711 (2011)

[He] F. Heß, Computing Riemann–Roch spaces in algebraic function fields and related topics. J. Symb. Comput. **33**(4), 425–445 (2002)

[JMV]   D. Jao, S.D. Miller, R. Venkatesan, Do all elliptic curves of the same order have the same difficulty of discrete log?, in *Advances of Cryptology-Asiacrypt 2005*. Lecture Notes in Computer Science, vol. 3788 (Springer, Berlin 2005), pp. 21–40

[JS]   D. Jao, V. Soukharev, A subexponential algorithm for evaluating large degree isogenies, in *Algorithmic Number Theory* (Springer Berlin 2010), pp. 219–233

[K]   D. Kohel, Endomorphism rings of elliptic curves over finite fields. Ph.D. thesis, Berkeley, 1996

[Le]   R. Lercier, Algorithmique des courbes elliptiques dans les corps finis. Thèse, LIX-CNRS, 1997

[LR]   D. Lubicz, D. Robert, Computing isogenies between abelian varieties. Compos. Math. **148**, 1483–1515 (2012)

[M1]   D. Mumford, *Abelian Varieties* (Oxford University Press, Oxford, 1970)

[M2]   D. Mumford, On the equations defining abelian varieties I–III. Invent. Math. **1**, 287–354 (1967); Invent. Math. **3**, 75–135 (1967); Invent. Math. **3**, 215–244 (1967)

[Oe]   J. Oesterlé, Versions effectives du théorème de Chebotarev sous l'hypothèse de Riemann généralisée. Astérisque **61**, 165–167 (1979)

[R]   K. Ribet, On modular representations of $G(\bar{\mathbb{Q}}|\mathbb{Q})$ arising from modular forms. J. Math. **100**, 431–476 (1990)

[Se1]   J.P. Serre, Propriétés galoisiennes des points d'ordre fini des courbes elliptiques. Invent. Math. **15**, 259–331 (1972)

[Se2]   J.P. Serre, Résumé des cours de 1985–1986 (Annuaire du Collège de France, 1986)

[S]   B. Smith, Isogenies and the Discrete Logarithm Problem in Jacobians of Genus 3 Hyperelliptic Curves, in *Advances in Cryptology: EUROCRYPT 2008, Istanbul*. Lecture Notes in Computer Science, vol. 4965 (2008)

[T]   J. Tate, Endomorphisms of abelian varieties over finite fields. Invent. Math. **2**, 134–144 (1966)

[V]   J. Vélu, Isogénies entre courbes elliptiques. C.R. Acad. Sci. Paris Ser. A **273**, 238–241 (1971)

Another Look at Security Theorems for 1-Key Nested MACs

Neal Koblitz and Alfred Menezes

Abstract We prove a security theorem without collision resistance for a class of 1-key hash function-based MAC schemes that includes HMAC and Envelope MAC. The proof has some advantages over earlier proofs: it is in the uniform model, it uses a weaker related-key assumption, and it covers a broad class of MACs in a single theorem. However, we also explain why our theorem is of doubtful value in assessing the real-world security of these MAC schemes. In addition, we prove a theorem assuming collision resistance. From these two theorems, we conclude that from a provable security standpoint, there is little reason to prefer HMAC to Envelope MAC or similar schemes.

1 Introduction

The purpose of our "Another Look" series of papers [14] is to examine the way the "provable security" paradigm is used in the cryptographic literature. In particular, we hope to foster a less credulous attitude toward some of the claims that are frequently made about "provable" security.

Starting in the early days of "practice-oriented provable security"—a term coined by Bellare and Rogaway [1, 4]—there has been an unfortunate tendency to exaggerate both the security guarantees that are proved and the efficiency advantages of the provably secure protocols. For example, in [5] the authors used the word "optimal" to advertise the OAEP version of RSA encryption (OAEP = "optimal asymmetric encryption padding"). Shortly after Victor Shoup [27] discovered that the security proof in [5] was fallacious, the claim of optimal efficiency was also reexamined. It now seems that Boneh–Rabin encryption [9] comes closer than OAEP to being both provably secure (in a limited sense) and optimally efficient; see Sect. 4 of [15].

N. Koblitz
Department of Mathematics, University of Washington, Box 354350, Seattle, WA 98195, USA
e-mail: koblitz@uw.edu

A. Menezes (✉)
Department of Combinatorics & Optimization, University of Waterloo, Waterloo, ON, Canada N2L 3G1
e-mail: ajmeneze@uwaterloo.ca

© Springer International Publishing Switzerland 2014
Ç.K. Koç (ed.), *Open Problems in Mathematics and Computational Science*,
DOI 10.1007/978-3-319-10683-0_4

69

Excessive enthusiasm in marketing protocol designs can also be seen in certain statements about hash-based key agreement and message authentication. According to a letter to the *AMS Notices* from Hugo Krawczyk [20], the designer of the HMQV key agreement protocol, "the HMQV work represents a prime example of the success of theoretical cryptography, not only in laying rigorous mathematical foundations for cryptography at large, but also in its ability to guide us in the design of truly practical solutions to real-world problems." Similarly, speaking of his hash-based message authentication code HMAC in an invited talk at Asiacrypt 2010 on "Cryptography: from theory to practice," Krawczyk proclaimed that with HMAC "balance [between engineering and theory was] regained and the rest is history."

One of the conclusions of the present chapter is that Krawczyk's claim of a unique benefit provided by HMAC cannot be justified by provable security considerations. Rather, very similar security results can be proved for a broad class of message authentication codes, including some (such as "Envelope MAC") that arguably are a little more efficient than HMAC.

Another theme that recurs in several of our papers, including the present one, is that the security definitions that are at the heart of any "proof" of security are often open to debate and are far from definitive (see [16] for more discussion of this). In [18] we found that even such a fundamental concept of computer science as the distinction between a uniform and nonuniform algorithm is frequently dealt with in a confusing and inconsistent manner in the cryptographic literature. In the present chapter, we argue that in the MAC setting, two of the basic definitions used by earlier authors—that of a pseudorandom function and that of security against related-key attacks—need to be replaced by more suitable versions.

As we have written on many occasions, starting with [15], we have no objection to formal arguments in cryptography provided that their significance is properly interpreted and they are not misnamed "proofs of security." Indeed, reductionist security arguments for hash functions, message authentication codes, and other symmetric and asymmetric cryptographic protocols can provide a type of baseline guarantee of a certain limited security feature. We show that a broad class of 1-key nested MACs have such a property. But the choice of which MAC in the class one wants to use cannot be made using reductionist security arguments but rather should be based on an ad hoc analysis of efficiency and security in the settings in which it will be deployed.

* * *

A common method of constructing a message authentication code (MAC) is the "nested" construction (NMAC). One first applies a keyed iterated hash function $h(K_1, M)$ (constructed from a compression function f) to the message M, and then one puts this hash value into a second keyed function $\tilde{f}(K_2, h(K_1, M))$ (where $\tilde{f}$ is also a compression function). For efficiency and ease of implementation, one usually wants the MAC to depend on a single key K, and so one sets $K_1 = K_2 = K$ or, more generally, $K_1 = g_1(K)$, $K_2 = g_2(K)$ for some functions g_1, g_2. Our main purpose is to prove a new security theorem without collision resistance that covers arbitrary constructions of this type. The theorem says, roughly speaking, that the MAC is a pseudorandom function (prf) provided that both $\tilde{f}$ and f

are pseudorandom functions and the functions $f, \tilde{f}, g_1, g_2$ satisfy a certain rather weak related-key assumption. This theorem is a generalized 1-key version of our Theorem 10.1 in [17].

The two most important examples of this type of MAC are the "hash-based message authentication code" (HMAC) [6] (standardized in [7, 21]) and Envelope MAC (also called "Sandwich MAC"; see [29] for a recent version). In these cases there are earlier security proofs without collision resistance in [2, 29], but unfortunately those proofs are not valid in the uniform model of complexity.[1] In other words, they use unconstructible adversaries and so have to assume that the cryptographic primitives withstand attack even by unconstructible adversaries. For this reason, as we explained in [17] (see also [8]), they do not give useful concrete bounds on the resources a prf-adversary would need in order to defeat the MAC. In contrast, our theorem is proved in the uniform model; this means that it needs much milder assumptions.

One of the five finalists in the NIST SHA-3 competition used Envelope MAC. The designers of the "Grøstl" construction wrote (Sect. 6.1 of [11]):

> We propose this envelope construction as a dedicated MAC mode using Grøstl. This construction has been proved to be a secure MAC under similar assumptions as HMAC.

Here the designers were referring to the proof in [29], but they were apparently unaware that Yasuda's proof is not valid in the uniform model and for that reason gives much weaker guarantees than one would expect. As we commented in [18], one of the drawbacks of results obtained in the nonuniform model is the possibility that they will be used by other authors who are unaware of the extremely limited nature of such results from a practice-oriented standpoint. In any case, in the present chapter we remove this gap in the security argument in [11] by supplying a uniform proof.

There is a second respect in which our theorem makes a milder assumption than earlier theorems of this type: our related-key assumption is weaker than the one defined in [2, 3]. This not only gives us a stronger theorem but also enables us to unify HMAC and Envelope MAC in a single theory.

Despite these advantages over earlier security theorems, the sad fact is that our main theorem by itself provides very little assurance about the real-world security of these MAC schemes. In Sect. 4 we recall some of the reasons for this.

In Sect. 5 we prove a second theorem, this time assuming collision resistance, that carries over the main result of [6] to 1-key nested MACs. Our two theorems together show that from the standpoint of security reductions, there is little difference between HMAC, Envelope MAC, and other similar constructions. We conclude that security theorems are not of much use in deciding which of the competing schemes—HMAC, Envelope MAC, or some other variant—has better security in practice.

---

[1]Fischlin [10] has a uniform proof of a security theorem for HMAC without collision resistance, but its usefulness is questionable because of the extremely large tightness gap in his result.

2 Statement of the Main Theorem

Let $f : \{0,1\}^c \times \{0,1\}^b \longrightarrow \{0,1\}^c$ and $\tilde{f} : \{0,1\}^c \times \{0,1\}^c \longrightarrow \{0,1\}^c$ be two compression functions. Here $b \geq 2c$ (typically $b = 512$ and $c = 128$ or 160), so that f compresses by a factor of at least 3, whereas $\tilde{f}$ compresses by a factor of 2. Let $g_i : \{0,1\}^c \longrightarrow \{0,1\}^c$, $i = 1, 2$. We suppose that all of these functions are publicly and efficiently computable.

By a (t, q)-adversary, we mean an adversary that makes $\leq q$ queries and has running time $\leq t$. Recall that f is said to be an (ϵ, t, q)-secure pseudorandom function (prf) if no (t, q)-adversary can distinguish between f with a hidden key and a random function with advantage $\geq \epsilon$. We say that f is strongly (ϵ, t, q) secure (see [17]) if such an adversary before any query is permitted to "reset" the oracle, by which we mean that in response to the adversary's request the oracle chooses either a new random key (if it is $f(K, .)$) or a new random function (if it is a random function $r'(.)$).

We now define the "related-key assumption" that we shall use in our main theorem.

Definition 1 In the above setting, we say that $(f, \tilde{f})$ is (ϵ, t, q)-secure against (g_1, g_2)-related-key attacks if no (t, q)-adversary has an advantage greater than or equal to ϵ in the following interaction with the oracle O_{rka}. First, the oracle chooses a random bit; if it is 0, the oracle chooses two random keys $K_1, K_2 \in \{0,1\}^c$; if it is 1, the oracle chooses one random key $K \in \{0,1\}^c$ and sets $K_1 = g_1(K)$, $K_2 = g_2(K)$. Each query of the adversary is a message M in either $\{0,1\}^b$ or $\{0,1\}^c$, to which the oracle responds with either $f(K_1, M)$ or $\tilde{f}(K_2, M)$, respectively. At the end, the adversary guesses the oracle's random bit.

We recall that in this situation the advantage of the adversary is defined as

Prob(adversary guesses 1 | oracle chose 1) − Prob(adversary guesses 1 | oracle chose 0),

where Prob(A|B) denotes the conditional probability of event A given event B.

This setting is general enough to include two of the best-known MAC constructions (see Fig. 1):

1. For HMAC, let IV be a fixed (and publicly known) initialization vector, and let ipad and opad be two fixed elements of $\{0,1\}^b$ (also publicly known). We let a superscript 0 on a bitstring in $\{0,1\}^c$ indicate that we are appending $b - c$ zero bits to it. We set $\tilde{f}(K, M) = f(K, M^0)$, $g_1(K) = f(\text{IV}, K^0 \oplus \text{ipad})$, $g_2(K) = f(\text{IV}, K^0 \oplus \text{opad})$.
2. For Envelope MAC, let IV be a fixed (and publicly known) initialization vector; let $\tilde{f}(K, M) = f(M, K^0)$, $g_1(K) = f(\text{IV}, K^0)$, and $g_2(K) = K$.

Remark 1 The above related-key assumption is weaker than the related-key assumption in [2, 3]. In that assumption, the oracle is required to simply give the adversary the two keys: K_1, K_2. In that case the adversary can of course compute

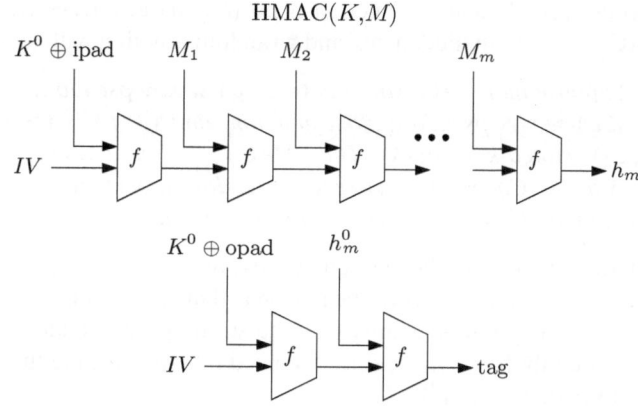

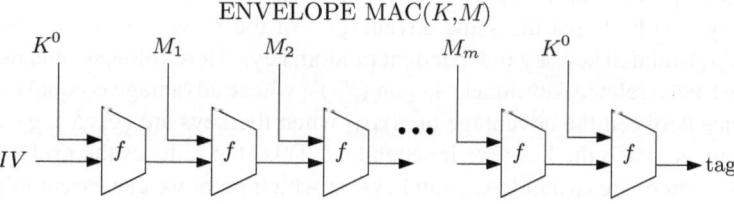

Fig. 1 HMAC and Envelope MAC

any number of desired values $f(K_1, M)$ or $\tilde{f}(K_2, M)$, limited only by the running time bound; in other words, the rka-adversary in our assumption is less powerful (because it has less information) than the rka-adversary in [2, 3]. Moreover, with the rka-assumption in [2, 3], we wouldn't have been able to include Envelope MAC in our theorem, because when $g_2(K) = K$, the adversary, if given K_1 and K_2, can trivially determine whether or not $K_1 = g_1(K_2)$.

In the above setting let $h : \{0, 1\}^c \times (\{0, 1\}^b)^* \longrightarrow \{0, 1\}^c$ denote the iterated hash function that, given a key $K \in \{0, 1\}^c$ and a message $M = (M_1, M_2, \ldots, M_m)$, $M_i \in \{0, 1\}^b$, successively computes $h_1 = f(K, M_1)$, $h_{i+1} = f(h_i, M_{i+1})$, $i = 1, 2, \ldots, m-1$ and sets $h(K, M) = h_m$. We define the message authentication code $\text{MAC}_{f, \tilde{f}, g_1, g_2}$ as follows:

$$\text{MAC}_{f, \tilde{f}, g_1, g_2}(K, M) = \tilde{f}(g_2(K), h(g_1(K), M)).$$

Notice that when $g_1(K) = f(\text{IV}, K^0 \oplus \text{ipad})$, and $g_2(K) = f(\text{IV}, K^0 \oplus \text{opad})$ this definition agrees with that of HMAC, and when $g_1(K) = f(\text{IV}, K^0)$ and $g_2(K) = K$, it agrees with that of Envelope MAC (see Fig. 1).

By a (t, q, n)-adversary, we mean an adversary that makes $\leq q$ queries of block length $\leq n$ and has running time $\leq t$. We say that $\text{MAC}_{f, \tilde{f}, g_1, g_2}$ is an

(ϵ, t, q, n)-secure pseudorandom function if no (t, q, n)-adversary can distinguish between $\text{MAC}_{f, \tilde{f}, g_1, g_2}$ with hidden key and a random function with advantage $\geq \epsilon$.

Theorem 1 *Suppose that f is a strongly (ϵ_1, t, q)-secure pseudorandom function, $\tilde{f}$ is an (ϵ_2, t, q)-secure pseudorandom function, and $(f, \tilde{f})$ is $(\epsilon_3, t, 2q)$-secure against (g_1, g_2)-related-key attacks. Then $\text{MAC}_{f, \tilde{f}, g_1, g_2}$ is a $(2n(\epsilon_1 + \binom{q}{2}2^{-c}) + \epsilon_2 + 2\epsilon_3, t - (qnT + Cq \log q), q, n)$-secure pseudorandom function. Here C is an absolute constant, and T denotes the time for one evaluation of f or $\tilde{f}$.*

Remark 2 In the statement of the theorem, the expression $2n(\epsilon_1 + \binom{q}{2}2^{-c}) + \epsilon_2 + 2\epsilon_3$ can be replaced by $2n\epsilon_1 + \epsilon_2 + 2\epsilon_3$. The reason is that, as explained in Remark 10.2 of [17], the generic key-guessing attack on the strong pseudorandomness property has advantage roughly $(qt/T)2^{-c}$; since we need $t > qnT$ for the theorem to have content, it follows that $\epsilon_1 \gg \binom{q}{2}2^{-c}$.

Before proving Theorem 1, we give an informal summary of the argument. The first step is to show that a prf-adversary A_{MAC} of $\text{MAC}_{f, \tilde{f}, g_1, g_2}$ is also a prf-adversary—with almost the same advantage—of the MAC obtained by replacing the (g_1, g_2)-related keys by independent random keys. Here "almost" means that we can construct a related-key attack A_{rka} on $(f, \tilde{f})$ whose advantage is equal to half the difference between the advantage of A_{MAC} when the keys are $(g_1(K), g_2(K))$ and its advantage when the keys are independent. This step reduces the problem to the case when there are two independent keys, at which point we can essentially follow the proof for NMAC in [17]. Namely, we show that a prf-adversary for the MAC succeeds only when either the prf property of the outer shell $\tilde{f}(K_2, .)$ is attacked (we call its adversary $A_{\tilde{f}}$) or else a collision is produced in the iterated hash function that's inside this shell. In the latter case we use the collision to construct a prf-adversary of f. Since there are two possible types of collisions that can occur and up to n iterations of the hash function, this leads to roughly $2n$ f adversaries. This intuitively explains why $2n\epsilon_1 + \epsilon_2 + 2\epsilon_3$ appears in the conclusion of the theorem. The term $2n\binom{q}{2}2^{-c}$ arises because of the possibility of random collisions between c-bit strings.

We shall give the actual proof in the next section. The above plausibility argument shows that the basic ideas in the proof are simple. However, the organization is a little intricate because of the need to proceed carefully with the reduction using all of the adversaries. We see no way to come up with a more concise self-contained proof, and we apologize to the reader for that.

3   Proof of the Main Theorem

Proof We will prove the following equivalent statement: if f is a strongly $((\frac{\epsilon - \epsilon_2 - 2\epsilon_3}{2n} - \binom{q}{2}2^{-c}), t + (qnT + Cq \log q), q)$-secure pseudorandom function, $\tilde{f}$ is an $(\epsilon_2, t + (qnT + Cq \log q), q)$-secure pseudorandom function, and $(f, \tilde{f})$ is $(\epsilon_3, t + (qnT + Cq \log q), 2q)$-secure against (g_1, g_2)-related-key attacks, then

$\mathrm{MAC}_{f,\tilde{f},g_1,g_2}$ is an (ϵ, t, q, n)-secure pseudorandom function. The proof starts by supposing that we have a (t, q, n)-adversary A_{MAC} that has advantage $\geq \epsilon$ in the pseudorandomness test for $\mathrm{MAC}_{f,\tilde{f},g_1,g_2}$, and then it proceeds to construct a $(t + (qnT + Cq\log q), 2q)$-adversary A_{rka} of the related-key property, a $(t + (qnT + Cq\log q), q)$-adversary $A_{\tilde{f}}$ of the pseudorandom property of $\tilde{f}$, and a $(t + (qnT + Cq\log q), q)$-adversary A_f of the pseudorandom property of f such that at least one of the following holds:

(i) A_{rka} has advantage $\geq \epsilon_3$ against the (g_1, g_2)-related key property of $(f, \tilde{f})$.
(ii) $A_{\tilde{f}}$ has advantage $\geq \epsilon_2$ in the pseudorandomness test for $\tilde{f}$.
(iii) A_f has advantage $\geq (\epsilon - \epsilon_2 - 2\epsilon_3)/(2n) - \binom{q}{2}2^{-c}$ in the strong pseudorandomness test for f.

Note that if any of these three conditions holds, we have a contradiction that proves the theorem.

For the ith message query M^i, we use the notation M^i_ℓ to denote its ℓth block, we let $M^{i,[m]} = (M^i_1, \ldots, M^i_m)$ be the truncation after the mth block, and we set $M^{i,(m)} = (M^i_m, M^i_{m+1}, \ldots)$, that is, $M^{i,(m)}$ is the message with the first $m-1$ blocks deleted. We say that a message is "non empty" if its block-length is at least 1.

Let h be the corresponding iterated function, and let $\tilde{f}h$ be the MAC that for a key $(K_1, K_2) \in \{0, 1\}^c \times \{0, 1\}^c$ is defined as follows: $\tilde{f}h(K_1, K_2, M) = \tilde{f}(K_2, h(K_1, M))$, where $M = (M_1, \ldots, M_m)$ is an m-block message, $m \leq n$. Note that $\mathrm{MAC}_{f,\tilde{f},g_1,g_2}(K, M) = \tilde{f}h(g_1(K), g_2(K), M)$. Let $r(M)$ denote a random function of messages, and let $r'(M_1)$ denote a random function of 1-block messages. In response to an input of suitable length, r' or r outputs a random c-bit string, subject only to the condition that if the same input is given a second time (in the same run of the algorithm), the output will be the same. In the test for pseudorandomness, the oracle is either a random function or the function being tested, as determined by a random bit (coin toss).

Now suppose that we have a (t, q, n)-adversary A_{MAC} that, interacting with its oracle O_{MAC}, has advantage $\geq \epsilon$ against the prf test for $\mathrm{MAC}_{f,\tilde{f},g_1,g_2}$. We use A_{MAC} to construct four adversaries $A_{\tilde{f}h}$, A_{rka}, $A_{\tilde{f}}$, and A_f. The last three are the adversaries in the above conditions (i)–(iii); the (t, q, n)-adversary $A_{\tilde{f}h}$ attacks the pseudorandomness property of $\tilde{f}h$. Each adversary makes at most the same number of queries as A_{MAC} (except that the related-key adversary can make up to $2q$ queries) and has a comparable running time. More precisely, the bound $t + (qnT + Cq\log q)$ on the running time of the adversaries A_{rka}, $A_{\tilde{f}}$, and A_f comes from the time required to run A_{MAC}, makes at most q computations of h values, and stores at most q values (coming from oracle responses and h computations) in lexicographical order and sorts them looking for collisions. (An adversary does not in all cases perform all of these steps; rather, this is an upper bound.)

The related-key adversary A_{rka} runs A_{MAC} and interacts with the related-key oracle O_{rka}, which chooses a random bit u. Recall that A_{rka}, after querying the oracle O_{rka} with at most $2q$ b-bit or c-bit messages, must guess whether the keys K_1 and

K_2 that O_{rka} is using are independent (i.e., $u = 0$) or are related by $K_i = g_i(K)$, $i = 1, 2$, for some K (i.e., $u = 1$).

The adversary A_{rka} randomly chooses a bit ℓ, and as A_{MAC} runs, A_{rka} responds to each query M^i as follows. If $\ell = 0$, its response to each query is a random c-bit string (except in the case when a query is repeated, in which case the response is also repeated). If $\ell = 1$, then it first queries O_{rka} with M_1^i and computes $H = h(O_{\mathrm{rka}}(M_1^i), M^{i,(2)})$, where $O_{\mathrm{rka}}(M_1^i)$ denotes the oracle's response. (If M^i is just a 1-block message, then H is set equal to $O_{\mathrm{rka}}(M_1^i)$.) Now A_{rka} makes a second query to O_{rka}—this time the c-bit query H—and responds to A_{MAC}'s query with $O_{\mathrm{rka}}(H)$.[2] At the end A_{rka} guesses that the random bit u chosen by O_{rka} is 1 if A_{MAC} guesses that the random bit ℓ chosen by A_{rka} (which is simulating an oracle) is 1; otherwise, it guesses that $u = 0$. (Note that A_{rka} guesses 0 if A_{MAC} stops or reaches time t without producing a guess; this could happen if A_{rka} is not properly simulating O_{MAC}, which would imply that $u = 0$.) Let δ denote the advantage of A_{rka}.

We also construct an adversary $A_{\tilde{f}h}$ that interacts with its oracle $O_{\tilde{f}h}$ and runs A_{MAC}. When A_{MAC} makes a query M^i, the adversary $A_{\tilde{f}h}$ queries $O_{\tilde{f}h}$ and sends A_{MAC} the response $O_{\tilde{f}h}(M^i)$. If A_{MAC} guesses that the oracle simulated by $A_{\tilde{f}h}$ is a random function, then $A_{\tilde{f}h}$ guesses that its oracle $O_{\tilde{f}h}$ is a random function; otherwise, $A_{\tilde{f}h}$ guesses that its oracle is $\tilde{f}h$ with hidden keys. In particular, note that if A_{MAC} stops or fails to produce a guess in time t—as may happen when $A_{\tilde{f}h}$ is not property simulating O_{MAC}—then $A_{\tilde{f}h}$ guesses that its oracle is $\tilde{f}h$ with hidden keys. (This makes sense, since if $O_{\tilde{f}h}$ were a random function, then the simulation of O_{MAC} would be correct.) Let γ denote the advantage of $A_{\tilde{f}h}$.

Returning to the description of A_{rka}, we see that there are two cases, depending on whether the random bit u of the oracle O_{rka} was (a) 1 (i.e., its keys are related) or (b) 0 (i.e., its keys are independent). In case (a) the interaction of A_{rka} with A_{MAC} precisely simulates O_{MAC}, and in case (b) it precisely simulates $O_{\tilde{f}h}$. (As we noted, in case (b) our original adversary A_{MAC} may stop or run for time t without producing a guess; in this case A_{rka} makes the guess 0.) Let

$$p_1 = \mathrm{Prob}(A_{\mathrm{MAC}} \text{ guesses } 1 \mid \ell = 1 \text{ and } u = 1).$$

$$p_2 = \mathrm{Prob}(A_{\mathrm{MAC}} \text{ guesses } 1 \mid \ell = 1 \text{ and } u = 0).$$

$$p_3 = \mathrm{Prob}(A_{\mathrm{MAC}} \text{ guesses } 1 \mid \ell = 0).$$

Note that when $\ell = 0$, there is no interaction with O_{rka}, and so the guess that A_{MAC} makes is independent of whether $u = 0$ or $u = 1$.

---

[2]The theorem's query bound for the related-key property is $2q$ because A_{rka} makes two queries for each query of A_{MAC}.

By assumption, A_{MAC} has advantage $\geq \epsilon$ in a prf test for $\text{MAC}_{f,\tilde{f},g_1,g_2}$; in other words, $p_1 - p_3 \geq \epsilon$. We also have $p_2 - p_3 = \gamma$. Subtracting gives $p_1 - p_2 \geq \epsilon - \gamma$. Next, the advantage δ of the related-key adversary A_{rka} is given by

$$\text{Prob}(A_{\text{rka}} \text{ guesses } 1 \mid u = 1) - \text{Prob}(A_{\text{rka}} \text{ guesses } 1 \mid u = 0)$$

$$= \text{Prob}(A_{\text{MAC}} \text{ guesses } 1 \mid u = 1) - \text{Prob}(A_{\text{MAC}} \text{ guesses } 1 \mid u = 0)$$

$$= \text{Prob}(A_{\text{MAC}} \text{ guesses } 1 \mid u = 1 \text{ and } \ell = 0) \cdot \text{Prob}(\ell = 0)$$

$$+ \text{Prob}(A_{\text{MAC}} \text{ guesses } 1 \mid u = 1 \text{ and } \ell = 1) \cdot \text{Prob}(\ell = 1)$$

$$- \text{Prob}(A_{\text{MAC}} \text{ guesses } 1 \mid u = 0 \text{ and } \ell = 0) \cdot \text{Prob}(\ell = 0)$$

$$- \text{Prob}(A_{\text{MAC}} \text{ guesses } 1 \mid u = 0 \text{ and } \ell = 1) \cdot \text{Prob}(\ell = 1)$$

$$= \frac{1}{2}(p_3 + p_1 - p_3 - p_2) = \frac{1}{2}(p_1 - p_2) \geq (\epsilon - \gamma)/2.$$

If condition (i) in the first paragraph of the proof does not hold, then $\delta < \epsilon_3$, in which case $\gamma > \epsilon - 2\epsilon_3$. For the remainder of the proof, we assume that the advantage of $A_{\tilde{f}h}$ satisfies this inequality, since otherwise (i) holds and we're done.

The rest of the proof closely follows the proof of Theorem 10.1 of [17]. We shall give the details rather than simply citing [17] because the present setting is slightly more general (with two pseudorandom compression functions f and $\tilde{f}$ rather than just one) and because there is some benefit in having a self-contained proof in one place.

We now construct an $\tilde{f}$-adversary $A_{\tilde{f}}$ and consider its advantage. As before, for any oracle O, we let $O(M)$ denote the response of O to the query M. The adversary $A_{\tilde{f}}$ is given an oracle $O_{\tilde{f}}$ and, using $A_{\tilde{f}h}$ as a subroutine, has to decide whether $O_{\tilde{f}}$ is $\tilde{f}(K_2, .)$ or a random function $r'(.)$ of 1-block messages. She chooses a random K_1 and presents the adversary $A_{\tilde{f}h}$ with an oracle that is either $\tilde{f}(K_2, h(K_1, .))$ or else a random function $r(.)$; that is, she simulates $O_{\tilde{f}h}$ (see below). In time $\leq t$ with $\leq q$ queries $A_{\tilde{f}h}$ is able with advantage $\gamma > \epsilon - 2\epsilon_3$ to guess whether $O_{\tilde{f}h}$ is $\tilde{f}h$ with hidden keys or a random function r. Here is how $A_{\tilde{f}}$ simulates $O_{\tilde{f}h}$: in response to a query M^i from $A_{\tilde{f}h}$, she computes $h(K_1, M^i)$, which she queries to $O_{\tilde{f}}$, and then gives $A_{\tilde{f}h}$ the value $O_{\tilde{f}}(h(K_1, M^i))$. Eventually (unless the simulation is imperfect, see below) $A_{\tilde{f}h}$ states whether it believes that its oracle $O_{\tilde{f}h}$ is $\tilde{f}h$ or r, at which point $A_{\tilde{f}}$ states the same thing for the oracle $O_{\tilde{f}}$—that is, if $A_{\tilde{f}h}$ said $\tilde{f}h$, then she says that $O_{\tilde{f}}$ must have been $\tilde{f}$, whereas if $A_{\tilde{f}h}$ said that $O_{\tilde{f}h}$ is r, then she says that $O_{\tilde{f}}$ is r'. Notice that if the oracle $O_{\tilde{f}}$ is $\tilde{f}(K_2, .)$, then the oracle $O_{\tilde{f}h}$ that $A_{\tilde{f}}$ simulates for $A_{\tilde{f}h}$ is $\tilde{f}h$ (with random key $K = (K_1, K_2)$); if the oracle $O_{\tilde{f}}$ is $r'(.)$, then the oracle that $A_{\tilde{f}}$ simulates for $A_{\tilde{f}h}$ acts as r with the important difference that if $h(K_1, M^i)$ coincides with an earlier $h(K_1, M^j)$ the oracle outputs the same value (even though $M^i \neq M^j$) rather than a second random value.[3] If the latter happens with negligible probability, then this algorithm $A_{\tilde{f}}$ is as successful in distinguishing $\tilde{f}$ from a random function as $A_{\tilde{f}h}$ is in distinguishing $\tilde{f}h$ from a random function. Otherwise, two sequences of f-adversaries $A_f^{(m)}$ and $B_f^{(m)}$ come into the picture, as described below.

The general idea of these adversaries is that they each use the oracle O_f in the pseudorandomness test for f to look for collisions between h values of two different messages M^i, M^j queried by $A_{\tilde{f}h}$. More precisely, the mth adversary in a sequence works not with all of a queried message but rather with the message with its first $m-1$ blocks deleted. If a collision is produced, then with a certain probability, O_f must be $f(K_2, .)$; however, one must also account for the possibility that O_f is $r'(.)$, and in the case of $A_f^{(m)}$, this brings in the next adversary in the sequence $A_f^{(m+1)}$.

First we make a remark about probabilities, which are taken over all possible coin tosses of the adversary, all possible keys, the oracle's "choice bit" (which determines whether it is the function being tested or a random function), and the coin tosses of the oracle in the case when it outputs a random function.[4] If the adversary's oracle is f or $\tilde{f}h$ with hidden keys, then the adversary's queries in general depend on the keys (upon which the oracle's responses depend) as well as the adversary's coin tosses. However, if the adversary's oracle is a random function—which is the situation when $A_{\tilde{f}}$ fails and the sequences of adversaries $A_f^{(m)}$ and $B_f^{(m)}$ are needed—then the oracle responses can be regarded simply as additional coin tosses, and the adversary's queries then depend only on the coin tosses and are independent of the keys. This is an important observation for understanding the success probabilities of the adversaries.

We define α_0 to be the probability, taken over all coin tosses of $A_{\tilde{f}h}$ (including those coming from random oracle responses) and all keys K_1, that the sequence of $A_{\tilde{f}h}$ queries M^i satisfies the following property:

There exist i and j, $j < i$, such that $h(K_1, M^i) = h(K_1, M^j)$.

For $m \geq 1$, we define α_m to be the probability, taken over all coin tosses of $A_{\tilde{f}h}$ and all q-tuples of keys $(K_1, K_2, \ldots, K_q)$, that the sequence of $A_{\tilde{f}h}$ queries M^i satisfies the following property:

(1_m) there exist i and j, $j < i$, such that $M^{i,(m+1)} \neq \emptyset$, $M^{j,(m+1)} \neq \emptyset$,

$$h(K_{\ell_i}, M^{i,(m+1)}) = h(K_{\ell_j}, M^{j,(m+1)}),$$

---

[3]If $A_{\tilde{f}h}$ fails to produce a guess about the oracle $O_{\tilde{f}h}$ in time t, as can happen if the simulation is imperfect, then $A_{\tilde{f}}$ guesses that $O_{\tilde{f}}$ is a random function. Note that the simulation is perfect if $O_{\tilde{f}}$ is $\tilde{f}$ with hidden key.

[4]The term "over all possible coin tosses" means over all possible runs of the algorithm with each weighted by 2^{-s}, where s is the number of random bits in a given run.

where for any index i for which $M^{i,(m+1)} \neq \emptyset$, we let ℓ_i denote the smallest index for which $M^{\ell_i,(m+1)} \neq \emptyset$ and $M^{i,[m]} = M^{\ell_i,[m]}$.

Finally, for $m \geq 1$, we define β_m to be the probability, taken over all coin tosses of $A_{\tilde{f}h}$ and all q-tuples of keys $(K_1, K_2, \ldots, K_q)$, that the sequence of $A_{\tilde{f}h}$ queries M^i satisfies the following property:

(2_m) there exist i and j such that $M^{i,(m+1)} = \emptyset$, $M^{j,(m+1)} \neq \emptyset$,

$$M^{i,[m]} = M^{j,[m]}, \qquad \text{and} \qquad h(K_i, M^{j,(m+1)}) = K_i.$$

We now return to the situation where with non-negligible probability α_0 the queries made by $A_{\tilde{f}h}$ lead to at least one collision $h(K_1, M^i) = h(K_1, M^j)$. Note that the advantage of the adversary $A_{\tilde{f}}$ is at least $\epsilon - 2\epsilon_3 - \alpha_0$. If condition (ii) fails, i.e., if this advantage is $< \epsilon_2$, it follows that $\alpha_0 > \epsilon - \epsilon_2 - 2\epsilon_3$. In the remainder of the proof, we shall assume that this is the case, since otherwise (ii) holds and we're done.

The first adversary in the sequence $A_f^{(m)}$ is A'_f, which is given the oracle O_f that is either $f(K_1, .)$ with a hidden random key K_1 or else $r'(.)$. As A'_f runs $A_{\tilde{f}h}$, giving random responses to its queries, she queries O_f with the first block M_1^i of each $A_{\tilde{f}h}$ query M^i. If $M^{i,(2)}$ is nonempty, she then computes $y_i = h(O_f(M_1^i), M^{i,(2)})$; if $M^{i,(2)}$ is empty, she just takes $y_i = O_f(M_1^i)$. If O_f is $f(K_1,.)$, then y_i will be $h(K_1, M^i)$, whereas if O_f is $r'(.)$, then y_i will be $h(L_i, M^{i,(2)})$ for a random key $L_i = O_f(M_1^i)$ if $M^{i,(2)}$ is nonempty and will be a random value L_i if $M^{i,(2)}$ is empty. As the adversary A'_f gets these values, she looks for a collision with the y_j values obtained from earlier queries M^j. If a collision occurs, she guesses that O_f is f with hidden key; if not, she guesses that O_f is $r'(.)$.

It is, of course, conceivable that even when O_f is $r'(.)$, there is a collision $h(L_i, M^{i,(2)}) = h(L_j, M^{j,(2)})$ with $M^{i,(2)}$ and $M^{j,(2)}$ nonempty. Note that $L_i = L_j$ if $M_1^i = M_1^j$, but L_i and L_j are independent random values if $M_1^i \neq M_1^j$. In other words, we have (1_1). Recall that the probability that this occurs is α_1.

It is also possible that even when O_f is $r'(.)$ there is a collision involving one or both of the random values L_i or L_j that is produced when $M^{i,(2)}$ or $M^{j,(2)}$ is empty. If both are empty, then the probability that $L_i = L_j$ is 2^{-c}. If, say, $M^{j,(2)}$ is nonempty, then in the case $M_1^i \neq M_1^j$, we again have probability 2^{-c} that $L_i = h(L_j, M^{j,(2)})$, whereas in the case $M_1^i = M_1^j$, we have (2_1) with $K_i = L_i$.

Bringing these considerations together, we see that the advantage of A'_f is $\geq \alpha_0 - \alpha_1 - \beta_1 - \binom{q}{2}2^{-c}$.

We next describe the sequence of adversaries $A_f^{(m)}$, $m \geq 2$. Let O_f again denote the prf-test oracle for f that $A_f^{(m)}$ can query. Like A'_f, he runs $A_{\tilde{f}h}$ once and gives random responses to its queries. As $A_{\tilde{f}h}$ makes queries, he sorts their prefixes (where we are using the word "prefix" to denote the first $m - 1$ blocks of a query that has block length at least m). If the ith query has block length at least m and if its prefix coincides with that of an earlier query, he records the index ℓ_i of the first query that has the same prefix; if it has a different prefix from earlier queries, he sets $\ell_i = i$. After running $A_{\tilde{f}h}$, he goes back to the first query M^{j_1} that has block-length at least m, and for all i for which $\ell_i = j_1$ (i.e., for all queries that have the same prefix as M^{j_1}), he queries M_m^i to O_f and computes $y_i = h(O_f(M_m^i), M^{i,(m+1)})$ if $M^{i,(m+1)}$ is nonempty and otherwise takes $y_i = O_f(M_m^i)$. Then he resets O_f and goes to the first j_2 such that M^{j_2} has block length at least m and a different prefix from M^{j_1}. For all i for which $\ell_i = j_2$, he queries M_m^i to O_f and computes $y_i = h(O_f(M_m^i), M^{i,(m+1)})$ if $M^{i,(m+1)}$ is nonempty and otherwise takes $y_i = O_f(M_m^i)$. He continues in this way until he's gone through all the queries. He then looks for two indices $j < i$ such that $y_j = y_i$. If he finds a collision, he guesses that O_f is f with hidden key; otherwise, he guesses that it is a random function.

The adversary $A_f^{(m)}$ takes advantage of the α_{m-1} probability of a collision of the form (1_{m-1}), and if such a collision occurs, he guesses that O_f is f with hidden key. The possibility that O_f is really $r'(.)$ is due to two conceivable circumstances—a collision of the form (1_m) or a collision among random values (either a collision between two random values L_i and L_j or between L_i and $h(L_j, M^{j,(m+1)})$) or else a collision of the form (2_m) with $K_i = L_i$—here the probability of such a collision is bounded by $\binom{q}{2}2^{-c}$ and by β_m, respectively).

Finally, the sequence of adversaries $B_f^{(m)}$, $m \geq 1$, is defined as follows. As usual, O_f denotes the prf-test oracle for f that $B_f^{(m)}$ can query. She runs $A_{\tilde{f}h}$ once and gives random responses to its queries. As $A_{\tilde{f}h}$ makes queries, she sorts their prefixes (where this time, we are using the word "prefix" to denote the first m blocks of a query that has block length at least m). She makes up a list of pairs $(i, S(i))$, where the ith query has block length exactly m and coincides with the prefix of at least one other query; in that case $S(i)$ denotes the set of indices $j \neq i$ such that $M^{j,[m]} = M^i$. After running $A_{\tilde{f}h}$, she chooses a message block Y that is different from all the blocks M_{m+1}^j of all queries M^j. She goes through all indices i with nonempty $S(i)$. For each such i, she queries Y to O_f, and for each $j \in S(i)$, she also queries M_{m+1}^j to O_f and computes $y_j = h(O_f(M_{m+1}^j), M^{j,(m+2)}, Y)$. She looks for a collision between $O_f(Y)$ and y_j for $j \in S(i)$. Before going to the next i, she resets O_f. If she finds a collision for any of the i, she guesses that O_f is f with hidden key; otherwise, she guesses that it is a random function. The advantage of this adversary is at least $\beta_m - q2^{-c}$, because if O_f is $f(K_i, .)$ and $h(K_i, M^{j,(m+1)}) = K_i$, then $h(O_f(M_{m+1}^j), M^{j,(m+2)}, Y) = f(K_i, Y) = O_f(Y)$, whereas if O_f is a random function, then $O_f(Y)$ has probability only 2^{-c} of coinciding with this h-value.

We thus have the following lower bounds for the advantages of the adversaries:

A_f': $\alpha_0 - \alpha_1 - \beta_1 - \binom{q}{2}2^{-c}$.

$A_f^{(m)}$, $m \geq 2$: $\alpha_{m-1} - \alpha_m - \beta_m - \binom{q}{2}2^{-c}$.

$B_f^{(m)}$, $m \geq 1$: $\beta_m - q2^{-c}$.

Trivially we have $\alpha_n = \beta_n = 0$, and so the adversaries go no farther than $A_f^{(n)}$ and $B_f^{(n-1)}$. The sum of all the advantages of the $2n - 1$ adversaries telescopes and is at least $\alpha_0 - (2n - 1)\binom{q}{2}2^{-c}$.

Since we have no way of knowing which of these adversaries has the greatest advantage, we make a random selection. That is, the adversary A_f we use to attack the pseudorandomness of f consists of randomly choosing one of the $2n - 1$ adversaries A'_f, $A_f^{(m)}$ ($2 \leq m \leq n$), $B_f^{(m)}$ ($1 \leq m \leq n - 1$) and running it. The advantage of the adversary A_f is the expectation obtained by summing the advantages of the $2n - 1$ adversaries with each one weighted by the probability $1/(2n - 1)$ that we choose the corresponding adversary. This advantage is at least $\frac{1}{2n-1}(\alpha_0 - (2n - 1)\binom{q}{2}2^{-c})) > (\frac{\epsilon-\epsilon_2-2\epsilon_3}{2n} - \binom{q}{2}2^{-c})$. Thus, returning to the first paragraph of the proof, we have shown that if conditions (i) and (ii) do not hold, then condition (iii) holds.   □

4 Interpretation

How useful is our theorem as a guarantee of real-world security? As in the case of Theorem 10.1 of [17], there are several reasons for skepticism concerning the practical assurance provided by Theorem 1:

1. In order to conclude that our MAC is an (ϵ, t, q, n)-secure pseudorandom function, we need the inner compression function f to be a strongly $(\epsilon/(2n), t, q)$-secure pseudorandom function. In other words, we have a tightness gap of about $2n$, which is large if, for example, we allow a block-length bound of 2^{20} or 2^{30}.[5]
2. Theorem 1 is in the single-user setting, and its security assurances could fail in the more realistic multiuser setting.
3. The three hypotheses in Theorem 1—pseudorandomness of the outer compression function, strong pseudorandomness of the inner compression function,[6] and the related-key property—are in general extremely difficult to evaluate. When the assumptions in a theorem cannot be evaluated in any convincing manner, we should not be surprised if practitioners view the theorem as having little value.

---

[5]The tightness gap in our theorem, bad as it is, is not nearly as extreme as the one in Fischlin's theorem [10], which establishes the secure-MAC property for NMAC and HMAC based on assumptions that are slightly weaker than the prf property. The gap in success probabilities in that theorem is roughly qn^2. In [10] this gap is compared to the q^2n gap in Bellare's Theorem 3.3 in [2]. However, any comparison based solely on success probabilities is misleading, since the factor q^2n in Bellare's theorem is multiplied by the advantage of a very low-resource adversary A_2 with running time $\leq nT$, much less than that of Fischlin's adversary. One must always include running time comparisons when evaluating tightness gaps, and this is not done in [10].

[6]Note that the inner compression function needs to be strongly (ϵ_1, t, q)-secure for a quite small value of ϵ_1, since the theorem loses content if $\epsilon_1 > 1/(2n)$.

5   Security Theorem with Collision Resistance

There are two types of security theorems that have been proved about nested MACs. Starting with Bellare's paper [2] (see also [10, 17, 24]), some authors have proved theorems without assuming collision resistance of the iterated hash function. The idea is that confidence in security of a MAC scheme should not depend upon the rather strong assumption that an adversary cannot find hash collisions. Our Theorem 1 continues this line of work.

On the other hand, if one is using a hash function that one strongly believes to be collision resistant and one wants to know that an adversary cannot forge message tags, then one can go back to the much earlier and more easily proved security theorem in [6]. The purpose of this section is to carry over the main result of [6] to our class of 1-key nested MACs.

An iterated hash function h is said to be (ϵ, t, q, n) weakly collision resistant if no (t, q, n)-adversary that queries $h(K, .)$ has success probability $\geq \epsilon$ of producing a collision $h(K, M') = h(K, M)$, where M and M' are distinct messages of block-length $\leq n$. (Here $h(K, .)$ is regarded as an oracle, i.e., a black box, and K is a hidden key.) A MAC is said to be (ϵ, t, q, n)-secure against forgery if no (t, q, n)-adversary has success probability $\geq \epsilon$ of producing a tag for an unqueried message of block length $\leq n$.

Theorem 2 *Suppose that the iterated hash function h coming from the compression function f is (ϵ_1, t, q, n) weakly collision resistant, the compression function $\tilde{f}$ is an $(\epsilon_2, t, q, 1)$-secure MAC, and $(f, \tilde{f})$ is $(\epsilon_3, t, 2q + 2)$-secure against (g_1, g_2)-related-key attacks. Then MAC$_{\tilde{f}, f, g_1, g_2}$ is an $(\epsilon_1 + \epsilon_2 + \epsilon_3, t - (q+1)nT, q, n)$-secure MAC, where T is the time required for one evaluation of f or $\tilde{f}$.*

Proof The proof is quite simple. Suppose that we are given a $(t - (q + 1)nT, q, n)$-adversary A_{MAC} that has probability $\geq \epsilon_1 + \epsilon_2 + \epsilon_3$ of forging a MAC$_{\tilde{f}, f, g_1, g_2}$-tag. Then we construct three adversaries—a (t, q)-adversary $A_{\tilde{f}}$, a (t, q, n)-adversary A_{wcr}, and a $(t, 2q + 2)$-adversary A_{rka}—such that at least one of the following is true:

(i) $A_{\tilde{f}}$ has probability $\geq \epsilon_2$ of forging a $\tilde{f}$-tag.
(ii) A_{wcr} has probability $\geq \epsilon_1$ of producing an h collision.
(iii) A_{rka} has advantage $\geq \epsilon_3$ against the (g_1, g_2)-related-key property of $(f, \tilde{f})$.

(It does not matter which of (i)–(iii) is true, since any one of the three would contradict the assumptions of the theorem.)

We first use A_{MAC} to construct both an adversary A_{rka} of the related-key property and an adversary $A_{\tilde{f}h}$ that can forge an $\tilde{f}h$ tag. (Recall that $\tilde{f}h$ denotes the MAC $\tilde{f}(K_2, h(K_1, M))$ with independent keys.) The adversary A_{rka} runs A_{MAC} and interacts with the oracle O_{rka}, which chooses a random bit u. After querying the oracle O_{rka} with at most $2q + 2$ b-bit or c-bit messages, A_{rka} must guess whether O_{rka} is using random keys (i.e., $u = 0$) or related keys (i.e., $u = 1$). As A_{MAC} runs, for each of its queries M^i the adversary A_{rka} queries the first block M_1^i to O_{rka}, then computes $H = h(O_{\text{rka}}(M_1^i), M^{i,(2)})$, and finally queries H to O_{rka}; it gives the value $O_{\text{rka}}(H)$ to A_{MAC} as the tag of the queried message. If in time $\leq t - (q+1)nT$ the adversary A_{MAC} forges a tag of an unqueried message,[7] then A_{rka} guesses that $u = 1$; otherwise, it guesses that $u = 0$. Let α denote the advantage of A_{rka}, where, by definition

$$\alpha = \text{Prob}(A_{\text{rka}} \text{ guesses } 1 \mid u = 1) - \text{Prob}(A_{\text{rka}} \text{ guesses } 1 \mid u = 0). \tag{1}$$

The time A_{rka} needs to perform these steps—that is, computing the values H, waiting for A_{MAC}, and verifying the forgery produced by A_{MAC}—that are bounded by $qnT + (t - (q+1)nT) + nT = t$.

We construct $A_{\tilde{f}h}$ as follows. It has an $\tilde{f}h$-oracle $O_{\tilde{f}h}$ that has hidden keys K_1, K_2. The adversary $A_{\tilde{f}h}$ runs A_{MAC}, responding to each of its queries M^i by querying $O_{\tilde{f}h}$ and giving A_{MAC} the response $O_{\tilde{f}h}(M^i)$. If A_{MAC} forges the $\tilde{f}h$ tag of an unqueried message $\tilde{M}$ (which $A_{\tilde{f}h}$ can verify with one further query to its oracle), then $A_{\tilde{f}h}$ has succeeded in forging the $\tilde{f}h$ tag of $\tilde{M}$. Let β denote the success probability of this adversary $A_{\tilde{f}h}$.

Note that the interaction of $A_{\tilde{f}h}$ with A_{MAC} is exactly the same as that of A_{rka} with A_{MAC} in the case $u = 0$. Thus, the second term on the right in the expression (1) for α is equal to β, whereas the first term is the success probability of A_{MAC}, which by assumption is at least $\epsilon_1 + \epsilon_2 + \epsilon_3$. We hence have $\alpha + \beta \geq \epsilon_1 + \epsilon_2 + \epsilon_3$, and this means that either $\alpha \geq \epsilon_3$ (which is the alternative (iii) above) or else $\beta \geq \epsilon_1 + \epsilon_2$.

We now use $A_{\tilde{f}h}$ to construct an $\tilde{f}$ tag-forging adversary $A_{\tilde{f}}$ and an h collision-finding adversary A_{wcr}. The former is constructed as follows. After choosing a random key K_1, $A_{\tilde{f}}$ runs $A_{\tilde{f}h}$. In response to each query, M^i from $A_{\tilde{f}h}$, $A_{\tilde{f}}$ computes $H = h(K_1, M^i)$, queries this H value to its oracle $O_{\tilde{f}} = \tilde{f}(K_2, .)$, and gives the value $\tilde{f}(K_2, H)$ to $A_{\tilde{f}h}$. With probability β in time $\leq t - (q+1)nT$, the adversary $A_{\tilde{f}h}$ finds a tag $\tilde{T} = \tilde{f}h(\tilde{M})$, where $\tilde{M}$ is different from all of the queried messages. The bound on the time $A_{\tilde{f}}$ needs to perform these steps—that is, computing the values $h(K_1, M^i)$ and waiting for $A_{\tilde{f}h}$—is $qnT + (t - (q+1)nT) = t - nT$. Then $A_{\tilde{f}}$ takes time $\leq nT$ to compute $\tilde{H} = h(K_1, \tilde{M})$, hoping that it is different from all H values that were queried to $O_{\tilde{f}}$, in which case it has succeeded in forging an $\tilde{f}$ tag. Meanwhile, the adversary A_{wcr}, which has an oracle O_{wcr} that responds to queries with $h(K_1, .)$ where K_1 is a hidden key, is constructed as follows. It chooses a random key K_2 and runs $A_{\tilde{f}h}$, responding to its queries M^i with $\tilde{f}(K_2, O_{\text{wcr}}(M^i))$. A_{wcr} looks for a collision between some

---

[7]Note that A_{rka} can verify that A_{MAC} has a valid forgery using the same procedure that was used to respond to its queries. This means that A_{rka} needs to be allowed two more queries of O_{rka}, and for this reason, the query bound for A_{rka} is $2q + 2$ rather than $2q$, and the time bounds have the term $(q+1)nT$ rather than qnT.

$O_{\text{wcr}}(M^i) = h(K_1, M^i)$ and $O_{\text{wcr}}(\tilde{M}) = h(K_1, \tilde{M})$, where $(\tilde{M}, \tilde{T})$ is the forgery produced by $A_{\tilde{f}h}$ in the event that the latter adversary succeeds in its task. Note that the success probability β of $A_{\tilde{f}h}$ is the sum of the success probability of $A_{\tilde{f}}$ and that of A_{wcr}. Since $\beta \geq \epsilon_1 + \epsilon_2$ if alternative (iii) does not hold, it follows that at least one of the above alternatives (i)–(iii) must hold.

This concludes the proof.   □

This theorem, like Theorem 1, provides only a very limited type of security assurance. Two of the three "reasons for skepticism" listed in Sect. 4 also apply to Theorem 2: it assumes the (unrealistic) single-user setting, and one of its hypotheses—the secure-MAC property for the compression function—is very difficult to evaluate in practice. On the positive side, at least Theorem 2 is tight, unlike Theorem 1. It's reasonable to regard Theorem 2 as providing a type of assurance that the "domain extender" feature of the MAC scheme is not likely to be a source of security breaches, provided that h is weakly collision resistant.

The proof of Theorem 2 is short, straightforward, and in some sense tautological. Some people would even say that it is "trivial," although we would prefer not to use such a pejorative word in connection with the proof of Theorem 2. But in any case, it seems to us that proofs of this sort merely confirm what is more or less intuitively obvious from the beginning. Such proofs cannot serve as a meaningful source of confidence in a protocol, and they certainly cannot be a substitute for extensive testing and concrete cryptanalysis.

6   HMAC vs. Envelope MAC Comparison

As discussed in [19], it often happens that a type of cryptography enjoys nearly universal acceptance more for reasons of historical happenstance than because of its intrinsic advantages over the alternatives. At present HMAC is widely deployed, whereas Envelope MAC languishes in relative obscurity. But the reasons for this seem to lie in the peculiarities of the history of Envelope MAC, and one can argue that, despite this history, it deserves serious consideration as a secure and practical MAC scheme.

An Envelope MAC scheme was first presented by Tsudik in [28]. His scheme used two independent c-bit keys, and he argued informally that this would give it $2c$ bits of security. However, Preneel and van Oorschot [25] showed that the keys can be recovered in 2^{c+1} steps if one knows approximately $2^{c/2}$ message-tag pairs. That is, Tsudik was wrong, and two independent keys do not give significantly more security than one key.

Soon after, Kaliski and Robshaw [12] and Piermont and Simpson [23] presented a 1-key variant of Envelope MAC, but it had a flaw. To explain this, for concreteness, we'll use MD5 as the underlying hash function with $c = 128$, $b = 512$. Let p denote a 384-bit padding, used to extend a key to fill a 512-bit block. Given a 128-bit key K and a message M of arbitrary length, the tag is $h(K \| p \| M \| K^0)$ (where the 0 superscript indicates that 0s are appended to fill out the last message block). Note that the second K may spill over into two blocks since the bitlength of M is not required to be a multiple of b. Preneel and van Oorschot [26] exploited this overlap to design a key-recovery attack that needs approximately 2^{64} message-tag pairs and has running time 2^{64}. Because a MAC based on MD5 would be expected to require exhaustive search—that is, roughly 2^{128} steps—for key recovery, this attack exposed a serious defect in the variant of Envelope MAC in [12,23]. The Preneel-van Oorschot attack gave Envelope MAC a bad name. However, the attack was possible only because of poor formatting of the second key block.

This flaw can be removed simply by ensuring that each key lies in its own block. This was done by Yasuda [29]. Yasuda also gave a security proof, along the lines of Bellare's HMAC security proof in [2]; in fact, he made crucial use of Bellare's Lemma 3.1. Like Bellare's proof, Yasuda's security theorem requires unconstructible adversaries and so is not valid in the uniform model of complexity. Our Theorem 1 gives a uniform proof for the version of 1-key Envelope MAC described by Yasuda.

As pointed out in [29], Envelope MAC has a minor efficiency advantage over HMAC because the iterated hash function needs to be applied just once. Generally, the accepted procedure when applying an iterated hash function is to append a block at the end of the message that gives the block length of the message. (With this modification, one can give a simple proof that collision resistance of f implies collision resistance of h.) In Envelope MAC, this is done just once, whereas in HMAC it needs to be done twice. Envelope MAC also has the advantage of simplicity—no need for ipad and opad.

In order to argue that HMAC is preferable, one thus has to make a persuasive case that it has better security. Our two theorems give the same security results in both cases, and the same building block (a compression function f) can be used in both. From this standpoint the only grounds for preferring HMAC would be if one of the following holds:

1. The prf assumption on $\tilde{f}$ in Theorem 1 is more credible for HMAC than for Envelope MAC. In the former case, the assumption is weaker than the prf assumption on f and in fact follows from it. In the latter case the assumption also seems to be weaker than the prf assumption on f in practice, but not in the formal sense; in Envelope MAC the prf assumption on $\tilde{f}$ is an additional condition that is not a consequence of the prf assumption on f.[8] One could claim that the need for a separate $\tilde{f}$ condition in Envelope MAC means that it is less secure than HMAC.

---

[8]In the prf test for f, the adversary gets the values $f(K, M)$ (with K a c-bit hidden key and M a b-bit queried message); in the test for $\tilde{f}$ in HMAC, he gets the values $f(K, M \| p)$ (with M a c-bit message and p a fixed $(b - c)$-bit padding); and in the test for $\tilde{f}$ in Envelope MAC, he gets the values $f(M, K \| p)$. Thus, the only difference is whether the key occurs in the first c bits or in the next c bits.

2. The secure-MAC assumption on $\tilde{f}$ in Theorem 2 is more credible for HMAC than for Envelope MAC. One would be claiming that it's harder to forge a tag if the key occurs in the first c bits than if it occurs in the next c bits.
3. The different choices of the pair of functions g_1, g_2 lead to a real difference in strength of the related-key assumptions. There would be a strong reason to prefer HMAC if one could argue that the choice $g_2(K) = K$ in Envelope MAC makes the related-key assumption less plausible.

However, we know of no evidence for any of the above three claims. To the best of our knowledge, no provable security theorem justifies preferring one of these two MACs over the other. Nor does any such theorem preclude the possibility that one would want to choose some other MAC with entirely different functions $\tilde{f}$, g_1, g_2.

Remark 3 Both HMAC and Envelope MAC offer the convenience of using only an off-the-shelf hash function with built-in IV. However, one can argue that for the outer compression function—which maps only from $\{0, 1\}^c \times \{0, 1\}^c$ rather than from the much larger set $\{0, 1\}^c \times \{0, 1\}^b$—it might be more efficient to use a specially chosen $\tilde{f}$. One can even argue that the inner compression function f needs to have better security than $\tilde{f}$ because it is iterated. That is why Theorem 1 has a $2n$ tightness gap with respect to the advantage bound ϵ_1 of an f adversary, but not with respect to the advantage bound ϵ_2 of an $\tilde{f}$ adversary. If one believes that security proofs should guide protocol design and that efficiency should not be sacrificed for security unless a provable security theorem gives grounds for doing so (in [13] Katz and Lindell argue forcefully for this viewpoint), then it is natural to conclude that $\tilde{f}$ *should* be less secure than f. Elsewhere (see [16]) we have raised doubts about this way of thinking, so our personal preference would be *not* to use a weaker $\tilde{f}$. But to someone who needs only short-term security, this might be an acceptable risk in order to gain a slight edge in efficiency.

Conclusion

The Importance of the "Right" Definitions

In their highly regarded textbook [13] on the foundations of cryptography, Katz and Lindell write that the "formulation of exact definitions" is Principle 1 in their list of "basic principles of modern cryptography" and that getting the definitions right is the "essential prerequisite for the... study of any cryptographic primitive or protocol." We agree with this statement, and in [16] we analyzed some of the difficulties and challenges that researchers in both symmetric and asymmetric cryptography have faced in trying to search for a consensus on what the "right" definitions are.

(continued)

In our study of 1-key nested MACs, we have encountered two instances where the standard accepted definitions are not, in our opinion, the natural and useful ones. First, as we explained in Sect. 9 of [17], in the context of iterated hash functions the usual definition of pseudorandomness needs to be replaced by a stronger definition in which the adversary is given the power to "reset" the oracle. In the second place, when analyzing the step from NMAC to 1-key versions such as HMAC and Envelope MAC, we believe that our definition of resistance to related-key attack is preferable to the one used by earlier authors.

We have given arguments justifying our use of these new definitions. Nevertheless, it would be arrogant in the extreme for us to claim that we have resolved the question or that our viewpoint is definitive. Cryptography is as much an art as a science, and to some extent decisions about which are the "right" definitions are matters of personal taste.

The Role of Mathematical Proofs

The NIST documents concerning the SHA-3 competition illustrate the limited role that provable security plays in evaluating real-world cryptosystems. The report [22] that explains the rationale for the selection of the winner devotes about 5 % of the section on security to security proofs. The report highlights the role of proofs in showing a hash function's "security against generic attacks—attacks that only exploit the domain extender and not the internals of the underlying building blocks" (p. 11). It notes that all five finalists have this sort of security proof. In other words, the security proofs are useful as a minimal type of assurance that basically says that concrete cryptanalysis should focus on the underlying building blocks rather than on the extension procedure. But the final decision about what to use must be based on extensive testing and concrete cryptanalysis. NIST makes it clear that provable security, although a necessary component in the security analysis, played no part in ranking the five finalists.[9]

In choosing a MAC scheme, provable security (which, as we argued in [15], is a misnomer) should play no greater role than it did in choosing SHA-3. All methods of the form in Theorem 1 for constructing MACs from compression functions are good domain extenders if they satisfy the

(continued)

---

[9]How can something be a necessary component, but play no role in the selection? By analogy, when one looks for an apartment, a functioning toilet is a requirement; however, one doesn't normally choose which apartment to rent based on which has the nicest toilet.

hypothesis of that theorem—of course, "good" only in the limited sense guaranteed by the conclusion of the theorem. As in the case of the SHA-3 competition, the final choice has to be made through ad hoc testing rather than mathematical theorems. In particular, the relative merits of HMAC and Envelope MAC cannot be determined from provable security considerations. The choice between the two (or a decision to go with a totally different $\tilde{f}, g_1, g_2$) is a judgment call.

References

1. M. Bellare, Practice-oriented provable-security, in *Proceedings of First International Workshop on Information Security (ISW '97)*. Lecture Notes in Computer Science, vol. 1396 (Springer, Berlin, 1998), pp. 221–231
2. M. Bellare, New proofs for NMAC and HMAC: security without collision resistance, in *Advances in Cryptology—Crypto 2006*. Lecture Notes in Computer Science, vol. 4117 (Springer, Heidelberg, 2006), pp. 602–619. Extended version available at http://cseweb.ucsd.edu/mihir/papers/hmac-new.pdf
3. M. Bellare, T. Kohno, A theoretical treatment of related-key attacks: RKA-PRPs, RKA-PRFs, and applications, in *Advances in Cryptology—Eurocrypt 2003*. Lecture Notes in Computer Science, vol. 2656 (Springer, Heidelberg, 2003), pp. 491–506
4. M. Bellare, P. Rogaway, Random oracles are practical: a paradigm for designing efficient protocols, in *Proceedings of First Annual Conference on Computer and Communications Security* (ACM, New York, 1993), pp. 62–73
5. M. Bellare, R. Rogaway, Optimal asymmetric encryption—how to encrypt with RSA, in *Advances in Cryptology—Eurocrypt '94*. Lecture Notes in Computer Science, vol. 950 (Springer, Heidelberg, 1994), pp. 92–111
6. M. Bellare, R. Canetti, H. Krawczyk, Keying hash functions for message authentication, in *Advances in Cryptology—Crypto '96*. Lecture Notes in Computer Science, vol. 1109 (Springer, Heidelberg, 1996), pp. 1–15. Extended version available at http://cseweb.ucsd.edu/mihir/papers/kmd5.pdf
7. M. Bellare, R. Canetti, H. Krawczyk, HMAC: Keyed-hashing for message authentication, Internet RFC 2104 (1997)
8. D. Bernstein, T. Lange, Non-uniform cracks in the concrete: the power of free precomputation, in *Advances in Cryptology—Asiacrypt 2013*. Lecture Notes in Computer Science, vol. 8270 (Springer, Heidelberg, 2013), pp. 321–340
9. D. Boneh, Simplified OAEP for the RSA and Rabin functions, in *Advances in Cryptology—Crypto 2001*. Lecture Notes in Computer Science, vol. 2139 (Springer, Heidelberg, 2001), pp. 275–291
10. M. Fischlin, Security of NMAC and HMAC based on non-malleability, in *Topics in Cryptology—CT-RSA 2008*. Lecture Notes in Computer Science, vol. 4064 (Springer, Heidelberg, 2008), pp. 138–154
11. P. Gauravaram, L. Knudsen, K. Matusiewicz, F. Mendel, C. Rechberger, M. Schläffer, S. Thomsen, Grøstl—a SHA-3 candidate (2011). Available at http://www.groestl.info/Groestl.pdf
12. B. Kaliski, M. Robshaw, Message authentication with MD5. CryptoBytes **1**(1), 5–8 (1995)
13. J. Katz, Y. Lindell, *Introduction to Modern Cryptography* (Chapman and Hall/CRC, Boca Raton, 2007)

14. N. Koblitz, A. Menezes. http://anotherlook.ca
15. N. Koblitz, A. Menezes, Another look at "provable security." J. Cryptol. **20**, 3–37 (2007)
16. N. Koblitz,A. Menezes, Another look at security definitions. Adv. Math. Commun. **7**, 1–38 (2013)
17. N. Koblitz, A. Menezes, Another look at HMAC. J. Math. Cryptol. **7**, 225–251 (2013)
18. N. Koblitz, A. Menezes, Another look at non-uniformity. Groups Complex. Cryptol. **5**, 117–139 (2013)
19. A.H. Koblitz, N. Koblitz, A. Menezes, Elliptic curve cryptography: the serpentine course of a paradigm shift. J. Number Theory **131**, 781–814 (2011)
20. H. Krawczyk, Koblitz's arguments disingenuous. Not. Am. Math. Soc. **54**(11), 1455 (2007)
21. National Institute of Standards and Technology, The keyed-hash message authentication code (HMAC). FIPS Publication 198 (2002)
22. National Institute of Standards and Technology, Third-round report of the SHA-3 cryptographic hash algorithm competition. Interagency Report 7896 (2012)
23. P. Piermont, W. Simpson, IP authentication using keyed MD5, IETF RFC 1828 (1995)
24. K. Pietrzak, A closer look at HMAC. Available at http://eprint.iacr.org/2013/212
25. B. Preneel, P. van Oorschot, MDx-MAC and building fast MACs from hash functions, in *Advances in Cryptology—Crypto '95*. Lecture Notes in Computer Science, vol. 963 (Springer, Heidelberg, 1995), pp. 1–14
26. B. Preneel, P. van Oorschot, On the security of iterated message authentication codes. IEEE Trans. Inf. Theory **45**, 188–199 (1999)
27. V. Shoup, OAEP reconsidered, in *Advances in Cryptology—Crypto 2001*. Lecture Notes in Computer Science, vol. 2139 (Springer, Heidelberg, 2001), pp. 239–259
28. G. Tsudik, Message authentication with one-way hash functions. ACM SIGCOMM Comput. Commun. Rev. **22**(5), 29–38 (1992)
29. K. Yasuda, "Sandwich" is indeed secure: how to authenticate a message with just one hashing, in *Information Security and Privacy—ACISP 2007*. Lecture Notes in Computer Science, vol. 4586 (Springer, Heidelberg, 2007), pp. 355–369

Non-extendable $\mathbb{F}_q$-Quadratic Perfect Nonlinear Maps

Ferruh Özbudak and Alexander Pott

Abstract Let q be a power of an odd prime. We give examples of non-extendable $\mathbb{F}_q$-quadratic perfect nonlinear maps. We also show that many classes of $\mathbb{F}_q$-quadratic perfect nonlinear maps are extendable. We give a short survey of some related results and provide some open problems.

1 Introduction

Let q be a power of an odd prime. Let n, m be positive integers with m dividing n. We use $\mathrm{Tr}_{\mathbb{F}_{q^n}/\mathbb{F}_{q^m}}$ and $\mathrm{Norm}_{\mathbb{F}_{q^n}/\mathbb{F}_{q^m}}$ to denote the relative trace and relative norm functions from $\mathbb{F}_{q^n}$ to $\mathbb{F}_{q^m}$ given by

$$\mathrm{Tr}_{\mathbb{F}_{q^n}/\mathbb{F}_{q^m}} = x + x^{q^m} + x^{2q^m} + \cdots + x^{q^{(n/m-1)m}}$$

and

$$\mathrm{Norm}_{\mathbb{F}_{q^n}/\mathbb{F}_{q^m}} = x \cdot x^{q^m} \cdot x^{2q^m} \cdots x^{q^{(n/m-1)m}}.$$

When $m = 1$, we denote $\mathrm{Tr}_{\mathbb{F}_{q^n}/\mathbb{F}_q}$ and $\mathrm{Norm}_{\mathbb{F}_{q^n}/\mathbb{F}_q}$ as Tr and Norm in short.

An arbitrary $\mathbb{F}_q$-quadratic form f on $\mathbb{F}_{q^n}$ is a map from $\mathbb{F}_{q^n}$ to $\mathbb{F}_q$ given by

$$f(x) = \mathrm{Tr}(a_0 x^2 + a_1 x^{q+1} + \cdots + a_{\lceil \frac{n}{2} \rceil} x^{\lceil \frac{n}{2} \rceil + 1})$$

where $a_0, a_1, \ldots, a_{\lceil \frac{n}{2} \rceil} \in \mathbb{F}_{q^n}$. We call such f an $\mathbb{F}_q$-quadratic map from $\mathbb{F}_{q^n}$ to $\mathbb{F}_q$ as well.

F. Özbudak (✉)
Department of Mathematics and Institute of Applied Mathematics, Middle East
Technical University, Dumlupınar Bulvarı No. 1, 06800 Ankara, Turkey
e-mail: ozbudak@metu.edu.tr

A. Pott
Faculty of Mathematics, Otto-von-Guericke University of Magdeburg, Universitätplatz 2,
39106 Magdeburg, Germany
e-mail: alexander.pott@ovgu.de

© Springer International Publishing Switzerland 2014
Ç.K. Koç (ed.), *Open Problems in Mathematics and Computational Science*,
DOI 10.1007/978-3-319-10683-0_5

An $\mathbb{F}_q$-quadratic map F on $\mathbb{F}_{q^n}$ is a map in Dembowski–Ostrom polynomial form given by

$$F(x) = \sum_{i,j=0}^{n-1} a_{i,j} x^{q^i + q^j}$$

where $a_{i,j} \in \mathbb{F}_{q^n}$.

Let $\{e_1, e_2, \ldots, e_n\}$ be an $\mathbb{F}_q$-basis of $\mathbb{F}_{q^n}$. Equivalently, an $\mathbb{F}_q$-quadratic map on $\mathbb{F}_{q^n}$ is a map given by

$$F(x) = \sum_{i=1}^{n} f_i(x) e_i. \tag{1}$$

where $f_i(x)$ is an $\mathbb{F}_q$-quadratic form on $\mathbb{F}_{q^n}$. Indeed if $\{e_1^*, e_2^*, \ldots, e_n^*\}$ is the trace-orthogonal basis (see [18]), then

$$f_i(x) = \mathrm{Tr}\left(e_i^* \sum_{i_1, i_2=0}^{n-1} a_{i_1, i_2} x^{q^{i_1} + q^{i_2}} \right)$$

is an $\mathbb{F}_q$-quadratic form given in (1) for $1 \leq i \leq n$.

We find it more convenient to represent the $\mathbb{F}_q$-quadratic map F from $\mathbb{F}_{q^n}$ to $\mathbb{F}_{q^n}$ (or $\mathbb{F}_q^n$) as

$$F(x) = \begin{bmatrix} f_1(x) \\ \vdots \\ f_n(x) \end{bmatrix}.$$

In general, for $1 \leq r \leq n$, we say that F is an $\mathbb{F}_q$-quadratic map from $\mathbb{F}_{q^n}$ to $\mathbb{F}_q^r$ if

$$F(x) = \begin{bmatrix} f_1(x) \\ \vdots \\ f_r(x) \end{bmatrix}.$$

where $f_1(x), \ldots, f_r(x)$ are $\mathbb{F}_q$-quadratic forms on $\mathbb{F}_{q^n}$. Choosing a basis $\{e_1, \ldots, e_n\}$ of $\mathbb{F}_{q^n}$ over $\mathbb{F}_q$, F can equivalently be considered as an $\mathbb{F}_q$-quadratic map from $\mathbb{F}_{q^n}$ to $\mathbb{F}_{q^r}$ given by

$$F(x) = f_1(x) e_1 + \cdots + f_r(x) e_r.$$

We give an important definition.

Definition 1.1 For $1 \leq r \leq n$, let F be an $\mathbb{F}_q$-quadratic map from $\mathbb{F}_{q^n}$ to $\mathbb{F}_q^r$. For $a \in \mathbb{F}_{q^n}^*$, let $D_{F,a}$ be the difference map from $\mathbb{F}_{q^n}$ to $\mathbb{F}_q^r$ given by

$$D_{F,a}(x) = F(x + a) - F(x) - F(a)$$

We call F is *perfect nonlinear* or (q^n, q^r)-*bent* if the cardinality of the set

$$\{x \in \mathbb{F}_{q^n} : D_{F,a}(x) = b\}$$

is equal to q^{n-r} for all $a \in \mathbb{F}_{q^n}^*$ and $b \in \mathbb{F}_q^r$. We also call F is (n, r)-*bent* if q is clear from the context. Moreover we say that F is *planar* if $n = r$ and F is *bent* if $r = 1$.

Definition 1.2 For $1 \leq r \leq n$, let $\begin{bmatrix} f_1 \\ \vdots \\ f_r \end{bmatrix}$ and $\begin{bmatrix} g_1 \\ \vdots \\ g_r \end{bmatrix}$ be $\mathbb{F}_q$-quadratic maps from

$\mathbb{F}_{q^n}$ to $\mathbb{F}_q^r$. We call that $\begin{bmatrix} f_1 \\ \vdots \\ f_r \end{bmatrix}$ and $\begin{bmatrix} g_1 \\ \vdots \\ g_r \end{bmatrix}$ are *equivalent* if there exists an $\mathbb{F}_q$-

linearized permutation polynomial $L(x) \in \mathbb{F}_{q^n}[x]$ and an invertible $r \times r$ matrix $[a_{ij}]$ with entries from $\mathbb{F}_q$ such that

$$[a_{ij}] \begin{bmatrix} f_1(L(x)) \\ \vdots \\ f_r(L(x)) \end{bmatrix} = \begin{bmatrix} g_1(x) \\ \vdots \\ g_r(x) \end{bmatrix}$$

for all $x \in \mathbb{F}_{q^n}$.

Remark 1.3 There are more general notions of equivalence for arbitrary finite fields and more general maps between them: extended affine equivalence and Carlet–Charpin–Zinoviev equivalence [6]. However the equivalence in Definition 1.2 gives the same results if we use these two other equivalence notions for the $\mathbb{F}_q$-quadratic maps from $\mathbb{F}_{q^n}$ to $\mathbb{F}_q^r$ in this paper (see [21]).

Definition 1.4 For $1 \leq r \leq n - 1$, let F be an $\mathbb{F}_q$-quadratic (q^n, q^r)-bent map. We call that F is *extendable* if there exists an $\mathbb{F}_q$-quadratic form f on $\mathbb{F}_{q^n}$ such that the map $\begin{bmatrix} F(x) \\ f(x) \end{bmatrix}$ is an (q^n, q^{r+1})-bent map. Otherwise we call that F is *non-extendable*.

Note that F is non-extendable if and only if G is non-extendable for any G equivalent to F. Moreover non-extendable $\mathbb{F}_q$-quadratic perfect nonlinear maps can be considered as a kind of "atomic" structures.

It is a difficult problem to characterize all $\mathbb{F}_q$-quadratic perfect nonlinear maps up to equivalence. In Sect. 2 we give a short survey on some of the results in this problem.

It seems also a difficult problem to characterize all $\mathbb{F}_q$-quadratic non-extendable $\mathbb{F}_q$-quadratic perfect nonlinear maps up to equivalence. In Sect. 3 we prove that many classes of $\mathbb{F}_q$-quadratic perfect nonlinear maps are extendable. In Sect. 4 we give some examples of non-extendable $\mathbb{F}_q$-quadratic perfect nonlinear maps.

There is a natural connection to finite semifields. We explain such a connection in Sects. 5 and 6. Throughout the chapter, we explain many open problems.

2 $\mathbb{F}_q$-Quadratic Perfect Nonlinear Maps

In this section we mainly consider $\mathbb{F}_q$-quadratic perfect nonlinear maps from $\mathbb{F}_{q^n}$ to $\mathbb{F}_{q^n}$. They are also called planar maps. The equivalence in Definition 1.2 of Sect. 1 becomes the following. Let $f, g : \mathbb{F}_{q^n} \to \mathbb{F}_{q^n}$ be $\mathbb{F}_q$-quadratic maps on $\mathbb{F}_{q^n}$. We say that f and g are *equivalent* if there exist $\mathbb{F}_q$-linearized permutation polynomials $L_1, L_2 \in \mathbb{F}_{q^n}[x]$ such that

$$f(x) = L_1 \circ g \circ L_2 \text{ for all } x \in \mathbb{F}_{q^n}.$$

It is a difficult problem to decide whether a given map is planar or not in general. All $\mathbb{F}_q$-quadratic monic monomial maps on $\mathbb{F}_{q^3}$ are

$$x^2, x^{q+1}, x^{2q}, x^{q^2+1}, x^{q^2+q}, x^{2q^2}.$$

It is easy to prove that all these $\mathbb{F}_q$-quadratic monomial maps are planar. Recently the authors characterized all $\mathbb{F}_q$-quadratic planar binomial maps on $\mathbb{F}_{q^3}$ explicitly in [16]. Up to multiplication with a nonzero constant in $\mathbb{F}_{q^3}$, there are 15 distinct $\mathbb{F}_q$-quadratic binomial maps on $\mathbb{F}_{q^3}$:

(1) $x^2 + ux^{q^2+q}$
(2) $x^{q+1} + ux^{2q^2}$
(3) $x^{2q} + ux^{q^2+1}$
(4) $x^2 + ux^{q+1}$
(5) $x^2 + ux^{2q}$
(6) $x^2 + ux^{q^2+1}$
(7) $x^2 + ux^{2q^2}$
(8) $x^{q+1} + ux^{2q}$
(9) $x^{q+1} + ux^{q^2+1}$
(10) $x^{q+1} + ux^{q^2+q}$
(11) $x^{2q} + ux^{q^2+q}$
(12) $x^{2q} + ux^{2q^2}$
(13) $x^{q^2+1} + ux^{q^2+q}$

(14) $x^{q^2+1} + ux^{2q^2}$
(15) $x^{q^2+q} + ux^{2q^2}$

Here u is an arbitrary nonzero element in $\mathbb{F}_{q^3}$. It is easier to decide whether the maps in the sublist $4, 5, \ldots, 15$ are planar or not. First we recall that for $u \in \mathbb{F}_{q^3}^*$, the polynomial maps

$$x^{q+1} + ux^2 \in \mathbb{F}_{q^3}[x] \text{ and } x^{q^2+1} + ux^2 \in \mathbb{F}_{q^3}[x] \tag{2}$$

are not planar (see Lemma 5.1 and page 643 in [12]). It is clear that the items $4, 6, 8, 11, 14$, and 15 are all equivalent to one of the maps in (2). For example, regarding the map in item 8, we have

$$x^{q+1} + ux^{2q} = (x^{q^2+1} + ux^2) \circ x^q \mod (x^{q^3} - x).$$

Therefore, the maps in items $4, 6, 8, 11, 14$, and 15 are not planar for any $u \in \mathbb{F}_{q^3}^*$. Next we consider the remaining maps in the sublist. The maps in items $5, 7$, and 12 are compositions of x^2 with a linearized polynomial map. For example, regarding the map in item 5, we have

$$x^2 + ux^{2q} = (x + ux^q) \circ x^2$$

Therefore the map in item 5 is planar if and only if the map $x \mapsto x + ux^{q^2}$ is a permutation map. Similarly it is easy to decide whether the maps in items 7 and 12 are planar or not depending on the corresponding linearized polynomials in terms of $u \in \mathbb{F}_{q^3}$. Moreover they are equivalent to x^2 if they are planar.

The maps in items $9, 10$, and 13 are compositions of x^{q+1} with a linearized polynomial map. For example, regarding the map in item 9, we have

$$x^{q+1} + ux^{q^2+1} \equiv (x + ux^{q^2}) \circ x^{q+1} \mod (x^{q^3} - x)$$

Therefore the map in item 9 is planar if and only if the map $x \mapsto x + ux^{q^2}$ is a permutation map. Similarly the maps in items 10 and 13 are planar if and only if the corresponding linearized polynomials give permutation maps, which depend on $u \in \mathbb{F}_{q^3}^*$. Moreover they are equivalent to x^{q+1} if they are planar.

It remains to consider the maps in items $1, 2$, and 3. These maps are quite different from the rest of the maps in the list. First the maps in items 2 and 3 are equivalent to the map in item 1 as

$$x^{q+1} + ux^{2q^2} \equiv (x^{q^2+q} + ux^2) \circ x^{q^2} \mod (x^{q^3} - x)$$

and

$$x^{2q} + ux^{q^2+1} \equiv (x^2 + ux^{q^2+q}) \circ x^q \mod (x^{q^3} - x).$$

Hence for $u \in \mathbb{F}_{q^3}^*$ let F_u be the binomial map given by

$$F_u(x) = x^{q^2+q} + ux^2 \in \mathbb{F}_{q^3}[x].$$

In [16] the authors prove that $F_u(x)$ is planar for some $u \in \mathbb{F}_{q^3}$ and equivalent to x^2. They also prove that $F_u(x)$ is planar for some $u \in \mathbb{F}_{q^3}$ and equivalent to x^{q+1}. Therefore the binomial form of $F_u(x)$ is different from the sublist of the binomial forms in $4, 5, \ldots, 15$ as they cannot be equivalent to both x^2 and x^{q+1} for some different choices of $u \in \mathbb{F}_{q^3}^*$. Moreover, it seems that proving planarity of $F_u(x)$ is quite difficult, and it has direct connections with arithmetic of some function fields [16]. Their main results is (see Theorem 4.5 in [16]):

Theorem 2.1 *Let q be a power of an odd prime and $u \in \mathbb{F}_{q^3}$. Let $F_u(x)$ be the polynomial $x^{q^2+q} + ux^2$ in $\mathbb{F}_{q^3}[x]$. If $q \equiv 1 \mod 3$, then let G and H be the subgroups of $\mathbb{F}_{q^3}^*$ with $|G| = q^2 + q + 1$ and $|H| = (q^2 + q + 1)/3$. Then the polynomial $F_u(x)$ is planar if and only if $q \equiv 1 \mod 3$ and*

$$u \in \left(-(G \setminus H) \cup \frac{1}{2}(G \setminus H) \right).$$

Moreover we have the followings. Assume that $q \equiv 1 \mod 3$.

- *The polynomial $F_u(x)$ is equivalent to x^2 if and only if $u \in \frac{1}{2}(G \setminus H)$.*
- *The polynomial $F_u(x)$ is equivalent to x^{q+1} if and only if $u \in -(G \setminus H)$.*

We note that one of the important ingredients in the proof of Theorem 4.5 in [16] comes from the theory of (commutative) finite semifields. We will give more information about it in Sects. 5 and 6 below.

We also note that there are many related problems stated either explicitly or implicitly in [16]. We would like to refer to Remark 3.9 in [16] and Corollary 4.6 in [16].

A special type of Dembowski–Ostrom is of the form

$$F(x) = L_1(x)L_2(x),$$

where both $L_1(x)$ and $L_2(x)$ are $\mathbb{F}_q$-linearized polynomials in $\mathbb{F}_{q^n}[x]$. In [15] the planarity of such products of linearized polynomials is investigated. An important tool they use is the link between the set

$$\mathcal{M}(L) : \{\alpha \in \mathbb{F}_q : L(x) + \alpha x \text{ is bijective on } \mathbb{F}_{q^n}\},$$

which is studied in finite geometry [1, 4, 23].

Another important tool they use is Hasse–Weil–Serre Theorem (see, Theorem 4 in [15] and Theorem 5.3.7 in [24]). Using Hasse–Weil–Serre Theorem, they prove the following result.

Theorem 2.2 *Let q be a power of an odd prime and $n \geq 2$ be an integer. Let $\mathrm{Tr}_n(x) : \mathbb{F}_{q^n} \to \mathbb{F}_q$ be the linearized polynomial $x \mapsto x + x^q + \cdots + x^{q^{n-1}}$. For $a \in \mathbb{F}_{q^n}$, if $n \geq 5$, then the mapping*

$$F(x) = x(\mathrm{Tr}_n(x) + ax)$$

is not planar.

There are planar maps of the form $x(\mathrm{Tr}_n(x) + ax)$ with $a \in \mathbb{F}_{q^n}$ for $n = 2$ and $n = 3$. The case of $n = 2$ was completely characterized in Theorem 2 of [15].

For the case $n = 3$, it was proved in [15] that

$$F(x) = x(\mathrm{Tr}_3(x) + ax) \tag{3}$$

is planar for $a \in \{-1, -2\}$. Moreover for certain values of $a \in \mathbb{F}_q$, it was proved that $F(x)$ in (3) is not planar on $\mathbb{F}_{q^3}$. Moreover they conjectured in Conjecture 1 and Conjecture 2 that there is no planar map of the form (3) for the remaining values of $a \in \mathbb{F}_{q^3}$. These conjectures were proved in [8] again using some facts from the theory of finite semifields.

The authors in [15] also conjectured that there is no planar map on $a \in \mathbb{F}_{q^4}$ of the form

$$F(x) = x(\mathrm{Tr}_4(x) + ax) \text{ with } a \in \mathbb{F}_{q^4}.$$

In [8] it is commented that their method for resolving the conjecture for the case $n = 3$ would not work for the case $n = 4$. Indeed the conjecture for the case $n = 4$ was proved in [25] using a different approach. They also proved the subcase for $n = 3$ where $a \in \{-1, -2\}$ using an elementary approach. It is still a problem whether an approach using exponential sums would work to prove the conjecture of the remaining situation of the subcase $n = 3$ that $a \in \mathbb{F}_{q^3} \setminus \mathbb{F}_q$. We also refer to Sect. 2 in [15] for another open problem.

Recall that we define the equivalence of $\mathbb{F}_q$-quadratic perfect nonlinear maps from $\mathbb{F}_{q^n}$ to $\mathbb{F}_q^r$ in Definition 1.2 of Sect. 1. Up to recently, such maps were classified completely only in the following cases:

- $n \geq 1, r = 1$ (nondegenerate $\mathbb{F}_q$-quadratic form),
- $n = 2, r = 2$ ([10], finite fields),
- $n = 3, r = 3$ ([20], finite fields or twisted finite field).

Namely, for $(n, r) = (2, 2)$ all $\mathbb{F}_q$-quadratic perfect nonlinear maps are equivalent to the map $x \mapsto x^2$. Moreover for $(n, r) = (3, 3)$ all $\mathbb{F}_q$-quadratic perfect nonlinear maps are equivalent to one of the following maps:

- $x \mapsto x^2$ (finite field),
- $x \mapsto x^{q+1}$ (twisted finite field).

Recently we proved in [21] that all $\mathbb{F}_q$-quadratic perfect nonlinear maps for $(n, r) = (3, 2)$ are equivalent. We used some results from algebraic geometry, in particular Bezout's Theorem in our proof. We also presented an explicit algorithm for finding the corresponding equivalence using a geometric method.

First we note that, up to equivalence, an $\mathbb{F}_q$-quadratic perfect nonlinear map from $\mathbb{F}_{q^3}$ to $\mathbb{F}_q^2$ has to be of the form

$$F(x) = \begin{bmatrix} \mathrm{Tr}(x^2) \\ \mathrm{Tr}(\theta x^2 + wx^{q+1}) \end{bmatrix}$$

for some $\theta, w \in \mathbb{F}_{q^3}$ (see Sect. 3 in [21]). However it is perfect nonlinear only for some $\theta, w \in \mathbb{F}_{q^3}$. We also provided a characterization result as follows (see Proposition 3.1 in [21]).

Proposition 2.3 *Let q be a power of an odd prime. For $\theta, w \in \mathbb{F}_{q^3}$ consider the $\mathbb{F}_q$-quadratic map $F : \mathbb{F}_{q^3} \to \mathbb{F}_q^2$ given by*

$$F(x) = \begin{bmatrix} \mathrm{Tr}(x^2) \\ \mathrm{Tr}(\theta x^2 + wx^{q+1}) \end{bmatrix}.$$

Then F is perfect nonlinear if and only if the polynomial

$$T^3 + A_2 T^2 + A_1 T + A_0 \in \mathbb{F}_q[T],$$

where

$$\begin{aligned} A_2 &= \mathrm{Tr}(2\theta), \\ A_1 &= \mathrm{Tr}((2\theta)^{q+1} - w^2), \\ A_0 &= \mathrm{Norm}(2\theta) - \mathrm{Tr}(2\theta w^{2q}) + 2\mathrm{Norm}(w) \end{aligned}$$

is irreducible over $\mathbb{F}_q$.

For some special $\mathbb{F}_q$-quadratic maps, it is easy to decide whether they are perfect nonlinear or not (also equivalence among them is much easier to find). The following was proved in Sect. 2 of [21].

Theorem 2.4 *Let $\{w_1, w_2\}, \{w_3, w_4\} \subseteq \mathbb{F}_{q^3}$ be $\mathbb{F}_q$-linearly independent sets. Let F_1 and F_2 be the $\mathbb{F}_q$-quadratic maps given by*

$$F_1(x) = \begin{bmatrix} \mathrm{Tr}(w_1 x^2) \\ \mathrm{Tr}(w_2 x^2) \end{bmatrix} \quad and \quad F_2(x) = \begin{bmatrix} \mathrm{Tr}(w_3 x^{q+1}) \\ \mathrm{Tr}(w_4 x^{q+1}) \end{bmatrix}$$

For $w \in \mathbb{F}_{q^3} \setminus \mathbb{F}_q$ let G and H be the $\mathbb{F}_q$-quadratic maps

$$G(x) = \begin{bmatrix} \mathrm{Tr}(x^2) \\ \mathrm{Tr}(wx^2) \end{bmatrix} \quad and \quad H(x) = \begin{bmatrix} \mathrm{Tr}(x^{q+1}) \\ \mathrm{Tr}(wx^{q+1}) \end{bmatrix}.$$

Then all of the maps F_1, F_2, G, and H are perfect nonlinear and equivalent to each other.

The main result of the [21] is the following theorem.

Theorem 2.5 *Let $F : \mathbb{F}_{q^3} \to \mathbb{F}_q^2$ be an $\mathbb{F}_q$-quadratic perfect nonlinear map. We can assume that there exist $\theta, w \in \mathbb{F}_{q^3}$ such that*

$$F(x) = \begin{bmatrix} \mathrm{Tr}(x^2) \\ \mathrm{Tr}(\theta x^2 + wx^{q+1}) \end{bmatrix}$$

without loss of generality (up to equivalence). Let $PG(2, \mathbb{F}_{q^3})$ denote the projective plane over $\mathbb{F}_{q^3}$. Let $\mathcal{T}_1$ be the set of consisting $(x_0 : x_1 : x_2) \in PG(2, \mathbb{F}_{q^3})$ such that

$$\mathcal{T}_1 : x_0^2 + x_1^2 + x_2^2 = 0.$$

Let $\mathcal{T}_2$ be the set of consisting of $(x_0 : x_1 : x_2) \in PG(2, \mathbb{F}_{q^3})$ such that

$$\mathcal{T}_2 : (x_0^2 \theta + x_1^2 \theta^{q^2} + x_2^2 \theta^q) + (x_0 x_1 w^{q^2} + x_0 x_1 w + x_1 x_2 w^q) = 0.$$

Let $E : PG(2, \mathbb{F}_{q^3}) \to PG(2, \mathbb{F}_{q^3})$ be the action defined as

$$(x_0 : x_1 : x_2) \mapsto (x_0^q : x_1^q : x_2^q).$$

Then we have the following:

(1) *There exist $Q, P_1 \in PG(2, \mathbb{F}_{q^3})$ such that*

$$\mathcal{T}_1 \cap \mathcal{T}_2 = \{Q, P_1, E(P_1), (E \circ E)(P_1)\},$$

$E(Q) = Q$ and $|\mathcal{T}_1 \cap \mathcal{T}_2| = 4$.
(2) *Put $P_1 = (a_0 : a_1 : a_2) \in PG(2, \mathbb{F}_{q^3})$ with $a_0, a_1, a_2 \in \mathbb{F}_{q^3}$. Let $L(x)$ be the $\mathbb{F}_q$-linearized polynomial in $\mathbb{F}_{q^3}[x]$ given by*

$$L(x) = a_0 x + a_1^q x^q + a_2^{q^2} x^{q^2}.$$

Then there exist $\delta_1, \delta_2 \in \mathbb{F}_{q^3}$ such that

$$\begin{bmatrix} \mathrm{Tr}(\delta_1 x^{q+1}) \\ \mathrm{Tr}(\delta_2 x^{q+1}) \end{bmatrix} \text{ is perfect nonlinear}$$

and

$$\begin{bmatrix} \mathrm{Tr}(\delta_1 x^{q+1}) \\ \mathrm{Tr}(\delta_2 x^{q+1}) \end{bmatrix} = F(L(x)) \text{ for all } x \in \mathbb{F}_{q^3}.$$

Note that Theorem 2.5 gives an explicit geometric method for finding an equivalence from $F(x)$ to an $\mathbb{F}_q$-quadratic perfect nonlinear map in the form

$$\begin{bmatrix} \mathrm{Tr}(\delta_1 x^{q+1}) \\ \mathrm{Tr}(\delta_2 x^{q+1}) \end{bmatrix}.$$

As it is easy to find equivalence for the $\mathbb{F}_q$-quadratic perfect nonlinear maps in between the maps F_1, F_2, G, and H in Theorem 2.4, we easily obtain a geometric algorithm for finding an equivalence from $F(x)$ to, for example, $G(x)$. We refer to Sect. 3 in [21] for such an explicit algorithm.

There is a consequence of Theorems 2.4 and 2.5 given in Sect. 3 below.

3   Extendable $\mathbb{F}_q$-Quadratic Perfect Nonlinear Maps

In this section we show that many classes of $\mathbb{F}_q$-quadratic perfect nonlinear maps are extendable.

Proposition 3.1 *Let $n \geq 2$ and $F : \mathbb{F}_{q^n} \to \mathbb{F}_q$ be an $\mathbb{F}_q$-quadratic perfect nonlinear (or bent) map. Then F is extendable.*

Proof Let $f_1(x) = F(x)$. Note that $f_1(x)$ is a nondegenerate $\mathbb{F}_q$-quadratic form on $\mathbb{F}_{q^n}$. We will show that there exist $\mathbb{F}_q$-quadratic forms $f_2(x), \ldots, f_n(x)$ on $\mathbb{F}_{q^n}$ such that

$$\begin{bmatrix} f_1(x) \\ f_2(x) \\ \vdots \\ f_n(x) \end{bmatrix}$$

is an $\mathbb{F}_q$-quadratic perfect nonlinear map (or planar map) from $\mathbb{F}_{q^n}$ to $\mathbb{F}_q^n$. This shows that F is extendable.

Assume first that n is odd. Let $\lambda \in \mathbb{F}_q^*$ be a nonsquare. Let $g_1(x) = \mathrm{Tr}(x^2)$ and $h_1(x) = \lambda \mathrm{Tr}(x^2)$. Note that both $g_1(x)$ and $h_1(x)$ are nondegenerate $\mathbb{F}_q$-quadratic forms on $\mathbb{F}_{q^n}$. Moreover the discriminants of $g_1(x)$ and $h_1(x)$ are not both square or both nonsquare in $\mathbb{F}_q^*$. Here we use the fact that n is odd and λ^n is a nonsquare in $\mathbb{F}_q^*$. It is well known that there are exactly two nondegenerate $\mathbb{F}_q$-quadratic forms on $\mathbb{F}_{q^n}$ (also for n is even) up to a choice of $\mathbb{F}_q$-basis of $\mathbb{F}_q^n$; they are determined by whether the discriminant is square or not in $\mathbb{F}_q^*$ (see, e.g., Theorem 4.9 in [11]). Hence, there exists an $\mathbb{F}_q$-linearized polynomial $L(x) \in \mathbb{F}_{q^n}[x]$ such that

$$f_1(L(x)) = g_1(x) \text{ or } h_1(x) \text{ for all } x \in \mathbb{F}_{q^n}.$$

As $x \mapsto x^2$ and $x \mapsto \lambda x^2$ are $\mathbb{F}_q$-quadratic perfect nonlinear maps from $\mathbb{F}_{q^n}$ to $\mathbb{F}_{q^n}$, there exist $\mathbb{F}_q$-quadratic forms $g_2(x), h_2(x), \ldots, g_n(x), h_n(x)$ on $\mathbb{F}_{q^n}$ such that

$$\begin{bmatrix} g_1(x) \\ g_2(x) \\ \vdots \\ g_n(x) \end{bmatrix} \quad \text{and} \quad \begin{bmatrix} h_1(x) \\ h_2(x) \\ \vdots \\ h_n(x) \end{bmatrix}$$

are $\mathbb{F}_q$-quadratic nonlinear maps from $\mathbb{F}_{q^n}$ to $\mathbb{F}_q^n$. Therefore

$$\begin{bmatrix} f_1(x) \\ g_2(L^{-1}(x)) \\ \vdots \\ g_n(L^{-1}(x)) \end{bmatrix} \quad \text{or} \quad \begin{bmatrix} f_1(x) \\ h_2(L^{-1}(x)) \\ \vdots \\ h_n(L^{-1}(x)) \end{bmatrix}$$

is an $\mathbb{F}_q$-quadratic perfect nonlinear map from $\mathbb{F}_{q^n}$ to $\mathbb{F}_q^n$ if $f_1(L(x)) = g_1(x)$ or $f_1(L(x)) = h_1(x)$, respectively. This completes the proof if n is odd.

Assume next then n is even. Let $\lambda \in \mathbb{F}_{q^n}^*$ be a nonsquare. Let $g_1(x) = \text{Tr}(x^2)$ and $h_1(x) = \text{Tr}(\lambda x^2)$. Note that $g_1(x)$ and $h_1(x)$ are again nondegenerate $\mathbb{F}_q$-quadratic forms on $\mathbb{F}_{q^n}$. Comparing with the proof of the case n is odd above, it is enough to show that the discriminants $\Delta(g_1)$ and $\Delta(h_1)$ are distinct modulo squares in $\mathbb{F}_q^*$. Let $\eta : \mathbb{F}_q^* \to \{1, -1\}$ be the quadratic character

$$\eta(u) = \begin{cases} 1 & \text{if } u \text{ is a square in } \mathbb{F}_q^*, \\ -1 & \text{if } u \text{ is not a square in } \mathbb{F}_q^*. \end{cases}$$

Let $Q : \mathbb{F}_{q^n} \to \mathbb{F}_q$ be an arbitrary nondegenerate $\mathbb{F}_q$-quadratic form on $\mathbb{F}_{q^n}$ with discriminant $\Delta \in \mathbb{F}_q^*$. Let $b \in \mathbb{F}_{q^*}$ be an arbitrary nonzero element. It is well known that (see, e.g., Theorem 6.26 in [18])

$$|\{x \in \mathbb{F}_{q^n} : Q(x) = b\}| = q^{n-1} - q^{\frac{n-2}{2}} \eta(-1)^{n/2} \Delta. \tag{4}$$

Note that we also have

$$\begin{aligned} 2q^{n-1} &= 2|\{y \in \mathbb{F}_{q^n} : \text{Tr}(y) = b\}| \\ &= |\{x \in \mathbb{F}_{q^n} : \text{Tr}(x^2) = b\}| + |\{x \in \mathbb{F}_{q^n} : \text{Tr}(\lambda x^2) = b\}| \end{aligned} \tag{5}$$

Here we use the fact that if $y \in \mathbb{F}_{q^n}^*$ is a square, then

$$|\{x \in \mathbb{F}_{q^n} : y = x^2\}| = 2 \text{ and } |\{x \in \mathbb{F}_{q^n} : y = \lambda x^2\}| = 0;$$

otherwise if $y \in \mathbb{F}_{q^n}^*$ is not a square, then

$$|\{x \in \mathbb{F}_{q^n} : y = x^2\}| = 0 \text{ and } |\{x \in \mathbb{F}_{q^n} : y = \lambda x^2\}| = 2.$$

Assume the contrary that $\eta(\Delta(g_1)) = \eta(\Delta(h_1))$. Using (4) and (5), we obtain that

$$2q^{n-1} = 2(q^{n-1} \pm q^{\frac{n-2}{2}})$$

which is a contradiction. Therefore $\eta(\Delta(g_1)) \neq \eta(\Delta(h_1))$, which completes the proof. □

The following is a corollary of Theorems 2.4 and 2.5 of Sect. 2.

Corollary 3.2 *Let* $F : \mathbb{F}_{q^3} \to \mathbb{F}_{q^2}$ *be an* $\mathbb{F}_q$-*quadratic perfect nonlinear map. Then* F *is extendable.*

Proof By Theorems 2.4 and 2.5 of Sect. 2, we know that F is equivalent to

$$\begin{bmatrix} \mathrm{Tr}(x^2) \\ \mathrm{Tr}(wx^2) \end{bmatrix} \text{ and } \begin{bmatrix} \mathrm{Tr}(x^{q+1}) \\ \mathrm{Tr}(wx^{q+1}) \end{bmatrix}$$

for $w \in \mathbb{F}_{q^3} \setminus \mathbb{F}_q$. Hence F is extendable to both of the maps $x \mapsto x^2$ and $x \mapsto x^{q+1}$ on $\mathbb{F}_{q^3}$. □

Example 3.3 Let $q = 3$ and $n = 4$. There are exactly 2 distinct $\mathbb{F}_q$-quadratic planar maps from $F : \mathbb{F}_{3^4}$ to $\mathbb{F}_{3^4}$ up to equivalence. They are the maps

$$x \mapsto x^4 + x^{10} - x^{36} \text{ and } x \mapsto x^2.$$

Using computer, we have verified that all $\mathbb{F}_3$-quadratic (q^4, q^2)-bent maps are extendable.

Example 3.4 Let $q = 3$ and $n = 5$. There are exactly 7 distinct $\mathbb{F}_q$-quadratic planar maps from $F : \mathbb{F}_{3^5}$ to $\mathbb{F}_{3^5}$ up to equivalence. They are the maps

- $x \mapsto x^2$.
- $x \mapsto x^{q+1}$.
- $x \mapsto x^{q^2+1}$.
- $x \mapsto x^{10} + x^6 - x^2$.
- $x \mapsto x^{10} - x^6 - x^2$.
- $x \mapsto x^{90} + x^2$.
- $x \mapsto -(x^3 - x) + D(x^3 - x) + \frac{1}{2}x^2$ with $D(x) = -x^{36} + x^{28} + x^{12} + x^4$.

Using computer, we have also verified that all $\mathbb{F}_3$-quadratic (q^5, q^2)-bent maps are extendable.

In the next section we give the first examples of non-extendable $\mathbb{F}_q$-quadratic perfect nonlinear maps in the literature as far as we know.

4 Non-extendable $\mathbb{F}_q$-Quadratic Perfect Nonlinear Maps

In this section we give some non-extendable $\mathbb{F}_q$-quadratic perfect nonlinear maps. We note that finding non-extendable $\mathbb{F}_q$-quadratic perfect nonlinear maps seems to be more difficult than finding $\mathbb{F}_q$-quadratic perfect nonlinear maps. The reason is that non-extendable maps require that they are impossible to extend, which is an extra condition.

Example 4.1 Let $q = 3$ and $n = 4$. Recall that, up to equivalence, there are exactly two $\mathbb{F}_q$-quadratic perfect nonlinear maps from $\mathbb{F}_{q^4}$ to $\mathbb{F}_{q^4}$, which are the maps given by the polynomials x^2 and $x^4 + x^{10} - x^{36}$. We also recall that all $\mathbb{F}_q$-quadratic perfect nonlinear maps $F : \mathbb{F}_{q^4} \to \mathbb{F}_{q^2}$ are extendable (see Sect. 3). In fact, up to equivalence, there are exactly 7 $\mathbb{F}_q$-quadratic perfect nonlinear maps from $\mathbb{F}_{q^4}$ to $\mathbb{F}_{q^2}$.

However there are non-extendable $\mathbb{F}_q$-quadratic perfect nonlinear maps from $\mathbb{F}_{q^4}$ to $\mathbb{F}_{q^3}$. There are exactly 18 $\mathbb{F}_q$-quadratic perfect nonlinear maps from $\mathbb{F}_{q^4}$ to $\mathbb{F}_{q^3}$ up to equivalence. Only 5 of them are extendable. We define them as $E_1, E_2, E_3, E_4,$ and E_5 below. Only E_5 extends to both x^2 and $x^4 + x^{10} - x^{36}$. The others extend only to $x^4 + x^{10} - x^{36}$. The remaining 13 are $\mathbb{F}_q$-quadratic perfect nonlinear maps from $\mathbb{F}_{q^4}$ to $\mathbb{F}_{q^3}$ are non-extendable. We define them as $NE_1, NE_2, \ldots, NE_{13}$ below.

Now we give these maps explicitly. Let w be a primitive element of $\mathbb{F}_{q^4}$ such that $w^4 + 2w^3 + 2 = 0$. Let $E_1, E_2, E_3, E_4, E_5 : \mathbb{F}_{q^4} \to \mathbb{F}_q^3$ be the maps given by

$$E_1(x) = \begin{bmatrix} \mathrm{Tr}(x^2) \\ \mathrm{Tr}(wx^{10}) \\ \mathrm{Tr}(w^2 x^{10} + w^5 x^4) \end{bmatrix}, \quad E_2(x) = \begin{bmatrix} \mathrm{Tr}(x^2) \\ \mathrm{Tr}(wx^{10}) \\ \mathrm{Tr}(w^2 x^2) \end{bmatrix}, \quad E_3(x) = \begin{bmatrix} \mathrm{Tr}(x^2) \\ \mathrm{Tr}(wx^{10}) \\ \mathrm{Tr}(w^8 x^2) \end{bmatrix},$$

$$E_4(x) = \begin{bmatrix} \mathrm{Tr}(x^2) \\ \mathrm{Tr}(w^{13} x^{10}) \\ \mathrm{Tr}(w^5 x^4) \end{bmatrix}, \quad E_5(x) = \begin{bmatrix} \mathrm{Tr}(x^2) \\ \mathrm{Tr}(wx^4) \\ \mathrm{Tr}(w^2 x^4) \end{bmatrix}.$$

These are the extendable maps.

Let $NE_1, NE_2, \ldots, NE_{13} : \mathbb{F}_{q^4} \to \mathbb{F}_{q^3}$ be the maps given by

$$NE_1(x) = \begin{bmatrix} \mathrm{Tr}(x^2) \\ \mathrm{Tr}(wx^{10}) \\ \mathrm{Tr}(wx^4) \end{bmatrix}, \quad NE_2(x) = \begin{bmatrix} \mathrm{Tr}(x^2) \\ \mathrm{Tr}(wx^{10}) \\ \mathrm{Tr}(w^5 x^4) \end{bmatrix}, \quad NE_3(x) = \begin{bmatrix} \mathrm{Tr}(x^2) \\ \mathrm{Tr}(wx^{10}) \\ \mathrm{Tr}(w^7 x^4) \end{bmatrix},$$

$$NE_4(x) = \begin{bmatrix} \mathrm{Tr}(x^2) \\ \mathrm{Tr}(wx^{10}) \\ \mathrm{Tr}(w^3 x^{10} + w^8 x^4 + wx^2) \end{bmatrix}, \quad NE_5(x) = \begin{bmatrix} \mathrm{Tr}(x^2) \\ \mathrm{Tr}(wx^{10}) \\ \mathrm{Tr}(w^3 x^{10} + w^{25} x^4 + wx^2) \end{bmatrix},$$

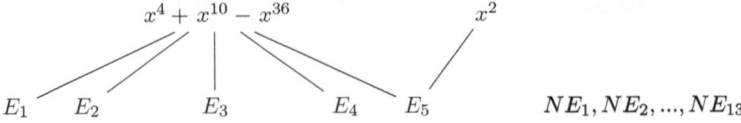

$$x^4 + x^{10} - x^{36} \qquad\qquad x^2$$

$$E_1 \quad E_2 \qquad\quad E_3 \qquad\qquad E_4 \quad E_5 \qquad\qquad NE_1, NE_2, ..., NE_{13}$$

Fig. 1 Lattice for extendable and non-extendable $\mathbb{F}_3$-quadratic perfect nonlinear maps on $\mathbb{F}_{3^4}$

$$NE_6(x) = \begin{bmatrix} \mathrm{Tr}(x^2) \\ \mathrm{Tr}(wx^{10}) \\ \mathrm{Tr}(w^3x^{10} + wx^4 + w^2x^2) \end{bmatrix}, \quad NE_7(x) = \begin{bmatrix} \mathrm{Tr}(x^2) \\ \mathrm{Tr}(wx^{10}) \\ \mathrm{Tr}(w^{30}x^4 + w^6x^2) \end{bmatrix},$$

$$NE_8(x) = \begin{bmatrix} \mathrm{Tr}(x^2) \\ \mathrm{Tr}(wx^{10}) \\ \mathrm{Tr}(w^{15}x^2) \end{bmatrix}, \quad NE_9(x) = \begin{bmatrix} \mathrm{Tr}(x^2) \\ \mathrm{Tr}(wx^{10}) \\ \mathrm{Tr}(w^4x^4 + w^{15}x^2) \end{bmatrix},$$

$$NE_{10}(x) = \begin{bmatrix} \mathrm{Tr}(x^2) \\ \mathrm{Tr}(w^{13}x^{10}) \\ \mathrm{Tr}(wx^4) \end{bmatrix}, \quad NE_{11}(x) = \begin{bmatrix} \mathrm{Tr}(x^2) \\ \mathrm{Tr}(wx^4) \\ \mathrm{Tr}(w^2x^{10} + wx^2) \end{bmatrix},$$

$$NE_{12}(x) = \begin{bmatrix} \mathrm{Tr}(x^2) \\ \mathrm{Tr}(wx^4) \\ \mathrm{Tr}(w^{14}x^{10} + w^8x^4 + wx^2) \end{bmatrix}, \quad NE_{13}(x) = \begin{bmatrix} \mathrm{Tr}(x^2) \\ \mathrm{Tr}(wx^4) \\ \mathrm{Tr}(w^4x^{10} + w^8x^4 + w^2x^2) \end{bmatrix}.$$

These are the non-extendable maps.

We summarize these results in Fig. 1 in the form of a lattice.

5 Semifields

In this section we give a short introduction to finite semifields. We refer to [3, 5, 7, 9, 13, 14, 17, 19, 20, 22, 26] for further information.

Recall that a field is a nonempty set $\mathbb{F}$ with two binary operations $+$ and $\times$ satisfying the following axioms:

- $(\mathbb{F}, +)$ is an abelian group with 0.
- $(\mathbb{F} \setminus \{0\}, \times)$ is an abelian group with 1.
- If $a, b, c \in \mathbb{F}$, then

$$a \times (b + c) = a \times b + a \times c \text{ and } (a + b) \times c = a \times c + b \times c.$$

If $\mathbb{F}$ is infinite, it is possible to weaken the commutativeness condition above. $\mathbb{F}$ is called a *skew field (or a division ring)* if $(\mathbb{F} \setminus \{0\}, \times)$ is a group with 1 (not necessarily abelian), and all the other conditions are satisfied.

Example 5.1 Let $\mathbb{R}$ be the field of real numbers. Let $\hat{i}, \hat{j}, \hat{k}$ be the symbols satisfying

$$\hat{i}^2 = \hat{j}^2 = \hat{k}^2 = -1, \ \hat{i}\hat{j} = -\hat{j}\hat{i} = \hat{k}, \ \hat{j}\hat{k} = -\hat{k}\hat{j} = \hat{i}, \ \hat{k}\hat{i} = -\hat{i}\hat{k} = \hat{j}. \quad (6)$$

Let $\mathbb{H}$ be the set consisting of

$$a + b\hat{i} + c\hat{j} + d\hat{k} \text{ with } a, b, c, d \in \mathbb{R}.$$

We define $+$ on $\mathbb{H}$ componentwise. We define $\times$ on $\mathbb{H}$ using (6), distributive law and multiplication on $\mathbb{R}$. Then $\mathbb{H}$ is called the Hamilton quaternions, and it is a skew field, which is not a field.

Note that $\mathbb{H}$ in Example 5.1 is an infinite set. The following important results of Wedderburn (see, e.g., [18]) says that we have no such an example for finite sets.

Theorem 5.2 *If $\mathbb{F}$ is a finite set and $\mathbb{F}$ is a skew field, then $\mathbb{F}$ must be a finite field.*

However, if $\mathbb{F}$ is finite, then it is possible to weaken associativity condition of $(\mathbb{F} \setminus \{0\}, \times)$ instead of the commutativity condition.

Definition 5.3 A finite *presemifield* $\mathbb{S}$ is a finite set with two binary operations $+$ and $*$ satisfying the following axioms:

- $(\mathbb{S}, +)$ is an abelian group with 0.
- If $a, b, c \in \mathbb{S}$, then

$$a * (b + c) = a * b + a * c \text{ and } (a + b) * c = a * c + b * c.$$

- If $a, b \in \mathbb{S}$ and $a * b = 0$, then $a = 0$ or $b = 0$.

If $\mathbb{S}$ is a presemifield and there exists $e \in \mathbb{S} \setminus \{0\}$ such that

$$e * a = a * e = a \text{ for all } a \in \mathbb{S},$$

then $\mathbb{S}$ is called *semifield*.

If $\mathbb{S}$ is presemifield (or semifield) and

$$a * b = b * a \text{ for all } a, b \in \mathbb{S},$$

then $\mathbb{S}$ is called *commutative presemifield* (or *commutative semifield*).

The additive group of a finite semifield is elementary abelian. Therefore the order of $\mathbb{S}$ is p^n for a prime p and integer $n \geq 1$. Here p is called the characteristic of $\mathbb{S}$. In fact there is no finite semifield $\mathbb{S}$, which is not a field, with $|\mathbb{S}| = p^2$ or $|\mathbb{S}| = 8$, where p is a prime. However, for each prime p and integer $n \geq 3$ with $(p, n) \neq (2, 3)$, there exists a semifield $\mathbb{S}$ which is not a finite field with $|\mathbb{S}| = p^n$.

Any finite presemifield can be represented by $(\mathbb{S}, +, *)$, where $\mathbb{S}$ is the underlying set of a finite field $\mathbb{F}_{p^n}$ and the addition of $\mathbb{S}$ coincides with the addition of $\mathbb{F}_{p^n}$. The notion to classify finite presemifield is isotopism:

Definition 5.4 Two finite presemifields $(\mathbb{S}, +, *)$ and $(\mathbb{S}, +, \star)$ of order p^n are *isotopic* if there exist linearized permutation polynomials L, M, N over $\mathbb{F}_{p^n}$ such that

$$M(x) \star N(y) = L(x * y) \text{ for all } x, y \in \mathbb{S}.$$

Note that two presemifields can be isotopic only if their orders are the same. Finite fields are classified up to isomorphism by their orders. However there are many non-isotopic semifields of the same order. It is a big open problem to classify all finite semifields up to isotopism.

There is not much restriction in considering semifields instead of presemifields. Indeed any presemifield is isotopic to a semifield. Let $(\mathbb{S}, +, *)$ be a presemifield. Choose any nonzero element $a \in \mathbb{S}$. Let $(\mathbb{S}, +, \star)$ be the semifield with the binary operation $\star : \mathbb{S} \times \mathbb{S} \to \mathbb{S}$ defined using

$$x * y = (x * a) \star (a * y).$$

Namely, if $\alpha, \beta \in \mathbb{S}$, then there exists uniquely determined $x, y \in \mathbb{S}$ such that $x * a = \alpha$ and $a * y = \beta$. Then $\alpha \star \beta$ is defined as

$$\alpha \star \beta = x * y.$$

It is now clear that $e = a * a$ is the identity of $\mathbb{S} \setminus \{0\}$ under $\star$, and hence $(\mathbb{S}, +, \star)$ is a semifield. This is known as Kaplansky's trick. Here if $(\mathbb{S}, +, *)$ is a commutative presemifield, then $(\mathbb{S}, +, \star)$ is a commutative semifield. In general isotopism does not preserve commutativity.

Let $(\mathbb{S}, +, *)$ be a semifield. The subsets

$$\mathcal{N}_l(\mathbb{S}) = \{a \in \mathbb{S} : (a * x) * y = a * (x * y) \text{ for all } x, y \in \mathbb{S}\}$$
$$\mathcal{N}_m(\mathbb{S}) = \{a \in \mathbb{S} : (x * a) * y = x * (a * y) \text{ for all } x, y \in \mathbb{S}\}$$
$$\mathcal{N}_r(\mathbb{S}) = \{a \in \mathbb{S} : (x * y) * a = x * (y * a) \text{ for all } x, y \in \mathbb{S}\}$$

are called the *left, middle,* and *right nucleus* of $\mathbb{S}$, respectively. They are in fact finite fields. The subset

$$\mathcal{N}(\mathbb{S}) = \mathcal{N}_l(\mathbb{S}) \cap \mathcal{N}_m(\mathbb{S}) \cap \mathcal{N}_r(\mathbb{S})$$

is called the *nucleus* (or *associative center*) of $\mathbb{S}$. The *center* $C(\mathbb{S})$ of $\mathbb{S}$ is defined by

$$C(\mathbb{S}) = \mathcal{N}(\mathbb{S}) \cap \{c \in \mathbb{S} : c * a = a * c \text{ for any } a \in \mathbb{S}\}.$$

It is known that $\mathbb{S}$ is a division algebra over its center. Moreover the nuclei are invariant under isotopism.

Let $f : \mathbb{F}_{p^n} \to \mathbb{F}_{p^n}$ be an $\mathbb{F}_p$-quadratic polynomial map. If f is planar, then we get a commutative presemifield $(\mathbb{S}_f, +, *)$ defined as

$$x * y = \frac{1}{2}(f(x + y) - f(x) - f(y)).$$

Conversely if $(\mathbb{S}, +, \star)$ is a finite commutative semifield of odd characteristic with order p^n, then there exists an $\mathbb{F}_p$-quadratic polynomial $f \in \mathbb{F}_{p^n}[x]$ such that $(\mathbb{S}_f, +, *)$ is isotopic to $(\mathbb{S}, +, \star)$. Therefore the classification of finite commutative semifields and classification of $\mathbb{F}_{p^n}$-quadratic planar maps on $\mathbb{F}_{p^n}$ are the same problem.

In the sections above, we mainly considered nondegenerate symmetric quadratic forms over $\mathbb{F}_q$. They are directly linked to symmetric bilinear forms (or symmetric matrices) over $\mathbb{F}_q$ as the characteristic is odd. Moreover planarity is related to commutative semifields as explained in this section. However, symmetric matrices are known to be connected symplectic semifields. In the next section, we explain a geometric connection among these concepts.

6 Knuth's Cubical Array

In this section we explain Knuth's cubical array. This also gives an important and well-known method to construct up to six non-isotopic semifields of order p^n starting from a given finite semifield of order p^n.

Let $F : \mathbb{F}_{q^n} \to \mathbb{F}_{q^n}$ be a perfect nonlinear and $\mathbb{F}_q$-quadratic map. For $a \in \mathbb{F}_{q^n}$, the corresponding difference map is

$$\begin{aligned} D_a : \mathbb{F}_{q^n} &\to \mathbb{F}_{q^n} \\ x &\mapsto F(x + a) - F(x) - F(a). \end{aligned}$$

Note that D_0 is the zero map and D_a is an $\mathbb{F}_{q^n}$-linear map for all $a \in \mathbb{F}_{q^n}$. The corresponding presemifield $(\mathbb{S}, +, \circ)$ has the operation

$$x \circ y = D_y(x).$$

Let $e_1, e_2, \ldots, e_n$ be a basis of $\mathbb{F}_{q^n}$ over $\mathbb{F}_{q^n}$. The presemifield is determined by the operations

$$D_{e_j}(e_i) = e_i \circ e_j = \sum_{k=1}^{n} a_{ijk} e_k \text{ for } 1 \leq i, j \leq n.$$

Here $(a_{ijk}) = \underset{1 \le i,j,k \le n}{(a_{ijk})}$ is Knuth's cubical array with entries from $\mathbb{F}_q$. Namely, up to a choice of basis, $(\mathbb{S}, +, \circ)$ is determined by Knuth's cubical array (a_{ijk}).

Note that $(a_{ijk}) = (a_{jik})$ as $\mathbb{S}$ is commutative in this chapter. In general $\mathbb{S}$ can be noncommutative.

For $1 \le i, j, k \le n$, the slice $M(\mathbb{S}, j)$ of the cubical array (a_{ijk}) is an $n \times n$ matrix with entries from $\mathbb{F}_q$ defined as

$$M(\mathbb{S}, j) = \underset{1 \le i,k \le n}{(a_{ijk})} .$$

The matrix $M(\mathbb{S}, j)$ is the matrix of the linear transformation

$$D_{e_j}(x) = y$$

via left multiplication

$$\underline{x} \in \mathbb{F}_q^n \mapsto \underline{y} = \underline{x} \cdot M(\mathbb{S}, j).$$

Here we have $\underline{x} = (x_1, x_2, \ldots, x_n)$, $\underline{y} = (y_1, y_2, \ldots, y_n)$, $x = \sum_{k=1}^{n} x_i e_i$, $y = \sum_{k=1}^{n} y_i e_i$, and $D_{e_j}(x) = y$. The matrices $M(\mathbb{S}, 1), M(\mathbb{S}, 2), \ldots, M(\mathbb{S}, n)$ are linearly independent over $\mathbb{F}_q$. In fact their linear span is a spread set, which is defined, for example, in [2]. A spread set also determines $\mathbb{S}$.

The matrices $M(\mathbb{S}, 1), M(\mathbb{S}, 2), \ldots, M(\mathbb{S}, n)$ are not necessarily symmetric. Recall that $M(\mathbb{S}, j)$ is obtained from the cubical array (a_{ijk}) by fixing j. Let us fix k instead. For $1 \le k \le n$, the slice $\hat{M}(\mathbb{S}, k)$ of the cubical array (a_{ijk}) is an $n \times n$ matrix with entries from $\mathbb{F}_q$ defined as

$$\hat{M}(\mathbb{S}, k) = \underset{1 \le i,j \le n}{(a_{ijk})} . \tag{7}$$

As $\mathbb{S}$ is commutative, $\hat{M}(\mathbb{S}, k)$ is symmetric for all k. In fact $\{\hat{M}(\mathbb{S}, k) : 1 \le k \le n\}$ corresponds to a basis of the spread set $\mathrm{span}\{\hat{M}(\mathbb{S}^{d*}, k) : 1 \le k \le n\}$ of another semifield $\mathbb{S}^{d*}$ obtained via Knuth operations $\mathbb{S} \mapsto \mathbb{S}^d$ and $\mathbb{S}^d \mapsto \mathbb{S}^{d*}$ that we explain below.

It is well known that any permutation of the indices of the cubical array (a_{ijk}) gives a cubical array corresponding to a (pre)semifield. These permutations also respect semifield isotopism. Since a presemifield is isotopic to a semifield, we assume that $\mathbb{S}$ is a semifield without loss of generality here.

$\mathbb{S} : (a_{ijk})$
$\mathbb{S}^* : (a_{jik})$
$\mathbb{S}^d : (a_{kji})$
$\mathbb{S}^{*d} : (a_{jki})$
$\mathbb{S}^{d*} : (a_{jki})$
$\mathbb{S}^{*d*} = \mathbb{S}^{d*d} : (a_{ikj})$

Fig. 2 Knuth's cubical array

$\mathbb{S} : (a_{ijk})$
$\mathbb{S}^d : (a_{kji})$
$\mathbb{S}^{d*} = \mathbb{S}^{d*d} : (a_{jki})$

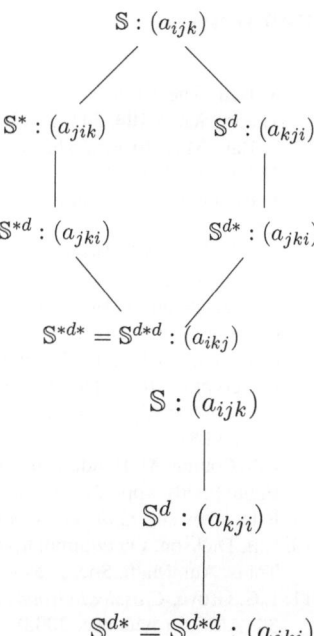

Fig. 3 Knuth's cubical array for commutative semifields

The permuted cubical array (a_{jik}) corresponds to the permutation (12) and gives the opposite semifield $\mathbb{S}^*$. The permuted cubical array (a_{kji}) corresponds to the permutation (13) and gives the dual semifield $\mathbb{S}^d$. In fact we have the lattices in Fig. 2 in general.

Moreover if $\mathbb{S}$ is commutative, then we only have the lattice in Figure 3.

As $\mathbb{S}$ is commutative, it is well known that $\mathbb{S}^{d*}$ is symplectic (see [2]). This means that the matrices (cf [13, 17])$M(\mathbb{S}^{d*}, 1), M(\mathbb{S}^{d*}, 2), \ldots, M(\mathbb{S}^{d*}, n)$ are all symmetric (see [13]). It is not difficult to observe that for the matrices $\hat{M}(\mathbb{S}, 1), \hat{M}(\mathbb{S}, 2), \ldots, \hat{M}(\mathbb{S}, n)$ obtained in (7), we have

$$\hat{M}(\mathbb{S}, 1) = M(\mathbb{S}^{d*}, 1), \hat{M}(\mathbb{S}, 2) = M(\mathbb{S}^{d*}, 2), \ldots, \hat{M}(\mathbb{S}, n) = M(\mathbb{S}^{d*}, n).$$

In other words slicing Knuth's cubical array (a_{ijk}) by fixing k and forming the symmetric matrices $\hat{M}(\mathbb{S}, 1), \hat{M}(\mathbb{S}, 2), \ldots, \hat{M}(\mathbb{S}, n)$ of the commutative semifield $\mathbb{S}$ corresponds to forming a basis $M(\mathbb{S}^{d*}, 1), M(\mathbb{S}^{d*}, 2), \ldots, M(\mathbb{S}^{d*}, n)$ of a spread set for the symplectic semifield $\mathbb{S}^{d*}$.

Acknowledgements Ferruh Özbudak is partially supported by TUBİTAK under Grant no. TBAG-112T011.

References

1. S. Ball, The number of directions determined by a function over a finite field. J. Combin. Theory Ser. A **104**, 341–35 (2003)
2. S. Ball, M.R. Brown, The six semifield planes associated with a semifield flock. Adv. Math. **189**(1), 68–87 (2004)
3. J. Bierbrauer, New semifields, PN and APN functions. Des. Codes Cryptogr. **54**, 189–200 (2010)
4. A. Blokhuis, A.E. Brouwer, T. Szonyi, The number of directions determined by a function f on a finite field. J. Combin. Theory Ser. A **70**, 349–353 (1995)
5. L. Budaghyan, T. Helleseth, New commutative semifields defined by new PN multinomials. Cryptogr. Commun. **3**, 1–16 (2011)
6. C. Carlet, P. Charpin, V. Zinoviev, Codes, bent functions and permutations suitable for DES-like cryptosystems. Des. Codes Cryptogr. **15**(2), 125–156 (1998)
7. R.S. Coulter, M. Henderson, Commutative presemifields and semifields. Adv. Math. **217**, 282–304 (2008)
8. R.S. Coulter, M. Henderson, On a conjecture on planar polynomials of the form $X\mathrm{Tr}_n(X)-uX$. Finite Fields Appl. **21**, 30–34 (2013)
9. P. Dembowski, *Finite Geometries* (Springer, Berlin, 1968)
10. L.E. Dickson, On commutative linear algebras in which division is always uniquely possible. Trans. Am. Math. Soc. **7**, 514–522 (1906)
11. L.C. Grove, *Classical Groups and Geometric Algebra*. Graduate Studies in Mathematics, vol. 39 (AMS, Providence, 2002)
12. X.-D. Hou, C. Sze, On certain diagonal equations over finite fields. Finite Fields Appl. **15**, 633–643 (2009)
13. W.M. Kantor, Commutative semifields and symplectic spreads. J. Algebra **270**(1), 96–114 (2003)
14. D.E. Knuth, Finite semifields and projective planes. J. Algebra **2**, 153–270 (1965)
15. M. Kyuregyan, F. Özbudak, Planarity of products of two linearized polynomials. Finite Fields Appl. **18**, 1076–1088 (2012)
16. M. Kyuregyan, F. Özbudak, A. Pott, Some planar maps and related function fields, in *Arithmetic, Geometry, Cryptography and Coding Theory*. Contemporary Mathematics, vol. 574, (American Mathematical Society, Providence, RI, 2012), pp. 87–114
17. M. Lavrauw, O. Polverino, Finite semifields, in *Current Research Topics in Galois Geometry*, Chap. 6, ed. by J. De Beule, L. Storme (NOVA Academic Publishers, New York, 2011)
18. R. Lidl, H. Niederrieter, *Finite Fields* (Cambridge University Press, Cambridge, 1997)
19. G. Marino, O. Polverino, R. Trombetti, On $\mathbb{F}_q$-linear sets of $PG(3,q^3)$ and semifields. J. Combin. Theory Ser. A **114**, 769–788 (2007)
20. G. Menichetti, On a Kaplansky conjecture concerning three-dimensional division algebras over a finite field. J. Algebra **47**, 400–410 (1977)
21. F. Özbudak, A. Pott, Uniqueness of $\mathbb{F}_q$-quadratic perfect nonlinear maps from $\mathbb{F}_{q^3}$ to $\mathbb{F}_q^2$. Finite Fields Appl. **29**, 49–88 (2014)
22. A. Pott, Y. Zhou, A new family of semifields with 2 parameters. Adv. Math. **234**, 43–60 (2013)
23. L. Rédei, *Lückenhafte Polynome über endlichen Körpern* (Birkhäuser, Basel, 1970)
24. H. Stichtenoth, *Algebraic Function Fields and Codes*, 2nd edn. (Springer, Berlin, 2009)
25. M. Yang, S. Zhu, K. Feng, Planarity of mappings $x\mathrm{Tr}(x) - \frac{\alpha}{2}x$ on finite fields. Finite Fields Appl. **23**, 1–7 (2013)
26. Z. Zha, G. Kyureghyan, X. Wang, Perfect nonlinear binomials and their semifields. Finite Fields Appl. **15**, 125–133 (2009)

Open Problems for Polynomials over Finite Fields and Applications

Daniel Panario

Abstract We survey open problems for univariate polynomials over finite fields. We first comment in some detail on the existence and number of several classes of polynomials. The open problems in that part of the survey are of a more theoretical nature. Then, we center on classes of low-weight (irreducible) polynomials. The conjectures here are more practically oriented. Finally, we give brief descriptions of a selection of open problems from several areas including factorization of polynomials, special polynomials (APN functions, permutations), and relations between rational integers and polynomials.

1 Background and Goals of This Chapter

1.1 Introduction

We introduce a series of open problems for univariate polynomials over finite fields. The list is incomplete but still provides several topics of current research on this type of polynomials.

This chapter is an extended transcription of the author's invited talk at the Open Problems in Mathematics and Computer Science Conference, Istanbul 2013. The topics presented in this chapter have been selected mainly from the *Handbook of Finite Fields* by Mullen and Panario [50]. Further problems on polynomials over finite fields are presented in that reference.

D. Panario (✉)

School of Mathematics and Statistics, Carleton University, Ottawa, ON, Canada K1S 5B6

e-mail: daniel@math.carleton.ca

© Springer International Publishing Switzerland 2014   111

Ç.K. Koç (ed.), *Open Problems in Mathematics and Computational Science*,

DOI 10.1007/978-3-319-10683-0_6

1.2 Background on Finite Fields

In order to make this survey self-contained, we briefly review fundamental notions and results in finite fields. Research on finite fields bridges the gap between several branches of mathematics like:

- algebra (field extensions and Galois theory);
- discrete mathematics and combinatorics (representing finite field elements as combinatorial objects; algorithms in finite fields; finite field constructions of combinatorial arrays);
- number theory (counting special finite field elements; analogies between polynomials over finite fields and rational integers).

Many projects undertaken in finite fields can be applied almost immediately to "real-world" problems. Finite fields are used extensively in:

- coding theory;
- public key cryptography;
- communications and electrical engineering;
- computer science.

The interested reader is referred to the handbooks of cryptography, coding theory, and combinatorial designs for further results and connections of finite fields to those areas [8, 11, 48, 53].

Next we recall the definition of finite fields as well as several basic results that we need in this chapter.

Definition 1 A *field* $(F, +, \cdot)$ is a set F together with binary operations "+" and "·" such that:

1. $(F, +)$ is an abelian group;
2. $(F \setminus \{0\}, \cdot)$ is an abelian group;
3. distributive laws hold, that is, for $a, b, c \in F$, we have

$$a \cdot (b + c) = a \cdot b + a \cdot c,$$
$$(b + c) \cdot a = b \cdot a + c \cdot a.$$

If $\#F$ is finite, then we say that F is a *finite field*.

It is well known that

$$\mathbb{Z}/(p) \text{ is a field if and only if } p \text{ is a prime.}$$

Up to isomorphisms, there exists exactly one finite field with $q = p^n$ elements, denoted by $\mathbb{F}_q$, for all primes p and positive integers n. The *characteristic* of the finite field $\mathbb{F}_q$ is p.

We also need the well-known fact that the multiplicative group of $\mathbb{F}_q$ is cyclic. The generators of this multiplicative group are *primitive* elements and play a fundamental role in many applications.

Polynomials As usual, a monic polynomial over $\mathbb{F}_q$ of degree n is of the form $x^n + a_{n-1}x^{n-1} + \cdots + a_1x + a_0$ with $a_i \in \mathbb{F}_q$ for $0 \le i < n$.

Irreducible polynomials are the most fundamental polynomials. A polynomial $f \in \mathbb{F}_q[x]$ is *irreducible* over $\mathbb{F}_q$ if $f = gh$ with $g, h \in \mathbb{F}_q[x]$ implies that g or h is in $\mathbb{F}_q$. Through unique factorization, irreducible polynomials play the rôle to polynomials that prime numbers play to rational integers; we comment on this relation in more detail in Sect. 4.4. Moreover, we can construct $\mathbb{F}_{q^n}$ by taking the quotient of $\mathbb{F}_q[x]$ by an irreducible polynomial f of degree n over $\mathbb{F}_q$, that is, $\mathbb{F}_{q^n} \cong \mathbb{F}_q[x]/(f)$. The finite field elements are represented as polynomials of degree less than n with coefficients in $\mathbb{F}_q$. In this extension field, addition is performed term-wise; multiplication is taken (mod f). There are other ways of representing elements over a finite field, but in this chapter we focus only on polynomials. One of the most prominent and practical of those other representations is normal bases; see Chap. 5 of [50] for an account on normal bases results.

1.3   Outline of the Chapter

We focus on open problems for *univariate polynomials over finite fields*.

- In Sect. 2, we comment in some detail on the existence and number of several classes of polynomials. The open problems here are more of a *theoretical* nature.
- Then, in Sect. 3, we center in classes of low-weight (irreducible) polynomials. The conjectures here are more *practically* oriented.
- Finally, in Sect. 4, we give brief descriptions of a selection of open problems from several areas including factorization of polynomials, special polynomials (APN functions, permutation), iterations of polynomials, and relations between integer numbers and polynomials.

To know more about basics of finite fields, the reader is referred to the textbook *Finite Fields* by Lidl and Niederreiter [44].

Remark Given the amount of work already done on the topics presented in this chapter, we are intentionally *vague* on the statement of the open problems. Including precise statements would make for a quite long presentation. However, the reader should check the references cited before each open problem to find exact results known so far. We note that for each open problem given in this chapter, there is still considerably work to be done to completely solve the problem.

2 Prescribed Coefficients

2.1 Irreducible Polynomials

The *number $I_q(n)$* of monic irreducible polynomials of degree n over $\mathbb{F}_q$ is

$$I_q(n) = \frac{1}{n} \sum_{d|n} \mu(d) q^{n/d} = \frac{q^n}{n} + O(q^{n/2}),$$

where $\mu : \mathbb{N} \to \mathbb{N}$ is the Möbius function, given by

$$\mu(n) = \begin{cases} 1 & \text{if } n = 1, \\ (-1)^k & \text{if } n \text{ is a product of } k \text{ distint primes,} \\ 0 & \text{otherwise.} \end{cases}$$

This is known for almost 150 years, but if we *prescribed some coefficient to some value*, can we characterize and count those type of irreducible polynomials? We briefly introduce in the following some results on this direction; a wealth of references to these and related problems on irreducible polynomials over finite fields can be found on Chap. 3 of [50].

2.2 Irreducibles with Prescribed Coefficients: Existence

The Hansen–Mullen conjecture [32] states the existence of irreducibles over $\mathbb{F}_q$ with *any* one coefficient *prescribed to a fixed value*. Wan [61] proved the Hansen–Mullen conjecture using Dirichlet characters and Weil bounds. Generalizations have been given for the existence of irreducibles with the trace and norm coefficients (i.e., the coefficients of x^{n-1} and x^0, respectively) prescribed as well as with few more coefficients prescribed; see Sect. 3.5 of [50].

On the other hand, there are also results for up to *half coefficients prescribed* [35] and variants, for example, prescribing precisely upper and lower coefficients of the polynomial:

$\diagdown\!\!\diagdown$ = coefficients prescribed to any value with total size of roughly $\frac{n}{2} - \log_q n$

However, *experiments* show that we could prescribe almost all coefficients and still obtain irreducible polynomials!

Open Problem 1 *Prefix some coefficients of a polynomial over a finite field to some values; prove that there exist irreducible polynomials with those coefficients prescribed to those values.*

We remark that the techniques used so far for *existence* results are from number theory (characters, bounds on character sums).

2.3 Irreducibles with Prescribed Coefficients: Number

There are far less results proved about the number of irreducible polynomials with coefficients prescribed than about existence. Results so far include exact estimates for the *number of irreducibles* with up to 2 coefficients (x^{n-1} and x^0, or x^{n-1} and x^{n-2}) prescribed over any finite field.

Over $\mathbb{F}_2$, in addition to the previous results, there are formulas for the number of irreducible polynomials

- with up to the three most significant coefficients ($x^{n-1}, x^{n-2}, x^{n-3}$) prescribed to any value,
- conjectures for the four most significant coefficients prescribed.

Open Problem 2 *Give exact (or asymptotic) counting for irreducible polynomials with prescribed coefficients.*

The techniques used for these estimates are more elementary than the ones for existence results, and they come from several areas of discrete mathematics.

The long-term goal here is to provide existence and counting results for irreducibles with *any* number of prescribed coefficients to *any* given values. This goal is completely out of reach at this time. Incremental steps seem doable, but it would be most interesting if new techniques were introduced to attack these problems.

2.4 Primitive Polynomials with Prescribed Coefficients

A polynomial $f \in \mathbb{F}_q[x]$ of degree $n \geq 1$ is *primitive* if it is the minimal polynomial of a primitive element of $\mathbb{F}_{q^n}$. These polynomials are fundamental in many applications in engineering involving sequences and LFSRs (linear feedback shift registers).

Primitive polynomials exist for any degree $n \geq 1$ and any finite field $\mathbb{F}_q$. As it is easy to check, the number of primitive polynomials of degree n over $\mathbb{F}_q$ is $\phi(q^n - 1)/n$, where ϕ denotes Euler's function. However, if we prescribe some coefficients, only some partial results are known. We comment on some results for this type of polynomials; more references can be found on Chap. 4 of [50].

The Hansen–Mullen conjecture for primitive polynomials asks whether primitive polynomials exist with *any coefficient prescribed* to a given value. This conjecture was proved for $n \geq 9$ by Cohen [7] and without restrictions by Cohen and Prešern [9]. There are generalizations to few prescribed coefficients but no exact results for the *number* of primitive polynomials with prescribed coefficients.

Open Problem 3 *Prefix some coefficients of a primitive polynomial over a finite field to some values; prove that there* exist *(or give the* number *of) primitive polynomials with those coefficients prescribed to those values.*

One can require for a polynomial to be primitive and also hold other algebraic conditions. Primitive normal polynomials are an important class of this type of polynomials. A polynomial is *primitive normal* if its roots form a normal basis and are primitive elements. We recall that an element α in $\mathbb{F}_{q^n}$ is *normal* if $\{\alpha, \alpha^q, \ldots, \alpha^{q^{n-1}}\}$ is a basis of $\mathbb{F}_{q^n}$ over $\mathbb{F}_q$.

The *existence* of primitive normal polynomials was established by Carlitz [5], for sufficiently large q and n, Davenport [13] for prime fields, and finally for all (q, n) by Lenstra and Schoof [41]. A proof without the use of a computer was later given Cohen and Huczynska [10]. Gauss sums, hybrids of additive and multiplicative characters sums, are employed in primitive normal results.

Hansen–Mullen also conjectured that *primitive normal polynomials with one prescribed coefficient* exist for all q and n. Fan and Wang [18] proved the conjecture for $n \geq 15$. There are generalizations for two coefficients (norm and trace) and for three coefficients but not much is known beyond those cases; see Sect. 4.2 of [50].

An element α in $\mathbb{F}_{q^n}$ is *completely normal* if α is a normal element of $\mathbb{F}_{q^n}$ over $\mathbb{F}_{q^d}$, for every subfield $\mathbb{F}_{q^d}$ (hence $d \mid n$). The minimal polynomial of α over $\mathbb{F}_q$ is a *completely normal polynomial*. Not much is known about these polynomials even in the case when no prescribed condition on the coefficients is included. Morgan and Mullen [49] conjectured that for any $n \geq 2$ and any prime power q there *exists* a completely normal primitive basis of $\mathbb{F}_{q^n}$ over $\mathbb{F}_q$.

Open Problem 4 *Prove the Morgan and Mullen conjecture for completely normal primitive bases.*

This conjecture is still wide open though major advances on this problem have been done by Hachenberger [30, 31]. The methods here are algebraic and allow the derivation of lower bounds.

3   Low-Weight Polynomials

3.1   *Introduction*

A particular important case of *prescribed coefficients* occurs when *most coefficients are set to zero*. The *weight* of a polynomial is the number of nonzero coefficients of the polynomial. Loosely speaking, a polynomial has low weight when "most" coefficients are zero.

This case is relevant in *practice* where we prefer sparse irreducible polynomials, like *trinomials* (polynomials with 3 monomials) or *pentanomials* (polynomials with 5 monomials) over $\mathbb{F}_2$, to construct the extension fields. These are, for example, the recommendations of IEEE [36]. Among same degree irreducible trinomials or pentanomials, for reasons of confidence, we choose polynomials following a *lowest lexicographical order* selection. However, Scott [56] shows that the irreducible with the optimal performance for a given implementation does not necessarily follow the lowest lexicographical order!

3.2 Conjectures and Open Problems

The state of affairs for low-weight polynomials is very poor. Indeed the following open problem presents several old questions that are still not mathematically answered.

Open Problem 5 • *What is the density of n's such that there is an irreducible trinomial of degree n over $\mathbb{F}_2$?*
• *Are there irreducible pentanomials over $\mathbb{F}_2$ for all n?*
• *Are there irreducible tetranomials over $\mathbb{F}_q$, $q \geq 3$, for all $n \geq 3$?*

Experimentally, there are only about 50 % of n with irreducible trinomials of degree n over $\mathbb{F}_2$. However, there seems to be a pentanomial for every n. There are tables of trinomials and pentanomials, using Magma, for the following values of q and n:

q	$n \leq$	q	$n \leq$	q	$n \leq$	q	$n \leq$
2	120, 000	3	50, 000	4, 5, 7	2, 000	$9 \leq q \leq 127$	1, 000

Those experiments suggest that the existential questions raised in Open Problem 5 should have a positive answer, but there are no proofs in sight for any of those problems.

A *sedimentary polynomial* is a polynomial over $\mathbb{F}_q$ of the form $f(x) = x^n + g(x)$ with g of degree close to $\log_q n$.

Open Problem 6 *Prove that for every positive integer n, there exists a polynomial g of degree at most $\log_q n + 3$ such that $f(x) = x^n + g(x)$ is irreducible over $\mathbb{F}_q$.*

These polynomials have been used, for example, by Coppersmith [12] to represent elements in $\mathbb{F}_{2^n}$ in a subexponential algorithm for discrete logarithm computations in finite fields. Again, experiments [27] seem to imply the existence of sedimentary polynomials as stated in Open Problem 6, but no proof of this fact is available at this time.

3.3   Reducibility of Low-Weight Polynomials

Swan [58] characterizes the parity of the number of irreducible factors of a trinomial over $\mathbb{F}_2$ relating it to the discriminant of the trinomial (due to Pellet and Stickelberger). Obviously, if the number of irreducible factors of a polynomial is even, the trinomial is then *reducible*.

In principle one could use this to provide reducibility conditions. However there is a main problem with this approach: the calculation of the discriminant of the polynomial is hard even when the polynomial has a moderate number of terms.

We exemplify in the following a typical result in this area.

Theorem 1 ([58]) *Let $n > k > 0$. Assume precisely one of n, k is odd. Then if r is the number of irreducible factors of $f(x) = x^n + x^k + 1 \in \mathbb{F}_2[x]$, then r is even in the following cases:*

- *n even, k odd, $n \neq 2k$ and $nk/2 \equiv 0, 1 \pmod 4$;*
- *n odd, k even, $k \nmid 2n$ and $n \equiv 3, 5 \pmod 8$;*
- *n odd, k even, $k \mid 2n$ and $n \equiv 1, 7 \pmod 8$.*

In other cases f has an odd number of factors.

The case where n and k are both odd can be covered using that the reverse of f has the same number of irreducible factors. If both n and k are even the trinomial is a square and has an even number of irreducible factors.

The next is an important practical consequence that asserts that there are no irreducible trinomials over $\mathbb{F}_2$ for the usual computer word sizes.

Corollary 1 *There are no irreducible trinomials over $\mathbb{F}_2$ with degree a multiple of 8.*

By now, over $\mathbb{F}_2$, we know the reducibility of few pentanomials, but not if irreducible pentanomials exist for all degrees. Over $\mathbb{F}_q, q > 2$, we know when binomials are reducible; we also have partial results for trinomials and tetranomials, as well as for some very special type of polynomials; see Sects. 3.4 and 3.5 in [50].

Open Problem 7 *Give new reducibility results for low-weight polynomials over finite fields.*

It would also be interesting to study the distribution of the discriminant for random polynomials over finite fields.

Open Problem 8 *Give expected value and distributional studies for the discriminant of a random polynomial over a finite field.*

Low-weight polynomials have several applications; see Sect. 14.9 of [50]. We only comment here on two of those. Consider a maximum-length shift-register sequence generated by a primitive polynomial f over a finite field. The set of its subintervals is *a linear code whose dual code is formed by all polynomials divisible by f* [51]. Since the minimum weight of a dual code is directly related to the strength of the corresponding orthogonal array [3], one can produce orthogonal arrays by studying divisibility of polynomials with low weight. (For information on orthogonal arrays, we refer to the book [33].)

To obtain orthogonal arrays of larger strength t (equivalently dual codes of minimum weight $t + 1$), we need conditions on when a low-weight polynomial divides another (low) t-weight polynomial. At this moment, we only know conditions for divisibility of trinomials and pentanomials over $\mathbb{F}_2$, and some similar cases over $\mathbb{F}_3$.

In addition, low-weight multiples of a public polynomial compromise the private key for the $\mathcal{TCHo}$ *cryptosystem*, and its security therefore rests on the difficulty of finding low-weight multiples [2, 34].

Open Problem 9 *Study the divisibility of a low-weight polynomial over a finite field by another low-weight polynomial over the same finite field.*

4 Potpourri of Polynomial Topics and Problems

4.1 Factorization of Polynomials

Given a monic univariate polynomial $f \in \mathbb{F}_q[x]$, the factorization problem asks to find monic distinct irreducible f_i and positive integers e_i, $1 \leq i \leq r$, such that $f = f_1^{e_1} \cdots f_r^{e_r}$. Much work has been done in this area; see [23, 24].

A standard method for this task uses three steps:

1. *Elimination of repeated factors* (ERF) replaces a polynomial by a square-free one which contains all the irreducible factors of the original polynomial with exponents reduced to 1.
2. *Distinct-degree factorization* (DDF) splits a square-free polynomial into a product of polynomials whose irreducible factors have all the same degree.
3. *Equal-degree factorization* (EDF) completely factors a polynomial whose irreducible factors have the same degree.

All efficient practical versions use a *probabilistic algorithm* for EDF. The next is a long-standing problem of a main theoretical interest.

Open Problem 10 *Find a polynomial time* deterministic *algorithm for factoring polynomials over finite fields.*

We remark that this problem is open even assuming the generalized Riemann hypothesis (GRH) [37]. So far, the techniques used to answer this question are mostly algebraic.

In terms of fast practical versions of the general methodology commented above based on three steps, it has been well known that the bottleneck is on the second step, the distinct-degree factorization stage. A key role on the algorithmic improvements for this problem has been played by the use of fast modular compositions and interval partitions [26, 39].

By now the algorithms are very efficient, at least to factor very large degree polynomials taken uniformly at random. Advances in the next problem may improve the current best algorithm versions when the polynomials to be factored are taken at random.

Open Problem 11 *Find the* best interval partition *for factoring a random polynomial over a finite field.*

The techniques used so far for this type of probabilistic analysis come from analytic combinatorics [20]. They proceed in two steps; first generating functions for the quantities of interest are derived, and then asymptotic analyses for the extraction of coefficient asymptotics are used. This general methodology was used in [21] for the complete analysis of the algorithms above for the factorization of polynomials over finite fields; for the latest results, see [25, 52].

4.2   Special Polynomials Over Finite Fields

4.2.1   PN and APN Functions

Definition 2 Let G_1 and G_2 be finite abelian groups of the same cardinality and $f : G_1 \to G_2$. We say that f is a *perfect nonlinear (PN) function* if

$$\Delta_{f,a}(x) = f(x + a) - f(x) = b$$

has exactly one solution for all $a \neq 0 \in G_1$ and all $b \in G_2$.

PN functions provide optimal resistance to linear and differential cryptographic attacks. However, PN functions *cannot exist in finite fields of characteristic* 2 (the most important for implementations). They were introduced as *planar functions* by Dembowski–Ostrom [14]; they are also known as *bent functions*; see Chap. 9 of [50].

We obtain an alternate definition for almost best-possible differential structure by slightly relaxing the condition on the definition of PN function.

Definition 3 Let G_1 and G_2 be finite abelian groups of the same cardinality and $f : G_1 \to G_2$. We say that f is an *almost perfect nonlinear function* if

$$\Delta_{f,a}(x) = f(x + a) - f(x) = b \tag{1}$$

has at most two solutions for all $a \neq 0 \in G_1$ and all $b \in G_2$.

As an example, the *inverse function* $f : x \mapsto x^{2^n-2}$ in $\mathbb{F}_{2^n}$ is *APN* if and only if n is odd. We remark that this function is used in the *Advanced Encryption Standard (AES)* but in that case $n = 8$. If n is even, then $\Delta_{f,a}$ is close to APN; indeed, it is *differential 4 uniform*, that is, Eq. (1) has at most four solutions for each a and all b.

In most applications, candidate functions for use in symmetric key cryptosystems must be permutations. Furthermore, for implementation purposes, functions over $\mathbb{F}_{2^e}$ with e even are preferred. There are no PN permutations in these fields. Hence, combining these criteria, the most desirable candidate functions for cryptographical applications are *APN permutations over* $\mathbb{F}_{2^e}$ *where* e *is even*. Currently, there is *only one* known *APN permutation over* $\mathbb{F}_{2^e}$, *when* e *is even*. This function for $\mathbb{F}_{2^6}$ was given by Dillon and collaborators [4].

Open Problem 12 *Find APN permutations over* $\mathbb{F}_{2^e}$, *when* e *is even*.

4.2.2 Permutation Polynomials Over Finite Fields

Definition 4 A polynomial $f \in \mathbb{F}_q[x]$ is a *permutation polynomial* (PP) of $\mathbb{F}_q$ if the function $f : c \to f(c)$ from $\mathbb{F}_q$ into itself induces a permutation.

There have been massive amount of work on PPs since the nineteenth century. Many results have appeared on the last 30 years, some of them due to the many cryptographic applications of PPs; see Chap. 8 of [50]. However, many questions are still not fully answered [42, 43] even though substantial work have been done on these problems.

Some well-known classes of PPs include monomials x^n when $\gcd(n, q-1) = 1$, Dickson polynomials $D_n(x, a) = \sum_{j=0}^{\lfloor n/2 \rfloor} \frac{n}{n-j} \binom{n-j}{j} (-a)^j x^{n-2j}$ when $\gcd(n, q^2 - 1) = 1$, and linearized polynomials $L(x) = \sum_{s=0}^{n-1} a_s x^{q^s} \in \mathbb{F}_{q^n}[x]$ when $\det(a_{i-j}^{q^j}) \neq 0, 0 \leq i, j \leq n-1$.

Open Problem 13 *1. Find new classes of PPs.*
2. Find PPs with some prescribed coefficients to some values.

Let $N_n(q)$ denote the number of PPs of $\mathbb{F}_q$ of degree n. It is easy to show that $N_1(q) = q(q-1)$, $N_n(q) = 0$ if n is a divisor of $q-1$ larger than 1, and $\sum N_n(q) = q!$ where the sum is taken over all $1 \leq n < q-1$ such that n is either 1 or is not a divisor of $q-1$. However, in general, $N_n(q)$ is still not known.

Open Problem 14 *Find the number* $N_n(q)$ *of PPs of degree* n.

If the polynomial is not a permutation, it is interesting to find its *value set*, that is, the distinct values that the function takes. The value sets of some functions have been studied, but, in general, similar questions as above have only been partially studied; see Sect. 8.3 of [50].

4.3 Iteration of Polynomials Over Finite Fields

Given a polynomial $f \in \mathbb{F}_q[x]$, one can define the functional graph of f as a directed graph on q nodes labelled by elements of $\mathbb{F}_q$ where there is an edge from a to b if and only if $f(a) = b$. This graph has one or more connected components, and each connected component contains one cycle with trees attached to some of the cycle nodes. The cycle may be of length 1, a fixed point. The graph of a polynomial f encodes characteristics of the map like the distribution and length of periodic points (points in the cycle of a connected component) and pre-periodic points (points in the trees, not in the cycle).

A key motivation on this area is to better understand Pollard's ρ-algorithm [55] for integer factorization. In the analysis of that algorithm, a composite integer m is given, and we are interested in the properties of the polynomial mappings $x \mapsto x^2 - 1 \pmod{p}$, where $p|m$. If p is prime, this mapping has the property that every image has at most 2 preimages. This type of property can also be desirable for mappings used in cryptographic hash functions. The quadratic mappings x^2 and $x^2 - 2$ has been studied in [60], but the shape of more generic quadratic maps has not been fully understood.

Some nonquadratic maps have also been studied. The dynamics of the maps x^n over finite fields has been analyzed [6,57]. Iterations of Chebyshev polynomials of the first kind has also been studied [28], as well as rational maps of the form $x + x^{-1}$ over small finite fields [59].

Open Problem 15 *Describe the graphs of iterations of polynomials over finite fields.*

It is not clear what is the proper heuristic model to describe Pollard's ρ-algorithm. A study of general random maps was executed by Flajolet and Odlyzko [19]. A model with restrictions on the number of preimages, that would in principle adapt better to Pollard's ρ-algorithm, is in [1,45]. However, a model that can fully explain Pollard's ρ-algorithm is still not available.

Open Problem 16 *Develop a heuristic model to completely describe Pollard's ρ-algorithm.*

4.4 Relations Between Integers and Polynomials

The unique factorization of polynomials into irreducibles allow the derivation of analogous results to the ones for the decomposition of integers into prime numbers. For example, one can study properties such as:

- expected number of irreducible factors of a polynomial (number of primes of an integer);
- probability of a factorization pattern;

- expected largest and smallest degree irreducible factor (largest and smallest prime);
- irreducibles (primes) in arithmetic progression.

A basic technique from analytic combinatorics allows the derivation of such results (see [21] and Sect. 11.1 of [50]). This technique also allows the study of the distribution of the factors in the gcd of several univariate polynomials over a finite field.

Some classical number theoretic problems have been successfully translated to polynomials. This is the case of the *twin prime conjecture*. As in the integer case, we consider two irreducible polynomials to be *twins* if they differ by as little as possible. We measure the size of a monic polynomial $f \in \mathbb{F}_q[x]$ of degree n with the *absolute value* $|f| = q^n$. Two polynomials f and g, both of degree n over $\mathbb{F}_q$, are twin irreducible polynomials if $|g - f| = 4$ for $q = 2$, or $|g - f| = 1$ otherwise.

The *twin irreducible polynomials conjecture* states that if q is fixed and the degree n tends to infinity, then there are infinitely many twin irreducible polynomials. This conjecture has been proved for all finite fields of order bigger than two [15, 54].

Open Problem 17 *Prove the twin irreducible polynomial conjecture in* $\mathbb{F}_2$.

Classical generalizations of the twin prime conjecture in the integer setting (if we consider more than two primes, or if the primes are not as close as possible) have not been given yet for polynomials over finite fields.

There have been some results about *additive* properties for polynomials related to the *Goldbach conjecture* and their generalizations (e.g., sum of three irreducibles); see [16, 17].

Many other classical problems from number theory have been treated for polynomials over finite fields. The most famous result is the polynomial version of the Generalized Riemann Hypothesis of Weil [62]; there is as well a polynomial analogue of Artin's conjecture on primitive roots [38]

On the other hand, several recent results in number theory have not been completely translated into polynomials over finite fields yet including, for example, the ones in the following.

Open Problem 18 *Give polynomial over a finite field versions for the following problems already studied for prime numbers:*

1. *divisors and shifted divisors in intervals [22, 40];*
2. *primes in small gaps [29, 47, 63];*
3. *sum of digits function [46].*

Some of these problems may be amenable to the techniques from analytic combinatorics commented above.

Conclusions
Polynomials over finite fields are fundamental in several theoretical areas of mathematics and in many practical applications in communications. We comment on some open problems for univariate polynomials over finite fields. The selection of topics is by no means complete, but we hope is representative of the intense current research in polynomials over finite fields.

References

1. J. Arney, E.A. Bender, Random mappings with constraints on coalescence and number of origins. Pac. J. Math. **103**, 269–294 (1982)
2. J.-P. Aumasson, M. Finiasz, W. Meier, S. Vaudenay, A hardware-oriented trapdoor cipher, in *Information Security and Privacy*. Lecture Notes in Computer Science, vol. 4586 (Springer, New York, 2007), pp. 184–199
3. R.C. Bose, On some connections between the design of experiments and information theory. Bull. Inst. Int. Stat. **38**, 257–271 (1961)
4. K.A. Browning, J.F. Dillon, M.T. McQuistan, A.J. Wolfe, An APN permutation in dimension six, in *Finite Fields: Theory and Applications*. Contemp. Math., vol. 518 (The American Mathematical Society, Providence, 2010), pp. 33–42
5. L. Carlitz, Primitive roots in a finite field. Trans. Am. Math. Soc. **73**, 373–382 (1952)
6. W. Chou, I. Shparlinski, On the cycle structure of repeated exponentiation modulo a prime. J. Number Theory **107**, 345–356 (2004)
7. S.D. Cohen, Primitive polynomials with a prescribed coefficient. Finite Fields Appl. **12**, 425–491 (2006)
8. H. Cohen, G. Frey, R. Avanzi, C. Doche, T. Lange, K. Nguyen, F. Vercauteren, *Handbook of Elliptic and Hyperelliptic Curve Cryptography*, Series on Discrete Mathematics and Its Applications (CRC Press, Boca Raton, 2006)
9. S.D. Cohen, M. Prešern, The Hansen–Mullen primitive conjecture: completion of proof, in *Number Theory and Polynomials*. London Math. Soc. Lecture Note Series, vol. 352 (Cambridge University Press, Cambridge, 2008), pp. 89–120
10. S.D. Cohen, S. Huczynska, The primitive normal basis theorem—without a computer. J. Lond. Math. Soc. 2nd Ser. **67**, 41–56 (2003)
11. C.J. Colbourn, J.H. Dinitz, *Handbook of Combinatorial Designs*, 2nd edn., Series on Discrete Mathematics and Its Applications (CRC Press, Boca Raton, 2007)
12. D. Coppersmith, Fast evaluation of logarithms in fields of characteristic two. IEEE Trans. Inf. Theory **30**, 587–594 (1984)
13. H. Davenport, Bases for finite fields. J. Lond. Math. Soc. 2nd Ser. **43**, 21–39 (1968)
14. P. Dembowski, T.G. Ostrom, Planes of order n with collineation groups of order n^2. Math. Z. **103**, 239–258 (1968)
15. G. Effinger, Toward a complete twin primes theorem for polynomials over finite fields, in *Finite Fields and Applications*. Contemp. Math., vol. 461 (The American Mathematical Society, Providence, 2008), pp. 103–110
16. G. Effinger, D.R. Hayes, *Additive Number Theory of Polynomials over a Finite Field*, Oxford Mathematical Monographs (Oxford University Press, New York, 1991)
17. G. Effinger, K. Hicks, G.L. Mullen, Integers and polynomials: comparing the close cousins $\mathbb{Z}$ and $\mathbb{F}_q[x]$. Math. Intelligencer **27**, 26–34 (2005)

18. S. Fan, X. Wang, Primitive normal polynomials with a prescribed coefficient. Finite Fields Appl. **15**, 682–730 (2009)
19. P. Flajolet, A.M. Odlyzko, Random mapping statistics, in *Advances in cryptology—EUROCRYPT '89*. Lecture Notes in Comput. Sci., vol. 434 (Springer, New York, 1990), pp. 329–354
20. Ph. Flajolet, R. Sedgewick, *Analytic Combinatorics* (Cambridge University Press, Cambridge, 2009)
21. P. Flajolet, X. Gourdon, D. Panario, The complete analysis of a polynomial factorization algorithm over finite fields. J. Algorithms **40**, 37–81 (2001)
22. K. Ford, The distribution of integers with a divisor in a given interval. Ann. Math. **168**, 367–433 (2008)
23. J. von zur Gathen, J. Gerhard, *Modern Computer Algebra*, 2nd edn. (Cambridge University Press, Cambridge/New York/Melbourne, 2003)
24. J. von zur Gathen, D. Panario, Factoring polynomials over finite fields: a survey. J. Symb. Comput. **31**, 3–17 (2001)
25. J. von zur Gathen, D. Panario, B. Richmond, Interval partitions and polynomial factorization. Algorithmica **63**, 363–397 (2012)
26. J. von zur Gathen, V. Shoup, Computing Frobenius maps and factoring polynomials. Comput. Complex. **2**, 187–224 (1992)
27. S. Gao, J. Howell, D. Panario, Irreducible polynomials of given forms, in *Finite Fields: Theory, Applications, and Algorithms*. Contemp. Math., vol. 225 (The American Mathematical Society, Providence, 1999), pp. 43–54 s
28. T.A. Gassert, Chebyshev action on finite fields. arXiv:1209.4396v3
29. D.A. Goldston, J. Pintz, C.Y. Yldrm, Primes in tuples I. Ann. of Math. **170**, 819–862 (2009)
30. D. Hachenberger, Primitive complete normal bases for regular extensions. Glasgow Math. J. **43**, 383–398 (2001)
31. D. Hachenberger, Primitive complete normal bases: existence in certain 2-power extensions and lower bounds. Discrete Math. **310**, 3246–3250 (2010)
32. T. Hansen, G.L. Mullen, Primitive polynomials over finite fields. Math. Comput. **59**, 639–643, S47–S50 (1992)
33. A.S. Hedayat, N.J.A. Sloane, J. Stufken, *Orthogonal Arrays, Theory and Applications*, Springer Series in Statistics (Springer, New York, 1999)
34. M. Herrmann, G. Leander, A practical key recovery attack on basic $\mathcal{TCHo}$, in *Public Key Cryptography—PKC 2009*. Lecture Notes in Comput. Sci., vol. 5443 (Springer, New York, 2009), pp. 411–424
35. C.-N. Hsu, The distribution of irreducible polynomials in $F_q[t]$. J. Number Theory **61**, 85–96 (1996)
36. IEEE, Standard specifications for public key cryptography, Standard P1363-2000, Institute of Electrical and Electronics Engineering, 2000, Draft D13 available at http://grouper.ieee.org/groups/1363/P1363/draft.html
37. G. Ivanyos, M. Karpinski, L. Rónyai, N. Saxena, Trading GRH for algebra: algorithms for factoring polynomials and related structures. Math. Comput. **81**, 493–531 (2012)
38. E. Jensen, M.R. Murty, Artin's conjecture for polynomials over finite fields, in *Number Theory*, Trends in Mathematics (Birkhauser, Basel, 2000), pp. 167–181
39. K.S. Kedlaya, C. Umans, Fast modular composition in any characteristic, in *49th Annual IEEE Symposium on Foundations of Computer Science* (2008), pp. 146–155
40. D. Koukoulopoulos, Divisors of shifted primes. Int. Math. Res. Not. IMRN **2010**, 4585–4627 (2010)
41. H.W. Lenstra, Jr., R.J. Schoof, Primitive normal bases for finite fields. Math. Comp. **48**, 217–231 (1987)
42. R. Lidl, G.L. Mullen, When does a polynomial over a finite field permute the elements of the field? Am. Math. Mon. **95**, 243–246 (1988)
43. R. Lidl, G.L. Mullen, When does a polynomial over a finite field permute the elements of the field? II. Am. Math. Mon. **100**, 71–74 (1993)

44. R. Lidl, H. Niederreiter, *Finite Fields*, vol. 20, 2nd edn., Encyclopedia of Mathematics and its Applications (Cambridge University Press, Cambridge, 1997)
45. A. MacFie, D. Panario, Random mappings with restricted preimages, in *Proceedings of LatinCrypt 2012*. Lecture Notes in Comput. Sci., vol.7533 (Springer, Berlin, 2012), pp. 254–270
46. C. Mauduit, J. Rivat, Sur un problème de Gelfond: la somme des chiffres des nombres premiers. Ann. Math. **171**, 1591–1646 (2010)
47. J. Maynard, Small gaps between primes. arXiv:1311.4600 (2013)
48. A.J. Menezes, P.C. van Oorschot, S.A. Vanstone, *Handbook of Applied Cryptography*, Series on Discrete Mathematics and its Applications (CRC Press, Boca Raton, 1997)
49. I.H. Morgan, G.L. Mullen, Completely normal primitive basis generators of finite fields. Utilitas Math. **49**, 21–43 (1996)
50. G.L. Mullen, D. Panario, *Handbook of Finite Fields*, Series on Discrete Mathematics and Its Applications (CRC Press, Boca Raton, 2013)
51. A. Munemasa, Orthogonal arrays, primitive trinomials, and shift-register sequences. Finite Fields Appl. **4**, 252–260 (1998)
52. D. Panario, What do random polynomials over finite fields look like? in *Finite Fields and Applications. Lecture Notes in Comput. Sci.*, vol. 2948 (Springer, Berlin, 2004), pp. 89–108
53. V.S. Pless, W.C. Huffman, R.A. Brualdi, *Handbook of Coding Theory* (North-Holland, Amsterdam, 1998)
54. P. Pollack, A polynomial analogue of the twin primes conjecture. Proc. Am. Math. Soc. **136**, 3775–3784 (2008)
55. J.M. Pollard, A Monte Carlo method for factorization, Nordisk Tidskr. Informationsbehandling (BIT) **15**, 331–334 (1975)
56. M. Scott, Optimal irreducible polynomials for GF(2^m) arithmetic, in *Software Performance Enhancement for Encryption and Decryption (SPEED 2007)*, Cryptology ePrint Archive (2007)
57. M. Sha, S. Hu, Monomial dynamical systems of dimension over finite fields. Acta Arith. **148**, 309–331 (2011)
58. R.G. Swan, Factorization of polynomials over finite fields. Pac. J. Math. **12**, 1099–1106 (1962)
59. S. Ugolini, Graphs associated with the map $X \mapsto X + X^{-1}$ in finite fields of characteristic three and five. J. Number Theory **133**, 1207–1228 (2013)
60. T. Vasiga, J. Shallit, On the iteration of certain quadratic maps over GF(p). Discrete Math. **277**, 219–240 (2004)
61. D. Wan, Generators and irreducible polynomials over finite fields. Math. Comput. **66**, 1195–1212 (1997)
62. A. Weil, *Sur les Courbes Algébriques et les Variétés qui s'en dÉduisent*, Actualités Sci. Ind., no. 1041; Publ. Inst. Math. Univ. Strasbourg, vol. 7 (Hermann et Cie., Paris, 1945/1948)
63. Y. Zhang, Bounded gaps between primes. Ann. Math. **170**, 1121–1174 (2014)

Generating Good Span n Sequences Using Orthogonal Functions in Nonlinear Feedback Shift Registers

Kalikinkar Mandal and Guang Gong

Abstract A binary span n sequence generated by an n-stage nonlinear feedback shift register (NLFSR) is in a one-to-one correspondence with a de Bruijn sequence and has the following randomness properties: period $2^n - 1$, balance, and ideal n-tuple distribution. A span n sequence may have a high linear span. However, how to find a nonlinear feedback function that generates such a sequence constitutes a long-standing challenging problem for about 5 decades since Golomb's pioneering book, *Shift Register Sequences*, published in the middle of the 1960s. In hopes of finding good span n sequences with large linear span, in this chapter we study the generation of span n sequences using orthogonal functions in polynomial representation as nonlinear feedback in a nonlinear feedback shift register. Our empirical study shows that the success probability of obtaining a span n sequence in this technique is better than that of obtaining a span n sequence in a random span n sequence generation method. Moreover, we analyze the linear span of new span n sequences, and the linear span of a new sequence lies between $2^n - 2 - 3n$ (near optimal) and $2^n - 2$ (optimal). Two conjectures on the linear span of new sequences are presented and are valid for $n \leq 20$.

1 Introduction

Nonlinear feedback shift registers (NLFSRs) are used to design many cryptographic primitives such as pseudorandom sequence generators (PRSGs), stream ciphers [11], and lightweight block ciphers [7] for providing security and privacy in communication systems. Ciphers based on NLFSRs are of great practical importance in many constrained environments, for instance, RFID tags and sensor networks due to their need for efficient hardware implementation and high throughput. In general, an arbitrary NLFSR cannot be used for generating keystreams in stream ciphers

K. Mandal (✉) • G. Gong
Department of Electrical and Computer Engineering, University of Waterloo, Waterloo, ON, Canada N2L 3G1
e-mail: kmandal@uwaterloo.ca; ggong@uwaterloo.ca

© Springer International Publishing Switzerland 2014
Ç.K. Koç (ed.), *Open Problems in Mathematics and Computational Science*,
DOI 10.1007/978-3-319-10683-0_7

127

because the randomness properties including the period of a sequence generated by that NLFSR are unknown and hard to determine.

A binary *de Bruijn sequence* is a binary sequence with period 2^n which satisfies the property that all n-tuples occur exactly once in one period. De Bruijn sequences have known randomness properties, namely, maximum period, balance property, ideal n-tuple distribution, and large linear span [3, 21, 34]. A *modified de Bruijn sequence* or *span n sequence* with period $2^n - 1$ is a pseudorandom sequence where each nonzero n-tuple occurs exactly once in one period of the sequence. This property is referred to as *the span n property* [21]. Often, de Bruijn sequences as well as span n sequences are generated recursively by an n-stage nonlinear feedback shift register. Only, m-sequences are a class of span n sequences generated by linear feedback shift registers.

A span n sequence can be constructed from a de Bruijn sequence by removing any one zero from the run of zeros of length n, and similarly, a de Bruijn sequence can be formed from a span n sequence by adding one zero to the run of zeros of length $n - 1$. The *linear span or linear complexity* of a sequence is the length of the shortest LFSR that produces the given sequence. We remember that "linear span" and "span n" are two different properties of a span n sequence. Note that by adding an extra zero to the run of zeros of length $n - 1$ to an m-sequence, the linear span of the resultant de Bruijn sequence varies between $2^{n-1} + n$ and $2^n - 1$ [3], but by removing any one zero from the run of zeros of length n from the resultant de Bruijn sequence, it becomes an m-sequence or a span n sequence with linear complexity n. So the lower bound of the linear span of the span n sequence drops to n [23]. This phenomenon suggests to study the randomness properties, particularly, the linear span property of span n sequences instead of de Bruijn sequences for cryptographic usages. Until recently, there is no known general construction of a nonlinear feedback function which generates a span n sequence, and this is open since the last 5 decades. Therefore, the generation of span n sequences by NLFSRs is a challenging problem.

Our objective is to produce span n/de Bruijn sequences using orthogonal functions as feedback functions in nonlinear feedback shift registers. An orthogonal feedback function has a trace representation and is composed of three parameters, namely, a decimation number, a primitive polynomial, and a t-tap position ($5 \leq t \leq n - 1$). In an NLFSR, a class of feedback functions is constituted by varying the decimation numbers and the polynomial bases of the finite fields. Finding span n sequences by using this class of feedback functions and all possible tap positions of the feedback functions is called a *structured search*. We show that a number of new span n sequences with a moderate n can be produced through the structured search. For $n \geq 10$, all the feedback functions of degree greater than or equal to two cannot be employed to search span n sequences. Using the structure search, on the other hand, one can employ a number of feedback functions with different degrees and a variable number of terms.

In this chapter, we present some new theoretical results on generating span n sequences and experimental results on finding the number of new span n sequences. The chapter is organized as follows. In Sect. 2, we provide some basic definitions of shift register sequences and their properties. Section 2.2 recalls the definitions of known orthogonal functions, and Sect. 3 introduces some known constructions of de Bruijn sequences. In Sect. 4, we describe the span n sequence generation technique using orthogonal functions and develop some properties of this technique, including an estimation of the number of orthogonal feedback functions used in this technique. Sections 5 and 6 present the experimental results on the number of span n sequences produced using orthogonal functions, and Sect. 7 presents an empirical success probability comparison of obtaining span n sequences using orthogonal functions. In Sect. 8, we analyze the linear span of newly produced span n sequences by the aforementioned orthogonal functions and present two conjectures on the linear span of the span n sequences produced by the orthogonal functions. Our empirical results show that the success probability of obtaining a span n sequence in the structured search is larger than that of generating a span n sequence in a random search. Our results show that the linear span of a new span n sequence lies in the range of $2^n - 2 - 3n$ (near optimal) and $2^n - 2$ (optimal). In Sect. 9, some applications of new span n sequences are shown, and in the section "Conclusions", we conclude the chapter.

2 Preliminaries

In this section, we define and explain the terms and mathematical functions that will be used in this chapter to produce span n sequences.

- $\mathbb{F}_2 = \{0, 1\}$: the Galois field with two elements.
- $\mathbb{F}_{2^t} = \{(x_0, x_1, \ldots, x_{t-1}) \mid x_i \in \mathbb{F}_2\}$—an extension field that is defined by a primitive element α with $p(\alpha) = 0$, where $p(x) = c_0 + c_1 x + \cdots + c_{t-1} x^{t-1} + x^t$ is a primitive polynomial of degree t (≥ 2) over $\mathbb{F}_2$.
- $\mathrm{Tr}(x) = x + x^2 + \cdots + x^{2^{t-1}}$: the trace function mapping from $\mathbb{F}_{2^t}$ to $\mathbb{F}_2$.
- $D_t = \{d : d$ is a coset leader with $\gcd(d, 2^t - 1) = 1\}$. The cardinality of D_t, denoted as $|D_t|$, is given by $\frac{\phi(2^t - 1)}{t}$, where $\phi(\cdot)$ is the Euler phi function.

2.1 Basic Definitions and Properties of Feedback Shift Registers

Usually, an n-stage linear or nonlinear feedback shift register is used to generate a periodic binary sequence $\mathbf{a} = \{a_i\}$, and the recurrence relation for the (N)LFSR is defined as [20]

$$a_{n+k} = a_k \oplus g(a_{k+1}, \ldots, a_{k+n-1}), \ a_i \in \mathbb{F}_2, \ k \geq 0$$

where $(a_0, a_1, \ldots, a_{n-1})$ is the *initial state* of the shift register, g is a Boolean function in $(n-1)$ variables, and $\oplus$ is the addition operation over $\mathbb{F}_2$. If the function g is an affine function, then the sequence **a** is called an LFSR sequence; otherwise, it is called an NLFSR sequence. The above recurrence relation is also known as a nonsingular recurrence relation.

The complementary binary sequence of binary sequence $\mathbf{b} = \{b_i\}_{i \geq 0}$, denoted as $\bar{\mathbf{b}}$, is defined by $\{\bar{b}_i\}_{i \geq 0}$, where $\bar{b}_i = b_i \oplus 1$. The *linear span or linear complexity* of a sequence is the length of the shortest LFSR that produces the sequence.

Definition 1 ([22]) The autocorrelation of a binary sequence $\{a_i\}$ with period N is defined as

$$C(\tau) = \sum_{i=0}^{N-1} (-1)^{a_{i+\tau} + a_i}.$$

Moreover, if $N = 2^n - 1$, the sequence $\{a_i\}$ has 2-level autocorrelation if

$$C(\tau) = \begin{cases} 2^n - 1 & \text{if } \tau \equiv 0 \pmod{2^n - 1} \\ -1 & \text{if } \tau \not\equiv 0 \pmod{2^n - 1}. \end{cases}$$

Property 1 The linear span of a de Bruijn sequence, denoted as LS_{db}, is bounded by [3]

$$2^{n-1} + n \leq LS_{db} \leq 2^n - 1. \tag{1}$$

On the other hand, the linear span of a span n sequence that is generated by an NLFSR, denoted as LS_s, is bounded by

$$2n < LS_s \leq 2^n - 2. \tag{2}$$

From this property, we say that a span n sequence has the optimal linear span if its linear span is equal to $2^n - 2$.

2.2 Review of the Trace Representation of 2-level Autocorrelation Sequences

An orthogonal function from $\mathbb{F}_{2^t}$ to $\mathbb{F}_2$ is in one-to-one correspondence with a binary sequence with (ideal) 2-level autocorrelation function, 2-level autocorrelation sequence in short. There are only very few known constructions on 2-level autocorrelation sequences, which constitutes another challenge problem for years. Interestingly, those functions possess good cryptographic properties. The reader is referred to Golomb and Gong's book [22] for the details about the constructions of 2-level autocorrelation sequences and their related cryptographic properties. In the following, for easy reference, we formally provide the definition of orthogonal functions and their known constructions from corresponding trace representation of 2-level autocorrelation sequences.

Definition 2 A function, say, $f(x)$, from $\mathbb{F}_{2^t}$ to $\mathbb{F}_2$ is called an orthogonal function if $\sum_{x \in \mathbb{F}_{2^t}} (-1)^{f(\lambda x) + f(x)} = 0$ for all $(1 \neq)\lambda \in \mathbb{F}_{2^t}$.

Let α be a primitive element of $\mathbb{F}_{2^t}$ and let $a_i = f(\alpha^i)$ where the binary sequence $\{a_i\}$ is called an evaluation of $f(x)$ and $f(x)$, the trace representation of $\{a_i\}$.

Property 2 With the above notation:

1. $f(x)$ is orthogonal if and only if its evaluation has 2-level autocorrelation.
2. If $f(x)$ is orthogonal, then $f(x^r)$ is orthogonal for all r with $(r, 2^t - 1) = 1$.

Let $C = \{r, 2r, \ldots, 2^{n_r-1}r\}$ where n_r is the smallest number such that $r2^{n_r} \equiv r \mod 2^t - 1$. Then C is called a (cyclotomic) coset consisting r modulo $2^t - 1$, and the smallest number in C is called the coset leaders of C. Let I consist of all coset leaders modulo $2^t - 1$.

2.2.1 Number Theory-Based Constructions

This type of the constructions includes Legendre sequences and Hall sextic residue sequences. Let $p = 2^t - 1$ be a prime number, u be a primitive element in $\mathbb{F}_p$, and $c = \frac{2^t - 2}{t}$.

Orthogonal Functions from Legendre Sequences (A1) Let

$$f(x) = \sum_{i=0, i \in I}^{c/2-1} \mathrm{Tr}(x^{u^{2i}}), x \in \mathbb{F}_{2^t}.$$

Or equivalently,

$$f(x) = \sum_{i \in I_0} \mathrm{Tr}(x^i), x \in \mathbb{F}_{2^t}$$

where $I_0 \subset I$ consist of all quadratic coset leaders modulo $2^t - 1$. Then $f(x)$ is an orthogonal function from $\mathbb{F}_{2^t}$ to $\mathbb{F}_2$ whose evaluation gives a Legendre sequence with 2-level autocorrelation.

Hall's "Sextic Residue Sequence" (A2) Additional to the Legendre sequences, $p = 4t - 1 = 4a^2 + 27$. Let

$$f(x) = \sum_{i=0, i \in I}^{c/6-1} \mathrm{Tr}(x^{u^{6i}}), x \in \mathbb{F}_{2^t}.$$

Then $f(x)$ is an orthogonal function from $\mathbb{F}_{2^t}$ to $\mathbb{F}_2$ whose evaluation gives a Hall's "sextic residue sequence" with 2-level autocorrelation function.

2.2.2 Finite Fields-Based Constructions

There are four types of constructions for 2-level autocorrelation sequences: m-sequences, hyperoval constructions, Welch–Gong transformation construction, and Kasami power function construction including three-term and five-term sequences.

Orthogonal Functions from m-Sequences Let

$$f(x) = \text{Tr}(x), x \in \mathbb{F}_{2^t},$$

then $f(x)$ is an orthogonal function whose evaluation gives an m-sequence with period $2^t - 1$, and the other m-sequences are given by $\text{Tr}(x^d)$ where $\gcd(d, 2^t - 1) = 1$.

Orthogonal Functions from Hyperoval Sequences There are three monomial hyperoval sequences with 2-level autocorrelation, namely, Segre type and Glynn type 1 and type 2. Except for Segre hyperoval sequences, the trace representation is not represented in a formula. Instead, it is described in terms of some relation which needs to be computed for different t.

Let $(1)^l$ denote a string of l consecutive 1s. Let $\mathcal{A}$ denote the set consisting of all strings of the form $(1)^{4a+1}0$ or $a \geq 0$ and $(1)^{4b}$, $b \geq 0$. Let $\mathcal{A}^*$ denote the set of all strings obtained by concatenating zero, one or more strings from $\mathcal{A}$. Let t be a prime and

$$\left\| \begin{array}{l} 01(\text{string in } \mathcal{A}^*)0(1)^{2s}, s \geq 0 \text{ or} \\ 011(\text{string in } \mathcal{A}^*)11 \end{array} \right\|. \tag{3}$$

The trace representation of a Segre hyperoval sequence of period $2^t - 1$, t odd, is given by

$$f(x) = \sum_{i \in T_{\text{Segre}}} \text{Tr}(x^i), x \in \mathbb{F}_{2^t}$$

where $T_{\text{Segre}} \subset I$ which are the collections of coset leaders of the all binary numbers given by (3) [5].

Let $T_{\text{Glynn}} \subset I$ be the collections of the coset leaders of solutions to

$$w(j) + w((k-1)j - w(kj) = 1, j = 1, \ldots, 2^t - 1$$

where $w(x)$ is the Hamming weight of binary number x. Then the trace representation of a Glynn hyperoval sequence of period $2^t - 1$, t odd, is given by

$$f(x) = \sum_{i \in T_{\text{Glynn}}} \text{Tr}(x^i), x \in \mathbb{F}_{2^t}$$

where $k = \sigma + \gamma$ for Glynn type 1 and $k = 3\sigma + 4$ for Glynn type 2 where $\sigma = 2^{(t+1)/2}$ and $\gamma = 2^{(3t+1)/4}$ [13].

Orthogonal Functions from Three-Term, Five-Term, and Welch–Gong Transformation Constructions In [38], it was conjectured that three-term and five-term sequences have 2-level autocorrelation as well as Welch–Gong transformation sequences discovered by Golomb, Gong, and Gaal. The validity of those conjectures is established later on by Dillon and Dobbertin in [8, 9].

Let $t = 2k - 1$ for some positive integer k and $t \geq 5$. Let

$$f(x) = \text{Tr}(x + x^{2^k+1} + x^{2^k-1}), x \in \mathbb{F}_{2^t}.$$

Then its evaluation gives three-term 2-level autocorrelation sequences.

Let t be a positive integer with $t \bmod 3 \not\equiv 0$ and $3k \equiv 1 \bmod t$ for some integer k. We define the function h from $\mathbb{F}_{2^t}$ to $\mathbb{F}_{2^t}$ by

$$h(x) = x + x^{q_1} + x^{q_2} + x^{q_3} + x^{q_4}$$

where

$$q_1 = 2^k + 1, q_2 = 2^{2k} + 2^k + 1, q_3 = 2^{2k} - 2^k + 1, q_4 = 2^{2k} + 2^k - 1.$$

(Note that $h(x)$ is a permutation over $\mathbb{F}_{2^t}$ [8].) Let

$$g(x) = \text{Tr}(h(x)) \text{ and } f(x) = \text{Tr}(h(x + 1) + 1)$$

where $f(x)$ is known as the *WG transformation*. The evaluations of $g(x)$ and $f(x)$ yield five-term sequences and WG transformation sequences.

Orthogonal Functions from Kasami Power Function Construction Let $\gcd(k, t) = 1$, $k < t$, $kk' \equiv 1$, and

$$f(x) = \text{Tr}(R(x)), x \in \mathbb{F}_{2^t}$$

where $R(x)$ is given by

$$R(x) = \sum_{i=1}^{k'} A_i(x) + V_{k'}(x)$$

where A_i and V_i are iteratively defined by

$$A_1(x) = x$$
$$A_2(x) = x^{2^k+1}$$
$$A_{i+2}(x) = x^{2^{(i+1)k}} A_{i+1}(x) + x^{2^{(i+1)k}-2^{ik}} A_i(x), i \geq 1$$

and

$$V_1(x) = 0$$
$$V_2(x) = x^{2^k-1}$$
$$V_{i+2}(x) = x^{2^{(i+1)k}} V_{i+1}(x) + x^{2^{(i+1)k}-2^{ik}} V_i(x), i \geq 1.$$

Orthogonal Functions from Subfield Constructions Let $1 < m \mid t, m \neq t$, and $g(x)$ be any orthogonal function from $\mathbb{F}_{2^m}$ to $\mathbb{F}_2$, listed in the above subsections, and let

$$f(x) = \text{Tr}_m^t(g(x)), x \in \mathbb{F}_{2^t}$$

where $\text{Tr}_m^t(x)$ is the trace function from $\mathbb{F}_{2^t}$ to $\mathbb{F}_{2^m}$, i.e.,

$$\text{Tr}_m^t(x) = x + x^{2^m} + \cdots + x^{2^{(l-1)m}}, x \in \mathbb{F}_{2^t}, l = t/m.$$

Then $f(x)$ is an orthogonal function from $\mathbb{F}_{2^t}$ to $\mathbb{F}_2$, and its evaluation is called a subfield 2-level autocorrelation sequences which includes GMW sequence for $g(x) = \text{Tr}_1^m(x^d)$ where $\text{Tr}(x)$ is the trace function from $\mathbb{F}_{2^m}$ to $\mathbb{F}_2$ and $\gcd(d, 2^m - 1) = 1$ and generalized GMW sequences for the rest of $g(x)$. Here we shorten them as GMW sequences.

2.2.3 Orthogonal Functions for Small Fields

In the following, we give the exponents explicitly for all known orthogonal functions of the form $f(x) = \sum_{i \in I} \text{Tr}(x^i)$ from $\mathbb{F}_{2^t}$ to $\mathbb{F}_2$ for $5 \leq t \leq 11$ in Tables 1, 2 and 3 where the monomial function $\text{Tr}(x)$ is not listed.

Table 1 Exponents in the orthogonal functions over $\mathbb{F}_{2^t}$, $5 \leq t \leq 9$

Orthogonal functions	Trace spectra	# of terms
$t = 5$		
T3	1, 3, 5	3
$t = 7$		
T3	1, 9, 13	3
T5	1, 5, 21, 13, 29	5
WG	1, 3, 7, 19, 29	5
QR	3, 5, 7, 23, 27, 29, 43, 55, 63	9
Hall	5, 27, 63	3
$t = 8$		
GMW	7, 13, 37, 11	4
T5	1, 9, 37, 29, 39	5
WG	13, 19, 21, 29, 39	5
$t = 9$		
T3	1, 17, 25	3
GMW	3, 17, 129	3
Segre	1,5,7, 9, 19, 25, 37, 77, 117	9
Glynn 1	1, 5, 9, 13, 19, 37, 43	7
Glynn 2	17, 23, 37, 43, 45, 75, 87	7

Table 2 Exponents in the orthogonal functions over $\mathbb{F}_{2^t}$, $t = 10$

Orthogonal functions	Trace spectra	# of terms
T5	1, 9, 57, 73, 121	5
WG	1, 3, 5, 7, 11, 13, 15, 35, 69, 71, 89, 105, 121	13
GMW1	3, 17	2
GMW2	5, 9	2
GMW3	7, 19, 25, 69	4
GMW4	11, 13, 21, 73	4
GMW5	1, 5, 7, 9, 19, 25, 69	7
GMW6	15, 23, 27, 29, 77, 85, 89, 147	8
GMW7	3, 7, 11, 13, 15, 21, 23, 27, 29, 73, 77, 85, 89, 147	14

We define the following set:

$$D_t^* = \{d \; : \; d \in D_t \text{ and } f_d(\cdot), \text{ is nonlinear and } f_d(x) \neq f_{d_1}(x), d \neq d_1 (\in D_t^*)\}.$$

For all decimation numbers in D_t^*, we take into account all distinct orthogonal functions obtained from an orthogonal function using decimations.

Table 3 Exponents in the orthogonal functions over $\mathbb{F}_{2^t}, t = 11$

Orthogonal functions	Trace spectra	# of terms
T3	1, 33, 49	3
T5	1, 17, 121, 137, 143	5
WG	21, 23, 29, 35, 37, 41, 71, 89, 139, 165, 213, 307, 415	13
Segre = B2	1, 5, 13, 21, 53, 77, 85, 205, 213, 309, 333, 341, 413, 423, 469	15
Glynn 1	1, 5, 9, 13, 19, 37, 43, 67, 69, 137, 163, 211, 293	13
Glynn 2	1, 5, 13, 17, 29, 37, 49, 61, 69, 81, 93, 101, 113, 125, 139, 147, 151, 157, 171, 173, 183	21
B3	1, 5, 7, 9, 19, 25, 81, 169, 295	9

3 Review of Known Constructions of (Modified) de Bruijn Sequences

There is a one-to-one correspondence between a de Bruijn sequence and a modified de Bruijn/span n sequence. When the construction of a feedback function that generates a span n sequence is known, the construction of a de Bruijn sequence can be known and vice versa. In this section, we provide some known de Bruijn and span n sequence generation techniques.

3.1 Known Constructions for de Bruijn Sequences

Problem of generating a de Bruijn sequence is easy to understand, but providing a solution for generating a de Bruijn sequence efficiently is a challenging problem. This problem is studied from algorithmic, graph theoretic, and algebraic technique points of view in the literature. In particular, generating a de Bruijn sequence using a feedback shift register is an algebraic technique, which exploits properties of a feedback function. In the following, we present some well-known approaches of constructing de Bruijn sequences.

3.1.1 Lempel's D-Morphism-Based Techniques for de Bruijn Sequences

Lempel in [26] proposed the concept of generating a de Bruijn sequence of period 2^{n+1} by first computing two D-morphic preimages of a de Bruijn sequence of period 2^n and then concatenating these two preimages at a conjugate pair. In this construction, it is assumed that the construction of the de Bruijn sequence of period 2^n is known. Later on, Annexstein in [1] and Chang et al. in [6] proposed two algorithms based on Lempel's D-homomorphism for producing de Bruijn sequences of long period. Games [18] proposed a generalized construction of

Lempel's construction in which a de Bruijn sequence of period 2^{n+1} is constructed from two different de Bruijn sequences of period 2^n using Lempel's conjugate.

In [36], Mykkeltveit et al. presented Lempel's construction in the form of a composited recurrence relation. Following Mykkeltveit et al.'s construction, Mandal and Gong in [28] refined and studied the composited construction, for producing strong composited de Bruijn sequences of arbitrarily long period from a span n sequence. For the properties and cycle structures of composited recurrence relations, see [27, 36]. Note that, in the composited construction, the feedback function of a de Bruijn sequence is a bit complicated, which contains a number of sum-of-product terms. Recently, Mandal and Gong in [29] analyzed composited de Bruijn sequences from D-morphic point of view and presented an iterative technique for computing the nonlinear feedback function of a composited de Bruijn sequence. In the composited construction one needs to know the construction of a feedback function of a span n sequence in order to generate a de Bruijn sequence of long period.

3.1.2 Algorithms for de Bruijn Sequence Generation

Fredricksen and Kessler in [16] proposed an algorithm based on lexicographic compositions for constructing de Bruijn sequences of period 2^n, and the amount of storage required in implementing the algorithm is linear in n. Fredricksen and Maiorana in [17] presented an algorithm for generating necklaces of length n in k colors, and a k-ary de Bruijn sequence of period k^n is produced by juxtaposing in order the periodic reductions of the necklaces.

Fredricksen [14] developed an algorithm to generate nonlinear de Bruijn sequences, and the algorithm requires $3n$ units of storage and outputs one bit in around n units of time. Fredricksen also exhibited that new de Bruijn sequences can be obtained from a de Bruijn sequence by cross-joining, and the number of such new de Bruijn sequences is 2^{2n-5}. The storage requirement for implementing the method is about $6n$ units. When this method is compared with Mandal and Gong's iterative technique (MG iterative technique) for composited de Bruijn sequences, MG iterative technique for the composited feedback function requires less amount of time as well as memory.

Etzion and Lempel [12] developed a construction of de Bruijn sequences with linear complexity $(2^{n-1} + n)$ for all $n \geq 3$. A detailed survey by Fredricksen of many other de Bruijn sequence generation techniques can be found in [15].

3.1.3 Cycle Joining Techniques for de Bruijn Sequence Generation

Cycle joining technique is one of the well-known methods of generating a de Bruijn sequence in which a de Bruijn sequence is constructed by joining a finite number of cycles produced by a feedback shift register. In this technique, first a feedback

function of a nonsingular feedback shift register is chosen, and then a different
feedback function for a de Bruijn sequence is constructed from the first feedback
function based on its cycle decomposition.

Jansen et al. [25] presented a cycle joining algorithm for generating de Bruijn
sequences where the feedback function of a de Bruijn sequence is the sum of
two functions; one function is the feedback function itself, and another function
is constructed from the feedback function for joining cycles. In [25], it is shown
that $O(2^{\frac{2n}{\log(2n)}})$ de Bruijn sequences of period 2^n can be produced when all
irreducible polynomials of degree n is taken in a feedback shift register. The storage
requirement for this method is $3n$ bits, and $4n$-unit of time is required to generate
each bit of a de Bruijn sequence. A storage-time comparison between this algorithm
and the MG iterative technique can be found in [6].

Yang and Dai in [40] proposed a construction of an m-ary de Bruijn sequence
based on joining the cycles using modification sets of a feedback function f. In the
construction, a nonlinear feedback function F of a de Bruijn sequence is constructed
from the feedback function f using the modification sets of f. The authors showed
that, when a circulating register is chosen, at least $2^{(\frac{m^n}{n}-mn)}$ feedback functions
that generate de Bruijn sequences can be constructed. However, this method is not
efficient for large values of n, since the method requires the cycles decomposition
of f to construct the function F, and for a large n, it is very hard to obtain the cycle
decomposition of f. Moreover, the feedback function would contain many product
terms for joining of the cycles.

Hauge and Helleseth [24] proposed a technique based on an irreducible polynomial and its adjacency graph to generate de Bruijn sequences. In this technique, a de
Bruijn sequence is obtained as maximum spanning trees from the adjacency graph of
a feedback function corresponding to an irreducible polynomial. The lower bound
for the number of de Bruijn sequences is determined in terms of the cyclotomic
numbers.

3.2 Known Techniques for Generating Modified de Bruijn Sequences

Most of the research efforts devoted on span n sequences have been concerned
about the number of span n sequences and the characteristics of nonlinear feedback
functions [21, 33, 34] including the number of terms in the feedback functions
[33, 35] and the weight of truth tables of the feedback functions [32, 33]. Mayhew
and Golomb reported the number of span n sequences for different values of the
linear span of span n sequences and for different values of the number of terms in
the feedback functions ($4 \leq n \leq 6$) [34, 35]. Mayhew reported the number of span
n sequences for different weight classes of the truth tables of the feedback functions
for $n = 6$ [33]. However, the task of finding the number of span n sequences for

different weight classes and for different values of the linear span is an unsolved problem for $n \geq 7$.

In [4], Chan et al. have considered the generation of quadratic m-sequence that uses very simple quadratic functions as the feedback function, which is the sum of a linear function in n variables and a quadratic term for any two variables and reported the number of span n sequences for $5 \leq n \leq 12$. Dubrova in [10] and Rachwalik et al. in [39] found a few quadratic m-sequences, i.e., span n sequence generated using quadratic feedback functions for $4 \leq n \leq 24$ and $25 \leq n \leq 27$, respectively. Gammel et al. have searched span n sequences while designing stream cipher Achterban:128/80 based on nonlinear feedback shift registers [19].

Note that the feedback functions of an NLFSR in [10, 19, 39] contain only a few terms and are of low algebraic degree. All the methods for finding the number of span n sequences and verifying the span n property of a sequence use an exhaustive search method which is an exponential time algorithm in n.

4 A New Construction

In this section we first describe the recurrence relation of nonlinear feedback shift registers whose feedback functions are orthogonal functions. In an n-stage NLFSR, the feedback function can also be regarded as a Boolean function in t variables where $5 < t \leq n - 1$. Our considered orthogonal feedback functions in t variables are balanced as the evaluation of the feedback function has 2-level autocorrelation and have even Hamming weight 2^{t-1}. Thus, the new span n sequences generated by a class of feedback functions belong to the weight class 2^{n-2}. Then we calculate the approximate number of feedback functions used in the structured search.

4.1 Description of Span n Sequence Generation Using Orthogonal Function

Let $\mathbf{a} = \{a_i\}$ be a binary sequence generated by an n-stage NLFSR whose nonlinear recurrence relation is defined as

$$a_{n+k} = a_k \oplus f_d(x_k), \ x_k = (a_{r_1+k}, a_{r_2+k}, \ldots, a_{r_t+k}) \in \mathbb{F}_{2^t}, d \in D_t^*,$$
$$0 < t < n, \ k \geq 0 \tag{4}$$

where $(r_1, r_2, \ldots, r_t)$ with $0 < r_1 < r_2 < \cdots < r_t \leq n - 1$ is called a t-*tap position* of the NLFSR, $f_d(x) = f(x^d)$, $f(x)$ is an orthogonal function, and $\oplus$ is the addition over $\mathbb{F}_2$. For a proper selection of a t-tap position and a feedback function $f_d(x)$, the binary sequence $\mathbf{a}$ can be a *span n sequence*. We note that for any choice of a t-tap position and a feedback function $f_d(x)$, the binary sequence may not be a span n sequence. The reason for choosing $t \leq (n-1)$ is to involve a small number of state variables in the feedback functions, which is benefited to the implementation of the NLFSR as well as the production of more feedback functions.

Let $\mathbf{b} = \{b_i\}$ be a binary sequence generated by the following recurrence relation

$$b_{n+k} = 1 \oplus b_k \oplus f_d(x_k), \ x_k = (b_{r_1+k}, \dots, b_{r_t+k}) \in \mathbb{F}_{2^t}, d \in D_t^*,$$

$$0 < t < n, \ k \geq 0. \tag{5}$$

Similarly, for a proper selection of a t-tap position and a feedback function $f_d(x)$, the complementary binary sequence $\bar{\mathbf{b}}$ of $\mathbf{b}$ can be a *span n sequence*, but the sequence $\mathbf{b}$ is not a span n sequence since it contains the all-zero state.

If the number of terms in the algebraic normal form representation of the function f_d is even, then the recurrence relations (4) and (5) cannot generate a span n sequence for any choice of a t-tap position, since for the all-one state, recurrence relation (4) generates the all-one sequence, and recurrence relation (5) contains the all-one n-tuple.

Proposition 1 *If $f_d(x) = 0$ for $x = (1, 1, \dots, 1) \in \mathbb{F}_{2^t}$, then recurrence relations (4) and (5) cannot generate span n sequences.*

In the recurrence relations (4) and (5), by varying three parameters, namely, the primitive polynomial $p(x)$, the decimation number d, and the t-tap position $(r_1, r_2, \dots, r_t)$, a number of new span n sequences can be produced, and that number mainly depends on the length n of the NLFSR and the number t of inputs to the function f_d. We call this searching technique a *structured search*, where an NLFSR has a compact representation in terms of feedback functions and tap positions. Note that we may not always obtain a span n sequence for a fixed value of t and for any length n of the NLFSR. A special case of the recurrence relation (4) with the trace function in $(n-1)$ variables as the feedback function is defined in [37].

A periodic reverse binary sequence is defined as follows [32, 35]: for a binary sequence $\{a_0, a_1, \dots, a_{2^n-2}\}$ with period $2^n - 1$, the reverse sequence of the binary sequence is defined by $\{a_{2^n-2}, a_{2^n-3}, \dots, a_1, a_0\}$. A reverse sequence of a span n sequence is also a span n sequence, which is not shift equivalent to the original one, and the reverse span n sequence can be generated by the same function but with a different t-tap position.

Proposition 2 ([32]) *Let $g(x_0, x_1, \dots, x_{n-1}) = x_0 \oplus f(x_1, \dots, x_{n-1})$ generates a span n sequence with period $2^n - 1$. Then the function $h(x_0, x_{n-1}, \dots, x_1) = x_0 \oplus f(x_{n-1}, \dots, x_1)$ generates a reverse span n sequence.*

Our span n sequences generated by recurrence relations (4) and (5) with a fixed $P(x)$ are uniquely determined by the following three parameters:

1. the decimation number d,
2. the primitive polynomial $p(x)$,
3. the t-tap position $(r_1, r_2, \dots, r_t)$.

Similarly, the reverse span n sequence of a span n sequence with parameters d, $p(x)$, and $(r_1, r_2, \ldots, r_t)$ is represented by the same decimation number d and the same primitive polynomial $p(x)$, but with a different t-tap position $(n - r_1, n - r_2, \ldots, n - r_t)$. For a fixed function $f_d(x)$, a span n sequence generated by $f_d(x)$ is different if the t-tap position is different. We now describe the span n sequence generation by the above structured search in the following example.

Example 1 The following example describes our span n sequence generation procedure for $t = 5$.

The WG transformation over $\mathbb{F}_{2^5}$ is given by

$$f(x) = \text{Tr}(x + (x + 1)^5 + (x + 1)^{13} + (x + 1)^{19} + (x + 1)^{21}).$$

After simplification, $f(x)$ can be written as

$$f(x) = \text{Tr}(x^{19}), \ x \in \mathbb{F}_{2^5},$$

which is degenerated into an m-sequence. For $t = 5$, the set of coset leaders is given by $D_t = \{1, 3, 5, 7, 11, 15\}$, and the coset leaders for which $f_d(x)$ is nonlinear is given by $D_t^* = \{1, 3, 7, 11, 15\}$, since for $d = 5$, the function $f_d(x)$ is linear. The d-th decimation of $f(x)$ is given by

$$f_d(x) = f(x^d) = \text{Tr}(x^{d'}), \ d' = (19 \cdot d) \bmod 2^t - 1, \ d \in D_t^*.$$

The n-stage nonlinear recurrence relation with a t-tap position is given by

$$a_{n+k} = a_k \oplus f_d(x_k), x_k = (a_{r_1+k}, \ldots, a_{r_5+k}) \in \mathbb{F}_{2^5}, \ k \geq 0.$$

The Boolean representation of $f(x) = \text{Tr}(x^{19})$ with defining polynomial $p(x) = 1 + x + x^2 + x^4 + x^5$ of $\mathbb{F}_{2^5}$ is as follows:

$$f(x_0, \ldots, x_4) = x_0 + x_3 + x_0x_1 + x_0x_2 + x_0x_3 + x_0x_4 + x_1x_2 + x_1x_3 + x_1x_4$$
$$+ x_2x_4 + x_0x_1x_3 + x_0x_1x_4 + x_0x_2x_3 + x_0x_3x_4 + x_1x_2x_4.$$

For the span n sequence with parameters $d = 1$, $p(x) = 1 + x + x^2 + x^4 + x^5$, $(r_1, r_2, r_3, r_4, r_5) = (1, 2, 3, 4, 5)$ in Table 4, the above recurrence relation can be written as

$$a_{7+k} = a_k + a_{1+k} + a_{4+k} + a_{1+k}a_{2+k} + a_{1+k}a_{3+k} + a_{1+k}a_{4+k} + a_{1+k}a_{5+k}$$
$$+ a_{2+k}a_{3+k} + a_{2+k}a_{4+a} + a_{2+k}a_{5+k} + a_{3+k}a_{5+k} + a_{1+k}a_{2+k}a_{4+k}$$
$$+ a_{1+k}a_{2+k}a_{5+k} + a_{1+k}a_{3+k}a_{4+k} + a_{1+k}a_{4+k}a_{5+k} + a_{2+k}a_{3+k}a_{5+k},$$
$$a_k \in \mathbb{F}_2, k \geq 0.$$

Table 4 Span n sequences generated using WG5 for $n = 7$

Decimation	Polynomial	t-tap position
By recurrence relation (4)		
d	$(c_0, c_1, c_2, c_3, c_4)$	$(r_1, r_2, r_3, r_4, r_5)$
1	1 1 1 0 1	1 2 3 4 5
1	1 1 0 1 1	1 3 4 5 6
7	1 0 0 1 0	1 2 3 4 6
7	1 0 1 0 0	1 2 4 5 6
7	1 0 1 1 1	2 3 4 5 6
11	1 0 0 1 0	1 2 4 5 6
11	1 1 1 1 0	1 2 4 5 6
11	1 1 1 0 1	1 2 4 5 6
15	1 1 1 1 0	1 2 4 5 6
By recurrence relation (5)		
1	1 1 1 1 0	1 2 3 4 5
1	1 1 1 0 1	1 3 4 5 6
1	1 0 1 0 0	1 3 4 5 6
7	1 0 1 1 1	1 2 3 4 5
7	1 0 1 0 0	1 2 3 4 5
7	1 1 0 1 1	1 2 3 5 6
15	1 1 1 1 0	1 2 3 4 5

The above generates the following span n sequence of period $2^7 - 1$

1111111000111001000100000110110000001001011011101011100001011110110101011001010000111100110001010100100111110100110100011001110.

For $n = 7$, all the span n sequences produced by recurrence relations (4) and (5) are presented in Table 4.

4.2 Approximate Number of Functions in the Search Space

Note that three parameters, namely, a decimation number d, a primitive polynomial $p(x)$, and a t-tap position, determine a nonlinear recurrence relation or a feedback function that may generate a span n sequence. In other words, each feedback function can be considered as a candidate span n sequence. For a fixed value of n and t, a search space is formed by including all possible combinations of these three parameters. In order to find span n sequences, an exhaustive search is performed over this search space. We determine the size of the search space or the number of candidate span n sequences in terms of n and t in the following proposition.

Proposition 3 *For any $n > t \geq 6$, the number of feedback functions in the search space of recurrence relations (4) and (5) is given by $C = \left(\frac{\phi(2^t-1)}{t}\right)^2 \binom{n-1}{t}$ if $|D_t^*| = \frac{\phi(2^t-1)}{t}$.*

Proof As in the recurrence relations, the first position is fixed for the sequence to be periodic, and any t-tap position is chosen from $n - 1$ positions ($n \geq 6$) to form a t-tap position; the number of distinct t-tap positions is given by $T = \binom{n-1}{t}$. Again, the total number of nonlinear feedback functions is given by $n_p \cdot |D_t^*|$, where $n_p = \frac{\phi(2^t-1)}{t}$ is the number of t degree primitive polynomials over $\mathbb{F}_2$ and $|D_t^*|$ is the number of decimation numbers for which the feedback function is nonlinear. Hence, for fixed n and t, the number of feedback functions in the search space is

$$C = n_p \cdot |D_t^*| \cdot T = \left(\frac{\phi(2^t-1)}{t}\right)^2 \binom{n-1}{t} \text{ if } |D_t^*| = \frac{\phi(2^t-1)}{t}.$$

$\square$

Proposition 4 *A feedback shift register defined by recurrence relations (4) and (5) produces the maximum number of span n sequences when about half the length of the shift register tap positions participate in the feedback functions.*

Proof Without loss generality, we assume that the number of terms in a feedback function is even. In a feedback shift register, the feedback functions are different for different t-tap positions. Thus, for a particular value of n and t and for a feedback function in t variables, the number of different feedback functions in n variables is equal to $N_{n,t} = \binom{n-1}{t}$ and $N_{n,t}$ is maximum when $t = \lceil \frac{n}{2} \rceil$ (for linear feedback functions, t is always odd and $t \approx \lceil \frac{n}{2} \rceil$). If the feedback functions in n variables that are candidate span n sequences are uniformly distributed over the set of all Boolean functions, then the FSR generates the maximum number of span n sequences when $t \approx \lceil \frac{n}{2} \rceil$. Hence, the assertion is established. $\square$

We note that an LFSR also produces the maximum number of span n sequences when $t \approx \lceil \frac{n}{2} \rceil$ (see Table 20). This property is also satisfied by the nonlinearly generated span n sequences using recurrence relations (4) and (5) (see Tables 6, 7, 8, 9, 10, 11, and 12). We now estimate the number of feedback functions in the search space for finding the maximum number of span n sequences. Assume that we use NLFSRs defined by recurrence relations (4) and (5) for $t = \lceil \frac{n}{2} \rceil$. Let N denote the number of span n sequences (including reverse span n sequences) obtained by recurrence relations (4) and (5). Then we have the following theorem.

Theorem 1 *An approximate number of candidate span n sequences or feedback functions in recurrence relations (4) and (5) is given by C_0, where $C_0 \approx \left(\frac{\phi(2^{\lceil \frac{n}{2} \rceil}-1)}{\lceil \frac{n}{2} \rceil}\right)^2 \cdot \frac{2^{n-1}}{\sqrt{\pi \cdot \frac{n-1}{2}}}$ and $C_0 \approx \frac{2^{2n-1}-2^{\frac{3n}{2}+1}}{\sqrt{\pi} \cdot (\lceil \frac{n}{2} \rceil)^{5/2}}$, if $2^t - 1$ is a Mersenne prime, and the success probability of obtaining such a span n sequence is given by $\frac{N}{C_0}$.*

Proof We recall that the size of the search space is

$$C = \left(\frac{\phi(2^t - 1)}{t}\right)^2 \binom{n-1}{t}, \quad \text{for } |D_t^*| = \frac{\phi(2^t - 1)}{t}.$$

Putting $t = \lceil \frac{n}{2} \rceil$ in the above formula, then we get

$$C_0 = \left(\frac{\phi(2^{\lceil \frac{n}{2} \rceil} - 1)}{\lceil \frac{n}{2} \rceil}\right)^2 \cdot \binom{n-1}{\lceil \frac{n}{2} \rceil}$$

$$= \left(\frac{\phi(2^{\lceil \frac{n}{2} \rceil} - 1)}{\lceil \frac{n}{2} \rceil}\right)^2 \cdot \binom{n-1}{\lfloor \frac{n-1}{2} \rfloor + 1}, \quad \text{for positive } n$$

$$= \left(\frac{\phi(2^{\lceil \frac{n}{2} \rceil} - 1)}{\lceil \frac{n}{2} \rceil}\right)^2 \cdot \frac{(n - \lfloor \frac{n-1}{2} \rfloor - 1) \cdot \binom{n-1}{\lfloor \frac{n-1}{2} \rfloor}}{(\lfloor \frac{n-1}{2} \rfloor + 1)}.$$

By Stirling's formula

$$\binom{m}{\lfloor \frac{m}{2} \rfloor} \sim \frac{2^m}{\sqrt{\pi m/2}},$$

the above equation can be written as

$$C_0 \sim \left(\frac{\phi(2^{\lceil \frac{n}{2} \rceil} - 1)}{\lceil \frac{n}{2} \rceil}\right)^2 \cdot \frac{\lfloor \frac{n-1}{2} \rfloor \cdot 2^{n-1}}{(\lfloor \frac{n-1}{2} \rfloor + 1) \cdot \sqrt{\pi \cdot \frac{n-1}{2}}}$$

$$\sim \left(\frac{\phi(2^{\lceil \frac{n}{2} \rceil} - 1)}{\lceil \frac{n}{2} \rceil}\right)^2 \cdot \frac{2^{n-1}}{\sqrt{\pi \cdot \frac{n-1}{2}}}.$$

$$\approx \frac{2^{2n-1} - 2^{\frac{3n}{2}+1}}{\sqrt{\pi} \cdot (\lceil \frac{n}{2} \rceil)^{5/2}}, \quad \text{if } 2^t - 1 \text{ is a Mersenne prime.}$$

Thus, the success probability of obtaining a span n sequence is equal to $\frac{N}{C_0}$. Hence, the result is proved. $\square$

5 Experimental Results on Span n Sequence Generation Using WG Transformations

In this section, we report the number of new span n sequences generated using WG transformations. We also present a heuristic method for searching WG span n sequences of long length. Table 5 provides a summary of the list of orthogonal functions used to produce span n sequences.

5.1 WG Span n Sequences

WG span n sequences are obtained by putting the WG transformation in recurrence relations (4) and (5) for different t and n. The span n sequences are generated by computer simulations. We consider the WG transformations over the field $\mathbb{F}_{2^t}$ for $t = 5, 7, 8, 10$, and 11. We denote by WG-t the WG transformations over the field $\mathbb{F}_{2^t}$. Table 6 presents the number of new span n sequences (new reverse span n sequences are not taken into account) produced by recurrence relations (4) and (5) for $6 \leq n \leq 20$. However, this method can be applied to generate span n sequences of long length. In Table 6, "×" denotes the recurrence relations that are not defined for such values of n and t, and $\sim$ represents those cases wherein the number of span n sequences is not yet determined. We present some instances of new span n sequences in the Appendix and all span n sequences in http://www. comsec.uwaterloo.ca/~kmandal/WG-Span-n/index.html.

A graphical representation of the number of new span n sequences is provided in Fig. 1, which shows that for different t the distribution of the number of new span n sequences has the following property: the number of span n sequences increases as n increases, and it reaches the maximum for some value of n, and thereafter the number of span n sequences decreases as n increases. At a quick glance, we can observe that the number of span n sequences is maximal close to $n = 2t$, which follows from the fact that the size of the search space is a multiple of a binomial coefficient (see Proposition 4). This fact reveals that there exists a trade off between n and t for obtaining the maximum number of span n sequences.

Table 5 Orthogonal functions used in the structured search

Parameter t	Orthogonal functions
$t = 5$	T1, T3
$t = 7$	T1, T3, T5, WG, Hall, QR
$t = 8$	T5, WG, GMW
$t = 9$	T1, T3, GMW, Segre, Glynn 1
$t = 10$	T1, T5, WG, GMWi, $i = 1 \ldots 7$
$t = 11$	T1, T3, T5, WG, Segre, Glynn 1, Glynn 2, B3

Table 6 Number of WG span n sequences

By recurrence relation (4)

t	n 6	7	8	9	10	11	12	13	14	15	16	17	18	19	20
5	0	9	7	14	8	11	17	11	13	10	3	7	7	0	1
7	×	×	3	25	42	63	108	138	138	125	126	111	83	86	63
8	×	×	×	3	9	18	34	76	96	104	106	108	110	90	79
10	×	×	×	×	×	5	40	107	246	373	627	819	999	~	~
11	×	×	×	×	×	×	31	204	574	1313	2539	4079	~	~	~
Total	0	9	10	42	59	97	230	536	1067	1925	3401	5124	–	–	–

By recurrence relation (5)

t	n 6	7	8	9	10	11	12	13	14	15	16	17	18	19	20
5	1	7	7	10	16	18	10	8	4	10	2	1	3	1	0
7	×	×	4	25	47	59	121	122	137	125	123	98	74	84	54
8	×	×	×	1	6	35	33	75	73	91	123	115	106	99	77
10	×	×	×	×	×	4	47	118	270	401	680	863	~	~	~
11	×	×	×	×	×	×	33	186	576	1350	2522	4010	~	~	~
Total	1	7	11	36	69	116	244	509	1060	1977	3450	5087	–	–	–

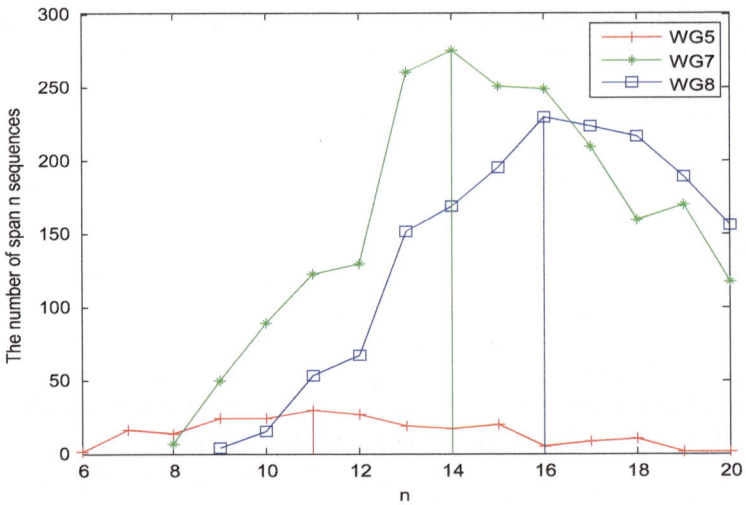

Fig. 1 Distribution of the number of span n sequences

Remark 1 There exist many span n sequences whose t-tap positions and the bases of the finite fields are the same, but their decimation numbers are different.

5.2 The Search Complexity Reduction for WG Span n Sequences

It is worth noticing that as t increases, the number of feedback functions in the search space increases exponentially. For large t, it is hard to find span n sequences by considering all functions in the search space. Thus, for large n and t, a search in a restricted search space can be performed to find span n sequences by imposing restrictions over decimation numbers and t-tap positions. Below we list a type of decimation numbers and t-tap positions that are observed for WG span n sequences. In some cases, we may not find any span n sequence. However, according to our observations, it is possible to obtain many span n sequences.

5.2.1 Observations on Decimation Numbers

We have performed a search on the following type of decimation numbers for different n

$$D_{\text{dec}} = \{d \,:\, d \in D_t^* \text{ and } d = 2^i - 1, i = 1, 2, \ldots, t - 1\}$$

for $t = 7$, 8, and 10, and the result shows that there exist many span n sequences whose decimation numbers in the recurrence relations (4) and (5) are of the above type. For this type of decimation numbers in the recurrence relations, the size of the search space is given by

$$C_{\text{dec}} = \frac{\phi(2^t - 1)}{t}(t - 1)\binom{n - 1}{t} \approx \phi(2^t - 1)\binom{n - 1}{t}.$$

Obviously, the reduced complexity C_{dec} is less than the original complexity C.

5.2.2 Observations on t-Tap Positions

Likewise, a search in the search space can be performed according to some pattern of t-tap positions for finding long period span n sequences. Assume that it is possible to fix, say, k tap positions ($1 \leq k \leq t$). Then, the total number of fixed tap positions in the recurrence relations is $(k + 1)$, and we only need to choose $(t - k)$ positions out of $(n - 1 - k)$ positions. So, for k fixed choices of tap positions, the search complexity is

$$C_{\text{tap}} = \left(\frac{\phi(2^t - 1)}{t}\right)^2 \binom{n - 1 - k}{t - k}.$$

Based on our observations on the t-tap positions for $t = 7, 8$, and 10, the following types of t-tap positions are effective when the slope of the curves in Fig. 1 increases gradually. For example, when $t = 7$, $n = 11, 12, 13$, and 14 and $t = 8$, $n = 13, 14, 15, 16, 17$, and 18, the t-tap positions are given by: $\{1, 2, 3, 4, \ldots\}$, $\{1, 2, 3, \ldots, n - 1\}$, $\{1, 2, \ldots, n - 2, n - 1\}$, $\{1, \ldots, n - 3, n - 2, n - 1\}$, where the numbers in the tap positions represent fixed positions in the t-tap positions (i.e., $k = 4$ fixed positions) and "$\ldots$" represents a combination of $(n - k - 1)$ tap positions. We performed a search according to the first pattern of t-tap position; the following span n sequence generated by a WG transformation has been found for $t = 13$ and $n = 24$.

Decimation d	Polynomial $(c_0, c_1, c_2, \ldots, c_{11}, c_{12})$	t-tap position $(r_1, r_2, \ldots, r_{12}, r_{13})$
1207	$(1, 0, 1, 0, 1, 0, 1, 0, 1, 0, 1, 0, 0)$	$(1, 2, 3, 4, 5, 6, 7, 10, 11, 12, 13, 15, 22)$

6 Experimental Results on Span n Sequences Generated by Other Orthogonal Functions

This section reports the number of span n sequences produced using three-term, five-term, monomial, Hall, quadratic residue, Glynn, Segre, GMW, and Kasami power functions. Explicit representations of these function are provided in Tables 1, 2, and 3.

6.1 Three-Term and Five-Term and Monomial Span n Sequences

Considering three-term and five-term functions in recurrence relations (4) and (5), a number of span n sequences can be obtained by the structured search. Tables 7 and 8 present the number of span n sequences for three-term functions and five-term functions, respectively. When $t = 5$, three-term functions and five-term functions degenerate to the same functions, as a result, the number of span n sequences obtained by three-term functions and five-term function are the same.

Table 9 presents the number of span n sequences produced using monomial functions for $6 \leq n \leq 20$. In tables, $\times$ denotes that the recurrence relation is not defined by the parameters t and n, and $\sim$ denotes that the cases are incomplete due to a huge number of functions in the search space. When $t = 5$, the WG transformations and monomial functions degenerate to the same functions.

Table 7 Number of three-term span *n* sequences

By recurrence relation (4)

t	*n*											
	6	7	8	9	10	11	12	13	14	15	16	17
5	1	3	9	8	9	8	4	3	5	2	3	1
7	×	×	6	25	51	89	103	150	131	128	127	123
9	×	×	×	×	8	52	104	223	391	549	710	770
11	×	×	×	×	×	×	35	190	624	1323	2580	4056
Total	1	3	15	33	68	149	246	566	1151	2002	3420	4950

By recurrence relation (5)

t	6	7	8	9	10	11	12	13	14	15	16	17
5	1	2	2	5	10	5	6	5	3	1	3	5
7	×	×	4	24	44	84	98	122	133	146	128	111
9	×	×	×	×	12	47	109	237	361	553	694	823
11	×	×	×	×	×	×	34	186	578	1416	2554	4007
Total	1	3	6	29	66	136	247	550	1075	2116	3379	4946

Table 8 Number of five-term span *n* sequences

By recurrence relation (4)

t	*n*													
	6	7	8	9	10	11	12	13	14	15	16	17	18	19
5	1	3	9	8	9	8	4	4	5	2	3	1	0	1
7	×	×	5	22	44	66	118	131	115	135	124	118	99	90
8	×	×	×	1	9	18	37	56	88	101	104	86	92	90
10	×	×	×	×	×	9	37	116	246	411	621	797	943	~
11	×	×	×	×	×	×	25	171	590	1443	2618	4194	~	~
Total	1	3	14	31	62	101	221	478	1044	2092	3470	5196	–	–

By recurrence relation (5)

t	6	7	8	9	10	11	12	13	14	15	16	17	18	19
5	1	2	2	5	10	5	6	5	3	1	3	5	0	1
7	×	×	8	19	43	74	108	138	138	127	117	102	84	91
8	×	×	×	0	6	22	38	54	66	116	89	106	83	93
10	×	×	×	×	×	7	47	119	223	443	627	861	~	~
11	×	×	×	×	×	×	20	172	609	1397	2558	4062	~	~
Total	1	2	10	24	59	108	219	488	1039	2084	3394	5136	–	–

6.2 Hall, QR, Segre, Glynn, and GMW Span n Sequences

In this section, we present the number of span *n* sequences produced by Hall, QR, Segre, Glynn, and GMW functions for $7 \leq n \leq 20$. We use the functions defined in Tables 1, 2, and 3 for Hall, quadratic residue, Glynn, Segre, and GMW functions in recurrence relations (4) and (5).

Table 9 Number of span n sequences generated by monomial functions

By recurrence relation (4)

t	n 6	7	8	9	10	11	12	13	14	15	16	17	18	19
5	0	9	7	14	8	11	17	11	13	10	3	7	7	0
7	×	×	6	17	41	76	79	118	108	99	125	78	88	72
9	×	×	×	×	10	43	120	258	410	519	662	788	~	~
11	×	×	×	×	×	×	26	188	604	1423	2491	4056	~	~
Total	0	9	13	31	59	130	242	575	1135	2051	3281	4929	–	–

By recurrence relation (5)

t	6	7	8	9	10	11	12	13	14	15	16	17	18	19
5	1	7	7	10	16	18	10	8	4	10	2	1	3	1
7	×	×	4	25	45	60	98	117	114	104	116	96	86	77
9	×	×	×	×	6	37	131	239	367	558	740	860	~	~
11	×	×	×	×	×	×	32	184	596	1403	2547	4074	~	~
Total	1	7	11	35	67	115	271	548	1081	2075	3405	5031	–	–

Table 10 Number of span n sequences generated by Hall functions and QR functions

By recurrence relation (4)

t		n 8	9	10	11	12	13	14	15	16	17	18	19	20
7	Hall	2	9	19	21	41	38	35	45	28	34	30	30	–
7	QR	0	4	4	5	14	27	16	9	18	14	12	6	6

By recurrence relation (5)

t		8	9	10	11	12	13	14	15	16	17	18	19	20
7	Hall	1	6	20	25	37	48	36	44	46	24	39	–	–
7	QR	0	3	6	7	13	12	13	18	16	13	14	10	8

For the range $7 \leq t \leq 11$, the Hall and QR functions with trace representations exist only for $t = 7$. Table 10 presents the number of span n sequences produced using recurrence relations (4) and (5) with Hall and QR functions for $8 \leq n \leq 20$. When all the decimated QR functions are considered, the class of 18 QR functions degenerates to two distinct QR orthogonal functions, and similarly, the class of 18 Hall functions degenerates to six distinct Hall orthogonal functions. Due to this reason, the number of span n sequences in Table 10 is smaller compared to other cases for $n = 7$.

When all the decimations are considered, Glynn 1 functions and Glynn 2 functions over $\mathbb{F}_{2^9}$ degenerate to the same class of orthogonal functions. Therefore, the number of span n sequences for Glynn 1 and Glynn 2 functions are the same in the structured search. However, for $t = 11$, the Glynn 1 class of functions and Glynn 2 class of functions are different. We provide the number of span n sequences produced by Glynn functions in Table 11, which also contains the number of span n sequences generated by Segre functions for $t = 9$ and 11. In Tables 10, 11, and 12,

Table 11 Number of span n sequences generated by Segre and Glynn functions

By recurrence relation (4)

t	OF	n								
		10	11	12	13	14	15	16	17	18
9	Segre	15	51	131	245	418	528	706	783	–
11	Segre	×	×	34	172	586	1413	2564	–	–
9	Glynn 1	11	52	129	253	415	584	673	790	–
11	Glynn 1	×	×	28	177	587	1418	2553	–	–
11	Glynn 2	×	×	30	185	595	1320	2646	–	–

By recurrence relation (5)

t	OF-t	10	11	12	13	14	15	16	17	18
9	Segre	7	48	108	264	371	521	692	–	–
11	Segre	×	×	37	153	627	1372	–	–	–
9	Glynn 1	6	49	126	248	397	529	709	–	–
11	Glynn 1	×	×	26	185	562	1351	–	–	–
11	Glynn 2	×	×	28	183	598	1340	–	–	–

Table 12 Number of span n sequences generated by GMW functions

By recurrence relation (4)

t	OF	n										# of terms
		9	10	11	12	13	14	15	16	17	18	
8	GMW	1	11	13	50	75	71	99	97	117	78	4
9	GMW	×	15	45	128	223	382	–	–	–	–	3
10	GMW1	×	×	7	37	114	236	424	606	810	–	2
10	GMW2	×	×	6	51	97	247	405	–	–	–	2
10	GMW3	×	×	5	33	119	255	415	672	865	–	4
10	GMW4	×	×	7	36	110	248	405	–	–	–	4
10	GMW5	×	×	10	39	147	261	411	645	853	–	7
10	GMW6	×	×	5	39	113	234	440	654	816	–	8
10	GMW7	×	×	10	39	118	236	422	664	888	–	14

By recurrence relation (5)

t	OF	9	10	11	12	13	14	15	16	17	18	# of terms
8	GMW	1	5	21	45	77	80	90	107	116	111	4
9	GMW	×	11	44	140	247	414	559	716	–	–	3
10	GMW1	×	×	7	34	117	257	414	609	–	–	2
10	GMW2	×	×	8	41	126	243	409	–	–	–	2
10	GMW3	×	×	7	44	122	257	411	641	–	–	4
10	GMW4	×	×	4	35	130	257	424	–	–	–	4
10	GMW5	×	×	6	43	113	239	407	638	–	–	7
10	GMW6	×	×	2	42	113	247	455	630	–	–	8
10	GMW7	×	×	5	51	133	258	429	643	–	–	14

"−" denotes the computation for the number of span n sequences is in progress and will be finished soon.

Table 12 presents the number of span n sequences produced by GMW functions in the structured search for $t = 8, 9$, and 10 and $9 \leq n \leq 19$. For the GMW functions over $\mathbb{F}_{2^8}$ and $\mathbb{F}_{2^9}$, there exists only one class of GMW functions. On the other hand, for the GMW functions over $\mathbb{F}_{2^{10}}$, there exist total seven distinct classes of orthogonal GMW functions with different number of terms in the trace representation. GMW span n sequences with $9 \leq n \leq 18$ are generated using recurrence relations (4) and (5) with GMWi functions, $1 \leq i \leq 7$. In Table 12, the term "# of terms" denotes the number of terms in the trace representation of a GMW function.

Remark 2 For a class of orthogonal functions in recurrence relations (4) and (5), each span n sequence is uniquely determined by a decimation number, a primitive polynomial, and a t-tap position. Unfortunately, we could not find any relation among these three parameters.

7   The Success Probability Comparison

In this section, an empirical success probability of obtaining a span n sequence using a orthogonal feedback function is presented. Note that the success probability of obtaining a randomly generated span n sequence is $\frac{1}{2^{n-3}}$ [33], where a random span n sequence is generated by randomly choosing a feedback function from the set of all Boolean functions in n variables and checking the condition for a span n sequence.

We compared the success probability of obtaining a span n sequence using WG transformations (including reverse sequences) in the structured search with a random span n sequence generation method for $t = 5, 7, 8$ (for $t \approx \lceil \frac{n}{2} \rceil$), 10, and 11 (for $13 \leq n \leq 17$), and the comparison shows that in the structured search, one can produce a span n sequence with a better success probability than that of a random span n sequence generation method. A comparison of success probability for $t = 5, 7$, and 8 is provided in Table 13. Furthermore, we compared the success probability of obtaining a span n sequences using three-term, five-term, and monomial functions in Table 13 for $t = 5, 7, 8, 9$. Table 13 illustrates that a span n sequence can be produced using any of three-term, five-term, and monomial functions with a better success probability. Our empirical comparisons also show that the success probability of obtaining a span n sequence using Hall, QR, Segre, Glynn, and GMW functions is greater than that of a random span n sequence generation method. We don't provide the success probability values due to the large number of cases.

Table 13 The success probability comparison for WG, three-term, five-term, and monomial span n sequences

WG span n sequences			
	$n = 2t$	Our approach	Randomly chosen
WG-5	10	$\frac{1}{2^{6.56}}$	$\frac{1}{2^7}$
WG-7	14	$\frac{1}{2^{9.98}}$	$\frac{1}{2^{11}}$
WG-8	16	$\frac{1}{2^{11.81}}$	$\frac{1}{2^{13}}$

Three-term span n sequences			
	$n \approx 2t$	Our approach	Randomly chosen
T3-5	10	$\frac{1}{2^{6.89}}$	$\frac{1}{2^7}$
T3-7	14	$\frac{1}{2^{10.04}}$	$\frac{1}{2^{11}}$
T3-9	17	$\frac{1}{2^{13.04}}$	$\frac{1}{2^{14}}$

Five-term span n sequences			
	$n = 2t$	Our approach	Randomly chosen
T5-5	10	$\frac{1}{2^{6.89}}$	$\frac{1}{2^7}$
T5-7	14	$\frac{1}{2^{10.10}}$	$\frac{1}{2^{11}}$
T5-8	16	$\frac{1}{2^{12.02}}$	$\frac{1}{2^{13}}$

Monomial span n sequences			
	$n \approx 2t$	Our approach	Randomly chosen
T1-5	10	$\frac{1}{2^{6.88}}$	$\frac{1}{2^7}$
T1-7	14	$\frac{1}{2^{10.29}}$	$\frac{1}{2^{11}}$
T1-9	17	$\frac{1}{2^{12.96}}$	$\frac{1}{2^{14}}$

8   Linear Span of New Span n Sequences

In this section, we analyze the linear span of new span n sequences produced by orthogonal functions and present two conjectures on linear span of span n sequences produced by orthogonal functions.

We study the linear span of new span n sequences generated using orthogonal functions. The linear span of a sequence is an important randomness property that is considered as an upper bound on sequence unpredictability because using only twice-linear span consecutive bits one can certainly predict the remaining bits of the sequence by the Berlekamp–Massey algorithm [2, 31]. Sequences with optimal linear complexity are of practical interests, since an attacker requires the whole sequence to decrypt the message in a stream cipher. There is no theoretical result on the linear span of span n sequences generated by a nonlinear feedback shift register. What we know is the bounds presented in Property 1 in Sect. 2.

We compute the linear span of new span n sequences by the Berlekamp–Massey algorithm, and our computational results show that the linear span of a new sequence lies in the range of $(2^n - 2 - 3n)$ (near optimal) and $(2^n - 2)$ (optimal). Table 14 presents a summary of the linear spans of WG span n sequences generated by the recurrence relations (4) and (5), respectively. Moreover, Tables 15, 16, and 17 exhibit a summary of the linear spans of the span n sequences generated by

Table 14 The bounds of the linear span of WG span n sequences

Range on n	t	Upper bound of LS	Lower bound of LS
By recurrence relation (4)			
$7 \leq n \leq 20$	5	$2^n - 2$	$2^n - 2 - 2n$
$8 \leq n \leq 20$	7	$2^n - 2$	$2^n - 2 - 2n$
$9 \leq n \leq 20$	8	$2^n - 2$	$2^n - 2 - 3n$
$11 \leq n \leq 17$	10	$2^n - 2$	$2^n - 2 - 3n$
$12 \leq n \leq 17$	11	$2^n - 2$	$2^n - 2 - 2n$
By recurrence relation (5)			
$7 \leq n \leq 20$	5	$2^n - 2$	$2^n - 2 - 2n$
$8 \leq n \leq 20$	7	$2^n - 2$	$2^n - 2 - 3n$
$9 \leq n \leq 20$	8	$2^n - 2$	$2^n - 2 - 3n$
$11 \leq n \leq 17$	10	$2^n - 2$	$2^n - 2 - 3n$
$12 \leq n \leq 16$	11	$2^n - 2$	$2^n - 2 - 3n$

Table 15 The bounds of the linear span of monomial span n sequences

Range on n	t	Upper bound of LS	Lower bound of LS
By recurrence relation (4)			
$7 \leq n \leq 19$	5	$2^n - 2$	$2^n - 2 - 2n$
$8 \leq n \leq 19$	7	$2^n - 2$	$2^n - 2 - 3n$
$8 \leq n \leq 17$	9	$2^n - 2$	$2^n - 2 - 3n$
$12 \leq n \leq 16$	11	$2^n - 2$	$2^n - 2 - 3n$
By recurrence relation (5)			
$7 \leq n \leq 19$	5	$2^n - 2$	$2^n - 2 - 2n$
$8 \leq n \leq 19$	7	$2^n - 2$	$2^n - 2 - 3n$
$8 \leq n \leq 17$	9	$2^n - 2$	$2^n - 2 - 3n$
$12 \leq n \leq 16$	11	$2^n - 2$	$2^n - 2 - 3n$

Table 16 The bounds of the linear span of three-term span n sequences

Range on n	t	Upper bound of LS	Lower bound of LS
By recurrence relation (4)			
$7 \leq n \leq 17$	5	$2^n - 2$	$2^n - 2 - 2n$
$8 \leq n \leq 17$	7	$2^n - 2$	$2^n - 2 - 3n$
$8 \leq n \leq 17$	9	$2^n - 2$	$2^n - 2 - 3n$
$12 \leq n \leq 17$	11	$2^n - 2$	$2^n - 2 - 3n$
By recurrence relation (5)			
$7 \leq n \leq 17$	5	$2^n - 2$	$2^n - 2 - 2n$
$8 \leq n \leq 17$	7	$2^n - 2$	$2^n - 2 - 2n$
$8 \leq n \leq 17$	9	$2^n - 2$	$2^n - 2 - 3n$
$12 \leq n \leq 17$	11	$2^n - 2$	$2^n - 2 - 2n$

monomial functions, three-term functions, and five-term functions, respectively, for different values of t, and Table 18 presents a summary of the linear span of span n sequences produced by other orthogonal functions. Our computational results also show that most of new sequences obtain the optimal linear span ($2^n - 2$), only very

Table 17 The bounds of the linear span of five-term span n sequences

Range on n	t	Upper bound of LS	Lower bound of LS
By recurrence relation (4)			
$7 \le n \le 19$	5	$2^n - 2$	$2^n - 2 - 2n$
$8 \le n \le 19$	7	$2^n - 2$	$2^n - 2 - 2n$
$9 \le n \le 19$	8	$2^n - 2$	$2^n - 2 - 3n$
$11 \le n \le 17$	10	$2^n - 2$	$2^n - 2 - 3n$
$12 \le n \le 16$	11	$2^n - 2$	$2^n - 2 - 2n$
By recurrence relation (5)			
$7 \le n \le 20$	5	$2^n - 2$	$2^n - 2 - 2n$
$8 \le n \le 20$	7	$2^n - 2$	$2^n - 2 - 3n$
$9 \le n \le 20$	8	$2^n - 2$	$2^n - 2 - 3n$
$11 \le n \le 17$	10	$2^n - 2$	$2^n - 2 - 2n$
$12 \le n \le 16$	11	$2^n - 2$	$2^n - 2 - 3n$

Table 18 The upper and lower bounds of the linear span of Hall, QR, GMW, Segre, and Glynn span n sequences

By recurrence relations (4) and (5)				
t	Function	Range on n	Upper bound	Lower bound
7	Hall	$8 \le n \le 19$	$2^n - 2$	$2^n - 2 - 2n$
	QR	$8 \le n \le 20$	$2^n - 2$	$2^n - 2 - 3n$
8	GMW	$9 \le n \le 18$	$2^n - 2$	$2^n - 2 - 2n$
9	Segre	$10 \le n \le 16$	$2^n - 2$	$2^n - 2 - 3n$
	Glynn	$10 \le n \le 16$	$2^n - 2$	$2^n - 2 - 3n$
	GMW	$10 \le n \le 16$	$2^n - 2$	$2^n - 2 - 3n$
10	GMW	$11 \le n \le 17$	$2^n - 2$	$2^n - 2 - 3n$
11	Segre	$12 \le n \le 16$	$2^n - 2$	$2^n - 2 - 3n$
	Glynn	$12 \le n \le 16$	$2^n - 2$	$> 2^n - 2 - 3n$

few span n sequences obtain the linear span ($2^n - 2 - 3n$), and in some cases all the linear spans are greater than ($2^n - 2 - 3n$).

Based on our observation on the linear span of new span n sequences produced by orthogonal functions, we have the following two conjectures. These two conjectures are valid and verified by our computational results for $n \le 20$.

Conjecture 1 Let the function g be an orthogonal function and $\mathbf{s} = \{s_i\}$ be a binary sequence generated by an n-stage NLFSR with $n > m$ whose feedback function is given by

$$f(x_0, x_1, \ldots, x_{n-1}) = c \oplus x_0 \oplus g(y)$$

where $c = 0/1$ and $y = (x_{r_1}, x_{r_2}, \ldots, x_{r_m})$, $y \in \mathbb{F}_{2^m}$, and $0 < r_1 < r_2 < \cdots < r_m < n$. If $\mathbf{s}$ or $\bar{\mathbf{s}}$ is a span n sequence, then the linear span of $\mathbf{s}$, denoted as LS_s, is bounded by

$$(2^n - 2 - 3n) \le LS_s \le (2^n - 2).$$

Conjecture 2 For a prime length of an NLFSR, the linear span of a span n sequence produced by the above feedback function with an orthogonal function takes one of the following three values $\{2^n - 2 - 2n, 2^n - 2 - n, 2^n - 2\}$.

9 Applications

Our span n sequences and span n sequences produced by the structured search in this chapter can be used in the following scenarios. In [28], Mandal and Gong analyzed the composited construction based on a span n sequence for generating long and strong de Bruijn sequences. Based on their analysis, the span n sequence to be used in the construction must have high linear span in order to produce strong de Bruijn sequences. Since our span sequences have optimal or near-optimal linear span, these span n sequences can be used in the composited construction for producing long and strong de Bruijn sequences. Mandal et al. [30] designed **Warbler**, a pseudorandom number generator for EPC C1 Gen2 RFID tags using NLFSRs where two span n sequences with optimal linear span are used to promise the randomness properties such as period and linear span of an output sequence. Our span n sequences or span n sequences produced by the structured search can be used to design lightweight pseudorandom number generators and stream ciphers. Thus, our span n sequences have an immediate application in cryptography, which can be found in [28, 30].

Conclusion
In this chapter, we have studied the span n sequence generation using orthogonal functions and presented some theoretical results on generating span n sequences and experimental results about the number of span n sequences produced by orthogonal functions. We used all known and well-studied orthogonal functions as nonlinear feedback functions in an NLFSR for $5 \leq t \leq 11$ and presented the number of span n sequences produced using orthogonal functions for $6 \leq n \leq 20$. Finally, we analyzed the linear span of new span n sequences produced by the orthogonal functions and gave a summary of the bounds of the linear span for each class of span n sequences. Interestingly, the linear span of a new span n sequence lies between the near optimal ($2^n - 2 - 3n$) and optimal ($2^n - 2$). We observed that the majority of span n sequences have an optimal linear span. According to our study, it is possible to obtain span n sequences of high linear span with a better probability of success using orthogonal feedback functions.

Appendix: A Upper and Lower Bounds of Linear Span of Span n Sequences

We present the upper and lower bounds of the linear span of new span n sequences generated using orthogonal functions for different n and t and give all new span n sequences generated using WG transformations for $t = 5$ (Tables 14, 15, 16, 17, 18, 19, 20, 21, and 22). All new span n sequences generated using WG transformations with $t = 7, 8, 10$, and 11 can be found in http://www.comsec. uwaterloo.ca/~kmandal/WG-Span-n/index.html.

Table 19 Span n sequences generated using WG7

Length n	Decimation d	Polynomial $(c_0, c_1, \ldots, c_5, c_6)$	t-tap position $(r_1, r_2, \ldots, r_6, r_7)$
8	5	1 1 0 0 0 0 0	1 2 3 4 5 6 7
9	1	1 0 1 1 1 1 1	1 2 3 4 5 6 7
10	27	1 1 1 1 0 1 1	1 2 3 4 5 6 7
11	1	1 1 1 1 0 1 1	1 2 3 5 8 9 10
12	1	1 0 1 1 1 0 0	1 2 4 5 8 10 11
13	9	1 1 0 0 1 0 1	1 2 3 4 5 6 8
14	43	1 1 1 0 1 1 1	1 2 3 4 5 6 7
15	31	1 1 0 0 0 0 0	1 2 3 4 7 12 14
16	27	1 1 1 1 0 1 1	1 2 3 5 6 8 14
17	1	1 0 1 1 1 0 0	1 2 3 4 7 9 13
18	1	1 0 1 1 1 0 0	1 2 3 4 6 9 16
19	3	1 1 1 1 1 1 0	1 2 3 5 7 15 17
20	31	1 1 1 1 1 1 0	1 2 3 7 8 12 15

Table 20 Tap-position distribution for an LFSR of length ≤ 20

# of taps	5	6	7	8	9	10	11	12	13	14	15	16	17	18	19	20
2	2	2	4	–	2	2	2	–	–	–	6	–	6	2	–	2
4	4	4	10	12	16	20	44	18	66	42	82	52	152	72	158	100
6	–	–	4	4	28	28	80	86	236	226	470	368	1050	718	1774	1104
8	–	–	–	–	2	10	50	36	264	338	720	812	2674	2296	6696	4522
10	–	–	–	–	–	–	–	4	60	140	450	648	2696	2910	10238	8436
12	–	–	–	–	–	–	–	–	4	12	66	156	1006	1470	6766	7000
14	–	–	–	–	–	–	–	–	–	–	6	12	122	284	1772	2460
16	–	–	–	–	–	–	–	–	–	–	–	–	–	24	190	354
18	–	–	–	–	–	–	–	–	–	–	–	–	–	–	–	22

Table 21 WG span n sequences generated using rec. rel. (4)

n	Decimation d	Polynomial $(c_0, c_1, c_2, c_3, c_4)$	Tap position $(r_1, r_2, r_3, r_4, r_5)$
8	1	1 0 1 0 0	1 2 4 5 7
	1	1 1 1 1 0	1 3 4 5 6
	1	1 1 1 1 0	2 4 5 6 7
	3	1 1 0 1 1	1 2 3 5 6
	7	1 0 1 1 1	1 2 3 5 7
	7	1 0 1 0 0	2 3 4 6 7
	15	1 1 1 1 0	2 3 4 6 7
9	1	1 1 1 0 1	1 2 5 6 8
	1	1 1 1 0 1	1 3 6 7 8
	1	1 1 1 1 0	2 3 5 7 8
	1	1 1 1 0 1	4 5 6 7 8
	3	1 1 0 1 1	1 2 4 5 6
	3	1 0 1 0 0	1 2 4 5 8
	3	1 0 1 0 0	2 4 6 7 8
	7	1 0 1 0 0	1 2 3 4 6
	11	1 1 1 0 1	1 4 6 7 8
	11	1 1 1 1 0	2 4 5 6 7
	11	1 1 1 1 0	2 4 5 6 8
	11	1 1 1 0 1	2 4 6 7 8
	15	1 1 1 1 0	1 2 3 4 6
	15	1 1 1 0 1	1 2 5 7 8
10	1	1 1 0 1 1	1 2 4 5 8
	1	1 1 1 0 1	1 3 4 6 7
	1	1 1 1 0 1	1 3 4 6 9
	3	1 1 0 1 1	1 2 3 4 8
	7	1 0 0 1 0	1 2 4 7 8
	11	1 0 1 1 1	1 2 3 4 5
	11	1 0 0 1 0	1 2 3 7 8
	11	1 1 1 1 0	1 4 5 8 9
11	1	1 1 1 0 1	1 2 7 8 10
	1	1 1 1 1 0	3 4 5 8 10
	1	1 1 1 0 1	6 7 8 9 10
	7	1 0 1 1 1	1 2 3 6 7
	7	1 0 0 1 0	1 3 7 8 10
	7	1 0 1 1 1	2 3 4 7 10
	7	1 1 0 1 1	2 3 7 9 10
	7	1 0 0 1 0	2 4 5 6 10
	7	1 1 0 1 1	3 4 5 8 9
	11	1 1 1 1 0	1 2 4 5 8
	11	1 1 1 0 1	1 3 4 6 10

(continued)

Table 21 (continued)

n	Decimation d	Polynomial $(c_0, c_1, c_2, c_3, c_4)$	Tap position $(r_1, r_2, r_3, r_4, r_5)$
12	1	1 1 1 1 0	2 3 4 5 6
	1	1 0 1 0 0	2 3 4 5 8
	1	1 1 1 0 1	2 3 5 7 9
	1	1 0 1 0 0	2 3 6 9 10
	1	1 1 1 0 1	4 6 9 10 11
	3	1 1 0 1 1	1 2 3 4 5
	3	1 1 0 1 1	2 5 7 8 10
	3	1 0 1 0 0	4 5 6 9 11
	7	1 0 1 0 0	1 2 4 7 8
	7	1 1 0 1 1	1 2 5 6 8
	11	1 0 0 1 0	1 3 4 6 10
	11	1 1 1 0 1	1 3 4 9 11
	11	1 1 1 1 0	1 4 5 8 9
	11	1 1 1 0 1	2 3 6 7 10
	11	1 1 1 1 0	3 5 7 8 9
	11	1 1 1 1 0	4 6 7 9 10
	15	1 1 1 1 0	1 2 4 7 8

Table 22 WG span n sequences generated using rec. rel. (4)

n	Decimation d	Polynomial $(c_0, c_1, c_2, c_3, c_4)$	Tap position $(r_1, r_2, r_3, r_4, r_5)$
13	1	1 0 1 0 0	1 3 4 5 9
	1	1 0 1 0 0	5 8 9 11 12
	3	1 1 0 1 1	5 6 10 11 12
	7	1 0 1 0 0	1 2 3 6 8
	7	1 1 0 1 1	3 5 7 10 12
	7	1 1 0 1 1	6 7 9 10 12
	11	1 0 0 1 0	1 2 3 5 10
	11	1 1 1 0 1	1 2 5 10 12
	11	1 1 1 0 1	1 5 6 10 12
	11	1 1 1 0 1	4 5 7 8 9
	15	1 1 1 1 0	1 2 3 6 8
14	1	1 0 1 0 0	1 3 5 7 9
	1	1 1 1 1 0	2 6 8 9 13
	1	1 1 1 0 1	3 4 6 8 10
	1	1 1 1 0 1	3 5 8 10 13
	3	1 1 0 1 1	1 8 10 11 13

(continued)

Table 22 (continued)

n	Decimation d	Polynomial $(c_0, c_1, c_2, c_3, c_4)$	Tap position $(r_1, r_2, r_3, r_4, r_5)$
	7	1 0 0 1 0	1 2 6 9 12
	7	1 0 0 1 0	1 3 10 12 13
	7	1 0 0 1 0	1 6 9 12 13
	7	1 0 1 0 0	3 5 7 8 9
	11	1 1 1 1 0	1 2 4 11 12
	11	1 1 1 1 0	1 2 9 10 11
	15	1 1 1 0 1	3 5 6 8 13
	15	1 1 1 1 0	3 5 7 8 9
15	1	1 1 1 0 1	4 5 12 13 14
	3	1 0 1 0 0	2 6 8 9 10
	3	1 0 1 0 0	4 5 6 7 14
	7	1 0 1 1 1	2 5 7 10 13
	7	1 0 1 1 1	2 5 8 11 14
	7	1 0 0 1 0	3 4 5 7 12
	11	1 0 0 1 0	2 3 6 7 13
	11	1 1 1 0 1	2 4 9 11 13
	11	1 0 1 1 1	2 9 10 11 12
	15	1 1 1 0 1	1 2 3 5 6
16	1	1 1 0 1 1	1 10 11 12 14
	1	1 1 1 0 1	1 10 11 12 14
	15	1 1 1 0 1	3 6 9 12 14
17	3	1 0 1 0 0	1 6 7 8 9
	3	1 1 0 1 1	4 7 8 9 12
	7	1 0 1 0 0	1 3 12 13 14
	7	1 1 0 1 1	1 4 10 11 13
	7	1 0 0 1 0	1 5 11 12 13
	11	1 1 1 0 1	1 3 6 12 13
	15	1 1 1 1 0	1 3 12 13 14
18	1	1 1 1 0 1	1 2 12 13 14
	3	1 1 0 1 1	4 7 8 10 15
	3	1 1 0 1 1	5 10 11 14 17
	7	1 0 0 1 0	1 2 5 7 11
	7	1 1 0 1 1	5 7 8 11 17
	11	1 0 0 1 0	1 8 9 11 15
	15	1 1 1 0 1	2 9 12 15 17
20	1	1 1 1 0 1	5 10 12 18 19

References

1. F.S. Annexstein, Generating de Bruijn sequences: an efficient implementation. IEEE Trans. Inf. Theory **46**(2), 198–200 (1997)
2. E.R. Berlekamp, *Algebraic Coding Theory*, Ch. 7 (McGraw-Hill, New York, 1968)
3. A.H. Chan, R.A. Games, E.L. Key, On the complexities of de Bruijn sequences. J. Combin. Theory Ser. A **33**(3) 233–246 (1982)
4. A.H. Chan, R.A. Games, J.J. Rushanan, On quadratic m-sequences, in *IEEE International Symposium on Information Theory*, vol. 364 (1994)
5. A.C. Chang, S.W. Golomb, G. Gong, P.V. Kumar, On the linear span of ideal autocorrelation sequences arising from the Segre hyperoval, in *Sequences and their Applications—Proceedings of SETA'98, Discrete Mathematics and Theoretical Computer Science* (Springer, London, 1999)
6. T. Chang, B. Park, Y.H. Kim, I. Song, An efficient implementation of the D-homomorphism for generation of de Bruijn sequences. IEEE Trans. Inf. Theory **45**(4), 1280–1283 (1999)
7. C. De Canniére, O. Dunkelman, M. Knežević, KATAN and KTANTAN—a family of small and efficient hardware-oriented block ciphers. in *Proceedings of the 11th International Workshop on Cryptographic Hardware and Embedded Systems*, LNCS, vol. 5747 (Springer, Heidelberg, 2009). pp. 272–288
8. J. Dillon, H. Dobbertin, New cyclic difference sets with singer parameters. Finite Fields Appl. **10**(3), 342–389 (2004)
9. H. Dobbertin, Kasami power functions, permutation polynomials and cyclic difference sets, in *Proceedings of the NATO-A.S.I. Workshop Difference Sets, Sequences and their Correlation Properties*, (Kluwer, Bad Windsheim/Dordrecht, 1999), pp. 133–158
10. E. Dubrova, A list of maximum period NLFSRs. Report 2012/166, Cryptology ePrint Archive (2012), http://eprint.iacr.org/2012/166.pdf
11. eSTREAM: The ECRYPT stream cipher project. http://www.ecrypt.eu.org/stream/
12. T. Etzion, A. Lempel, Construction of de Bruijn sequences of minimal complexity. IEEE Trans. Inf. Theory **30**(5), 705–709 (1984)
13. R. Evan, H.D.L. Hollman, C. Krattenthaler, Q. Xiang, Gauss sums, Jacobi sums and p-ranks of cyclic difference sets. J. Combin. Theory Ser. A, **87**(1), 74–119 (1999)
14. H. Fredricksen, A class of nonlinear de Bruijn cycles. J. Combin. Theory Ser. A **19**(2), 192–199 (1975)
15. H. Fredricksen, A survey of full length nonlinear shift register cycle algorithms. SIAM Rev. **24**(2), 195–221 (1982)
16. H. Fredricksen, I. Kessler, Lexicographic compositions and de Bruijn sequences. J. Combin. Theory Ser. A **22**, 17–30 (1977)
17. H. Fredricksen, J. Maiorana, Necklaces of beads in k colors and k-ary de Bruijn sequences. Discrete Math. **23**(3), 207–210 (1978)
18. R.A. Games, A generalized recursive construction for de Bruijn sequences. IEEE Trans. Inf. Theory **29**(6), 843–850 (1983)
19. B.M. Gammel, R. Göttfert, O. Kniffler, Achterbahn-128/80 (2006), http://www.ecrypt.eu.org/stream/p2ciphers/achterbahn/achterbahn_p2.pdf
20. S.W. Golomb, *Shift Register Sequences* (Aegean Park Press, Laguna Hills, 1981)
21. S.W. Golomb, On the classification of balanced binary sequences of period $2^n - 1$. IEEE Trans. Inf. Theory, 26(6), 730–732 (1980)
22. S.W. Golomb, G. Gong, *Signal Design for Good Correlation: for Wireless Communication, Cryptography, and Radar* (Cambridge University Press, New York, 2004)
23. G. Gong, Randomness and representation of span n sequences, in *Proceedings of the 2007 International Conference on Sequences, Subsequences, and Consequences, SSC'07* (Springer, Heidelberg, 2007), pp. 192–203
24. E.R. Hauge, T. Helleseth, De Bruijn sequences, irreducible codes and cyclotomy. Discrete Math. **159**(1–3), 143–154 (1996)

25. C.J.A. Jansen, W.G. Franx, D.E. Boekee, An efficient algorithm for the generation of de Bruijn cycles. IEEE Trans. Inf. Theory **37**(5), 1475–1478 (1991)
26. A. Lempel, On a homomorphism of the de Bruijn graph and its applications to the design of feedback shift registers. IEEE Trans. Comput. **C-19**(12), 1204–1209 (1970)
27. K. Mandal, Design and analysis of cryptographic pseudorandom number/sequence generators with applications in RFID. Ph.D. Thesis, University of Waterloo, 2013
28. K. Mandal, G. Gong, in *Cryptographically Strong de Bruijn Sequences with Large Periods*, ed. by L.R. Knudsen, H. Wu SAC 2012. LNCS, vol. 7707 (Springer, Heidelberg, 2012), pp. 104–118
29. K. Mandal, G. Gong, Cryptographic D-morphic analysis and fast implementations of composited De Bruijn sequences. Technical Report CACR 2012–27, University of Waterloo (2012)
30. K. Mandal, X. Fan, G. Gong, in *Warbler: A Lightweight Pseudorandom Number Generator for EPC Class 1 Gen 2 RFID Tags*, ed. by N.W. Lo, Y. Li. Cryptology and Information Security Series—The 2012 Workshop on RFID and IoT Security (RFIDsec'12 Asia), vol. 8 (IOS Press, Amsterdam, 2012), pp. 73–84
31. J.L. Massey, Shift-register synthesis and BCH decoding. IEEE Trans. Inf. Theory **15**(1), 122–127 (1969)
32. G.L. Mayhew, Weight class distributions of de Bruijn sequences. Discrete Math. **126**, 425–429 (1994)
33. G.L. Mayhew, Clues to the hidden nature of de Bruijn sequences. Comput. Math. Appl., **39**(11), 57–65 (2000)
34. G.L. Mayhew, S.W. Golomb, Linear Spans of modified de Bruijn sequences. IEEE Trans. Inf. Theory **36**(5), 1166–1167 (1990)
35. G.L. Mayhew, S.W. Golomb, Characterizations of generators for modified de Bruijn sequences. Adv. Appl. Math. **13**, 454–461 (1992)
36. J. Mykkeltveit, M.-K. Siu, P. Tong, On the cycle structure of some nonlinear shift register sequences. Inf. Control **43**(2), 202–215 (1979)
37. J.L.-F. Ng, Binary nonlinear feedback shift register sequence generator using the trace function, Master's Thesis, University of Waterloo, 2005
38. J.S. No, S.W. Golomb, G. Gong, H.K. Lee, P. Gaal, New binary pseudorandom sequences of period $2^n - 1$ with ideal autocorrelation. IEEE Trans. Inf. Theory **44**(2), 814–817 (1998)
39. T. Rachwalik, J. Szmidt, R. Wicik, J. Zablocki, Generation of nonlinear feedback shift registers with special-purpose hardware. Cryptology ePrint Archive, Report 2012/314 (2012), http:// eprint.iacr.org/
40. J.-H. Yang, Z.-D. Dai, Construction of m-ary de Bruijn sequences (extended abstract), in *Advances in Cryptology—AUSCRYPT'92*, LNCS (Springer, Heidelberg, 1993), pp. 357–363

Open Problems on the Cross-correlation of m-Sequences

Tor Helleseth

Abstract Pseudorandom sequences are important for many applications in communication systems, in coding theory, and in the design of stream ciphers. Maximum-length linear sequences (or m-sequences) are popular in sequence designs due to their long period and excellent pseudorandom properties. In code-division multiple-access (CDMA) applications, there is a demand for large families of sequences having good correlation properties. The best families of sequences in these applications frequently use m-sequences in their constructions. Therefore, the problem of determining the correlation properties of m-sequences has received a lot of attention since the 1960s, and many interesting theoretical results of practical interest have been obtained. The cross-correlation of m-sequences is also related to other important problems, such as almost perfect nonlinear functions (APN) and almost bent functions (AB), and to the nonlinearity of S-boxes in many block ciphers including AES. This chapter gives an updated survey of the cross-correlation of m-sequences and describes some of the most important open problems that still remain in this area.

1 Introduction

Let $\{u_t\}$ and $\{u_t\}$ be sequences of period ε with symbols from the finite field GF(p) with p elements. Let ω be a primitive complex pth root of unity. The cross-correlation between the two sequences at shift τ is defined to be

$$C_{u,v}(\tau) = \sum_{t=0}^{\varepsilon-1} \omega^{u_{t+\tau}-v_t}.$$

If the two sequences are cyclically equivalent (i.e., only differ by a cyclic shift), the correlation is denoted autocorrelation instead of cross-correlation.

T. Helleseth (✉)
Department of Informatics, The Selmer Center, University of Bergen, Thormøhlensgate 55, 5008 Bergen, Norway
e-mail: Tor.Helleseth@ii.uib.no

© Springer International Publishing Switzerland 2014
Ç.K. Koç (ed.), *Open Problems in Mathematics and Computational Science*,
DOI 10.1007/978-3-319-10683-0_8

In a code-division multiple-access (CDMA) system, each user is assigned a sequence from a family of sequences. The quality of the communication depends on the selection of a family of sequences with good parameters.

Let $\mathscr{F}$ be a family of M cyclically distinct sequences of the same period ε:

$$\mathscr{F} = \{\{s_t^{(i)}\} \mid 1 \le i \le M\}.$$

The most important parameters for evaluating the quality of the family are $(M, \varepsilon, \theta_{\max})$ where $\theta_{\max}$ is the maximum value of the absolute magnitude of the (nontrivial) auto- and cross-correlation between any two sequences in the family, i.e.,

$$\theta_{\max} = \max\{|C_{s^{(i)},s^{(j)}}(\tau)| \mid i \ne j \text{ or } \tau \ne 0\}.$$

Many of the best sequence families can be constructed from linear recursions. To generate a sequence $\{s_t\}$ with symbols from GF(p), one can use a linear recursion of degree n and generate each symbol from the previous n symbols such that

$$s_{t+n} + c_{n-1}s_{t+n-1} + \cdots + c_0 s_t = 0, \quad c_i \in \text{GF(p)}, c_0 \ne 0.$$

The initial state $(s_0, s_1, \ldots, s_{n-1})$ and the linear recursion uniquely determine the sequence $\{s_t\}$. Thus, the linear recursion generates p^n distinct sequences corresponding to the p^n initial states $(s_0, s_1, \ldots, s_{n-1})$. Clearly, some of these generated sequences may be cyclically equivalent.

The characteristic polynomial of the linear recursion is defined to be

$$f(x) = \sum_{i=0}^{n} c_i x^i.$$

The period of the sequences generated by the recursion with characteristic polynomial $f(x)$ is completely determined by the polynomial. It is a well-known fact that all these sequences will have period e where e is the smallest positive integer such that $f(x) \mid x^e - 1$. Furthermore, at least one of these sequences will have e as its smallest period.

Let $f(x)$ be a primitive polynomial, i.e., an irreducible polynomial with a zero α being a generator for the multiplicative group of $GF(p^n)$. Then the factorization of the primitive polynomial $f(x)$ is given by

$$f(x) = \prod_{i=0}^{n-1}(x - \alpha^{p^i}).$$

Since the generator α has order $p^n - 1$, then $f(x) \mid x^{p^n-1} - 1$ and any nonzero sequence generated by the recursion with characteristic polynomial $f(x)$ has period $p^n - 1$. This is maximum possible for a linear recursion of degree n, and any such sequence is therefore called a maximum-length sequence (or m-sequence).

During a period of the m-sequence, each nonzero consecutive n-tuple occurs exactly once during its period. In particular this implies that the m-sequence is as balanced as it can be for a sequence of period $p^n - 1$ since all nonzero symbols occur p^{n-1} times, while the 0 element occurs $p^{n-1} - 1$ time.

The trace function Tr_n from GF(p^n) to GF(p) is defined by

$$Tr_n(x) = \sum_{i=0}^{n-1} x^{p^i}.$$

The m-sequence can be written as

$$s_t = Tr_n(c\alpha^t),$$

where the $p^n - 1$ different nonzero values of $c \in$ GF(p^n)$^* =$ GF(p^n)$\setminus\{0\}$ correspond to all possible shifts of the m-sequence.

Starting with one m-sequence of period $p^n - 1$, all other m-sequences of the same period can be obtained by decimating the sequence. The decimated sequence of $\{s_t\}$ is the sequence $\{s_{dt}\}$ which is an m-sequence if and only if $\gcd(d, p^n - 1) = 1$. The sequence and its decimated sequence are cyclically distinct if and only if $d \not\equiv p^i \pmod{p^n - 1}$ for $i = 0, 1, \ldots, n - 1$. The number of cyclically distinct m-sequences is $\phi(p^n - 1)/n$, where ϕ is Euler's ϕ function, and equals the number of primitive polynomials of degree n. For further results on linear recursions, the reader is referred to the classical book by Golomb [10].

Example 1 Let $p = 3$ and consider the linear recursion

$$s_{t+3} + 2s_{t+2} + s_t = 0.$$

The characteristic polynomial of the recursion is $f(x) = x^3 + 2x^2 + 1$. This is a primitive polynomial, and using the initial state (011), the recursion generates the m-sequence (01110211210100222012212020...) of period $\varepsilon = 26$. The recursion clearly generates all cyclic shifts of this sequence since all nonzero initial states are present in the m-sequence. In addition the recursion generates the all-zero sequence using the initial state (000). It is easily verified that decimating the sequence above by $d \equiv 3^i \pmod{26}$ gives the same sequence, while decimation by any $d \not\equiv 3^i \pmod{26}$ with $\gcd(d, 26) = 1$ gives a cyclically distinct m-sequence of period 26.

2 Correlation of m-Sequences

The cross-correlation at shift τ between two m-sequences that differ by a decimation d will be denoted by $C_d(\tau)$. The problem to determine the values and the number of occurrences of each value of the cross-correlation $C_d(\tau)$ between two m-sequences when τ runs through all $p^n - 1$ shifts has been studied for almost 50 years.

The simplest case to consider is the autocorrelation function of an m-sequence. One reason for the popularity of m-sequences is due to their two-valued autocorrelation and their importance in synchronization applications.

Theorem 1 *The autocorrelation function $C_1(\tau)$ of an m-sequence having period $\varepsilon = p^n - 1$ takes the value -1 for any shift $\tau \not\equiv 0 \pmod{p^n - 1}$ and the value $p^n - 1$ for any shift $\tau \equiv 0 \pmod{p^n - 1}$.*

Proof Let $\tau \not\equiv 0 \pmod{p^n - 1}$, then since the characteristic polynomial of the m-sequence $\{s_t\}$ also generates the sequence $\{s_{t+\tau} - s_t\}$, it follows that this is some shift of the m-sequence. Hence,

$$C_1(\tau) = \sum_{t=0}^{p^n-2} \omega^{s_{t+\tau} - s_t} = \sum_{t=0}^{p^n-2} \omega^{s_{t+\delta}} = -1$$

since $s_t - s_{t+\tau} = s_{t+\delta}$ for some δ depending on τ and the m-sequence is balanced (except for a "missing" 0) having p^{n-1} of each nonzero element and $p^{n-1} - 1$ zeros during a period of the sequence.

Some basic results useful for the analysis of $C_d(\tau)$ can be found in Helleseth [12].

Lemma 1 *The following properties hold for the cross-correlation $C_d(\tau)$:*

1. *If $dd' \equiv 1 \pmod{p^n - 1}$ or $d' \equiv dp^i$ for some integer i, then $C_d(\tau)$ and $C_{d'}(\tau)$ have the same correlation values with the same number of occurrences.*
2. *The value of $C_d(\tau)$ is a real number.*
3. *The sum of the cross-correlation values is determined by*

$$\sum_{\tau=0}^{p^n-2} (C_d(\tau) + 1) = p^n.$$

4. *The square sum of the cross-correlation values is determined by*

$$\sum_{\tau=0}^{p^n-2} (C_d(\tau) + 1)^2 = p^{2n}.$$

5. *The higher-order power sums of the cross-correlation are given by*

$$\sum_{\tau=0}^{p^n-2} (C_d(\tau))^r = -(p^n-1)^{r-1} + 2(-1)^{r-1} + a_r p^{2n}$$

where a_r is the number of solutions of the equations

$$x_1 + x_2 + \cdots + x_{r-1} + 1 = 0$$
$$x_1^d + x_2^d + \cdots + x_{r-1}^d + 1 = 0$$

and $x_i \in GF(p^n)^*$ for $i = 1, 2, \ldots, r-1$.

The lemma above is useful to determine the number of occurrences of each value in $C_d(\tau)$ when there are rather few, say r, values that have already been determined. Then one can determine the complete distribution of the cross-correlation if one can find a_i for $2 < i < r$.

In Helleseth [12], the previous lemma was applied to prove a result first mentioned without proof in Golomb [11].

Theorem 2 *If $d \notin \{1, p, \ldots, p^{n-1}\}$ then $C_d(\tau)$ takes on at least three different values when $\tau = 0, 1, \ldots, p^n - 2$.*

Proof Suppose that $C_d(\tau)$ takes on only the two values x and y that occur r and $p^n - 1 - r$ times, respectively, in $C_d(\tau)$ when τ runs through all shifts $\tau = 0, 1, \ldots, p^n - 2$. Then using (3) and (4) in Lemma 1 leads to two equations in three unknowns x, y, and r. Eliminating r leads to the equation

$$(p^n x - (x+1))(p^n y - (y+1)) = p^{2n}(2 - p^n).$$

For $p = 2$ this is a Diophantine equation that can be shown to have no valid integer solutions (i.e., except $x = -1$ and $y = p^n - 1$ or $x = p^n - 1$ and $y = -1$ corresponding to the autocorrelation). For the nonbinary case when $p > 2$, similar divisibility properties in the ring $Q[\omega]$, where Q denotes the rational number field, imply that two-valued cross-correlation is impossible except in the autocorrelation case, i.e., when $d \equiv p^i \pmod{p^n - 1}$.

The cross-correlation between any two m-sequences $\{s_t\}$ and $\{s_{dt}\}$ with symbols from $GF(p)$ of the same period $\varepsilon = p^n - 1$ can be written as an exponential sum. After a suitable shift, we can assume without loss of generality that $s_t = Tr_n(\alpha^t)$, and we therefore obtain

$$C_d(\tau) = \sum_{t=0}^{\varepsilon-1} \omega^{s_{t+\tau} - s_{dt}} = \sum_{t=0}^{\varepsilon-1} \omega^{Tr_n(\alpha^{t+\tau} - \alpha^{dt})} = \sum_{x \in GF(p^n)^*} \omega^{Tr_n(cx - x^d)}$$

where $c = \alpha^\tau$. Finding the values and the number of occurrences of each value in the cross-correlation function $C_d(\tau)$ for τ in $\{0, 1, \ldots, p^n - 2\}$ is equivalent to determine the distribution of this exponential sum for any $c \neq 0$.

Since a two-valued cross-correlation is only possible when $d \equiv p^i \pmod{p^n - 1}$, it was natural that the early research on the cross-correlation had a strong focus on finding decimations leading to three-valued cross-correlation. The following sections will survey known cases where the cross-correlation takes on three or four values.

The mathematical techniques used to prove these results are rather different for different decimations and give interesting connections between the cross-correlation, exponential sums, and the solutions of special equations over finite fields.

Note that when we in the following find a decimation d with a correlation distribution, then, due to (1) in Lemma 1, the correlation distribution is the same for the decimations $dp^i \pmod{p^n - 1}$ for any i and for the inverse decimation by $d^{-1} \pmod{p^n - 1}$.

3   Three-Valued Cross-Correlation

3.1   Binary Sequences

There are more decimations leading to three-valued cross-correlation when $p = 2$ than in the case $p > 2$. First we consider three-valued cross-correlation in the case of binary sequences.

The pioneering result on three-valued cross-correlation was due to Gold [9] in 1968. Gold considered binary sequences and showed that $d = 2^k + 1$ for n odd and $\gcd(n, k) = 1$ gave a three-valued cross-correlation. Note that the condition n odd was later relaxed to $n/\gcd(n, k)$ odd which still implies that $\gcd(d, 2^n - 1) = 1$.

In 1968 Golomb [11] was the first to conjecture that $d = 2^{2k} - 2^k + 1$ leads to a three-valued cross-correlation when $n/\gcd(n, k)$ is odd, and he mentioned in this paper that this result was first proved by Welch, who never published his proof. Later in 1971 Kasami [19] published a proof in his famous paper on the weight distribution of several subcodes of the second-order Reed–Muller code.

Theorem 3 *Let $e = \gcd(n, k)$ and let n/e be odd. Let $d = 2^k + 1$ or $d = 2^{2k} - 2^k + 1$. Then $C_d(\tau)$ has the following distribution:*

$$-1 + 2^{\frac{n+e}{2}} \quad occurs \quad 2^{n-e-1} + 2^{\frac{n-e-2}{2}} \quad times.$$
$$-1 \qquad\qquad occurs \quad 2^n - 2^{n-e} - 1 \quad times.$$
$$-1 - 2^{\frac{n+e}{2}} \quad occurs \quad 2^{n-e-1} - 2^{\frac{n-e-2}{2}} \quad times.$$

The proof of these decimations use, a simple squaring technique combined with arguments to determine the number of solutions of some linearized polynomial. In the case $d = 2^k + 1$, one can compute rather directly, using simple properties of the trace function, that

$$(C_d(\tau) + 1)^2 = 2^n (1 + (-1)^{Tr_n(c+1)}).$$

It follows that $C_d(\tau)$ can only take the values $-1, -1 \pm 2^{\frac{n+1}{2}}$, and the distribution can be determined from (3) and (4) in Lemma 1.

In the case $d = 2^{2k} - 2^k + 1$ when n is odd and $\gcd(n, k) = 1$, a similar squaring argument gives

$$(C_d(\tau) + 1)^2 = 2^n N$$

where N is either 0 or equal to the number of zeros in GF(2^n) of the linearized polynomial

$$L(z) = z^{2^{6k}} + c^{2^{3k}} z^{2^{4k}} + c^{2^{2k}} z^{2^{2k}} + z.$$

In this case a more detailed argument shows that there is only 1 or 2 solutions in GF(2^n) of $L(z) = 0$ and that $C_d(\tau)$ can only take on the values $-1, -1 \pm 2^{\frac{n+1}{2}}$.

The generalization to the general case when $\gcd(n, k) = e > 1$ and n/e is odd is rather straightforward but more cumbersome. An elegant method for counting the solutions of the equation above is given by Bracken [1].

The following theorem provides a list of all decimations known to give three-valued cross-correlation in the binary case.

Theorem 4 *The cross-correlation $C_d(\tau)$ is three-valued and the correlation distribution is known for the following values of d:*

1. $d = 2^k + 1$, *where* $n/\gcd(n, k)$ *is odd.*
2. $d = 2^{2k} - 2^k + 1$, *where* $n/\gcd(n, k)$ *is odd.*
3. $d = 2^{\frac{n}{2}} + 2^{\frac{n+2}{4}} + 1$, *where* $n \equiv 2 \pmod 4$.
4. $d = 2^{\frac{n+2}{2}} + 3$, *where* $n \equiv 2 \pmod 4$.
5. $d = 2^{\frac{n-1}{2}} + 3$, *where* n *is odd.*
6.

$$d = \begin{cases} 2^{\frac{n-1}{2}} + 2^{\frac{n-1}{4}} - 1, & when \ n \equiv 1 \pmod 4 \\ 2^{\frac{n-1}{2}} + 2^{\frac{3n-1}{4}} - 1, & when \ n \equiv 3 \pmod 4. \end{cases}$$

Comments Case (1) is the celebrated result proved by Gold [9]. Case (2) is the result first proved by Welch (see Golomb [11] and Kasami [19]). Cases (3) and (4) were proved by Cusick and Dobbertin [4] in 1996. Case (5) was a long-standing conjecture by Welch (see Golomb [11]) that was proved 30 years later by Canteaut et al. [2]. Case (6) is a consequence of the results by Dobbertin [6] and Hollmann and Xiang [16]. Cases (3), (4), and (6) were all conjectured in 1972 by Niho [23].

Since the conjectures (3), (4), and (6) by Niho [23] in 1972, which all have been proved, no new decimations of m-sequences have been found to give a three-valued cross-correlation, and it is widely believed that the list of decimations of binary m-sequences in the theorem above is complete.

Open Problem 1 Show that Theorem 4 contains all decimations with three-valued correlation between binary m-sequences.

This appears to be a very hard open problem. All decimations known to have only three values have their three values of the form $-1, -1 \pm 2^r$ for some r. Even to show that any three-valued decimation must have three such values is not known.

3.2 Nonbinary Sequences

For nonbinary sequences there are three-valued decimations that are analogous to the Gold as well as to the Kasami and Welch decimations. These are given in the following result due to Trachtenberg [25], for n odd, in his Ph.D. thesis from 1970. The result is generalized by Helleseth [12] (or actually in his master thesis from 1971) to the case when $n/\gcd(n,k)$ is odd. The generalization is rather straightforward using the properties of the subfield $GF(p^k)$ of $GF(p^n)$.

Theorem 5 *Let p be an odd prime. Then the following decimations have three-valued cross-correlation.*

1. $d = \frac{p^{2k}+1}{2}$ *where $n/\gcd(n,k)$ is odd.*
2. $d = p^{2k} - p^k + 1$ *where $n/\gcd(n,k)$ is odd.*

These are the only decimations for $p > 3$ that are known to give three-valued cross-correlation.

Open Problem 2 Show that Theorem 5 contains all decimations with three-valued cross-correlation between p-ary m-sequences when $p > 3$ is an odd prime.

For the ternary case there is an additional decimation with three-valued cross-correlation given in the following result by Dobbertin et al. [8].

Theorem 6 *Let $p = 3$ and $d = 2 \cdot 3^{\frac{n-1}{2}} + 1$ where n is an odd positive integer. Then the cross-correlation $C_d(\tau)$ is three-valued and has the following distribution:*

$$-1 + 3^{\frac{n+1}{2}} \quad \text{occurs } \tfrac{1}{2}(3^{n-1} + 3^{\frac{n-1}{2}}) \text{ times.}$$
$$-1 \qquad\qquad \text{occurs } 3^n - 3^{n-1} - 1 \quad \text{times.}$$
$$-1 - 3^{\frac{n+1}{2}} \quad \text{occurs } \tfrac{1}{2}(3^{n-1} - 3^{\frac{n-1}{2}}) \text{ times.}$$

There are numerical observations of the cross-correlation between ternary sequences that give decimations with three-valued cross-correlation and that have not yet been proved. If the following open problem, conjectured by Dobbertin et al. [8], is settled, this would explain all the known decimations of ternary

m-sequences with three-valued cross-correlation. Actually, a solution of the problem would complete the explanation of all currently known three-valued cross-correlation decimations for any p.

Open Problem 3 Let $p = 3$ and $d = 2 \cdot 3^r + 1$ where n is odd and

$$r = \begin{cases} \frac{n-1}{4} & \text{if } n \equiv 1 \pmod 4, \\ \frac{n-1}{4} & \text{if } n \equiv 3 \pmod 4. \end{cases}$$

Show that $C_d(\tau)$ has three-valued cross-correlation.

If $n / \gcd(n, k)$ is odd, Theorems 4 and 5 imply the existence of decimations with three-valued cross-correlation.

In the remaining cases when $n = 2^i$ for some positive integer $i \geq 2$, there are no known decimations having three-valued cross-correlation. It was conjectured by Helleseth [12] that in these cases, any decimation gives at least four cross-correlation values. This conjecture has recently been proved in the binary case by Katz [20]. The general case to settle the conjecture for all other values of p is still open.

Open Problem 4 Show that $C_d(\tau)$ is at least four-valued when $n = 2^i$ for all values of the prime p.

4 Four-Valued Cross-Correlation

One of the main contributions leading to new decimations with four-valued cross-correlation is due to Niho [23]. For his method to be applicable, then $n = 2k$ has to be even and d must be of the special form $d = s(2^k - 1) + 1$.

The main idea is to reduce the problem to compute the number of solutions of some special equations that depend on s.

The next theorem provides a list, in historical order, of all the decimations that have been proved to give four-valued cross-correlation. An important observation is that all the results in (1)–(4) are covered by the last case (5).

Theorem 7 *Let $v_2(i)$ be the highest power of 2 dividing the integer i. The cross-correlation $C_d(\tau)$ is four-valued and the correlation distribution is known for the following values of d:*

1. $d = 2^{\frac{n}{2}+1} - 1$, *where $n \equiv 0 \pmod 4$.*
2. $d = (2^{\frac{n}{2}} + 1)(2^{\frac{n}{4}} - 1) + 2$, *where $n \equiv 0 \pmod 4$.*
3. $d = \frac{2^{(n/2+1)r} - 1}{2^r - 1}$ *$(0 < r < n/2, \gcd(n, r) = 1)$ for $n \equiv 0 \pmod 4$.*
4. $d = \frac{2^{2k} + 2^{s+1} - 2^{k+1} - 1}{2^s - 1}$, *where $n = 2k$ and $2s | k$.*
5. $d = (2^k - 1)s + 1$, $s \equiv 2^r(2^r \pm 1)^{-1} \pmod{2^k + 1}$, *where $v_2(r) < v_2(k)$.*

Comments The first two cases in Theorem 7 are due to Niho [23] in his Ph.D. thesis. Case (3) is due to Dobbertin [5]. Case (4) was proved by Helleseth and Rosendahl [15]. The final case (5) is proved by Dobbertin et al. [7] and contains all the four previous cases.

Sketch of proof Since the Niho decimations have played a significant role in the cross-correlation of m-sequences, we will provide a short outline of the proof.

The main idea behind the proof of Theorem 7 is very simple and uses that any nonzero element $x \in \mathrm{GF}(2^n)$ can be written uniquely as $x = yz$ where $y \in \mathrm{GF}(2^k)$ and $z \in U$ where

$$U = \{z \in \mathrm{GF}(2^n) \mid z^{2^k+1} = 1\}.$$

In particular, since $d = s(2^k - 1) + 1$, it follows that $d \equiv 1 \pmod{2^k - 1}$ and therefore $d \equiv -2s + 1 \pmod{2^k + 1}$. Hence, $y^d = y$ and $z^d = z^{-2s+1}$ and the cross-correlation can be written as

$$C_d(\tau) = \sum_{x \in \mathrm{GF}(2^n)^*} (-1)^{\mathrm{Tr}_m(cx+x^d)}$$

$$= \sum_{y \in \mathrm{GF}(2^n)^*, z \in U} (-1)^{\mathrm{Tr}_n(cyz+yz^{-2s+1})}$$

$$= \sum_{y \in \mathrm{GF}(2^n)^*, z \in U} (-1)^{\mathrm{Tr}_k(yh(z))}$$

$$= (2^k - 1)N + (2^k + 1 - N)(-1)$$

$$= -1 + (N - 1)2^k.$$

Here N is the number of solutions $z \in U$ of the equation $h(z) = 0$ where

$$h(z) = cz + z^{-2s+1} + c^{2^k} z^{-1} + z^{2s-1}$$

which is equivalent to N being the number of solutions $z \in U$ to

$$p(z) = z^{2s-1} + c^{1/2} z^s + c^{2^{k-1}} z^{s-1} + 1 = 0.$$

Case (1) is one of the two decimations in Niho's thesis shown to be four-valued. In this case $s = 2$, i.e., $d = 2(2^k - 1) + 1$ and

$$p(z) = z^3 + c^{1/2} z^2 + c^{2^{k-1}} z + 1 = 0$$

which has at most three solutions for z. Hence, $N = 0, 1, 2, 3$ are the only possibilities leading to at most a four-valued cross-correlation with values in the set

$$\{-1 - 2^k, -1, -1 + 2^k, -1 + 2^{k+1}\}.$$

The correlation distribution follows from Lemma 1 using (3)–(5) and finding a_3.

In the other cases (2)–(5) in Theorem 7, we have

$$p(z) = z^{2^r+1} + az^{2^r} + bz + 1 = 0$$

which is known to have $0, 1, 2$ or $2^{\gcd(r,n)} + 1$ solutions in $GF(2^n)$. A more detailed analysis shows that the number of solutions in U has these four possibilities and the four-valued cross-correlation distribution can be found as above.

There are numerical results that give decimations with four values that are not explained by this list. However, one believes that case (5) in Theorem 7 contains all four-valued cases when d is of the Niho form $d = s(2^k - 1) + 1$. The following conjecture was stated in Dobbertin et al. [7].

Open Problem 5 Any binary decimation of Niho type $d = s(2^k - 1) + 1, n = 2k$ with four-valued cross-correlation is of the form $d = (2^k-1)s+1, s \equiv 2^r(2^r \pm 1)^{-1}$ (mod $2^k + 1$), where $v_2(r) < v_2(k)$.

In the nonbinary case there are a few families known with four-valued cross-correlation. The following decimation in Helleseth [12] is the only known four-valued decimation that works for any prime p.

Theorem 8 *Let p be an odd prime and $d = 2 \cdot p^{\frac{n}{2}+1} - 1$, where $n \equiv 0$ (mod 4). Then the cross-correlation $C_d(\tau)$ is four-valued and the distribution is known.*

Recently new ternary decimations with four-valued cross-correlation have been found by Zhang et al. [26].

Theorem 9 *Let $p = 3$, $n = 3k$, and $\gcd(k, 3) = 1$. If $d = 3^k + 1$ or $d = 3^{2k} + 2$. Then if r is odd, the cross-correlation $C_d(\tau)$ is four-valued (and six-valued for r even) and the distribution is known. (The distribution is conjectured to be the same if $\gcd(k, 3) = 3$).*

5   The −1 Conjecture

For binary sequences the cross-correlation values are obviously always integers. For $p > 2$, this may not always be the case even though the values of $C_d(\tau)$ are always real numbers. This follows from the definition of the cross-correlation function and the fact that the second half of an m-sequence is the negative of the first half. It was shown in Helleseth [12] that $C_d(\tau)$ is an integer for all τ if and only if $d \equiv 1$ (mod $p - 1$).

Numerical results reveal that for $p = 2$, the cross-correlation always has -1 as one of its values. For $p > 2$ this happens for all decimations d where $d \equiv 1 \pmod{p-1}$. This was conjectured by Helleseth [12] and this is still an open problem.

Open Problem 6 Show that if $d \equiv 1 \pmod{p-1}$, then -1 always occurs as a value in $C_d(\tau)$.

It is trivial to reformulate the conjecture as a result of the number of common solutions of a special equation system.

Lemma 2 *Let* $q = p^n$ *and* α *be a primitive element in* GF(q). *Let* N *be the number of solutions* $x_i \in$ GF(q) *of the equation system:*

$$x_0 + \alpha x_1 + \alpha^2 x_2 + \cdots + \alpha^{q-2} x_{q-2} = 0$$
$$x_0^d + x_1^d + x_2^d + \cdots + x_{q-2}^d = 0.$$

The -1 *conjecture holds if and only if* $N = q^{q-3}$ *for all d where* $\gcd(d, p^n - 1) = 1$ *and* $d \equiv 1 \pmod{p-1}$.

Proof Let N denote the number of common solutions of the two equations above. Then N can be expressed by the following exponential sum.

$$q^2 N = \sum_{x_0, x_1, \ldots, x_{q-2} \in \mathrm{GF}(q)} \sum_{z_1, z_2 \in \mathrm{GF}(q)} \omega^{Tr_n(z_1(x_0 + \alpha x_1 + \cdots + \alpha^{q-2} x_{q-2}) + z_2(x_0^d + x_1^d + \cdots + x_{q-2}^d))}$$

$$= \sum_{z_1, z_2 \in \mathrm{GF}(q)} \prod_{i=0}^{q-2} \sum_{x \in \mathrm{GF}(q)} \omega^{Tr_n(z_1 \alpha^i x + z_2 x^d)}$$

$$= q^{q-1} + (q-1) \prod_{c \in \mathrm{GF}(q)} \sum_{x \in \mathrm{GF}(q)} \omega^{Tr_n(cx + x^d)}$$

$$= q^{q-1} + \prod_{\tau=0}^{q-2} (C_d(\tau) + 1)$$

since the contribution from $z_1 = z_2 = 0$ is q^{q-2}, and $z_1 = 0$ or $z_2 = 0$ contributes 0 if not both are zero. Furthermore, $z_1/z_2^{d^{-1}}$ runs through all nonzero elements in GF(q) $q-1$ times when z_1 and z_2 run through all nonzero elements in the field. Hence, $C_d(\tau) = -1$ for some τ if and only of $N = q^{q-3}$.

Another old problem on the cross-correlation between m-sequences that is more than 30-year-old is the following conjecture due to Sarwate and Pursley [24].

Open Problem 7 Let n be even. Show that $|C_d(\tau) + 1| \geq 2^{\frac{n}{2}+1}$.

For related surveys on m-sequences the reader is referred to [13, 14]. Other interesting results and open problems on this topic can be found in [18, 22].

6   Relations to APN and AB Functions

The cross-correlation of m-sequences has some interesting relations to almost perfect nonlinear mappings (APN) and almost bent functions (AB).

An *almost perfect nonlinear* function f is a mapping $f : GF(2^n) \mapsto GF(2^n)$ such that

$$f(x + a) + f(x) = b$$

has at most two solutions for any $a \neq 0$, $b \in GF(2^n)$. The function is said to be Δ-uniform if the maximum number of solutions is Δ, such that an APN function is the same as being 2-uniform.

The *Walsh transform* of f is defined by

$$\lambda_f(a, b) = \sum_{x \in GF(2^n)} (-1)^{Tr(af(x)+bx)},$$

where $a, b \in GF(2^n)$.

A function f is *almost bent* (AB) if

$$\{\lambda_f(a, b) : a, b \in GF(2^n)\} = \{0, \pm 2^{(n+1)/2}\}.$$

It has been shown by Chaubaud and Vaudenay [3] that AB implies APN. APN functions and AB functions are of significant importance in the design of S-boxes in block ciphers.

Monomial AB functions where $f(x) = x^d$ can be obtained from Gold sequences and several of the decimations with three-valued cross-correlation.

Theorem 10 *The known monomial AB functions $f(x) = x^d$ are*

1. *Gold:* $d = 2^k + 1$, *where* $\gcd(n, k) = 1$.
2. *Kasami:* $d = 2^{2k} - 2^k + 1$, *where* $\gcd(n, k) = 1$.
3. *Welch:* $d = 2^{\frac{n-1}{2}} + 3$, *where n is odd.*
4. *Niho:*

$$d = \begin{cases} 2^{\frac{n-1}{2}} + 2^{\frac{n-1}{4}} - 1, & \text{if } n \equiv 1 \pmod 4 \\ 2^{\frac{n-1}{2}} + 2^{\frac{3n-1}{4}} - 1, & \text{if } n \equiv 3 \pmod 4. \end{cases}$$

Note that each of these cases corresponds to decimations with three-valued cross-correlation where the values are restricted to the set $\{0, \pm 2^{(n+1)/2}\}$. Thus, each corresponding monomial function $f(x) = x^d$ is AB. Dobbertin [6] conjectured that these are the only monomial AB functions.

Open Problem 8 Show that Theorem 10 contains all monomial $f(x) = x^d$ AB functions.

Since a monomial AB function is an APN function, the monomial functions $f(x) = x^d$ with d in Theorem 10 are also APN functions. In addition there are two more decimations leading to APN functions and which are not AB. The known monomial APN functions are given in the following theorem.

Theorem 11 *The known monomial APN functions $f(x) = x^d$ are*

1. *Gold: $d = 2^k + 1$, where $\gcd(n, k) = 1$.*
2. *Kasami: $d = 2^{2k} - 2^k + 1$, where $\gcd(n, k) = 1$.*
3. *Welch: $d = 2^{\frac{n-1}{2}} + 3$, where n is odd.*
4. *Niho:*

$$d = \begin{cases} 2^{\frac{n-1}{2}} + 2^{\frac{n-1}{4}} - 1, & if\ n \equiv 1\ (\text{mod } 4) \\ 2^{\frac{n-1}{2}} + 2^{\frac{3n-1}{4}} - 1, & if\ n \equiv 3\ (\text{mod } 4). \end{cases}$$

5. *Inverse: $d = 2^n - 2 \equiv -1\ (\text{mod } 2^n - 1)$, where n is odd.*
6. *Dobbertin: $d = 2^{4k} + 2^{3k} + 2^{2k} + 2^k - 1$, where $n = 5k$.*

Dobbertin [6] conjectured that these are the only monomial APN functions.

Open Problem 9 Show that Theorem 11 contains all monomial $f(x) = x^d$ APN functions.

It is easy to show that the cross-correlation values between two m-sequence obey $C_d(\tau) \equiv -1\ (\text{mod } 4)$. The cross-correlation between $\{s(t)\}$ and its reverse sequence $\{s(-t)\}$ corresponds to the famous Kloosterman sum defined by

$$C_{-1}(\tau) = \sum_{x \in GF(p^n)^*} \omega^{Tr(ax + x^{-1})}.$$

A well-known bound for the Kloosterman sum is

$$|C_{-1}(\tau)) + 1| \le 2p^{n/2}.$$

For $p = 2$ it was shown by Lachuad and Wolfmann [21] that $C_{-1}(\tau)$ takes on all possible values $\equiv -1\ (\text{mod } 4)$ that obey this bound.

The S-box used in the Advanced Encryption Standard (AES) is a permutation based on $f(x) = x^{-1}$ for $n = 8$. The correlation between x^{-1} and all affine functions take on the same values as $|C_{-1}(\tau)|$ when $\tau = 0, 1, \ldots, p^n - 2$. The S-box is 4-uniform (not APN) which is the best known uniformity for $n = 8$. The S-box is not AB but the correlation (and nonlinearity) is the best known for $n = 8$.

Conclusion

The cross-correlation of m-sequences is a challenging mathematical problem that has many important applications in communication systems. This chapter presents an updated overview of this problem and presented some of the remaining open problems that still exist in this area. Finally, a few connections have been given to AB and APN functions that are important in the design and analysis of S-boxes in block ciphers.

Acknowledgements This research was supported by the Norwegian Research Council.

References

1. C. Bracken, Designs, Codes, *Spin Models and the Walsh Transform*, Ph.D. thesis, Department of Mathematics, National University Ireland (NUI), Maynooth, 2004
   In this Ph.D. thesis one can find a nice proof of the number of solutions of a linearized polynomial playing an important role in the proof of the 3-valued crosscorrelation with the Kasami–Welch exponent $d = 2^{2k} - 2^k + 1$.
2. A. Canteaut, P. Charpin, H. Dobbertin, Binary m-sequences with three-valued crosscorrelation: a proof of Welch's conjecture. IEEE Trans. Inf. Theory **46**(1), 4–8 (2000)
   The more than 30 year old conjecture by Welch on a decimation with 3-valued crosscorrelation between two m-sequences is proved in this paper.
3. F. Chabaud, S. Vaudenay, Links between differential and linear cryptanalysis, in *Advances in Cryptology-EUROCRYPT'94* (Springer, New York, 1995), pp. 356–365
   The paper gives important relations between differential and linear analysis and shows in particular that AB functions are APN functions.
4. T.W. Cusick, H. Dobbertin, Some new three-valued crosscorrelation functions for binary m-sequences. IEEE Trans. Inf. Theory **42**(4), 1238–1240 (1996)
   The authors prove two conjectures due to Niho on two decimation that (for n even) give 3-valued crosscorrelation.
5. H. Dobbertin, One-to-one highly nonlinear power functions on $GF(2^n)$. Appl. Algebra Eng. Commun. Comput. **9**(2), 139–152 (1998)
   The author finds a new decimation with 4-valued crosscorrelation, the first new one since Niho's Ph.D. thesis from 1972.
6. H. Dobbertin, Almost perfect nonlinear power functions on $GF(2^n)$: the Niho case. Inf. Comput. **151**(1–2), 57–72 (1999)
   The author shows that two decimations conjectured by Niho to have 3-valued crosscorrelation for odd m give almost perfect nonlinear functions. This was an important step in order to later complete the proof of these conjectures in [16].
7. H. Dobbertin, P. Felke, T. Helleseth, P. Rosendahl, Niho type cross-correlation functions via Dickson polynomials and Kloosterman sums. IEEE Trans. Inf. Theory **52**(2), 613–627 (2006)
   Dickson polynomials were used for the first time to find the crosscorrelation between m-sequences. The paper also settled the correlation distribution of many new decimations with 4-valued crosscorrelation.
8. H. Dobbertin, T. Helleseth, P. Vijay Kumar, H. Martinsen, Ternary m-sequences with three-valued crosscorrelation function: two new decimations of Welch and Niho type. IEEE Trans. Inf. Theory **47**(4), 1473–1481 (2001)

The importance of this paper is that is found the first new nonbinary decimations with three values since the constructions 30 years earlier by Trachtenberg.

9. R. Gold, Maximal recursive sequences with 3-valued recursive cross-correlation functions. IEEE Trans. Inf. Theory **14**(1), 154–156 (1968)
   This pioneering paper defined the Gold decimation and proved that it had a 3-valued crosscorrelation and determined the complete correlation distribution. This was the basis for the important Gold sequences.
10. S.W. Golomb, *Shift Register Sequences* (Holden-Day, San Francisco, 1967)
    This is a classical book on linear and nonlinear recursions.
11. S.W. Golomb, Theory of transformation groups of polynomials over GF(2) with applications to linear shift register sequences. Inf. Sci. **1**(1), 87–109 (1968)
    The author states (without proof) that the crosscorrelation between binary m-sequences takes on at least three values. The Welch conjecture that two special decimations have 3-valued crosscorrelation was published here for the first time.
12. T. Helleseth, Some results about the cross-correlation function between two maximal linear sequences. Discrete Math. **16**(3), 209–232 (1976)
    This paper contains many basic results on the crosscorrelations of m-sequences. The first nonbinary decimation is found giving a four-valued crosscorrelation between two m-sequences. The distributions of several decimations are completely settled. The -1 conjecture is stated in this paper.
13. T. Helleseth, Crosscorrelation of m-sequences, exponential sums and Dickson polynomials. IEICE Trans. Fundamentals **E93A**(11), 2212–2219 (2010)
    Presents a survey on the crosscorrelation between binary m-sequences having at most 5-valued crosscorrelation with a focus on the many connections between exponential sums and Dickson polynomials.
14. T. Helleseth, P.V. Kumar, Sequences with low correlation, in *Handbook in Coding Theory*, eds. by V.S. Pless, W.C. Huffman, ch. 21 (Elsevier Science B.V., Amsterdam, 1998), pp.1765–1853
    This is a survey of sequences with low correlation that contains constructions and analysis of many important sequence families and some of their relations to coding theory.
15. T. Helleseth, P. Rosendahl, New pairs of m-sequences with 4-level cross-correlation. Finite Fields Appl. **11**(4), 674–683 (2005)
    This paper introduced new decimations with 4-valued cross correlation.
16. H.D.L. Hollmann, Q. Xiang, A proof of the Welch and Niho conjectures on cross-correlations of binary m-sequences. Finite Fields Appl. **7**(2), 253–286 (2001)
    This paper completed the proof of two decimations, for odd m, that were conjectured by Niho to lead to 3-valued crosscorrelation.
17. A. Johansen, T. Helleseth, A family of m-sequences with five-valued cross correlation. IEEE Trans. Inf. Theory **55**(2), 880–887 (2009)
    The distribution of the crosscorrelation of pairs of m-sequences with decimations giving five-valued crosscorrelation was found using techniques involving Dickson polynomials.
18. A. Johansen, T. Helleseth, A. Kholosha, Further results on m-sequences with five-valued cross correlation. IEEE Trans. Inf. Theory **55**(12), 5792–5802 (2009)
    This paper extends the results in [17] to other decimations with five-valued crosscorrelation. Some results depend on open conjectures on some exponential sums.
19. T. Kasami, The weight enumerators for several classes of subcodes of the 2nd order binary Reed–Muller codes. Inf. Control **18**(4), 369–394 (1971)
    The author determined the weight enumerator of some subcodes of the 2nd order Reed–Muller. A consequence of these results is a proof of the Kasami–Welch decimation leading to 3-valued crosscorrelation. This decimation was also proved by Welch (unpublished).
20. D. Katz, Weil sums of binomials, three-level cross-correlation and a conjecture by Helleseth. J. Combin. Theory A **119**(8), 1644–1659 (2012)
    The paper gives a solution of the conjecture of Helleseth that for $n = 2^i$ and $p = 2$ the crosscorrelation takes on at least 4 values.

21. G. Lachaud, J. Wolfmann, The weights of the orthogonals of the extended quadratic binary Goppa codes. IEEE Trans. Inf. Theory **36**(3), 686–692 (1990)
The paper shows that the Kloosterman sums takes on all possible values $\equiv -1 \pmod 4$ within its bound.
22. J. Lahtonen, G. McGuire, H.N. Ward, Gold and Kasami–Welch functions, quadratic forms, and bent functions. Adv. Math. Commun. **1**(2), 243–250 (2007)
Provides a local result on $C_d(0)$ for the Kasami–Welch decimation.
23. Y. Niho, *Multi-valued Cross-Correlation Functions Between Two Maximal Linear Recursive Sequences*, Ph.D. thesis, University of Southern California, Los Angeles, 1972
This thesis gave the complete crosscorrelation distribution of several decimations with 4-valued cross correlation. Furthermore, many conjectures on the cross correlation distribution of sequences with few values in the crosscorrelation were given. This Ph.D. thesis had a significant influence on later research on the crosscorrelation.
24. D. Sarwate, M. Pursley, Crosscorrelation properties of pseudorandom and related sequences. Proc. IEEE, **68**(5), 593–619 (1980)
This is a classical and excellent survey of the crosscorrelation between m-sequences.
25. H.M. Trachtenberg, *On the Cross-Correlation Functions of Maximal Linear Recurring Sequences*, Ph.D. thesis, University of Southern California, Los Angeles, 1970
The main result is the two families of decimation giving three-valued crosscorrelation. These are the only decimations that work for all nonbinary m-sequences.
26. T. Zhang, S. Li, T. Feng, G. Ge, Some new results on the cross correlation of m-sequences. arXiv:1309.7734 [cs.IT]
This recent paper gives new ternary decimations with four-valued crosscorrelation.

Open Problems on With-Carry Sequence Generators

Andrew Klapper

Abstract Pseudorandom sequences are used in a wide range of applications in computing and communications, including cryptography. It is common to use linear feedback shift registers (LFSRs) to generate such sequences, either directly or as components in more complex structures. Much of the analysis of such sequences is done using the algebra of polynomials and power series over finite fields. The subjects of this chapter are feedback with carry shift registers (FCSRs) and algebraic feedback shift registers (AFSRs, generalizations of both LFSRs and FCSRs), sequence generators that are analogous to LFSRs, but whose state update involves arithmetic with a carry. Their analysis is based on algebraic structures with carry, such as the integers and the N-adic numbers. After a brief review of the basics on LFSRs, FCSRs, and AFSRs, we describe several open problems. These include: given part of a sequence, how to find an optimal generator of the sequence; how to construct sequences that cannot be generated by short LFSRs, FCSRs, or AFSRs; and the analysis of various statistical properties related to these generators.

1 Introduction

The subject of this chapter is the generation of "pseudorandom" sequences using very high-speed devices. Here pseudorandom means that various statistical properties hold such as (in the binary case) a balance in the numbers of zeroes and ones. Such sequences play critical roles in many applications in communications and computing. Following are some important examples.

1. Cryptography: stream ciphers scramble messages by combining them with sequences that are unpredictable from short prefixes.
2. CDMA: large families of uncorrelated sequences minimize interference and allow a collection of channels to be shared by users (see Sect. 5.2).

A. Klapper (✉)
Department of Computer Science, University of Kentucky, 307 Marksbury Building,
Lexington, KY 40506-0633, USA
e-mail: klapper@cs.uky.edu

© Springer International Publishing Switzerland 2014
Ç.K. Koç (ed.), *Open Problems in Mathematics and Computational Science*,
DOI 10.1007/978-3-319-10683-0_9

3. Radar ranging and GPS: peaks in autocorrelations of a sequence allow delay to be measured.
4. Quasi-Monte Carlo: integrals are approximated by sampling integrands at points determined by pseudorandom sequences.
5. Built in self-test: test patterns are determined by pseudorandom sequences.
6. Wear leveling of storage media: pseudorandom sequences are used to remap the memory locations in a way that distributes the wear evenly across the whole disk.

For some 60 years linear feedback shift registers (LFSRs) (described in Sect. 3) have been used as generators (or components of generators) of pseudorandom sequences for these and other applications. In the form of linear equations modulo N, they have been studied by mathematicians since at least the 1920s [4]. The primary mathematical tools for analyzing these sequences are finite fields and particularly polynomials and power series over finite fields. A great deal is known about these sequences, but there is still much that is unknown.

More recently (since 1993 [6, 20]), researchers have been studying feedback with carry shift registers (FCSRs), a "with-carry" analog of LFSRs (described in Sect. 3). So far they have found a smaller number of applications—cryptanalysis of the summation combiner, quasi-Monte Carlo integration, and the F-FCSR stream cipher. One advantage they have is that the state change is nonlinear, which makes stream ciphers based on them resistant to algebraic attacks.

Much less is known about sequences generated by FCSRs (and algebraic feedback shift registers (AFSRs), a generalization). The purpose of this chapter is to describe some of the open problems in this area. The main focus is on properties of sequences that are of interest cryptographically.

Throughout this chapter, the book by Goresky and the author [10] serves as a reference.

2   Stream Ciphers

In this section we discuss one important application of pseudorandom sequences. The main problem of practical cryptography is how to send a message securely in real time. The common techniques of public key cryptography are too slow for large transmissions (such as video on demand). For example, RSA encrypts by computing $E(m) = m^e \mod pq$, where p and q are perhaps 500 bit primes. This is much too slow to encrypt large data sets in real time.

The alternative is to use symmetric key cryptography—*block or stream ciphers*. The trade-off between these two approaches is that the fastest stream ciphers are somewhat faster than the fastest block ciphers, but stream ciphers seem to be more vulnerable to attack. In this section we are interested in stream ciphers. In their simplest form, a sender and receiver agree on a pseudorandom sequence generator (PSG) G (publicly) and a small shared seed s (privately, perhaps by a slow key agreement protocol). G, initialized with s, generates a pseudorandom sequence

Fig. 1 Structure of a stream cipher

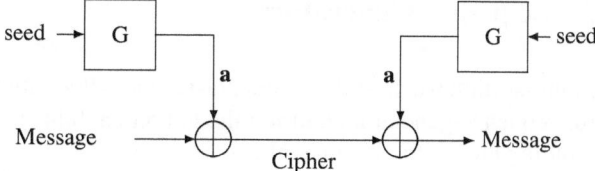

$G(s) = \mathbf{a} = a_0, a_1, \ldots \in \{0, 1\}^\infty$. A message $m = m_0, m_2, \ldots \in \{0, 1\}^\infty$ is encrypted by computing $c_i = m_i \oplus a_i$. See Fig. 1.

Sequence generators used in stream ciphers or other applications mentioned in the introduction must have various properties, depending on the applications. They must operate in (nearly) real time. They must resist known cryptanalytic attacks. They must have good statistical properties, such as the following:

Large period: A sequence $\mathbf{a} = a_0, a_1, \ldots$ is periodic if $\forall i : a_i = a_{i+p}$. It is eventually periodic if $\forall i > t : a_i = a_{i+p}$ for some t. The period, p, must be large for use in a stream cipher.

Balance: In one period the numbers of occurrences of different symbols must be nearly equal.

Uniform distribution of small subsequences: For any r, in one period the numbers of occurrences of different blocks of length r must be nearly equal.

Uncorrelated with shifts: Let $\mathbf{a}$ be a binary sequence with period p. The *autocorrelation* of $\mathbf{a}$ with shift t is

$$\mathscr{A}_{\mathbf{a}}(t) = \sum_{i=0}^{p-1} (-1)^{a_i + a_{i+t}}.$$

If t is not a multiple of p, this integer should be close to zero.

Unpredictable from a short prefix: It should not be possible to determine $\mathbf{a}$ knowing only $a_0, \ldots, a_{k-1}$ for small k using any known methods (e.g., using the Berlekamp–Massey algorithm). This is a critical requirement for stream ciphers.

Since we do not know what requirements will arise in the future, it is useful to have a large pool of high-quality pseudorandom sequences available.

Note that the approach to security described here is different from the complexity theory approach. In that approach one defines a *cryptographically strong pseudorandom bit generator* (CSPRBG) to be a sequence generator whose output is indistinguishable from a truly random sequence generator by any polynomial time probabilistic distinguisher. Unfortunately this is a strong constraint, and all known CSPRBGs are unable to approach real-time operation (and in fact the security of known CSPRBGs depends on the assumed intractability of certain computational problems such as quadratic residuosity).

3   Sequence Generators

In this section we describe simple, fast devices that satisfy many of the requirements
for sequence generators (but not the unpredictability). They are commonly used as
building blocks for stream ciphers.

LFSRs, FCSRs, and AFSRs (described in the next three subsections) are special
cases of a general model for sequence generators. A PSG is a (not necessarily finite)
state machine with output in an alphabet Σ, $G = (S, \Gamma, \delta)$, where the set S is the
state space, $\Gamma : S \to S$ is the state change function, and $\delta : S \to \Sigma$ is the output
function. Such a PSG generates a pseudorandom sequence from a given initial state
$\sigma \in S$ by iterating the state change forever. That is

$$\mathbf{a} = G(\sigma) = (\delta(\sigma), \delta(\Gamma(\sigma)), \delta(\Gamma^2(\sigma)), \cdots)$$

It is often desirable that for any given initial state σ, the set of states $\{\Gamma^i(\sigma) : i =
0, 1, 2, \ldots\}$ be finite. This implies that $G(\sigma)$ is eventually periodic.

In what follows, we are concerned with families of PSGs. We may be interested,
for example, in finding the most efficient PSG G that generates a given sequence $\mathbf{a}$,
where G is in a given family $\mathscr{G}$ of PSGs. In the next few subsections, we describe
some interesting families of PSGs.

3.1   LFSRs

A LFSR of length r over a field F is a finite state PSG whose state set is F^r and
whose state change function is determined by a set of coefficients $g_1, \ldots, g_r \in
F$ [10, p. 23]. If the current state is $(a_0, a_1, \ldots, a_{r-1})$, then the next state is
$(a_1, \ldots, a_{r-1}, a_r)$, where $a_r = g_r a_0 + \cdots + g_1 a_{r-1}$. The output function is
$\delta(a_0, a_1, \ldots, a_{r-1}) = a_0$. See Fig. 2.

There is a large literature on LFSRs. Some of their salient properties are the
following. We assume that the field $F = \mathbb{F}_q$ is finite, so the set of states is finite,
and the output is eventually periodic:

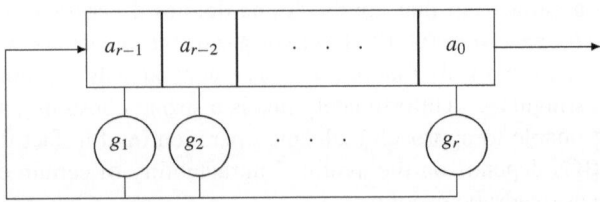

Fig. 2 A length r LFSR

1. The *connection polynomial* is $g(x) = -1 + g_1 x + \cdots + g_r x^r$. The generating function of the output sequence **a** is $a(x) = a_0 + a_1 x + a_2 x^2 + \cdots$. There is a polynomial $u(x)$, uniquely determined by the initial state $(a_0, \ldots, a_{r-1})$, so that $a(x) = u(x)/q(x)$.

2. The sequence **a** is eventually periodic. It's periodic if and only if $\deg(u) < \deg(g)$. The maximum possible period is $|F|^r - 1$. This is achieved when $g(x)$ is a *primitive polynomial*, meaning that a root of g is a primitive element in $\mathbb{F}_{q^r}$. In this case a is called an *m-sequence* [10, p. 208]. These sequences are the most commonly used LFSR sequences.

3. M-sequences have many good statistical properties. Their shifted autocorrelations are all -1. They are as balanced as possible for their period, and the distribution of subblocks of fixed size is as uniform as possible. They have the *run property* [10, p. 172] and the *shift and add property* [10, p. 191].

4. Let E be the unique degree r extension field of F. Let Tr be the trace function from E to F. If the connection polynomial $g(x)$ is irreducible, and the sequence **a** is periodic, then it can be expressed as $a_i = Tr(A\alpha^i)$ where α is a root of $g(x)$ and $A \in E$ corresponds to the initial state. More generally, if **a** is periodic, then it can be expressed as

$$a_i = (Ax^{-i} \mod g) \mod x,$$

meaning (1) compute the element $v \equiv Ax^{-i} \mod g$ with $\deg(v) < r$; and (2) a_i is the constant term of v [10, p. 48].

We can form a family of PSGs by fixing F and considering all LFSRs with entries in F.

3.2   FCSRs

Let $N \geq 2$ be an integer and $S = \{0, 1, \ldots, N-1\}$. A FCSR of length r based on N is a PSG whose state set is $S^r \times \mathbb{Z}$ and whose state change function is determined by a set of coefficients $g_1, \ldots, g_r \in \mathbb{Z}$ [10, p. 70]. If the current state is $(a_0, a_1, \ldots, a_{r-1}; z)$, then the next state is $(a_1, \ldots, a_{r-1}, a_r; z')$, where $a_r + Nz' = g_r a_0 + \cdots + g_1 a_{r-1} + z$. Here the addition and multiplication are in $\mathbb{Z}$. The output function is $\delta(a_0, a_1, \ldots, a_{r-1}; z) = a_0$. See Fig. 3.

FCSRs have many properties that parallel properties of LFSRs. Now, however, the algebra of polynomials and power series is replaced by the algebra of integers and *N-adic numbers*, which we briefly review [10, p. 72].

An N-adic number is an infinite expression

$$a = \sum_{i=0}^{\infty} a_i N^i,$$

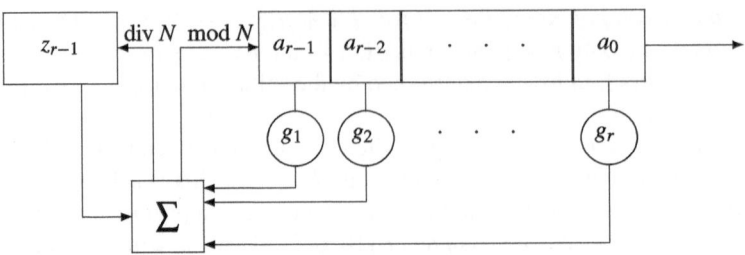

Fig. 3 A length r FCSR

where $a_i \in S$. Addition of N-adic numbers is addition with carry. That is,

$$\sum_{i=0}^{\infty} a_i N^i + \sum_{i=0}^{\infty} b_i N^i = \sum_{i=0}^{\infty} c_i N^i$$

if there are integers $d_0 = 0, d_1, d_2, \ldots$ so that for all $i \geq 0$ we have $a_i + b_i + d_i = c_i + Nd_{i+1}$. Similarly, we have

$$\sum_{i=0}^{\infty} a_i N^i \sum_{i=0}^{\infty} b_i N^i = \sum_{i=0}^{\infty} c_i N^i$$

if there are integers $d_0 = 0, d_1, d_2, \ldots$ so that for all $i \geq 0$ we have

$$\sum_{j=0}^{i} a_j b_{i-j} + d_i = c_i + Nd_{i+1}.$$

The set of N-adic numbers is thus an algebraic ring, denoted by $\mathbb{Z}_N$. Note that

$$-1 = (N-1) + (N-1)N + (N-1)N^2 + \cdots$$

(because adding 1 to the right-hand side gives 0). It can be seen that a sequence $\mathbf{a} = a_0, a_1, \ldots \in S^{\infty}$ is eventually periodic if and only if its associated N-adic number

$$a = \sum_{i=0}^{\infty} a_i N^i$$

is a rational number u/g with $\gcd(g, N) = 1$.

The following are some properties of FCSRs and their output sequences:

1. The *connection integer* of an FCSR is $g = -1 + g_1 N + \cdots + g_r N^r$. The associated N-adic number of the output sequence $\mathbf{a}$ is $a = a_0 + a_1 N +$

$a_2 N^2 + \cdots$. There is an integer u (uniquely determined by the initial state $(a_0, \ldots, a_{r-1}; z)$) so that $a = u/g$ [10, p. 80].

2. The sequence **a** is eventually periodic [10, p. 88]. This is equivalent to saying that the carry z is bounded in any infinite execution of the FCSR. The sequence **a** is periodic iff $-g \leq u \leq 0$. The period is at most $g-1$. The period equals $g-1$ when g is prime and N is a *primitive root* modulo g, meaning that the multiplicative order of N modulo g is $g-1$. In this case **a** is called an ℓ-*sequence* [10, p. 264]. These sequences are the most interesting FCSR sequences. It is unknown whether for a fixed N, there are infinitely many primes g such that N is primitive modulo g (Artin's conjecture). However, Hooley showed that if a certain generalized Riemann hypothesis holds, then for every N there are infinitely many primes g so that N is primitive modulo g [14]. Moreover, it is known that there are at most two values of N for which Artin's conjecture fails, although it is unknown what these values are [13].

3. ℓ-sequences have many good statistical properties. If $N = 2$, then their shifted *arithmetic* autocorrelations (defined in Sect. 5.2) are all 0 [10, p. 172]. They are as balanced as possible for their period and the distribution of subblocks of fixed size is as uniform as possible. They have the *arithmetic shift and add property* [10, p. 204].

4. If **a** is periodic, then it can be expressed as

$$a_i = (AN^{-i} \mod g) \mod N,$$

for some $A \in \mathbb{Z}$, meaning (1) compute the element $v \equiv AN^{-i} \mod g$ with $0 \leq v < g$; and (2) $a_i = v \mod N \in S$ [10, p. 87].

We can form a family of PSGs by fixing N and considering all N-ary FCSRs.

3.3 AFSRs

In this section we recall some details on AFSR, a generalization of both LFSRs FCSRs [10, p. 96]. Let R be an algebraic ring. Let $\pi \in R$ be neither a unit and nor a zero divisor, and assume that $R/(\pi)$ is finite. Let $S \subseteq R$ be a complete set of representatives for $R/(\pi)$. An AFSR of length r based on π is a PSG whose state set is $S^r \times R$ and whose state change function is determined by a set of coefficients $g_0, \ldots, g_r \in R$ with g_0 invertible modulo π. If the current state is $(a_0, a_1, \ldots, a_{r-1}; z)$, then the next state is $(a_1, \ldots, a_{r-1}, a_r; z')$, where $g_0 a_r + N z' = g_r a_0 + \cdots + g_1 a_{r-1} + z$. Here the addition and multiplication are in R. The output function is $\delta(a_0, a_1, \ldots, a_{r-1}; z) = a_0$. See Fig. 4.

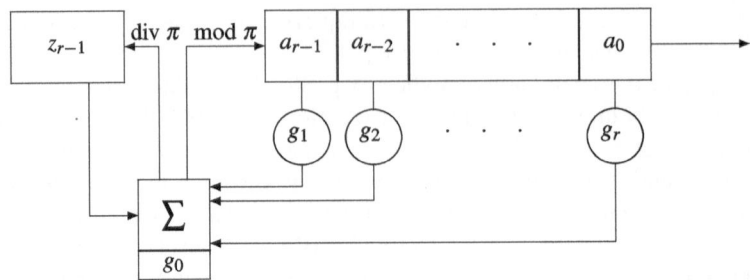

Fig. 4 A length r AFSR

Much of the analysis of AFSRs is based on the algebra of π-adic numbers, which we briefly recall [10, p. 98]. A π-adic number is an infinite expression

$$a = \sum_{i=0}^{\infty} a_i \pi^i,$$

where $a_i \in S$. Addition of π-adic numbers is again addition with carry. That is,

$$\sum_{i=0}^{\infty} a_i \pi^i + \sum_{i=0}^{\infty} b_i \pi^i = \sum_{i=0}^{\infty} c_i \pi^i$$

if there are elements $d_0 = 0, d_1, d_2, \ldots \in R$ so that for all $i \geq 0$ we have $a_i + b_i + d_i = c_i + \pi_{i+1}$. Multiplication is defined similarly. The set of π-adic numbers is $d_i = c_i + \pi d_{i+1}$ thus an algebraic ring, denoted by R_π.

In the case when $R = F[x]$, F a finite field, $\pi = x$, $S = F$, $g_0 = 1$, and $z = 0$, we obtain LFSRs (the carries are all 0 in this case). In the case when $R = \mathbb{Z}$, $\pi = N > 1$, $S = \{0, 1, \ldots, N - 1\}$, and $g_0 = 1$, we obtain FCSRs. Other special cases that have been studied include the case when $R = F[x]$ and $\deg(\pi) > 1$ [10, p. 250], and d-FCSRs, where $R = \mathbb{Z}[\pi]$ and $\pi = N^{1/d}$ with N square free [10, p. 133]. In the latter case, addition in R is addition with carry where the carry jumps d places ahead.

It is not in general the case that the output from an AFSR is eventually periodic. However, it is known that if R is a ring of integers in a number field, then the output is always eventually periodic iff for every embedding of the fraction field of R in $\mathbb{C}$ the complex norm of π is greater than 1. This is the case, for example, for d-FCSRs.

Following are some properties of AFSRs and their output sequences.

1. The *connection element* of an AFSR is $g = -g_0 + g_1\pi + \cdots + g_r\pi^r$. The associated π-adic number of the output sequence **a** is $a = a_0 + a_1\pi + a_2\pi^2 + \cdots$. There is an integer u (uniquely determined by the initial state $(a_0, \ldots, a_{r-1}; z)$) so that $a = u/q$.

2. The sequence **a** is eventually periodic if R is a ring of integers in a number field and the complex norm of π is greater than 1 under every embedding of the fraction field of R in $\mathbb{C}$. Otherwise there are AFSRs that do not produce eventually periodic output. There is in general no known condition on the numerator u characterizing the periodic output sequences, even in the case when all output sequences are eventually periodic. However, for d-FCSRs, we have the following.

Let $\pi^d = 2$. We denote by P the parallelepiped in $R = \mathbb{Z}[\pi]$ which is spanned by the d linearly independent vectors $-g, -g\pi, \dots, -g\pi^{d-1}$,

$$P = \left\{ \sum_{i=0}^{d-1} v_i g \pi^i \mid v_i \in \mathbb{Q} \text{ and } -1 \le v_i \le 0 \right\} \subset \mathbb{Q}[\pi].$$

Let $\Delta = P \cap \mathbb{Z}[\pi]$ be the set of points of the integer lattice $\mathbb{Z}[\pi]$ in P.

Theorem 3.1 ([9]) *Suppose $g \in \mathbb{Z}[\pi]$ is a unit modulo π. Let **a** be an output sequence from a d-FCSR with connection element g and let a be the π-adic number associated with **a**. Suppose that $a = u/g$. Then **a** is periodic if and only if $u \in \Delta$.*

The maximum possible period is $|R/(q)| - 1$. This is achieved when π is a *primitive element* modulo g, meaning that the multiplicative order of π modulo g is $|R/(q)| - 1$. In this case **a** is called a π-adic ℓ-*sequence*.

3. The statistical properties of π-adic ℓ-sequences are not well understood, except in some special cases. For example, for d-FCSRs with $d = 2$, we have the following. Let $\mathcal{N}_{\mathbb{Q}}^F$ denote the rational norm function on F.

Theorem 3.2 ([16]) *Let $\pi^2 = N \ge 2 \in \mathbb{Z}$ with N square free. Let F be the fraction field of $R = \mathbb{Z}[\pi]$. Suppose that $g = y + z\pi \in R$, with $y, z \in \mathbb{Z}$, is invertible modulo π, that $h = \mathcal{N}_{\mathbb{Q}}^F(g)$ is a prime integer, and that π is primitive modulo g. Let **a** be an ℓ-sequence defined over $\mathbb{Z}[\pi]$ with connection element g. If $s \in \mathbb{Z}^+$ is even, then the number K of occurrences of any s-tuple in one period of **a** satisfies*

$$\left| K - \frac{h}{N^s} \right| \le \frac{h}{N^{s/2}|z|} + \frac{|y| + |z|}{N^{s/2}} + 2.$$

If $(h/N)^{1/2} \le |z| \le h^{1/2}$, then

$$\left| K - \frac{h}{N^s} \right| \le \frac{(N^{1/2} + (N-1)^{1/2} + 1)h^{1/2}}{N^{s/2}} + 2.$$

If $N = 2$, then

$$\left| K - \frac{h}{2^s} \right| \le 3 \left(\frac{h}{2^s} \right)^{1/2} + 2.$$

4. It has only been shown that there is an exponential representation of periodic AFSR sequences under special conditions.

We can form a family of PSGs by fixing R, π, and S and considering all AFSRs based on these ingredients. We also may want to impose constraints on the g_is, such as requiring that they be in S. Note that if we let the g_is be arbitrary elements of R, then we can take $r = 1$ and $g = g_1\pi + g_0$ with $g_0 \in S$. Thus any π-adic number u/g can be generated by an AFSR of length one.

4   Register Synthesis Problem

Let $\mathscr{G}$ be a family of PSGs. Suppose that given part of a sequence **a** we can find the most efficient (in some sense) $G = (S, \Gamma, \delta) \in \mathscr{G}$ and $\sigma \in S$ so that $G(\sigma) = \mathbf{a}$. If G is efficient enough, then we have cryptanalyzed **a** [10, p. 295]. Let us make this more precise.

A *register synthesis algorithm* for the family $\mathscr{G}$ is an algorithm T that on input $a_0, a_1, \ldots, a_{n-1}$, a prefix of **a**, outputs $G = (S, \Gamma, \delta) \in \mathscr{G}$ and initial state $\sigma \in S$ so that

1. $G(\sigma) = a_0, a_1, \ldots, a_{n-1}, ?, ?, \cdots$.
2. If n is large enough, $G(\sigma) = \mathbf{a}$ (convergence).
3. T runs in polynomial time in n.

To measure the effectiveness of such an algorithm, we first need a notion of *size* of a sequence generator G in a family $\mathscr{G}$. This should at least approximate the amount of space needed to store the states that occur in an infinite execution of G. Then we define the $\mathscr{G}$-*complexity* $\lambda_{\mathscr{G}}(\mathbf{a})$ of a sequence **a** to be the minimum size of a generator in $\mathscr{G}$ that outputs **a**. We typically measure the effectiveness of a $\mathscr{G}$-synthesizing algorithm in terms of $\lambda_{\mathscr{G}}(\mathbf{a})$: for some slowly growing function μ, if the prefix length n is at least $\mu(\lambda_{\mathscr{G}}(\mathbf{a}))$, then T outputs G, σ with $G(\sigma) = \mathbf{a}$. In all cases we know, $\mu(\lambda)$ is linear in λ.

As a consequence, if $\lambda_{\mathscr{G}}(\mathbf{a})$ is small and $\mathscr{G}$ has an effective register synthesis algorithm, then **a** is cryptographically insecure.

We later use the notion of the $\mathscr{G}$-complexity of a finite sequence, the minimum size of a generator in $\mathscr{G}$ that outputs $a_0, a_1, \ldots, a_{n-1}$ as its first n output symbols. We denote the $\mathscr{G}$-complexity by $\lambda_{\mathscr{G}}(a_0, a_1, \ldots, a_{n-1})$.

4.1   LFSR Synthesis

LFSR synthesis amounts to solving a system of linear equations in the coefficients g_i. There is an efficient algorithm due to Berlekamp and Massey in 1969 [10, p. 296], [22]. This algorithm exploits the special structure of the equations and runs in time $O(n^2)$. Given $a_0, a_1, \ldots, a_{n-1}$, the goal is to find relatively prime polynomials $u(x)$ and $g(x)$ so that

$$a(x) = \sum_{i=0}^{\infty} a_i x^i = \frac{u(x)}{g(x)}.$$

Then $g(x)$ is the connection polynomial of a minimal size LFSR that generates **a**, and $u(x)$ determines the initial state. The algorithm proceeds iteratively—at the ith iteration it finds the minimal degree polynomials $u_i(x), g_i(x)$ so that

$$a(x) \equiv \frac{u_i(x)}{g_i(x)} \mod x^i.$$

The approximation $u_i(x)/g_i(x)$ is found by computing a linear combination of two earlier approximations: if $a(x) \not\equiv u_i(x)/g_i(x) \mod x^i$, then

$$(u_{i+1}(x), g_{i+1}(x)) = (u_i(x), g_i(x)) + bx^{i-m}(u_m(x), g_m(x))$$

for a certain index m and a certain b.

To measure the effectiveness, we define the *linear complexity* of **a**. If $a(x) = u(x)/g(x)$ and $\gcd(u(x), g(x)) = 1$, then

$$\lambda_{\mathrm{lin}}(\mathbf{a}) = \max(\deg(u) + 1, \deg(g))$$

$$= \text{the length of the smallest LFSR that generates } \mathbf{a}.$$

It can be seen that for the Berlekamp–Massey algorithm, $\mu(\lambda) = 2\lambda$. That is, if the sequence **a** has linear complexity λ and the input to the Berlekamp–Massey algorithm is $a_0, \ldots, a_{2\lambda-1}$, then the output is the precise rational representation of the generating function of **a**.

4.2 FCSR and AFSR Synthesis

A first attempt at solving the FCSR synthesis problem is to use the Berlekamp–Massey algorithm but using integer linear combinations instead of $\mathbb{F}_q$ linear combinations when finding a new approximation. This doesn't work—the propagation of carries interferes with convergence.

Instead, for $N = 2$, Goresky and Klapper developed an FCSR synthesis algorithm based on work of Mahler and De Weger. This *Rational Approximation Algorithm* iteratively finds a minimal basis for the kth approximation lattice,

$$L_k = \{(u, g) : g\alpha \equiv u \mod 2^k\},$$

by taking linear combinations of earlier bases [10, p. 334].

Subsequently Xu and Klapper solved the problem for any N [10, p. 348], [21, 28]. They modified the Berlekamp–Massey algorithm so that when a new rational approximation is needed, one is found that works for the next three symbols of $\mathbf{a}$. This means that the effect of the carry is overwhelmed by the growth in the number of terms accounted for. This algorithm also works for some classes of AFSRs: if the base ring R is the ring of integers of a number field $F = \mathbb{Q}[N^{1/d}]$ (d-FCSRs) or R is the ring of integers of certain quadratic extensions of $\mathbb{Q}$.

A third approach is due to Arnault et al. [1], [10, p. 338]. For any N, a modified Euclidean algorithm is used. The idea is, given $a_0, \dots, a_{n-1}$, to run the extended Euclidean algorithm on input

$$\left(N^n, \sum_{i=0}^{n-1} a_i N^i \right)$$

until the terms are less than $N^{n/2}$. If n is large enough, this is guaranteed to succeed.

To measure the effectiveness of these algorithms, we must have a clear notion of the size of an FCSR (or more generally of an AFSR). For LFSRs, it is clear that the number of cells is the size, and it can be seen that this is the same as $\max(\deg(u(x)) + 1, \deg(g(x)))$ if $u(x)/g(x)$ is the corresponding rational representation of the generating function of $\mathbf{a}$. For FCSRs, the "engineering" definition of size would be the number of cells plus the maximum number of N-ary digits needed to store the carry in an infinite execution of the FCSR. We call this the N-adic span. Unfortunately we know of no reasonable algebraic definition of N-adic span. Instead, we define the N-adic complexity of $\mathbf{a}$ to be

$$\lambda_N(\mathbf{a}) = \log_N(\max(|u|, |g|)),$$

where the N-adic number associated with $\mathbf{a}$ has rational representation u/g and $\gcd(u, g) = 1$. It can be seen that N-adic span and N-adic complexity differ only by a small amount.

The situation is more complicated for AFSRs based on a ring R and an element $\pi \in R$. There may be multiple competing choices for a size function. For example, represent $z \in R$ as $\sum_{i=0}^{k} z_i \pi^k$, and let the size of z be k. Or let the size of z be the log (to an appropriate base) of the rational norm of z. The former definition is inadequate in general since not all elements can be represented this way. The latter fails to distinguish the sizes of z and uz where u is a nontrivial unit.

All three algorithms have time complexity $O(n^2)$. For Goresky and Klapper's rational approximation algorithm, we have $\mu(\lambda) = \lceil 2\lambda \rceil + 2$. For Xu and Klapper's modified Berlekamp–Massey algorithm applied to $R = \mathbb{Z}$, we have $\mu(\lambda) = \lceil 6\lambda \rceil + 27$. For the Euclidean approximation algorithm, we have $\mu(\lambda) = \lceil 2\lambda \rceil + 3$.

This leaves the following open questions:

1. How can we build efficient generators of sequences that have large $\lambda_{\mathscr{G}}$ for all "reasonable" $\mathscr{G}$?

2. Are there other "interesting" families $\mathscr{G}$ of PSGs with good register synthesis algorithms?
3. Are there families $\mathscr{G}$ of PSGs that provably have no register synthesis algorithm? Or even just no algorithm with $\mu(\lambda)$ linear?
4. Can we find effective register synthesis algorithms for other classes of AFSRs?

4.3 Combined and Filtered Generators

In this section we consider two approaches to reducing vulnerability of stream ciphers to synthesis attacks. The general idea is to introduce some nonlinearity to the PSG while maintaining the good statistical properties.

The first approach is to use a set of n simple PSGs, such as LFSRs and FCSRs and to combine their outputs with a nonlinear *combiner* function $H(x_1, \ldots, x_n)$. How can we choose H to maximize security? In particular, how can we choose H to make the linear or N-adic complexity large? In the binary case, suppose the underlying PSGs generate sequences $\mathbf{a}^1, \ldots, \mathbf{a}^n$. Let the overall output sequence be $\mathbf{b} = b_0, b_1, \ldots$, with $b_i = H(a_i^1, \ldots, a_i^n)$. Then Key showed that the linear complexity of $\mathbf{b}$ satisfies

$$\lambda_{\text{lin}}(\mathbf{b}) \leq H(\lambda_{\text{lin}}(a^1), \ldots, \lambda_{\text{lin}}(a^n)), \tag{1}$$

where we treat H as a polynomial with integer coefficients that happen to be 0s and 1s [15]. Moreover, Key showed that if the a^j are m-sequences and their periods are pairwise relatively prime, then we have equality in equation (1).

This leaves several related questions:

1. Can we express or bound the 2-adic complexity of $\mathbf{b}$ in terms of the 2-adic complexities of the $\mathbf{a}^i$? Similarly for various π-adic complexities.
2. Are there conditions under which both the linear and 2-adic complexities of $\mathbf{b}$ are large? All π-adic complexities?
3. What if we add a small amount of memory to the combiner? Rueppell investigated the *summation combiner*, where H is binary addition with carry [26]. He gave a heuristic argument that the linear complexity should be large, but gave no actual proof. To our knowledge, no proof has yet been found. On the other hand, it is known that the 2-adic complexity of $\mathbf{b}$ is the sum of the 2-adic complexities of the $\mathbf{a}^i$, so the summation combiner is vulnerable to an FCSR synthesis attack. In fact it was this that motivated the invention of FCSRs.

We must point out that even if we achieve large linear and 2-adic complexities (or even large π-adic complexity for all π), this does not make these sequences secure. There are other attacks. For example, combiners tend to be vulnerable to *correlation attacks* [23].

A second approach to reducing vulnerability of stream ciphers to synthesis attacks is to use a single LFSR or FCSR for the state and state change function, but use a nonlinear function $F(a_0, \ldots, a_{r-1})$ as the output function. Regarding LFSRs, Key [15] showed that, in the binary case, if $d = \deg(F)$, then

$$\lambda_{\text{lin}}(\mathbf{b}) \leq \sum_{i=0}^{d} \binom{n}{i}. \tag{2}$$

The following questions are open:

1. How can F be chosen to achieve equality in inequality (2)?
2. What is the N-adic complexity of a filtered FCSR?
3. What is the linear complexity of an ℓ-sequence or a filtered FCSR?
4. What is the N-adic complexity of an m-sequence or a filtered LFSR?

Similar questions can be asked about π-adic complexity, where π is an element of a ring R. More generally, it is an open problem how to efficiently generate sequences that have both large linear complexity and large 2-adic complexity (or π-adic complexity for any π).

We mention here some additional motivation for studying FCSRs. There is a type of attack on filtered LFSR generators known as *algebraic cryptanalysis* [5]. The basic idea is to treat each monomial in the filter function F as a variable. Knowing some ciphertext and plaintext gives the attacker some keystream and thus gives an equation in these metavariables. More known keystream gives more equations using the composition of F with iterations of the state change. If the degree of F is small (or if there is a low degree multiple of F), then the number of metavariables is small, and if there are enough equations, we can solve for the metavariables.

Critical to this attack is the fact that F composed with the state change has the same degree as F. However, if we replace the LFSR with an FCSR, this is no longer the case and algebraic cryptanalysis no longer works.

5 Statistics of Sequences

In this section we consider various open questions on statistical properties of shift register sequences.

5.1 Average Behavior

We would like to understand the average behavior of the $\mathcal{G}$-complexity of sequences. Deviation from the average can be used as a measure of nonrandomness (the NIST test suite does this with linear complexity [25]). Moreover, if the average is large, then we know that randomly chosen sequences are likely to have large $\mathcal{G}$-complexity. This is important because many stream ciphers are designed to be hard to analyze. For such ciphers, it is likely to be impossible to determine the $\mathscr{G}$-complexity.

But averaged over what? We can use Haar measure on infinite sequences. However, the eventually periodic sequences are countable, so have measure zero. For most $\mathscr{G}$ of interest only eventually periodic sequences have finite $\mathscr{G}$-complexity, so the average $\mathscr{G}$-complexity is infinite. This tells us nothing.

Instead, we can consider two ways of averaging:

1. $E_n^{\text{fin},\mathscr{G}}$ = average $\mathscr{G}$-complexity over all finite, length n sequences.
2. $E_n^{\text{per},\mathscr{G}}$ = average $\mathscr{G}$-complexity over all infinite period n sequences.

Note that these are different. In the first case, the $\mathscr{G}$-complexity of a finite sequence is the minimum $\mathscr{G}$-complexity over all infinite extensions of $(a_0, \ldots, a_{n-1})$, not just the period n extensions. Thus

$$E_n^{\text{fin},\mathscr{G}} \le E_n^{\text{per},\mathscr{G}}.$$

The following averages are known:

1. $E_n^{\text{fin,lin}} = n/2 + O(1/q)$ for sequences over $\mathbb{F}_q$.
2. $E_n^{\text{per,lin}} \gtrsim n - m/(q-1)$. for sequences over $\mathbb{F}_q$, q a power of p, $n = p^v m$, $\gcd(m, p) = 1$ (the exact value can be expressed in terms of cyclotomic numbers).
3. $E_n^{\text{per},N\text{-adic}} \in n - O(\log(n))$ (the exact value can be expressed in terms of the prime factorization of $N^n - 1$).

This leaves open the determination of $E_n^{\text{fin},N\text{-adic}}$.

For AFSRs over $R = \mathbb{Z}[\pi]$, we know that if R is a UFD with $\pi^2 = -N < 0$ or $\pi^d = N > 0$ and n is a multiple of 4 in the former case and is arbitrary in the latter case, then $E_n^{\text{per},\mathscr{G}} \in n - O(\log(n))$ [19]. The average finite π-complexity is unknown, as are both averages for any other R. Note that in these cases, there are reasonable definitions of the size of an AFSR. For example, if $\pi^2 = -N$ and F is the field of fractions of R, then we define the size of an element $u + v\pi \in R$, $u, v \in \mathbb{Z}$, to be

$$\phi(u + v\pi) = \log_N(\mathscr{N}_{\mathbb{Q}}^F(u + v\pi)) = \log_N(u^2 + Nv^2).$$

Then the size of an AFSR with a given initial state is $\max(\phi(f), \phi(g))$ if the AFSR has connection element g and outputs a sequence whose associated π-adic number has rational representation f/g. If $\pi^d = N > 0$, then we let the size of an element be

$$\phi\left(\sum_{i=0}^{d-1} u_i \pi^i\right) = \max(d \log_N |u_i| + i),$$

and we extend this to sequences similarly. In both cases it can be seen that this notion of size approximates the number of N-ary digits needed to represent the state. It seems that the first step in extending these results to more general R is to find a suitable notion of size.

5.2   Correlations

Let $\mathbf{a}$ and $\mathbf{b}$ be binary sequences of period T. The classical notion of the cross-correlation of $\mathbf{a}$ and $\mathbf{b}$ is

$$\mathscr{C}_{\mathbf{a},\mathbf{b}}(t) = \sum_{i=0}^{T-1}(-1)^{a_i+b_{i+t}}$$

$$= \#\text{zeros} - \#\text{ones in one period of } \mathbf{a} + \text{shift}_t(\mathbf{b}).$$

If $\mathbf{a} = \mathbf{b}$, then the cross-correlation is called the *autocorrelation* of $\mathbf{a}$, denoted $\mathscr{A}_{\mathbf{a}}(t)$.

The cross-correlation is used in code division multiple access (CDMA) communications. Each user has a sequence $\mathbf{a}$ that determines how the user's signal is distributed across a set of T channels. Typically it is necessary that the sequences used by two users have low cross-correlation to prevent interference. Thus the capacity of the system is limited by the size of a family of sequences with low pairwise cross-correlations.

Unfortunately, there are various known constraints on this size. One such constraint is the *Welch bound* [27]. Let S be a set of n binary sequences of period T. Let $\mathscr{C}_{\max}$ be the maximum cross-correlation between distinct sequences in S (including shifts of sequences and including shifted autocorrelations). Then

$$\mathscr{C}_{\max}^2 \geq \frac{T^2(n-1)}{nT-1}.$$

Thus, for example, if $n = T^{1/2}$, then

$$\mathscr{C}_{\max}^2 \geq T\left(1 - \frac{T-1}{T^{3/2}-1}\right) \sim T.$$

There is an analogous notion for with-carry algebra. Let $-_2$ denote subtraction with borrow of binary sequences. That is, to compute $\mathbf{a} -_2 \mathbf{b}$, find the associated 2-adic numbers a and b, subtract them in $\mathbb{Z}_2$, and extract the sequence of coefficients of the result. Note that if $\mathbf{a}$ and $\mathbf{b}$ have period T, then $\mathbf{a} -_2 \mathbf{b}$ is only eventually periodic (in fact it is periodic from the Tth term on). We define the *arithmetic cross-correlation* to be

$$\mathscr{C}_{\mathbf{a},\mathbf{b}}^A(t) = \#\text{zeros} - \#\text{ones in one period of } \mathbf{a} -_2 \text{shift}_t(\mathbf{b}).$$

We next define a set of sequences whose pairwise arithmetic cross-correlations are identically zero. The d-*fold decimation* of a sequence $\mathbf{a} = a_0, a_1, a_2, \ldots$ is the sequence $\mathbf{a}^d = a_0, a_d, a_{2d} \ldots$.

Theorem 5.1 *Suppose 2 is a primitive root modulo the prime number g. Let $\mathbf{a}$ be an ℓ-sequence with connection integer g. Suppose that $\gcd(g, d) = \gcd(g, e) = 1$ and that $\mathbf{a}^d$ is not a shift of $\mathbf{a}^e$. Then for all t, $\mathscr{C}^A_{\mathbf{a}^d, \mathbf{a}^e}(t) = 0$.*

It follows that the set $S_g = \{\mathbf{a}^d : \gcd(g, d) = 1\}$ is a set of sequences with identically zero arithmetic cross-correlations. This is in stark contrast to the classical setting. Two questions remain. First, is there an application of this remarkable fact? Second, how large is S? That is, how many shift distinct decimations of an ℓ-sequence are there.

Conjecture 5.2 If $d \not\equiv e \mod q$ and $q > 13$, then $\mathbf{a}^d$ is not a shift of $\mathbf{a}^e$.

If true, this would give us sets S_g of period $g - 1$ sequences with zero arithmetic cross-correlations and $|S_g| = \phi(g - 1)$. Note that $\phi(g - 1)$ can be as large as $(g - 3)/2$.

It is known that the conjecture is true:

1. For $13 < q < 8 \times 10^9$ (by brute force search)
2. For some special cases ($d = 1$, $e = q - 2$ [8]; $d = 1$, $g \equiv 1 \mod 4$, and $e = (q + 1)/2$ [11, 12])
3. For $q > 4.92 \times 10^{34}$ [2, 3, 12]

Bourgain's et al. result is based on recent deep results on bounds for certain exponential sums.

5.3 Asymptotic Complexity

Let $\mathscr{G}$ be a family of PSGs. Typically a sequence $\mathbf{a}$ is eventually periodic if and only if it can be generated by some $G \in \mathscr{G}$. Let us call such a $\mathscr{G}$ *periodic*. The "if" part certainly holds if for any $G = (S, \Gamma, \delta)$ and $\sigma \in S$, the set of states $\{\Gamma^i(\sigma) : i \in \mathbb{N}\}$ is finite. The "if and only if" holds for LFSRs, FCSRs, and AFSRs based on ring $R = \mathbb{Z}[\pi]$ and π if $|\pi| > 1$ for every embedding of the fraction field of R in $\mathbb{C}$.

Suppose $\mathbf{a}$ is not eventually periodic. How can we understand the $\mathscr{G}$-complexity of $\mathbf{a}$? One way is to consider finite prefixes of $\mathbf{a}$ and study the growth in their $\mathscr{G}$ complexities as the length increases.

Let $\lambda_{\mathscr{G},n}(\mathbf{a}) = \lambda_{\mathscr{G}}(a_0, a_1, \ldots, a_{n-1})$. The $\mathscr{G}$-*complexity profile* of $\mathbf{a}$ is the sequence $\Lambda_{\mathscr{G}}(\mathbf{a}) = (\lambda_{\mathscr{G},1}(\mathbf{a}), \lambda_{\mathscr{G},2}(\mathbf{a}), \ldots)$. Assume for the remainder of this subsection that $\mathscr{G}$ is a family of periodic PSGs. Then

$$\lim_{n \to \infty} \lambda_{\mathscr{G},n}(\mathbf{a}) = \infty,$$

so the limit tells us nothing. We assume further that for all $\mathbf{a}$, we have $\lambda_{\mathscr{G},n}(\mathbf{a}) \leq n + o(n)$. This is the case for LFSRs, FCSRs, and AFSRs since these families contain pure cycling registers ($g_1 = \cdots = g_{r-1} = 0, g_r = 1$). In this case we can normalize by defining

$$\delta_{\mathscr{G},n}(\mathbf{a}) = \lambda_{\mathscr{G},n}(\mathbf{a})/n \in [0, 1 + o(1)].$$

Then we can ask about the limiting behavior of $\delta_{\mathscr{G},n}(\mathbf{a})$ as n tends to infinity.

However, it is not in general the case that this sequence has a single limit point. Rather, it has a set of accumulation points $T(\mathbf{a}) \subseteq [0, 1]$. It is this set we want to study. When does there exist a single limit point of the $\delta_{\mathscr{G},n}(\mathbf{a})$? In general what is the structure of $T(\mathbf{a})$?

The first question was answered by Niederreiter for linear complexity [24]. The answer is that generically a single limit exists and that limit is $1/2$. More precisely, recall that there is a natural measure on the set $L = \{0, 1\}^{\infty}$ of infinite binary sequences, called Haar measure. This is simply the infinite product of the uniform measure on $\{0, 1\}$. This is very nearly the uniform measure on the real unit interval $[0, 1]$. Niederreiter showed that there is a set $U \subseteq L$ with measure one such that if $\mathbf{a} \in U$ then $T(\mathbf{a}) = [1/2, 1/2]$.

It is an open problem to prove this for any other family of PSGs.

Next we mention a theorem that partially answers the second question.

Theorem 5.3 ([17]) *If $\lambda_{\mathscr{G},n} \leq \lambda_{\mathscr{G},n+1}$, then $T(\mathbf{a}) = [B, C] \subseteq [0, 1]$.*

But what are the possible values of B and C?

Conjecture 5.4 For all $\mathbf{a}$, we have $T(\mathbf{a}) = [B, 1 - B]$. For every $B \in [0, 1/2]$ there are uncountably many sequences $\mathbf{a}$ for which $T(\mathbf{a}) = [B, 1 - B]$.

The following are the cases when the conjecture is known to be true:

1. LFSRs [7].
2. 2^k-ary and 3^k-ary FCSRs [17].
3. N-ary FCSRs if $B < \log_N(2)$ [17].
4. π-adic AFSRs with $\pi^2 = -2$ if $B < \log_2(4/3)$ [18].

All other cases are open.

Acknowledgements This material is based upon work supported by the National Science Foundation under Grant No. CCF-0514660. Any opinions, findings, and conclusions or recommendations expressed in this material are those of the author and do not necessarily reflect the views of the National Science Foundation.

References

1. F. Arnault, T. Berger, A. Necer, Feedback with carry shift registers synthesis with the Euclidean algorithm. IEEE Trans. Inf. Theory **50**, 910–917 (2004)
   This paper modifies the extended Euclidean algorithm to find a minimal FCSR generating a sequence given a sufficiently long prefix of the sequence.
2. J. Bourgain, T. Cochrane, J. Paulhus, C. Pinner, Decimations of ℓ-sequences and permutations of even residues mod p. SIAM J. Discrete Math. **23**, 842–857 (2009)
3. J. Bourgain, T. Cochrane, J. Paulhus, C. Pinner, On the parity of k-th powers mod p, a generalization of a problem of Lehmer. Acta Arith. **147**, 173–203 (2011)
   These two papers show that if p is large enough and 2 is primitive modulo p, then all decimations of an ℓ-sequence with connection integer p are cyclically distinct. It is conjectured that this is true for all primes $p > 13$.
4. R. Carmichael, Sequences of integers defined by recurrence relations, Q. J. Pure Appl. Math. **48**, 343–372 (1920)
   This is one of the first papers to study integer linear recurrences modulo an integer.
5. N. Courtois, Fast algebraic attacks on stream ciphers with linear feedback, in *Advances in Cryptology: Crypto 2003*, ed. by D. Boneh, Lecture Notes in Computer Science, vol 2729 (Springer, Berlin, 2003), pp. 177–194
   Courtois' seminal paper describes an attack on stream ciphers based on finding low degree multiples of the polynomials that express the output from a keystream generator in terms of the state bits. If such multiples can be found, then the problem of recovering the state from the output can be solved by solving a system of linear equations in the monomials of low degree.
6. R. Couture, P. L'Écuyer, Distribution properties of multiply-with-carry random number generators. Math. Comput. **66**, 591–607 (1997)
   Couture and L'Écuyer invented multiply with carry sequences, generated by linear recurrences with carry. These are equivalent to FCSR sequences, which were invented independently at about the same time by Goresky and Klapper.
7. Z. Dai, S. Jiang, K. Imamura, G. Gong, Asymptotic behavior of normalized linear complexity of ultimately non-periodic sequences. IEEE Trans. Inf. Theory **50**, 2911–2915 (2004)
   Let **a** be an infinite, binary, eventually aperiodic sequence. The authors show that the set of accumulation points of the normalized linear complexities of prefixes of **a** is an interval of the form $[B, 1 - B]$.
8. M. Goresky, A. Klapper, Arithmetic cross-correlations of FCSR sequences. IEEE Trans. Inf. Theory **43**, 1342–1346 (1997)
   It is shown that the arithmetic cross-correlations of cyclically distinct binary ℓ-sequences are identically zero.
9. M. Goresky, A. Klapper, Periodicity, correlation, and distribution properties of d–FCSR sequences. Des. Codes Cryptogr. **33**, 123–148 (2004)
   d-FCSRs are a variant of FCSRs base on the algebra of $\mathbb{Z}[2^{1/d}]$. In this paper various statistical properties of maximum period d-FCSR sequences are considered.
10. M. Goresky, A. Klapper, *Algebraic Shift Register Sequences* (Cambridge University Press, Cambridge, 2012), http://www.cs.uky.edu/~klapper/algebraic.html
    This is an extensive monograph on sequence generators based on abstract algebra. Topics studied include statistical analysis, maximum period sequences, and the register synthesis problem: the problem of finding a minimal generator of a particular type for a sequence given a short prefix.
11. M. Goresky, A. Klapper, R. Murty, On the distinctness of decimations of ℓ-sequences, in *Sequences and Their Applications—SETA '01*, eds. by T. Helleseth, P.V. Kumar, K. Yang (Springer, Berlin, 2002), pp. 197–208

12. M. Goresky, A. Klapper. R. Murty, I. Shparlinski, On decimations of ℓ-sequences. SIAM J. Discrete Math. **18**, 130–140 (2004)

These two papers give a partial solution to the conjecture that all decimations of an ℓ-sequence are distinct if the connection integer is greater than 13. Exponential sum techniques are used.

13. D. Heath-Brown, Artin's conjecture for primitive roots. Q. J. Math. Oxford Ser. **37**(1), 27–38 (1986)

14. C. Hooley, On Artin's conjecture. J. Reine Angew. Math. **22**, 209–220, (1967)

These two papers give a partial solution to Artin's conjecture (that for any integer N there are infinitely many primes for which N is primitive), assuming a generalized Riemann hypothesis.

15. E. Key, An Analysis of the structure and complexity of nonlinear binary sequence generators. IEEE Trans. Inf. Theory **22**(1), 732–736 (1976)

Key analyzed the linear complexities of sequences generated by nonlinear combiners and LFSRs with nonlinear output functions.

16. A. Klapper, Distributional properties of d–FCSR sequences. J. Complexity **20**, 305–317 (2004)

Let **a** be a maximum period sequence generated by a length m p-ary d-FCSR (an AFSR based on a ring $\mathbb{Z}[p^{1/d}]$). We study the variation in the number of occurrences of blocks of length $s \leq m$. If $d = 2$, we see that the variation is bounded by a constant times the square root of the average number of occurrences of blocks of length s.

17. A. Klapper, The asymptotic behavior of 2-adic complexity. Adv. Math. Commun. **1**, 307–319 (2007)

Let **a** be an infinite, binary, eventually aperiodic sequence. We show that the set of accumulation points of the normalized 2-adic complexities of prefixes of **a** is an interval of the form $[B, 1 - B]$.

18. A. Klapper, The asymptotic behavior of π-adic complexity with $\pi^2 = -2$, in *Sequences, Subsequences, and Consequences*, eds. by S. Golomb, G. Gong, T. Helleseth, H.-Y. Song, Lecture Notes in Computer Science, vol 4893 (Springer, Berlin, 2007), pp. 134–147

Let **a** be an infinite, binary, eventually aperiodic sequence. We show that the set of accumulation points of the normalized π-adic complexities of prefixes of **a** is an interval of the form $[B, 1 - B]$ in some cases.

19. A. Klapper, Expected π-adic complexity of sequences. IEEE Trans. Inf. Theory **56**, 2486–2501 (2010)

This paper computes the average π-adic complexity of sequences of fixed period.

20. A. Klapper, M. Goresky, Feedback shift registers, 2-adic span, and combiners with memory. J. Cryptol. **10**, 111–147 (1997)

In this paper FCSRs were introduced and many basic properties were worked out. Parts of this analysis were based on the algebra of 2-adic numbers.

21. J. Xu, A. Klapper, Feedback with carry shift registers over $\mathbb{Z}/(n)$, in *Proceedings of International Conference on Sequences and their Application (SETA), Singapore, December 1998*, eds. by C. Ding, T. Helleseth, H. Niederreiter (Springer, Berlin, 1999), pp. 379–392

This paper generalizes Xu and Klapper's algorithm [28] to AFSRs over certain number fields (including d-FCSRs).

22. J.L. Massey, Shift register synthesis and BCH decoding. IEEE Trans. Inf. Theory **15**(1), 122–127 (1969)

Jim Massey showed here how Berlekamp's decoding algorithm could be used as an efficient solution to the LFSR synthesis problem. The idea is to process one symbol at a time. When the rational approximation needs updating (i.e., a discrepancy occurs), a new approximation that is correct for the new symbol is found as a linear combination of two earlier approximation.

23. W. Meier, O. Staffelbach, Correlation properties of combiners with memory in stream ciphers. J. Cryptol. **5**, 67–86 (1992)

This paper describes an effective attack on stream ciphers that combine several m-sequences with a nonlinear combiner, endowed with a small amount of extra memory. The basis is the analysis of the combining function to find a correlation between state bits and output bits.

24. H. Niederreiter, The probabilistic theory of linear complexity, in *Advances in Cryptology—Eurocrypt 88*, ed. by C. Günther, Lecture Notes in Computer Science, vol 330 (Springer, Berlin, 1988), pp. 191–209
   Niederreiter showed that with probability 1 the limit of the normalized linear complexity of a sequence exists and equals $1/2$. The proof of this fact uses a relationship between continued fractions and linear complexity and uses the theory of dynamical systems.
25. NIST, Statistical test suite for random and pseudorandom number generators for cryptographic applications, http://csrc.nist.gov/groups/ST/toolkit/rng/index.html
   This is a resource with useful tools for measuring statistical randomness of sequences.
26. R. Rueppel, *Analysis and Design of Stream Ciphers* (Springer, Berlin, 1986)
   This book studies several aspects of stream ciphers and statistical properties of sequences, including nonlinear combiners with memory such as the summation combiner. It was largely based on Ruepppel's Ph.D. dissertation.
27. L.R. Welch, Lower bounds on the maximum correlation of signals. IEEE Trans. Inf. Theory **20**(1), 397–399 (1974)
   Here Welch derived a fundamental constraint on the size of sequence families with low pairwise correlations.
28. A. Klapper, J. Xu, Register synthesis for algebraic feedback shift registers based on non-primes. Des. Codes Cryptogr. **31**, 227–250 (2004)
   This paper presents a solution to the FCSR synthesis problem for n-ary FCSRs with n arbitrary. This algorithm modifies the Berlekamp–Massey algorithm—when a discrepancy is found, the rational approximation is amended to account for several new sequence symbols instead of just one.

Open Problems on Binary Bent Functions

Claude Carlet

Abstract This chapter gives a survey of the recent results on Boolean bent functions and lists some open problems in this domain. It includes also new results. We recall the definitions and basic results, including known and new characterizations of bent functions; we describe the constructions (primary and secondary; known and new) and give the known infinite classes, in multivariate representation and in trace representation (univariate and bivariate). We also focus on the particular class of rotation symmetric (RS) bent functions and on the related notion of bent idempotent: we give the known infinite classes and secondary constructions of such functions, and we describe the properties of a recently introduced transformation of RS functions into idempotents.

1 Introduction

Boolean functions are functions from $\mathbb{F}_2^n$ to $\mathbb{F}_2$, where n is some positive integer called the number of variables. They play a role in almost all the domains of computer science. We are more interested here in their relationship with error-correcting codes and private-key cryptography. Multi-output Boolean functions are functions from $\mathbb{F}_2^n$ to $\mathbb{F}_2^m$ for some m (to specify that $m = 1$, we can speak of single-output Boolean function, but even without such precision, "Boolean function" without writing "vectorial" will imply single output).

A binary error-correcting code of a given length N is a subset C of $\mathbb{F}_2^N$. Each information to be sent over a noisy channel is encoded before transmission by the sender into an element of C (a codeword); if d is the minimum Hamming distance between two distinct codewords (called the minimum distance of the code), such encoding allows theoretically the receiver to correct up to $e = \left[\frac{d-1}{2}\right]$ binary errors in the transmission of a codeword (e is called the error correction capability of the

C. Carlet (✉)
LAGA, Universities of Paris 8 and Paris 13, CNRS, UMR 7539, France

Department of Mathematics, University of Paris 8, 2 rue de la liberté,
93526 Saint-Denis cedex 02, France
e-mail: claude.carlet@univ-paris8.fr

© Springer International Publishing Switzerland 2014
Ç.K. Koç (ed.), *Open Problems in Mathematics and Computational Science*,
DOI 10.1007/978-3-319-10683-0_10

code). When the code has length $N = 2^n$, the codewords can be interpreted as Boolean functions (some order on the elements of $\mathbb{F}_2^n$ being chosen beforehand). Reed-Muller codes [49] are examples of such codes defined as sets of Boolean functions. Any code of length N can be viewed itself as the support of an N-variable function, and some notions on codes (e.g., the dual distance) correspond to notions on Boolean functions (e.g., correlation immunity), but we shall not address this here.

Private-key cryptosystems (stream ciphers, which encrypt at a bit level, and block ciphers, which encrypt block by block) allow exchanging confidentially large-size data over a public channel. With such conventional symmetric ciphers, it is necessary to possess the secret encryption key (resp. the secret decryption key, which is in general equal to the encryption key, and is supposed to have been safely shared in advance between the sender and the receiver) for being able to encrypt (resp. decrypt) messages.[1] For reasons of speed, these cryptosystems involve linear operations, and for reasons of resistance to attacks, they need also to involve some amount of nonlinearity, often brought by single-output or multi-output Boolean functions. There are several ways of quantifying how nonlinear (i.e., different from linear functions) a given Boolean function can be. The main parameter for such quantification is the so-called *nonlinearity*, equal to the minimum Hamming distance between the function and all affine functions (i.e., sums of a linear function and a binary constant). The nonlinearity of any n-variable Boolean function is bounded above by $2^{n-1} - 2^{n/2-1}$. The functions achieving this bound with equality exist only for n even since the nonlinearity is an integer (in fact, they exist if and only if n is even). They are called *bent*. Such bent functions are not directly used in stream ciphers because bentness makes impossible the function to be balanced (i.e., to have output uniformly distributed over $\mathbb{F}_2$), and this induces a statistical correlation between the plaintext and the ciphertext; bentness also implies other cryptographic weaknesses. But bent functions can be involved in the substitution boxes (S-boxes) of block ciphers, whose role is also to bring some amount of nonlinearity (allowing to resist the differential and linear attacks), and the study of bent functions and those of Boolean functions for stream ciphers and of S-boxes for block ciphers are closely related.

Bent functions are involved in codes such as the Kerdock codes (see Sect. 3.4). They are also related to combinatorics (e.g., difference sets; see Sect. 2.2.1), design theory (any difference set can be used to construct a symmetric design), and sequence theory (see [16]).

Bent functions have been studied in numerous papers. A survey on Boolean bent functions is given in [11, Sect. 8.6] (and complements on non-binary bent functions, that we do not address here, can be found in [56]). The purpose of this chapter is to complete this survey on Boolean bent functions with results which appeared after the publication of [11] (and with a few original results) and to focus on open problems.

---

[1] In public-key cryptography, the only key which must be kept secret is the decryption key, but as far as we know, bent functions play no big role in such ciphers.

2 Boolean Bent Functions: Definitions and Basic Results

2.1 Representations of Affine Boolean Functions

Affine Boolean functions (i.e., sums of linear functions and constants) over $\mathbb{F}_2^n$ can be written in *multivariate representation*, that is, writing the input x as a vector $(x_1, \ldots, x_n)$ of $\mathbb{F}_2^n$, in the form

$$a_1 x_1 + \cdots + a_n x_n + \epsilon = a \cdot x + \epsilon \text{ (sums mod 2)}$$

for some $a = (a_1, \ldots, a_n) \in \mathbb{F}_2^n$, $\epsilon \in \mathbb{F}_2$.

Other representations exist. The vector space $\mathbb{F}_2^n$ can be endowed with the structure of the field $\mathbb{F}_{2^n}$ (since we know that this field is an n-dimensional vector space over $\mathbb{F}_2$). This allows to use the trace function $\mathrm{Tr}_1^n(x) = x + x^2 + x^{2^2} + \cdots + x^{2^{n-1}}$ as a basic linear form over $\mathbb{F}_{2^n}$, and any affine function of $x \in \mathbb{F}_{2^n}$ can then be written in the so-called *univariate trace representation*:

$$\mathrm{Tr}_1^n(ax) + \epsilon \; ; \; a \in \mathbb{F}_{2^n}, \; \epsilon \in \mathbb{F}_2 \text{ (sum mod 2)}.$$

The multivariate and univariate trace representations are at two opposite extremes. A representation which is intermediate and happens to be important in the framework of bent functions is the *bivariate trace representation*. For n even, we identify $\mathbb{F}_2^n$ with $\mathbb{F}_{2^{n/2}} \times \mathbb{F}_{2^{n/2}}$, and every affine Boolean function of $(x, y) \in \mathbb{F}_{2^{n/2}} \times \mathbb{F}_{2^{n/2}}$ can then be expressed as

$$\mathrm{Tr}_1^{n/2}(ax + ay) + \epsilon \; ; \; a, b \in \mathbb{F}_{2^{n/2}}, \; \epsilon \in \mathbb{F}_2 \text{ (sum mod 2)}.$$

2.2 Corresponding Expressions for the Walsh Transform of Boolean Functions, Nonlinearity, and Bentness

The Walsh transform of a Boolean function calculates the correlations between the function and linear Boolean functions.

In multivariate representation, that is, over $\mathbb{F}_2^n$, it can be expressed as

$$\widehat{\chi_f}(a) = \sum_{x \in \mathbb{F}_2^n} (-1)^{f(x) + a \cdot x}; \; a \in \mathbb{F}_2^n (\text{sum in } \mathbb{Z}),$$

where "·" is an inner product in $\mathbb{F}_2^n$ (for instance, we can take the usual inner product, introduced in Sect. 2.1).

In univariate trace representation (over $\mathbb{F}_{2^n}$), we take

$$\widehat{\chi_f}(a) = \sum_{x \in \mathbb{F}_{2^n}} (-1)^{f(x) + \mathrm{Tr}_1^n(ax)}; \ a \in \mathbb{F}_{2^n}.$$

In bivariate trace representation (over $\mathbb{F}_{2^m} \times \mathbb{F}_{2^m}$; $m = n/2$), we take:

$$\widehat{\chi_f}(a, b) = \sum_{x, y \in \mathbb{F}_{2^m}} (-1)^{f(x,y) + \mathrm{Tr}_1^m(ax+by)}.$$

The *Hamming distance* between two functions equals by definition the *Hamming weight* of their difference (i.e., their sum):

$$d_H(f, g) = w_H(f + g) = |\{x \in \mathbb{F}_2^n \ / \ f(x) \neq g(x)\}.$$

The *nonlinearity* $nl(f)$ of a Boolean function f equals by definition the minimum Hamming distance between f and affine functions. It is a simple matter to show that

$$nl(f) = 2^{n-1} - \frac{1}{2} \max_{a \in \mathbb{F}_2^n} |\widehat{\chi_f}(a)|$$

(note that this equality is valid whatever is the choice of the inner product "$\cdot$"). Because of the easily proved Parseval relation

$$\sum_{a \in \mathbb{F}_2^n} \widehat{\chi_f}^2(a) = 2^{2n},$$

the mean of $\widehat{\chi_f}^2(a)$, equals 2^n and this implies the so-called *covering radius bound*:

$$nl(f) \leq 2^{n-1} - 2^{n/2-1}.$$

The bound is tight for every even n; the functions achieving it with equality are called *bent*; the Walsh transforms of bent functions take values $\pm 2^{n/2}$, only. The *dual* $\tilde{f}$ of a bent function f is the function defined on $\mathbb{F}_2^n$ by

$$\widehat{\chi_f}(u) = 2^m (-1)^{\tilde{f}(u)}, u \in \mathbb{F}_2^n, \ m = n/2.$$

It is also a bent function, and the mapping $f \mapsto \tilde{f}$ is an isometry [11]. Bounds involving the algebraic degrees of bent functions (see definition below) and their duals and relations between the degree $n/2$ terms in the algebraic normal form (ANF, see also below) of bent functions and their duals are recalled in [11]. Self-dual bent functions are studied in [21].

2.2.1 Characterizations of Bent Functions

- Any Boolean function f is bent if and only if for any nonzero vector a, the so-called *derivative* $D_a f(x) = f(x) + f(x + a)$ (sum mod 2) is balanced (i.e., has Hamming weight 2^{n-1} or equivalently satisfies $\sum_{x \in \mathbb{F}_2^n} (-1)^{D_a f(x)} = 0$) [59], since we have $\sum_{a \in \mathbb{F}_2^n} \left((-1)^{a \cdot b} \sum_{x \in \mathbb{F}_2^n} (-1)^{D_a f(x)} \right) = \widehat{\chi_f}^2(b)$ and $\sum_{b \in \mathbb{F}_2^n} \widehat{\chi_f}^2(b)(-1)^{a \cdot b} = 2^n \sum_{x \in \mathbb{F}_2^n} (-1)^{D_a f(x)}$. The fact that $\sum_{b \in \mathbb{F}_2^n} (-1)^{a \cdot b}$ equals 2^n if $a = 0$ and is null otherwise completes the proof of the equivalence.

  Bent functions are also called *perfect nonlinear* and are equivalently the indicators of *difference sets* in elementary Abelian 2-groups [28].
- An n-variable Boolean function f is bent if and only if $\sum_{a \in \mathbb{F}_2^n} \widehat{\chi_f}^4(a)$ equals 2^{3n} (which, by the Cauchy–Schwarz inequality and the Parseval relation, is the minimum possible value).
- It is shown in [11] that:

Theorem 1 *A pair of n-variable Boolean functions f and f' satisfies, for every $a, b \in \mathbb{F}_2^n$, the relation $\sum_{x \in \mathbb{F}_2^n} (-1)^{D_a f'(x) + b \cdot x} = \sum_{x \in \mathbb{F}_2^n} (-1)^{D_b f(x) + a \cdot x}$ if and only if f and f' are bent and are the duals of each other, up to the addition of constant 1. Moreover, for any bent function f, we have $\sum_{x \in \mathbb{F}_2^n} (-1)^{D_a f(x) + b \cdot x} = \sum_{x \in \mathbb{F}_2^n} (-1)^{D_b f(x) + a \cdot x} = 0$ when $a \cdot b = 1$.*

- For n-variable Boolean functions f and f', we have $\sum_{x,y \in \mathbb{F}_2^n} (-1)^{f(x) + f'(y) + x \cdot y} = 2^{-n} \sum_{x,y,z \in \mathbb{F}_2^n} (-1)^{f(x) + (x+z) \cdot y} \widehat{\chi_f}(z) = \sum_{x \in \mathbb{F}_2^n} (-1)^{f(x)} \widehat{\chi_{f'}}(x) \leq \sum_{x \in \mathbb{F}_2^n} |\widehat{\chi_{f'}}(x)| \leq \sqrt{2^n \sum_{x \in \mathbb{F}_2^n} \widehat{\chi_{f'}}^2(x)} = 2^{3n/2}$, and this bound $\sum_{x,y \in \mathbb{F}_2^n} (-1)^{f(x) + f'(y) + x \cdot y} \leq 2^{3n/2}$ is achieved with equality if and only if both inequalities above are equalities, that is, if and only if $|\widehat{\chi_{f'}}(x)|$ is constant and $(-1)^{f(x)}$ is the sign of $\widehat{\chi_{f'}}(x)$ for every x, that is, if and only if f' is bent and $\widetilde{f'} = f$. This gives one more characterization of bent functions:

Theorem 2 *For every pair of n-variable Boolean functions f and f', we have*

$$\sum_{x,y \in \mathbb{F}_2^n} (-1)^{f(x) + f'(y) + x \cdot y} \leq 2^{3n/2},$$

with equality if and only if f and f' are bent and are the duals of each other.

In particular, a Boolean function f is bent and self-dual if and only if the so-called *Rayleigh quotient* $\sum_{x,y \in \mathbb{F}_2^n} (-1)^{f(x) + f(y) + x \cdot y}$ equals $2^{3n/2}$; this particular result was given in [21], but as far as we know, the general characterization of Theorem 2 is new. It has the interest of characterizing bent functions with a single character sum.

More characterizations are given in [11].

2.2.2 Number of Bent Functions

The number of bent functions is known only for $n \leq 8$ [43]. For larger values of n, only bounds are known. In particular, an upper bound exists [11, 17]. Comparing the bound and the actual value for $n = 8$ shows that the bound is weak (but it could not be improved during the last 10 years).

Open Problem 1: Bound more efficiently the number of n-variable bent functions (i.e., improve upon the bound of [17]).

2.2.3 Other Properties

We shall not list here all the properties of bent functions. We refer to the survey [11], in particular for the relationship with the sum-of-square indicator, with nonhomomorphicity, with codes, for the description of super-classes of Boolean functions and of bent sequences and for normal extensions of bent functions.

2.3 Equivalence of Boolean Functions

It seems elusive to determine all bent functions. This has been done for $n \leq 8$ only, and doing it for $n = 8$ was a difficult work [43]. An important subject of research is then to find constructions of bent functions leading to infinite classes of bent functions, or to find directly such infinite classes, after computer investigation. When a class is obtained, it remains to see if at least some of its elements are really new. Indeed, given a bent function, some simple transformations allow to obtain other bent functions; we say then that the known function and the functions we can obtain from it are equivalent (when the correspondence results in an equivalence relation); an infinite class is new if some of its elements are inequivalent to all previously known bent functions. We describe now the relevant notions of equivalence for bent functions.

The *automorphism group* of the set of bent functions

$$\{\sigma \text{ permutation of } \mathbb{F}_2^n \text{ s.t. } f \circ \sigma \text{ bent, } \forall f \text{ bent}\}$$

is the *general affine group*: $\sigma(x) = x \times A + a$ (A invertible matrix over $\mathbb{F}_2$) [11]. Two functions f and $f \circ \sigma$ are then called *affinely equivalent* (and $f(x)$ and $f(x \times A)$ are called linearly equivalent).

If f is bent and ℓ is affine, then $f + \ell$ is bent (sum mod 2). Two functions f and $f \circ \sigma + \ell$ are called *EA-equivalent*, and a class of bent functions is called *complete* if it is globally invariant under EA-equivalence. The completed version of a class is the set of all functions EA-equivalent to the functions in the class.

Another notion of equivalence called CCZ-equivalence [3, 20] exists and is more general than EA-equivalence for vectorial functions, but it is known from [1, 2] that for Boolean functions and for bent (Boolean or vectorial) functions, CCZ-equivalence coincides with EA-equivalence.

Note that X.-D. Hou and P. Langevin have made an observation reported in [11] showing that under some condition on a permutation σ, composing a bent function with σ may give another bent function. But this cannot be viewed as an equivalence since the condition on σ is hard to achieve.

2.4 Representations of Boolean Functions

Each representation of affine functions generalizes to a representation of general Boolean functions, useful for studying bent functions.

2.4.1 Multivariate Representation

Any Boolean function $f : \mathbb{F}_2^n \mapsto \mathbb{F}_2$ has a unique *ANF*:

$$f(x_1, \ldots, x_n) = \sum_{I \subseteq \{1, \ldots, n\}} a_I \left(\prod_{i \in I} x_i \right), \ a_I \in \mathbb{F}_2 \text{ (sum mod 2)}.$$

The algebraic degree of a Boolean function f is the global degree of its ANF. The *rth-order Reed-Muller code* of length 2^n equals by definition the set of n-variable Boolean functions (identified with binary vectors of length 2^n) of algebraic degrees at most r.

An important property of bent functions is that *their algebraic degrees are bounded above by $n/2$ [59]*.

A function f is called *quadratic* if it has algebraic degree 2, *cubic* if it has algebraic degree 3. We are able to characterize the ANF of bent functions only for quadratic functions: a quadratic Boolean function is bent if and only if it is affinely equivalent to $x_1x_2 + x_3x_4 + \cdots + x_{n-1}x_n$ or $x_1x_2 + x_3x_4 + \cdots + x_{n-1}x_n + 1$. Another characterization exists: according to the first result recalled in Sect. 2.2.1 and since all derivatives of a quadratic function are affine and are then balanced if and only if they are nonconstant, any quadratic function f is bent if and only if the so-called linear kernel of f, equal to the set of elements a such that $D_a f$ is constant, equals $\{0\}$. See more in [11].

Another representation of Boolean functions called the numerical normal form (NNF) and a related notion of degree called numerical degree exist (see the details in [11]), allowing characterizing bent functions. As far as we know, no recent result has been found on this relationship.

Open Problem 2 : The following question is posed by Tokareva in [63]: do the sums of two bent functions cover all Boolean functions of algebraic degrees at most $n/2$? Intuitively, the reply to this question would seem negative, for general n. However, the reply is shown by Tokareva to be yes for $n = 4, 6$ and for several classes of bent functions; it seems difficult to prove that the reply is no, in general: the usual parameters and properties of Boolean functions (ANF, NNF and numerical degree, generalized degree, divisibility of the Fourier transform or of the coefficients of the NNF, other properties of the Fourier or Walsh transform values) do not seem to allow discriminating sums of two bent functions from other Boolean functions of degrees at most $n/2$. Note that a positive reply to Tokareva's question would automatically imply a lower bound on the number of bent functions.

2.4.2 Univariate and Bivariate Representations: Trace Representations

Every function from $\mathbb{F}_{2^n}$ to $\mathbb{F}_{2^n}$ has a unique *univariate representation*

$$f(x) = \sum_{j=0}^{2^n-1} a_j \, x^j; \quad a_j, x \in \mathbb{F}_{2^n} \tag{1}$$

since the mapping $\sum_{j=0}^{2^n-1} a_j \, X^j \mapsto \sum_{j=0}^{2^n-1} a_j \, x^j$ from polynomials to functions is linear injective and therefore bijective.

Function f is Boolean if and only if:

$$a_0, a_{2^n-1} \in \mathbb{F}_2 \text{ and } a_{2j} = (a_j)^2, \forall j \in \mathbb{Z}/(2^n - 1)\mathbb{Z} \setminus \{0\}.$$

The univariate representation can then be written as a *trace representation*:

$$f(x) = \sum_{j \in \Gamma_n} \text{Tr}_1^{o(j)}(a_j x^j) + a_{2^n-1} x^{2^n-1}, \tag{2}$$

where:

- $a_{2^n-1} \in \mathbb{F}_2$ equals the Hamming weight of f modulo 2,
- Γ_n is the set of integers obtained by choosing one element in each cyclotomic coset of 2 mod $2^n - 1$,
- $o(j)$ is the size of the cyclotomic coset containing j,
- and $a_j \in \mathbb{F}_{2^{o(j)}}$.

In both univariate and trace representations (1), resp. (2), the algebraic degree equals: $\max\{w_2(j); \ j \, | \, a_j \neq 0\}$, where $w_2(j)$ is the Hamming weight of the binary expansion of j [11] (in particular, the univariate trace representation of a bent function does not involve the term x^{2^n-1}).

We have also $f(x) = \mathrm{Tr}_1^n(P(x))$ for some polynomial $P(x)$, but this representation is not unique and does not allow a general (simple) expression for the algebraic degree.

The bivariate representation is based on the identification $\mathbb{F}_2^n \approx \mathbb{F}_{2^m} \times \mathbb{F}_{2^m}$ and has the form:

$$f(x, y) = \sum_{0 \le i,j \le 2^m-1} a_{i,j} x^i y^j; \; a_{i,j} \in \mathbb{F}_{2^n}. \tag{3}$$

The existence of such representation for every function from $\mathbb{F}_{2^m} \times \mathbb{F}_{2^m}$ to $\mathbb{F}_{2^n}$ and the values of the coefficients $a_{i,j}$ in (3) can be derived from the univariate representation: choosing a basis (α, β) of $\mathbb{F}_{2^n}$ over $\mathbb{F}_{2^m}$ allows to express the input of f in the form $\alpha x + \beta y$, and expanding $(\alpha x + \beta y)^i$ in (1) gives (3) after reduction modulo $x^{2^m} + x$ and $y^{2^m} + y$. The uniqueness comes from the fact that the number of polynomials (3) equals the number $(2^n)^{2^m \times 2^m}$ of functions from $\mathbb{F}_{2^m} \times \mathbb{F}_{2^m}$ to $\mathbb{F}_{2^n}$ and that a surjective mapping from a finite set to a set of the same size is a bijection.

The function is Boolean if and only if the expression $(\sum_{0 \le i,j \le 2^m-1} a_{i,j} x^i y^j)^2$ equals $\sum_{0 \le i,j \le 2^m-1} a_{i,j} x^i y^j$ in the quotient algebra $\mathbb{F}_{2^n}[x, y]/(x^{2^m} + x, y^{2^m} + y)$, that is, if $a_{i,j}^2 = a_{2i,2j}$ for every (i, j), where $2i$ (resp. $2j$) is replaced by $2i - (2^m - 1)$ if $2i \ge m$ (resp. if $2j \ge m$).

This condition implies in particular that $a_{i,j}^{2^m} = a_{i,j}$, that is, $a_{i,j} \in \mathbb{F}_{2^m}$ (which can be directly deduced from $(\sum_{0 \le i,j \le 2^m-1} a_{i,j} x^i y^j)^2 \equiv \sum_{0 \le i,j \le 2^m-1} a_{i,j} x^i y^j$ [mod $(x^{2^m} + x, y^{2^m} + y)$]).

It also implies that the bivariate representation of any Boolean function over $\mathbb{F}_{2^m} \times \mathbb{F}_{2^m}$ can be written in *bivariate trace representation*:

$$f(x, y) = \sum_{(i,j) \in \Gamma_m'} \mathrm{Tr}_1^{o(i,j)}(a_{i,j} x^i y^j) + x^{2^m-1} \sum_{j \in \Gamma_m} \mathrm{Tr}_1^{o(j)}(a_{2^m-1,j} y^j)$$

$$+ y^{2^m-1} \sum_{i \in \Gamma_m} \mathrm{Tr}_1^{o(i)}(a_{i,2^m-1} x^i) + a_{2^m-1,2^m-1} x^{2^m-1} y^{2^m-1}, \tag{4}$$

where:

- $a_{2^m-1,2^m-1} \in \mathbb{F}_2$ equals the Hamming weight of f modulo 2,
- Γ_m' is a set obtained by choosing one ordered pair in each cyclotomic class of 2 modulo $2^m - 1$ in $\{0, \ldots, 2^m - 2\} \times \{0, \ldots, 2^m - 2\}$,
- Γ_m is a set obtained by choosing one element in each cyclotomic coset of 2 modulo $2^m - 1$ (in $\{0, \ldots, 2^m - 2\}$),
- $o(i, j)$ is the size of the equivalence class containing (i, j), and $o(i)$ is the size of the cyclotomic coset containing i,
- and $a_{i,j} \in \mathbb{F}_{2^{o(i,j)}}$, $a_j \in \mathbb{F}_{2^{o(j)}}$.

The algebraic degree of $f(x, y)$ equals $\max_{(i,j)| a_{i,j} \ne 0}(w_2(i) + w_2(j))$ in (3) and (4). In particular, the bivariate trace representation of a bent function does not involve any of the terms $x^{2^m-1}y^{2^m-1}$, $y^{2^m-1}\mathrm{Tr}_1^{o(i)}(a_{i,2^m-1}x^i)$, $i \neq 0$ and $x^{2^m-1}\mathrm{Tr}_1^{o(j)}(a_{2^m-1,j}y^j)$, $j \neq 0$).

As in the case of univariate representation, the simpler representation $f(x, y) = \mathrm{Tr}_1^m(P(x, y))$ (where $P(x, y)$ is a polynomial over $\mathbb{F}_{2^m}$) exists but is not unique.

3   Known Infinite Classes of Boolean Bent Functions

In all the rest of this chapter, n is even and we denote $n/2$ by m.

3.1   Basic Constructions in Multivariate Representation

In the *Maiorana–McFarland* (MM) construction, the input is represented in the form (x, y) where $x, y \in \mathbb{F}_2^m$. Given $\pi : \mathbb{F}_2^m \mapsto \mathbb{F}_2^m$ and $g : \mathbb{F}_2^m \mapsto \mathbb{F}_2$, the function:

$$f(x, y) = x \cdot \pi(y) + g(y); \ x, y \in \mathbb{F}_2^m$$

is bent if and only if π is a permutation on $\mathbb{F}_2^m$.

The dual function equals : $\tilde{f}(x, y) = y \cdot \pi^{-1}(x) + g(\pi^{-1}(x))$.

The truth table of an MM function is the concatenation of the truth tables of the affine functions $x \mapsto x \cdot \pi(y) + g(y)$ (in m variables). Completed MM class is the *widest known class* of bent functions. It covers all bent functions for $n \leq 6$ and all quadratic bent functions; but, for $n = 8$, it has size negligible with respect to all bent functions.

Generalizations of MM construction (see [11]):

- MM class has been modified by the addition of indicators of flats
- it has been generalized in diverse ways:

  - concatenations of quadratic functions,
  - concatenations of indicators of flats,
  - more complex concatenations (these constructions are more efficient for designing resilient functions than bent functions, though),

- it has also been generalized into a secondary construction (see Sect. 4).
- A construction including MM and PS_{ap} (see below) was given by Dobbertin.

3.2 Known Infinite Classes of Bent Functions in Univariate Trace Form

We list these known classes:

- $f(x) = \mathrm{Tr}_1^n\left(ax^{2^j+1}\right)$, where $a \in \mathbb{F}_{2^n} \setminus \{x^{2^j+1}; x \in \mathbb{F}_{2^n}\}$, $\frac{n}{\gcd(j,n)}$ even (the bentness of such function is directly deduced from the characterization recalled above of bent quadratic functions). This class has been generalized by Yu and Gong in [65] to functions of the form $\mathrm{Tr}_1^n(\sum_{i=1}^{m-1} a_i x^{2^i+1}) + c_m Tr_1^m(a_m x^{2^m+1})$. Being quadratic, these functions belong to the completed MM class.

- $f(x) = \mathrm{Tr}_1^n\left(ax^{2^{2j}-2^j+1}\right)$, where $a \in \mathbb{F}_{2^n} \setminus \{x^3; x \in \mathbb{F}_{2^n}\}$, $\gcd(j,n) = 1$ [29].

- $f(x) = \mathrm{Tr}_1^n\left(ax^{(2^{n/4}+1)^2}\right)$, where $n \equiv 4 \pmod 8$, $a = a'b^{(2^{n/4}+1)^2}$, $a' \in w\mathbb{F}_{2^{n/4}}$, $w \in \mathbb{F}_4 \setminus \mathbb{F}_2$, $b \in \mathbb{F}_{2^n}$ (Leander [44], see also [24]); the functions in this class belong to the completed MM class.

- $f(x) = \mathrm{Tr}_1^n\left(ax^{2^{n/3}+2^{n/6}+1}\right)$, where $6 \mid n$, $a = a'b^{2^{n/3}+2^{n/6}+1}$, $a' \in \mathbb{F}_{2^m}$, $\mathrm{Tr}_{m/3}^m(a') = 0$, $b \in \mathbb{F}_{2^n}$ [6]; the functions in this class belong to the completed MM class.

- $f(x) = \mathrm{Tr}_1^n\left(a[x^{2^i+1} + (x^{2^i} + x + 1)\mathrm{Tr}_1^n(x^{2^i+1})]\right)$, where $n \geq 6$, m does not divide i, $\frac{n}{\gcd(i,n)}$ even, $a \in \mathbb{F}_{2^n} \setminus \mathbb{F}_{2^i}$, $\{a, a+1\} \cap \{x^{2^i+1}; x \in \mathbb{F}_{2^n}\} = \emptyset$ [2]. These functions belong to the completed MM class when $a \in \mathbb{F}_{2^m}$.

- $f(x) = \mathrm{Tr}_1^n\big(a\big[(x + \mathrm{Tr}_3^n\big(x^{2(2^i+1)} + x^{4(2^i+1)}\big)$

$$+\mathrm{Tr}_1^n(x)\mathrm{Tr}_3^n\big(x^{2^i+1} + x^{2^{2i}(2^i+1)}\big)\big)^{2^i+1}\big]\big),$$

where $6 \mid n$, m does not divide i, $\frac{n}{\gcd(i,n)}$ even, $b + d + d^2 \notin \{x^{2^i+1}; x \in \mathbb{F}_{2^n}\}$ for every $d \in \mathbb{F}_{2^3}$ [2]. These functions are EA-inequivalent to functions in MM.

- *Niho bent functions* [31] whose restrictions to the cosets $u\mathbb{F}_{2^m}$ are linear. These functions can be written:

$$f(x) = \mathrm{Tr}_1^m(ax^{2^m+1}) + \mathrm{Tr}_1^n(bx^d); \; b = 0 \text{ or } a = b^{2^m+1} \in \mathbb{F}_{2^m}^\star.$$

The values of d are such that:

1. $d = (2^m - 1)3 + 1$ (the original condition $\exists u \in \mathbb{F}_{2^n}$ s.t. $b = u^5$ if $m \equiv 2 \pmod 4$ has been shown not useful by Helleseth–Kholosha–Mesnager [40]),
2. $d = (2^m - 1)\frac{1}{4} + 1$ (m odd),
3. $d = (2^m - 1)\frac{1}{6} + 1$ (m even).

Classes 1 and 3 are not EA-equivalent to MM functions [5]. Class 2 is in completed MM class. In classes 1 (for $m \not\equiv 2 \pmod 4$), 2, and 3, we can up to EA-equivalence fix $b = 1$.

Extension of the second class [45]:

$$\mathrm{Tr}_1^m(x^{2^m+1}) + \mathrm{Tr}_1^n\Big(\sum_{i=1}^{2^{r-1}-1} t^{s_i} \Big),$$

where

- $r > 1$ and $\gcd(r, m) = 1$,
- $s_i = (2^m - 1)\left(\frac{i}{2^r}[\mod 2^m + 1]\right) + 1, i \in \{1, \ldots, 2^{r-1} - 1\}$.

- *Dillon's and generalized Dillon's functions* [23, 28, 46]: let $\gcd(r, 2^m + 1) = 1$ and $a \in \mathbb{F}_{2^n}^*$, then

$$f(x) = \mathrm{Tr}_1^n(ax^{r(2^m-1)})$$

is bent if and only if

$$K(a^{2^m+1}) = 0,$$

where $K(u) = \sum_{x \in \mathbb{F}_{2^m}} \mathrm{Tr}_1^m\left(ux + \frac{1}{x}\right), u \in \mathbb{F}_{2^m}$, is a Kloosterman sum.
This class has been generalized to functions:

$$\mathrm{Tr}_1^n\left(\sum_{r \in R} a_r x^{r(2^m-1)}\right),$$

$$\mathrm{Tr}_1^n\left(ax^{r(2^m-1)} + bx^{\left(\frac{q+1}{e} - l\right)(2^m-1)} + cx^{\left(\frac{q+1}{e} + l\right)(2^m-1)}\right) + \mathrm{Tr}_1^\ell\left(\epsilon x^{\frac{2^n-1}{e}}\right),$$

$$\mathrm{Tr}_1^n\left(\sum_{i \in D} ax^{(ri+s)(2^m-1)}\right) + \mathrm{Tr}_1^\ell\left(\epsilon x^{\frac{2^n-1}{e}}\right),$$

where $\ell | n$ and $e | 2^\ell - 1$.

Two explicit classes are given in [37]: $\sum_{i=1}^{2^{m-1}-1} \mathrm{Tr}_1^n\left(\beta x^{i(2^m-1)}\right)$, where $\beta \in \mathbb{F}_{2^m} \setminus \mathbb{F}_2$ and $\sum_{i=1}^{2^{m-2}-1} \mathrm{Tr}_1^n\left(\beta x^{i(2^m-1)}\right)$, where m is odd, $\beta \in \mathbb{F}_{2^m}^*$, $\mathrm{Tr}_1^m\left(\beta^{(2^m-4)^{-1}}\right) = 0$.

- *Mesnager's functions* [52,53]: let $\gcd(r, 2^m + 1) = 1$; m odd > 3; $a \in \mathbb{F}_{2^n}^*, b \in \mathbb{F}_{2^2}^*$, then

$$f(x) = \mathrm{Tr}_1^n(ax^{r(2^m-1)}) + \mathrm{Tr}_1^2(bx^{\frac{2^n-1}{3}})$$

is bent if and only if

$$K(a^{2^m+1}) = 4.$$

This class can be extended to the case m even, but no necessary and sufficient condition is known in this case.

Generalizations exist by Mesnager [51] and Mesnager–Flori [55] to functions using trace functions in other subfields, including the class of functions $f(x) = \mathrm{Tr}_1^n(ax^{r(2^m-1)}) + \mathrm{Tr}_1^4(bx^{\frac{2^n-1}{5}})$.

Open Problem 3: Characterize the cases of bentness for Mesnager's functions when m is even.

- *Mesnager's functions, second class* [50]: let m odd , $a \in \mathbb{F}_{2^n}^*, b \in \mathbb{F}_{2^2}^*$, then

$$f(x) = \mathrm{Tr}_1^n(ax^{3(2^m-1)}) + \mathrm{Tr}_1^2(bx^{\frac{2^n-1}{3}})$$

is bent if and only if $m \not\equiv 3 \ [\mathrm{mod}\ 6]$ and $\left(\mathrm{Tr}_1^m(a^{\frac{2^m+1}{3}}) = 0 \text{ and } K\left(a^{2^m+1}\right) = 4\right)$ or $\left(\mathrm{Tr}_1^m(a^{\frac{2^m+1}{3}}) = 1 \text{ and } a \notin \mathbb{F}_{2^m} \text{ and } K\left(a^{2^m+1}\right) + C\left(a^{2^m+1}\right) = 4\right)$, where $C(u) = \sum_{x \in \mathbb{F}_{2^m}} \mathrm{Tr}_1^m\left(u(x^3 + x)\right)$.

- Finally, bent functions have been also obtained by Dillon and McGuire [30] as the restrictions of functions on $\mathbb{F}_{2^{n+1}}$, with $n+1$ odd, to a hyperplane of this field: these functions are the Kasami functions $tr_n\left(x^{2^{2k}-2^k+1}\right)$, and the hyperplane has equation $tr_n(x) = 0$. The restriction is bent under the condition that $n + 1 = 3k \pm 1$.

Remark 1 These classes are small for each n. Moreover, many of them belong to completed MM class, when viewed in multivariate representation, and their bentness may then seem easily explained. However, finding bent functions in univariate trace form is in general difficult and presents theoretical interest, since it gives more insight on bent functions. Note also that the output to such functions is often faster to compute thanks to their particular form.

Open Problem 4 : find more univariate classes. It has been checked that all *monomial* bent functions $\mathrm{Tr}_1^n(ax^i)$ are covered by the four first classes of bent functions described above and by Dillon's class for $n \leq 20$.

3.3 Known Infinite Classes of Bent Functions in Bivariate Trace Form

- MM class can be viewed in bivariate form. Its elements are the functions $f(x, y) = \mathrm{Tr}_1^m(x\,\pi(y)) + g(y); \ x, y \in \mathbb{F}_{2^m}$, where π is a permutation on $\mathbb{F}_{2^m}$.

- *Dillon's PS$_{ap}$ class*: $f(x, y) = g(xy^{2^m-2}) = g\left(\frac{x}{y}\right)$, where g is any balanced Boolean function on $\mathbb{F}_{2^m}$. The dual equals $g\left(\frac{y}{x}\right)$.

  This class is much larger than the classes above but much smaller than the MM class. It contains, up to EA-equivalence, the generalized Dillon's functions and the Mesnager functions. The functions in this class are, when viewed in univariate form, those bent functions whose restrictions to the cosets $u\mathbb{F}_{2^m}^*$ are constant. They are *hyperbent*: $f(x^s)$ is bent for every s co-prime with $2^n - 1$ (see more in [14, 64] on hyperbent functions).

Open Problem 5 : Find hyperbent functions EA-inequivalent to PS_{ap} functions in more than 4 variables (a sporadic example exists in 4 variables [14]).

PS_{ap} class is included in the more general Dillon's PS class (see [28]), which has itself been generalized to the GPS class (see [11]), which covers all bent functions up to EA-equivalence [38].

- An isolated class [11]: $f(x, y) = \mathrm{Tr}_1^m(x^{2^i+1} + y^{2^i+1} + xy)$, $x, y \in \mathbb{F}_{2^m}$, where n is co-prime with 3 and i is co-prime with m.
- *Functions related to Dillon's H class and o-polynomials*: In his thesis [28], J. Dillon had introduced several classes of functions. Some of these classes were merely constructions since the functions in them needed to satisfy some conditions difficult to achieve. One of such classes was class H, whose elements had linear restrictions to the lines through the origin of the $\mathbb{F}_{2^m}$-vector space $\mathbb{F}_{2^m} \times \mathbb{F}_{2^m}$. J. Dillon had not found inside it functions which were not already in previously defined classes. More than 35 years later, the condition for a function to be in class H was connected with a classical notion in finite geometry.

Definition 1 Let m be any positive integer. A permutation polynomial G over $\mathbb{F}_{2^m}$ is called an o-polynomial if, for every $\gamma \in \mathbb{F}_{2^m}$, the function $z \in \mathbb{F}_{2^m} \mapsto$
$$\begin{cases} \frac{G(z+\gamma)+G(\gamma)}{z} & \text{if } z \neq 0 \\ 0 \text{ if } z = 0 \end{cases} \quad \text{is a permutation.}$$

Theorem 3 ([18]) *Any Boolean function of the form*

$$g(x, y) = \begin{cases} \mathrm{Tr}_1^m\left(xH\left(\frac{y}{x}\right)\right) & \text{if } x \neq 0 \\ \mathrm{Tr}_1^m(\mu y) & \text{if } x = 0 \end{cases} \quad ,$$

that is, having linear restrictions to the lines through the origin of the two-dimensional vector space $\mathbb{F}_{2^m} \times \mathbb{F}_{2^m}$, is bent if and only if $G(z) := H(z) + \mu z$ is an o-polynomial.

The class of such functions is denoted by $\mathcal{H}$; its elements are EA-equivalent to functions in class H.

The known o-polynomials provide then the following bent functions:

1. m odd:

    - $f(x, y) = \mathrm{Tr}_1^m(x^{-5} y^6)$;
    - $f(x, y) = \mathrm{Tr}_1^m(x^{\frac{5}{6}} y^{\frac{1}{6}})$;

2. $m = 2k - 1$:

    - $f(x, y) = \mathrm{Tr}_1^m(x^{-3 \cdot (2^k+1)} y^{3 \cdot 2^k + 4})$;
    - $f(x, y) = \mathrm{Tr}_1^m(x^{2^m - 3 \cdot 2^{k-1} + 2} y^{3 \cdot 2^{k-1} - 2})$;

3. $m = 4k - 1$:

- $f(x, y) = \mathrm{Tr}_1^m (x^{2^m - 2^k - 2^{2k}} y^{2^k + 2^{2k}})$;
- $f(x, y) = \mathrm{Tr}_1^m (x^{2^{3k-1} - 2^{2k} + 2^k} y^{2^m - 2^{3k-1} + 2^{2k} - 2^k})$;

4. $m = 4k + 1$:

- $f(x, y) = \mathrm{Tr}_1^m (x^{2^m - 2^{2k+1} - 2^{3k+1}} y^{2^{2k+1} + 2^{3k+1}})$;
- $f(x, y) = \mathrm{Tr}_1^m (x^{2^{3k+1} - 2^{2k+1} + 2^k} y^{2^m - 2^{3k+1} + 2^{2k+1} - 2^k})$;

5. $m = 2k - 1$:

- $f(x, y) = \mathrm{Tr}_1^m (x^{1 - 2^k} y^{2^k} + x^{-(2^k + 1)} y^{2^k + 2} + x^{-3 \cdot (2^k + 1)} y^{3 \cdot 2^k + 4})$;

- $f(x, y) = \mathrm{Tr}_1^m \left(\dfrac{x \left(\frac{y}{x} + 1 + \frac{y^{2^k}}{x^{2^k}} \right) \frac{y^{2^{k-1}}}{x^{2^{k-1}}}}{\frac{y^{2^k}}{x^{2^k}} + \frac{y^2}{x^2} + 1} \right)$;

6. m odd :

- $f(x, y) = \mathrm{Tr}_1^m (x^{\frac{5}{6}} y^{\frac{1}{6}} + x^{\frac{3}{6}} y^{\frac{3}{6}} + x^{\frac{1}{6}} y^{\frac{5}{6}}) = D_5((y/x)^6)$;
- $f(x, y) = \mathrm{Tr}_1^m \left(x \left[D_{\frac{1}{5}} \left(\frac{y}{x} \right) \right]^6 \right)$; $D_{\frac{1}{5}}$ Dickson polynomial.

7. Four functions related to the two more o-polynomials:

- $\dfrac{\delta^2 (z^4 + z) + \delta^2 (1 + \delta + \delta^2)(z^3 + z^2)}{z^4 + \delta^2 z^2 + 1} + z^{1/2}$, where $\mathrm{Tr}_1^m (1/\delta) = 1$ and, if $m \equiv 2$ [mod 4], then $\delta \notin \mathbb{F}_4$;

- $\dfrac{\mathrm{Tr}_m^n (v')(z+1) + \mathrm{Tr}_m^n \left[(vz + v^{2^m})^r \right] \left(z + \mathrm{Tr}_m^n (v) z^{1/2} + 1 \right)^{1-r}}{\mathrm{Tr}_m^n (v)} + z^{1/2}$, where m is even, $r = \pm \frac{2^m - 1}{3}$, $v \in \mathbb{F}_{2^{2m}}$, $v^{2^m + 1} = 1$, and $v \neq 1$.

Class 1 of Niho bent functions corresponds to the so-called *Subiaco hyperovals*, related to the first of the two classes of o-polynomials recalled above in 7 [40] *Classes 2 and 3* correspond to *Adelaide hyperovals* [39] related to the second of the two classes of o-polynomials recalled above in 7 .

Open Problem 6 : Clarify the relation between EA-equivalence of Niho bent functions and equivalence of the corresponding o-polynomials/hyperovals.
Open Problem 7 : Find more univariate Niho bent functions having simple expression, related to the two last classes of o-polynomials.

- *Bent functions associated to AB functions*:

Definition 2 The *nonlinearity* of a *vectorial function* $F : \mathbb{F}_2^n \to \mathbb{F}_2^n$ (resp. $\mathbb{F}_{2^n} \to \mathbb{F}_{2^n}$) equals:

$$nl(F) = \min\{nl(b \cdot F); \ b \in \mathbb{F}_2^n \setminus \{0\}\} \ (\text{resp. } \min\{nl(\mathrm{Tr}_1^n (bF)); \ b \in \mathbb{F}_{2^n}^*\}).$$

F is *almost bent (AB)* if $nl(F) = 2^{n-1} - 2^{\frac{n-1}{2}}$, which is the best possible value (see more in [12]).

Table 1 Known AB power functions x^d on $\mathbb{F}_{2^m}$

Functions	Exponents d	Conditions
Gold	$2^i + 1$	$\gcd(i, m) = 1, 1 \le i < m/2$
Kasami	$2^{2i} - 2^i + 1$	$\gcd(i, m) = 1, 2 \le i < m/2$
Welch	$2^k + 3$	$m = 2k + 1$
Niho	$2^k + 2^{\frac{k}{2}} - 1, k$ even	$m = 2k + 1$
	$2^k + 2^{\frac{3k+1}{2}} - 1, k$ odd	

Theorem 4 ([20]) *Let F be a function from $\mathbb{F}_2^m$ to itself. Then F is AB if and only if $\gamma_F : (\mathbb{F}_{2^m})^2 \to \mathbb{F}_2$ defined by*

$$\gamma_F(a, b) = 1 \Leftrightarrow \begin{cases} a \ne 0 \\ \exists x \mid F(x) + F(x + a) = b \end{cases}$$

is bent. The dual function satisfies

$$\widetilde{\gamma_F}(a, b) = 1 \Leftrightarrow \begin{cases} b \ne 0 \\ \widehat{\chi_{b \cdot F}}(a) \ne 0 \end{cases} .$$

The known AB power functions $F(x) = x^d$, $x \in \mathbb{F}_{2^m}$ are given in Table 1.

The associated bent functions γ_F are studied in [4]. We give them below:

Gold : $\gamma_F(a, b) = \mathrm{Tr}_1^m \left(\frac{b}{a^{2^i+1}} \right)$ with $\frac{1}{0} = 0$;

Kasami, Welch, Niho : $F(x+1) + F(x) = q(x^{2^s} + x)$ (q permutation determined by Dobbertin, $\gcd(s, m) = 1$);

$F(x + 1) + F(x) = b$ has solutions if and only if $\mathrm{Tr}_1^m(q^{-1}(b)) = 0$.

Then: $\gamma_F(a, b) = \begin{cases} \mathrm{Tr}_1^m(q^{-1}(b/a^d)) + 1 & \text{if } a \ne 0, \\ 0 & \text{otherwise.} \end{cases}$

– Kasami: $s = i$, $q(x) = \dfrac{x^{2^i + 1}}{\sum_{j=1}^{i'} x^{2^{ji}} + \alpha \mathrm{Tr}_1^m(x)} + 1$, where

$$i' \equiv 1/i \mod m, \alpha = \begin{cases} 0 \text{ if } i' \text{ is odd} \\ 1 \text{ otherwise} \end{cases} .$$

– Welch: $s = k$, $q(x) = x^{2^{k+1}+1} + x^3 + x + 1$.
– Niho: $s = k/2$ if k is even and $s = (3k + 1)/2$ if k is odd, $q(x) = \begin{cases} \frac{1}{g(x^{2^s-1})+1} + 1 & \text{if } x \notin \mathbb{F}_2 \\ 1 & \text{otherwise} \end{cases}$ where

$$g(x) = x^{2^{2s+1}+2^{s+1}+1} + x^{2^{2s+1}+2^{s+1}-1} + x^{2^{2s+1}+1} + x^{2^{2s+1}-1} + x$$

The functions γ_F associated to Kasami, Welch, and Niho functions with $m = 7, 9$ are neither in the completed MM class nor in the completed PSap class.

The other known infinite classes of AB functions are quadratic; their associated γ_F belong to the completed MM class.

Open Problem 8 : Find infinite classes of bent functions whose bivariate trace representation (4) involves several values of $o(i, j)$.

3.4 Bent Functions in Hybrid Form and Kerdock Codes

A known infinite class of quadratic bent functions is defined over $\mathbb{F}_{2^{n-1}} \times \mathbb{F}_2$ as

$$f_u(x', x_n) = \sum_{i=1}^{\frac{n}{2}-1} \mathrm{Tr}_1^{n-1}((ux')^{2^i+1}) + x_n Tr_1^{n-1}(ux'), \text{ where } u \in \mathbb{F}_{2^{n-1}}^*, x' \in \mathbb{F}_{2^{n-1}}.$$

The difference (i.e., the sum) of two functions of such form corresponding to two distinct values of u is bent as well.

It is easily shown that any code of length 2^n (i.e., any set of Boolean functions) equal to the union of at least two cosets of the first-order Reed-Muller code RM(1,n) has minimum distance bounded above by $2^{n-1} - 2^{m-1}$, with equality if and only if all the differences between the elements of two distinct cosets are bent functions.

The Kerdock code of length 2^n ($n \geq 4$ even) equals the union of all the cosets $f_u + RM(1, n)$ where u ranges over $\mathbb{F}_{2^{n-1}}$. It is an optimal code (it was shown by Delsarte that no code exists with better parameters, e.g., with smaller length, same size, and same minimum distance, or larger size, same length, and same minimum distance, or larger minimum distance, same length, and same size).

Open Problem 9 : Find a code with the same parameters as the Kerdock code (for instance, find a set of $2^{n-1} - 1$ bent functions in n variables whose pairwise sums are also bent) and which is not equivalent to a subcode of the second-order Reed-Muller code.

3.5 Determining the Duals of Bent Functions in Univariate Form

Lemma 1 *Let (u, v) be an autodual basis of $\mathbb{F}_{2^n}$ over $\mathbb{F}_{2^m}$. Let f be bent over $\mathbb{F}_{2^n}$ and $g(x, y) = f(ux + vy)$, $x, y \in \mathbb{F}_{2^m}$.*
Then:

$$\widehat{\chi_f}(au + bv) = \widehat{\chi_g}(a, b).$$

The dual of Niho class 2 of bent functions [31] has been determined in [18]. Let v be such that $\text{Tr}_m^n(v) = 1$ and $b^4 = a^2 v^{2^m-1}$. Then

$$\tilde{f}(a^{\frac{1}{2}}x) =$$

$$\text{Tr}_1^m\left(\left(v^{\frac{2^m+1}{2}} + 1 + \text{Tr}_m^n(v^{2^m}x)\right)\left(\frac{\text{Tr}_m^n(vx) + v^{\frac{2^m+1}{2}}}{\text{Tr}_m^n(v^{-1})}\right)^{\frac{1}{3}}\right)$$

has algebraic degree $\frac{m+3}{2}$ (hence, $\tilde{f}$ is EA-inequivalent to f).

The dual of the Kholosha-Leander extension of class 2 has also been determined [5].

Open Problem 10 : Determine the duals of Niho bent functions 1 and 3.

The duals of Mesnager's functions have been determined [50, 52, 53].

4 Secondary Constructions of Bent Functions

We call secondary a construction of bent functions from already known bent functions, in the same numbers of variables or not (while primary constructions, like Maiorana-McFarland construction, build bent functions from scratch).

4.1 A Maiorana-McFarland-Like Construction

Let r, s be two positive integers such that $n = r + s$ is even and $r < s$. Let $\phi : \mathbb{F}_2^s \mapsto \mathbb{F}_2^r$ be such that $\phi^{-1}(a)$ is an $(s - r)$-dimensional affine subspace of $\mathbb{F}_2^s$, for every $a \in \mathbb{F}_2^r$, and let g be a Boolean function on $\mathbb{F}_2^s$ whose restriction to $\phi^{-1}(a)$ is bent, for every $a \in \mathbb{F}_2^r$.

Then $f(x, y) = x \cdot \phi(y) + g(y)$; $x \in \mathbb{F}_2^r, y \in \mathbb{F}_2^s$, is bent on $\mathbb{F}_2^n$.

4.2 Adding the Indicator of a Flat

The condition for preserving bentness when adding the indicator of an affine subspace of $\mathbb{F}_2^n$ is given in [7] and recalled in [11].

4.3 A Secondary Construction Which Does Not Increase the Number of Variables

Let f_1, f_2, and f_3 be three Boolean functions on $\mathbb{F}_2^n$. Let $s_1 = f_1 + f_2 + f_3$ and $s_2 = f_1 f_2 + f_1 f_3 + f_2 f_3$. Then

$$\widehat{\chi_{f_1}} + \widehat{\chi_{f_2}} + \widehat{\chi_{f_3}} = \widehat{\chi_{s_1}} + 2\,\widehat{\chi_{s_2}}. \tag{5}$$

If f_1, f_2, and f_3 are bent, then:

- if s_1 is bent and if $\tilde{s}_1 = \tilde{f}_1 + \tilde{f}_2 + \tilde{f}_3$, then s_2 is bent, and $\tilde{s}_2 = \tilde{f}_1 \tilde{f}_2 + \tilde{f}_1 \tilde{f}_3 + \tilde{f}_2 \tilde{f}_3$;
- if $\widehat{s_{2_\chi}}(a)$ is divisible by 2^m for every a (e.g., if s_2 is bent), then s_1 is bent [10].

Open Problem 11 : Deduce significantly new and large classes of bent functions from this construction (classes are found in [10], but they are a little peculiar).

4.4 Rothaus' Construction

Let f_1, f_2, f_3 be bent functions on $\mathbb{F}_2^n$ such that $f_1 + f_2 + f_3$ is bent as well, then the function on $\mathbb{F}_2^{n+2}$:

$$f(x, x_{n+1}, x_{n+2}) =$$
$$f_1(x) f_2(x) + f_1(x) f_3(x) + f_2(x) f_3(x) + x_{n+1} x_{n+2}$$
$$+[f_1(x) + f_2(x)] x_{n+1} + [f_1(x) + f_3(x)] x_{n+2}$$

is bent [59].

4.4.1 Designing the Initial Functions in the Rothaus Construction

Definition 3 A permutation π on $\mathbb{F}_2^m$ is called an *orthomorphic permutation* if the function $\pi + Id$, where $Id(x) = x$, is also a permutation.

Theorem 5 ([22]) *Let π be a permutation on $\mathbb{F}_2^m$. Let ϕ and ψ be orthomorphic permutations on $\mathbb{F}_2^m$. Let g_1, g_2, g_3 be three m-variable Boolean functions. Let*

$$\pi_1 = \pi; \quad \pi_2 = \phi \circ \pi_1; \quad \pi_3 = \psi \circ (\pi_1 + \pi_2)$$

then the four following MM functions are bent:

$$h_1(x, y) = x \cdot \pi_1(y) + g_1(y),$$
$$h_2(x, y) = x \cdot \pi_2(y) + g_2(y),$$
$$h_3(x, y) = x \cdot \pi_3(y) + g_3(y),$$
$$h_1 + h_2 + h_3.$$

Open Problem 12 : Find more general constructions of initial functions for the Rothaus construction.

4.5   The Indirect Sum and Its Generalizations

A very general secondary construction of bent functions with initial conditions was given in [8]:

Theorem 6 *Let f be a Boolean function on $\mathbb{F}_2^{r+s} = \mathbb{F}_2^r \times \mathbb{F}_2^s$, where r, s are even, such that, for any $y \in \mathbb{F}_2^s$, the function on $\mathbb{F}_2^r$:*

$$f_y : x \mapsto f(x, y)$$

is bent. Then f is bent if and only if for any element u of $\mathbb{F}_2^r$, the function

$$\vartheta_u : y \mapsto \widetilde{f_y}(u)$$

is bent on $\mathbb{F}_2^s$, and the dual of f is the function $\tilde{f}(u, v) = \widetilde{\vartheta_u}(v)$.

The Rothaus construction, which uses four bent functions whose sum is null, is a particular case of this construction, corresponding to $r = 2$ (indeed, any function $(x_{n+1}, x_{n+2}) \mapsto a_0 + x_{n+1}x_{n+2} + a_1x_{n+1} + a_2x_{n+2}$ is bent and has dual $a_0 + x_{n+1}x_{n+2} + a_2x_{n+1} + a_1x_{n+2} + a_1a_2$). Another particular case, called the *indirect sum*, uses four bent functions as well but is built differently and does not need any initial condition:

Corollary 1 ([9]) *Let f_1, f_2 be bent on $\mathbb{F}_2^r$ (r even) and g_1, g_2 be bent on $\mathbb{F}_2^s$ (s even). Define*

$$h(x, y) = f_1(x) + g_1(y) + (f_1 + f_2)(x)(g_1 + g_2)(y); \quad x \in \mathbb{F}_2^r, \ y \in \mathbb{F}_2^s.$$

Then h is bent and

$$\tilde{h}(x, y) = \tilde{f}_1(x) + \tilde{g}_1(y) + (\tilde{f}_1 + \tilde{f}_2)(x)(\tilde{g}_1 + \tilde{g}_2)(y); \quad x \in \mathbb{F}_2^r, \ y \in \mathbb{F}_2^s.$$

Indeed, any function $x \mapsto f_1(x) + a_0 + a_1(f_1 + f_2)(x)$ is bent and has dual $\tilde{f}_1(x) + a_0 + a_1(\tilde{f}_1 + \tilde{f}_2)(x)$. The name of "indirect sum" comes from the name of the well-known *direct sum*, which corresponds simply to $h(x, y) = f_1(x) + g_1(y)$, and that the indirect sum generalizes.

A generalization of the indirect sum needing initial conditions is given in [22]:

Theorem 7 *Let* $f_1, f_2,$ *and* f_3 *be bent on* $\mathbb{F}_2^r$. *Let* $g_1, g_2,$ *and* g_3 *be bent on* $\mathbb{F}_2^s$. *Let* $v_1 = f_1 + f_2 + f_3$ *and* $v_2 = g_1 + g_2 + g_3$. *If* v_1 *and* v_2 *are bent and if* $\widetilde{v_1} = \widetilde{f_1} + \widetilde{f_2} + \widetilde{f_3}$, *then* $f(x, y) =$

$$f_1(x) + g_1(y) + (f_1 + f_2)(x)(g_1 + g_2)(y) + (f_2 + f_3)(x)(g_2 + g_3)(y)$$

is a bent function in $r + s$ *variables.*

The indirect sum is a particular case of this construction: it corresponds to the case $f_2 = f_3$ and/or $g_2 = g_3$. The Rothaus construction is also a particular case.

Case of application: Let $p(x)$ and $\theta(x)$ be r-variable bent functions such that there exists $a \in \mathbb{F}_2^r$ nonzero such that $D_a p = D_a \theta$. We can take $f_1(x) = p(x), f_2(x) = p(x + a), f_3(x) = \theta(x)$.

Another generalization of the indirect sum is also given in [22]:

Theorem 8 *Let* $f_0, f_1, f_2,$ *and* f_3 *be bent functions on* $\mathbb{F}_2^r$ *and* $g_0, g_1, g_2,$ *and* g_3 *be bent functions on* $\mathbb{F}_2^s$.

Let $v_j = f_j + f_{(j+1)\bmod 4} + f_{(j+2)\bmod 4}$ *and* $\varepsilon_j = g_j + g_{(j+1)\bmod 4} + g_{(j+2)\bmod 4}$, *where* $j = 0, 1, 2, 3$. *If* v_j *and* ε_j *are bent functions and if for every* $j = 0, 1, 2, 3$, *we have* $\widetilde{v_j} = \widetilde{f_j} + \widetilde{f_{(j+1)\bmod 4}} + \widetilde{f_{(j+2)\bmod 4}}$, *then* $f(x, y) =$

$$f_0(x) + g_0(y) + (f_0 + f_1)(x)(g_0 + g_1)(y) + (f_1 + f_2)(x)(g_1 + g_2)(y) +$$
$$(f_2 + f_3)(x)(g_2 + g_3)(y) \text{ is bent.}$$

Case of application: under the same conditions as in the case of application of Theorem 7 (let $p(x)$ and $\theta(x)$ be r-variable bent functions such that there exists a nonzero vector $a \in \mathbb{F}_2^r$ such that $D_a p = D_a \theta$), we can take $f_0(x) = p(x), f_1(x) = p(x + a), f_2(x) = \theta(x)$ and $f_3(x) = \theta(x + a)$.

Open Problem 13 : Generalize the indirect sum without initial condition.
Open Problem 14 : Find more general cases of application of Theorems 7 and 8; deduce new classes.

4.5.1 A Modification of the Indirect Sum

A modified indirect sum is introduced in [66]:

Theorem 9 *Let* n *and* m *be two positive even numbers and* $\mu \in \{1, 2, \ldots, n\}$, $\rho \in \{1, 2, \ldots, m\}$. *For* $x = (x_1, \ldots, x_n) \in \mathbb{F}_2^n$ *and* $y = (y_1, \ldots, y_m) \in \mathbb{F}_2^m$, *let* $x' = (x_1, \ldots, x_{\mu-1}, x_{\mu+1}, \ldots, x_n) \in \mathbb{F}_2^{n-1}$ *and* $y' = (y_1, \ldots, y_{\rho-1}, y_{\rho+1}, \ldots, y_m) \in \mathbb{F}_2^{m-1}$. *Let* f *be an* n-*variable bent function and* g *an* m-*variable bent function. We* *consider the restrictions of f and g: $f_0(x') = f(x_1, \ldots, x_{\mu-1}, 0, x_{\mu+1}, \ldots, x_n)$, $f_1(x') = f(x_1, \ldots, x_{\mu-1}, 1, x_{\mu+1}, \ldots, x_n)$, $g_0(y') = g(y_1, \ldots, y_{\rho-1}, 0, y_{\rho+1}, \ldots, y_m)$, $g_1(y') = g(y_1, \ldots, y_{\rho-1}, 1, y_{\rho+1}, \ldots, y_m)$, and we define*

$$h(x', y') = f_0(x) \oplus g_0(y) \oplus (f_0 \oplus f_1)(x)(g_0 \oplus g_1)(y).$$

Then h is a bent function in $n + m - 2$ variables. Further, the dual of h is obtained from the functions $\widetilde{f_0}(x') = \tilde{f}(x_1, \ldots, x_{\mu-1}, 0, x_{\mu+1}, \ldots, x_n)$, $\widetilde{f_1}(x') = \tilde{f}(x_1, \ldots, x_{\mu-1}, 1, x_{\mu+1}, \ldots, x_n)$, $\widetilde{g_0}(y') = \tilde{g}(y_1, \ldots, y_{\rho-1}, 0, y_{\rho+1}, \ldots, y_m)$ and $\widetilde{g_1}(y') = \tilde{g}(y_1, \ldots, y_{\rho-1}, 1, y_{\rho+1}, \ldots, y_m)$, by the same formula as h is obtained from $f_0, f_1, g_0,$ and g_1.

4.5.2 A New Secondary Construction

We have seen just before Theorem 2 that, given two Boolean functions f and f', we have $\sum_{x,y \in \mathbb{F}_2^n} (-1)^{f(x)+f'(y)+x \cdot y} = \sum_{x \in \mathbb{F}_2^n} (-1)^{f(x)} \widehat{\chi_{f'}}(x)$. If f' is bent, then we deduce $\sum_{x,y \in \mathbb{F}_2^n} (-1)^{f(x)+f'(y)+x \cdot y} = 2^m \sum_{x \in \mathbb{F}_2^n} (-1)^{f(x)+\tilde{f'}(x)}$. This implies that, for every $a, b \in \mathbb{F}_2^n$, we have $\sum_{x,y \in \mathbb{F}_2^n} (-1)^{f(x)+f'(y)+x \cdot y + a \cdot x + b \cdot y} = \sum_{x,y \in \mathbb{F}_2^n} (-1)^{f(x)+a \cdot x + f'(y)+(x+b) \cdot y} = \sum_{x,y \in \mathbb{F}_2^n} (-1)^{f(x+b)+a \cdot (x+b) + f'(y)+x \cdot y}$ equals $2^m (-1)^{a \cdot b} \sum_{x \in \mathbb{F}_2^n} (-1)^{f(x+b)+a \cdot x + \tilde{f'}(x)}$. Denoting $\tilde{f'}$ by g, we deduce:

Proposition 1 *Let f be any n-variable Boolean function and g be any n-variable bent function. Then the $2n$-variable function $f(x) + \tilde{g}(y) + x \cdot y$ is bent if and only if $f(x + b) + g(x)$ is bent for every b (or equivalently $f(x) + g(x + b)$ is bent for every b).*

Note that the bent function $f(x, y) = \mathrm{Tr}_1^m(x^{2^i+1} + y^{2^i+1} + xy)$, $x, y \in \mathbb{F}_{2^m}$ (recalled in Sect. 3.3), where n is co-prime with 3 and i is co-prime with m, looks like the function of Proposition 1, but its bentness is not a case of application of Proposition 1 since $\mathrm{Tr}_1^m(x^{2^i+1})$ is never bent (its linear kernel containing always 1).

If f is quadratic, then the condition "$f(x + b) + g(x)$ is bent for every b" in Proposition 1 simplifies in "$f(x) + g(x)$ is bent" since $f(x) + f(x + b)$ is affine. We have then, denoting $f + g$ by h:

Corollary 2 *If two n-variable bent functions $g(x)$ and $h(x)$ differ by a quadratic function, then the $2n$-variable function $(g + h)(x) + \tilde{g}(y) + x \cdot y$ is bent.*

Examples of cases of application are:

- Any pairs of quadratic bent functions; for instance, pairs of functions involved in Kerdock codes (see Sect. 3.4);
- Maiorana-McFarland functions: let π and π' be permutations on $\mathbb{F}_2^m$ differing by an affine function (for instance, let π be an orthomorphic permutation and $\pi' = \pi + Id$) and let u, v be two Boolean functions on $\mathbb{F}_2^m$ differing by a quadratic

function, then we can take $g(x_1, x_2) = x_1 \cdot \pi(x_2) + u(x_2)$ and $h(x_1, x_2) = x_1 \cdot \pi'(x_2) + v(x_2)$; $x_1, x_2 \in \mathbb{F}_2^m$; these functions can be nonquadratic.

Note that Proposition 1 could have also been proved as a corollary of Theorem 6. In fact, Theorem 6 (or more precisely its version obtained by exchanging the roles of x and y) allows proving a slightly more general result:

Theorem 10 *Let f be any n-variable Boolean function, let g be any n-variable bent function and let ϕ be any mapping from $\mathbb{F}_2^n$ to itself. Then the $2n$-variable function $f(x) + \tilde{g}(y) + \phi(x) \cdot y$ is bent if and only if $f(x) + g(\phi(x) + b)$ is bent for every b.*

Indeed, for every fixed $x \in \mathbb{F}_2^n$, function $y \mapsto f(x) + \tilde{g}(y) + \phi(x) \cdot y$ is bent and the value of the dual of this function at $u \in \mathbb{F}_2^n$ equals $f(x) + g(y + \phi(x))$.

Note that if g is quadratic and ϕ is an affine permutation, then $f(x) + g(\phi(x) + b)$ is bent if and only if $f(x) + g(\phi(x))$ is bent. Let then $h(x)$ be another bent function; then taking $f(x) = g(\phi(x)) + h(x)$, the condition in Theorem 10 is satisfied. We have then:

Corollary 3 *Let g be any quadratic bent function and ϕ any affine permutation. Let h be any bent function; then the $2n$-variable function $g(\phi(x)) + h(x) + \tilde{g}(y) + \phi(x) \cdot y$ is bent.*

This gives one more case of application of Corollary 2 since the two bent functions $g(\phi(x)) + h(x) + \tilde{g}(y) + \phi(x) \cdot y$ and $h(x) + \tilde{g}(y)$ differ by a quadratic function.

Another case of application of Theorem 10 is when, for every $b \in \mathbb{F}_2^n$, the set $\{\phi(x) + b; \ x \in \mathbb{F}_2^n\}$ is either included in the support of g or disjoint from it, and f is bent.

Corollary 4 *Let f, g be two n-variable bent Boolean functions and let ϕ be any mapping from $\mathbb{F}_2^n$ to itself such that $Im(\phi) = \{\phi(x); \ x \in \mathbb{F}_2^n\}$ is either included in or disjoint from any translate of $\mathrm{supp}(g)$. Then the $2n$-variable function $f(x) + \tilde{g}(y) + \phi(x) \cdot y$ is bent.*

Denoting, for every $E, F \subseteq \mathbb{F}_2^n$, the set $\{x + x'; \ x \in E, x' \in F\}$ by $E + F$, the condition on $Im(\phi)$ is equivalent to $(Im(\phi) + Im(\phi)) \cap (\mathrm{supp}(g) + (\mathbb{F}_2^n \setminus \mathrm{supp}(g))) = \emptyset$.

The construction of Theorem 10 can be turned into a construction of bent functions from arbitrary functions, which generalizes the Maiorana-McFarland construction. Indeed, given two n-variable Boolean functions f and g and a mapping ϕ from $\mathbb{F}_2^n$ to itself, let us define

$$h(x, y) = f(x) + g(y) + \phi(x) \cdot y. \tag{6}$$

This $2n$-variable Boolean function has Walsh transform:

$$\widehat{\chi_h}(a, b) = \sum_{x,y \in \mathbb{F}_2^n} (-1)^{f(x)+g(y)+\phi(x) \cdot y + a \cdot x + b \cdot y} = \sum_{x \in \mathbb{F}_2^n} (-1)^{f(x)+a \cdot x} \widehat{\chi_g}(\phi(x) + b).$$

If g is an affine function, then without loss of generality, up to EA-equivalence, we can take $g(y) = 0$; the support of $\widehat{\chi_g}$ is then $\{0\}$, the value of $\widehat{\chi_g}$ at 0 equals 2^n, and h is bent if and only if ϕ is a permutation. This is the Maiorana-McFarland construction.

If g has Walsh support of a pair, then without loss of generality, up to EA-equivalence, we can take $g(y) = y_1 y_2$, and we have then $\mathrm{supp}(\widehat{\chi_g}) = \{x \in \mathbb{F}_2^n \mid x_3 = \ldots = x_n = 0\}$ and $\widehat{\chi_g}(x_1, x_2, 0, \ldots, 0) = 2^{n-1}(-1)^{x_1 x_2}$; then h is bent if and only if, for every $a, b \in \mathbb{F}_2^n$, we have, denoting by $\phi_i(x)$ the ith coordinate of $\phi(x)$:

$$\sum_{x \in \mathbb{F}_2^n \mid \phi_3(x) = b_3, \ldots, \phi_n(x) = b_n} (-1)^{f(x) + a \cdot x + (\phi_1(x) + b_1)(\phi_2(x) + b_2)} = \pm 2.$$

For instance, if:

- every pre-image by the mapping $x \mapsto (\phi_3(x), \ldots, \phi_n(x))$ is a two-dimensional affine subspace of $\mathbb{F}_2^n$,
- $\phi_1(x)$ and $\phi_2(x)$ are constant on every such pre-image,
- and the restriction of f to such pre-image has odd Hamming weight,

then h is bent. But such construction is close to that of Sect. 4.1 (in which y would be replaced by (x, y_1, y_2), x by $(y_3, \ldots, y_n)$, $g(y)$ by $f(x) + y_1 y_2$, $\phi(y)$ by $(\phi_3(x), \ldots, \phi_n(x))$ and with $2n$ in the place of n and $r = n + 2$). The only difference is with the terms $\phi_1(x) y_1 + \phi_2(x) y_2$ which are present here and are not in the construction of Sect. 4.1.

Open Problem 15 : Find secondary constructions based on new ideas, if possible without initial conditions.

5 Rotation Symmetric (RS) Bent Functions and Idempotent Bent Functions

Rotation symmetric (RS) Boolean functions, which have been originally introduced by Filiol and Fontaine in [32, 33] under the name of idempotent functions and soon after studied by Pieprzyk and Qu [58] under their final name, have received some attention since their introduction. RS structure allowed obtaining Boolean functions in odd numbers of variables beating the best known nonlinearities [41].

RS functions also have the interest of needing less space to be stored and of allowing faster computation of the Walsh transform.

There have been recent developments on RS and idempotent bent functions.

5.1 RS Bent Functions

A Boolean function f is RS if it is invariant under the cyclic shift:

$$f(x_{n-1}, x_0, x_1, \ldots, x_{n-2}) = f(x_0, x_1, \ldots, x_{n-1}).$$

In other words, the support of an RS function is a cyclic (but not necessarily linear) code. See more on RS functions in [27, 34, 61, 62]

The dual of an RS bent function is an RS bent function.

The next lemma on quadratic RS functions is more or less known, but, as far as we know, it has never been published.

Lemma 2 Let $f(x) = \sum_{0 \le i < j \le n-1} a_{i,j} x_i x_j + \ell(x)$ be any quadratic Boolean function, where $a_{i,j} \in \mathbb{F}_2$ and ℓ is affine. Let M be the associated matrix (see [49]), whose term located at row i and column j equals $a_{i,j}$ if $i < j$, $a_{j,i}$ if $i > j$ and 0 if $i = j$. Then f is RS if and only if M is circulant (i.e., each row of M is a cyclic shift of the previous row) and ℓ is RS.

Proof If f is RS, then ℓ is RS and $a_{i+k, j+k} = a_{i,j}$ for every $i < j$ and every k, where the indices are taken in $\mathbb{Z}/n\mathbb{Z}$. This equality applied for $1 \le k \le n - j - 1$ shows that the part of M located at the right of the diagonal is circulant; applied for $n - j \le k \le n - i - 1$, it shows that $a_{i',j'} = a_{i'+k, j'+k}$ for every $i' > j'$ and every k and the part of M located at the left of the diagonal is circulant. And we have also $a_{i,n-1} = a_{i+1,0}$. Hence, M is circulant. The converse is similar. $\qquad\square$

5.1.1 Infinite Classes of RS Bent Functions

The situation of RS bent functions is very similar to that of bent functions in trace forms: many of the known classes belong to completed MM class, and their bentness may then seem easily explained. However, finding RS bent functions is difficult and has theoretical and practical interest.

- Quadratic RS bent functions have been characterized in [35] by the fact that some related polynomial $P(X)$ over $\mathbb{F}_2$ such that $X^n P(\frac{1}{X}) = P(X)$ is co-prime with $X^n + 1$: given $c_1, \ldots, c_m$ in $\mathbb{F}_2$, the function:

$$\sum_{i=1}^{m-1} c_i \left(\sum_{j=0}^{n-1} x_j x_{i+j} \right) + c_m \left(\sum_{j=0}^{m-1} x_j x_{m+j} \right)$$

is bent if and only if the polynomial $\sum_{i=1}^{m-1} c_i (X^i + X^{n-i}) + c_m X^m$ is co-prime with $X^n + 1$. This condition is equivalent to the fact that the linearized polynomial $L(X) = \sum_{i=1}^{m-1} c_i (X^{2^i} + X^{2^{n-i}}) + c_m X^{2^m}$, which we can write $\sum_{i=0}^{n-1} c_i X^{2^i}$ by setting $c_{n-i} = c_i$, is a permutation polynomial (a necessary condition is that $L(1) \neq 0$, i.e., $c_m = 1$).

– An example of such polynomial is with $c_i = 0$ for $i \neq m$. Another example is
with $c_i = 1$ for $i = 1, \ldots, n-1$ since $L(X)$ equals then $X + \mathrm{Tr}_1^n(X)$ which is
a permutation polynomial since n is even (equivalently, $\sum_{i=1}^{n-1} X^i$ is co-prime
with $X^n + 1$). This provides the two infinite classes

$$\sum_{j=0}^{m-1} x_j x_{m+j}$$

and

$$\sum_{i=1}^{m-1} \left(\sum_{j=0}^{n-1} x_j x_{i+j} \right) + \left(\sum_{j=0}^{m-1} x_j x_{m+j} \right)$$

of quadratic RS bent functions.

– More examples can be found. For instance, let k be such that $2^k - 2$ divides n
and $2^k - 1$ is co-prime with n. Then $\left(\frac{X^{2^k-1}+1}{X+1} \right)^{\frac{n}{2^k-2}} + X^n + 1$ has the form

$\sum_{i=1}^{m-1} c_i (X^i + X^{n-i}) + c_m X^m$ (indeed, $\left(\frac{X^{2^k-1}+1}{X+1} \right)^{\frac{n}{2^k-2}}$ is self-reciprocal,

has degree n, and is normalized) and is co-prime with $X^n + 1$ (indeed, the

zeroes of $\left(\frac{X^{2^k-1}+1}{X+1} \right)^{\frac{n}{2^k-2}}$ in the algebraic closure of $\mathbb{F}_2$ are the elements of

$\mathbb{F}_{2^k} \setminus \mathbb{F}_2$, and for any $\xi \in \mathbb{F}_{2^k} \setminus \mathbb{F}_2$, we have $\xi^n + 1 \neq 0$, since $\xi \mapsto \xi^n$ is a
permutation of $\mathbb{F}_{2^k}^*$). Taking for example $k = 2$, we have $\left(X^2 + X + 1 \right)^m +$

$X^n + 1 = \sum_{\substack{0 \leq u,v,w \leq m \\ u+v+w=m, 2u+v \notin \{0,n\}}} \frac{m!}{u!v!w!} X^{2u+v}$, and for n not divisible by 3, the

following function is RS bent:

$$\sum_{\substack{0 \leq u,v,w \leq m \\ u+v+w=m, 2u+v \in \{1,\ldots,m-1\}}} \frac{m!}{u!v!w!} \left(\sum_{j=0}^{n-1} x_j x_{2u+v+j} \right) + \left(\sum_{j=0}^{m-1} x_j x_{m+j} \right),$$

where the coefficients are taken modulo 2.

– If n is a power of 2, then according to [60, Proposition 3.1], the func-
tion $\sum_{i=1}^{m-1} c_i \left(\sum_{j=0}^{n-1} x_j x_{i+j} \right) + c_m \left(\sum_{j=0}^{m-1} x_j x_{m+j} \right)$ is bent if and only if
$\sum_{i=0}^{n-1} c_i = 1$, that is, $c_m = 1$. Note that this can also be proved slightly
differently: given some normal element α of $\mathbb{F}_{2^n}$ (i.e., some element of $\mathbb{F}_{2^n}$
such that $(\alpha, \alpha^2, \ldots, \alpha^{2^{n-1}})$ is a normal basis, i.e., $\alpha, \alpha^2, \ldots, \alpha^{2^{n-1}}$ are linearly
independent), the condition on L is equivalent to the fact that $\sum_{i=0}^{n-1} c_i \alpha^{2^i}$ is
also a normal element (see more in [57]); according to [56, Corollary 5.2.9],
an element α of $\mathbb{F}_{2^n}$ is normal if and only if $\mathrm{Tr}_1^n(\alpha) = 1$ (and $\alpha \notin \mathbb{F}_2$, but this

is implied by $\mathrm{Tr}_1^n(\alpha) = 1$ since n is even). Hence, $\sum_{i=0}^{n-1} c_i \alpha^{2^i}$ is normal if and only if $\mathrm{Tr}_1^n(\sum_{i=0}^{n-1} c_i \alpha^{2^i}) = 1$, that is, $c_m = 1$. See more at Sect. 5.2.1.

Open Problem 16 : Find more infinite classes of quadratic RS bent functions, valid for every even n.

- Two infinite classes of cubic RS bent functions belonging to the completed MM class were found recently:

$$- \sum_{i=0}^{n-1} (x_i x_{t+i} x_{m+i} + x_i x_{t+i}) + \sum_{i=0}^{m-1} x_i x_{m+i}, \text{ where } m/\gcd(m,t) \text{ is odd [35]};$$

$$- \sum_{i=0}^{n-1} x_i x_{i+r} x_{i+2r} + \sum_{i=0}^{2r-1} x_i x_{i+2r} x_{i+4r} + \sum_{i=0}^{m-1} x_i x_{i+m}, \text{ where } m = 3r \text{ [36]}.$$

Open Problem 17 : Find more classes of cubic RS bent functions.

5.2 Univariate RS Functions (Idempotents), Bivariate Expressions

Definition 4 Let $f(x)$ be a Boolean function on $\mathbb{F}_{2^n}$. We say that f is an idempotent if

$$f(x) = f(x^2), \quad \text{for all } x \in \mathbb{F}_{2^n}.$$

A function $f(x) = \sum_{j=0}^{2^n-1} a_j x^j$ or $f(x) = \sum_{j \in \Gamma_n} \mathrm{Tr}_1^{o(j)}(a_j x^j) + a_{2^m-1} x^{2^m-1}$ is an idempotent if and only if every coefficient a_j belongs to $\mathbb{F}_2$.

For any Boolean function $f(x)$ over $\mathbb{F}_{2^n}$ and every normal basis $(\alpha, \alpha^2, \dots, \alpha^{2^{n-1}})$ of $\mathbb{F}_{2^n}$, the function

$$(x_0, \dots, x_{n-1}) \mapsto f\left(\sum_{i=0}^{n-1} x_i \alpha^{2^i}\right)$$

is RS if and only if f is an idempotent.

Remark 2 This property leads to a notion of equivalence between RS functions: if two RS functions are linked as above to the same idempotent, through the choices of two normal bases, these two RS functions can be considered as equivalent (note that this is a subcase of linear equivalence). More precisely, given a normal element α, another normal element can be written $\alpha' = \sum_{j \in \mathbb{Z}/n\mathbb{Z}} c_j \alpha^{2^j}$ where $x \mapsto \sum_{j \in \mathbb{Z}/n\mathbb{Z}} c_j x^{2^j}$ is a permutation (if n is a power of 2, then this condition is equivalent to $\sum_{j \in \mathbb{Z}/n\mathbb{Z}} c_j = 1$); the two functions $g(x_0, \dots, x_{n-1}) = f\left(\sum_{i=0}^{n-1} x_i \alpha^{2^i}\right)$ and $g'(x_0, \dots, x_{n-1}) = f\left(\sum_{i=0}^{n-1} x_i \alpha'^{2^i}\right)$ are related by the relation $g'(\ldots, x_j, \ldots) = g(\ldots, \sum_{i \in \mathbb{Z}/n\mathbb{Z}} x_i c_{j-i}, \ldots)$. In other words, g' is deduced from g by multiplying the input by a nonsingular circulant matrix (so we can call *circulant-equivalence* this equivalence). We can check that the rotation symmetry of g is equivalent to that of g' since the shift on the input of g corresponds to the inverse shift on the input of g' and vice versa.

The related equivalence classes can be large: if, for instance, n is a power of 2, then we have recalled above that any element $\alpha \in \mathbb{F}_{2^n}$ is normal if and only if $\mathrm{Tr}_1^n(\alpha) = 1$; there are then 2^{n-1} normal bases, and an equivalence class can potentially have a size near $\frac{2^{n-1}}{n}$.

Linear equivalence between RS functions is more general than the equivalence above. Even equivalence under permutation of the variables is. For instance, the 8-variable RS functions $\sum_{i=0}^{7} x_i x_{i+1} x_{i+2} x_{i+5}$ and $\sum_{i=0}^{7} x_i x_{i+1} x_{i+3} x_{i+4}$ are equivalent under permutation and not under circulant-equivalence; see [26, Remark 1.10]. Refer more generally to [26] and the references therein for linear and affine equivalences of RS functions.

Remark 3 Knowing an infinite class of idempotent bent functions is not equivalent to knowing an infinite class of RS bent functions, since there is no expression valid for an infinite number of values of n of the decomposition of $\left(\sum_{i=0}^{n-1} x_i \alpha^{2^i} \right)^j$ over the normal basis $(\alpha, \alpha^2, \ldots, \alpha^{2^{n-1}})$, except for j null or equal to a power of 2.

5.2.1 Known Bent Idempotents

- The function $f'(z) = \mathrm{Tr}_1^m(z^{2^m+1})$ and the function $f'(z) = \mathrm{Tr}_1^m(z^{2^m+1}) + \sum_{i=1}^{m-1} \mathrm{Tr}_1^n(z^{2^i+1})$ are bent quadratic idempotents. More generally, given $c_1, \ldots, c_m$ in $\mathbb{F}_2$, the function equal to $c_m T r_1^m(x^{2^m+1}) + \sum_{i=1}^{m-1} c_i T r_1^n(x^{2^i+1})$ is bent if and only if $\gcd(\sum_{i=1}^{m-1} c_i(X^i + X^{n-i}) + c_m X^m, X^n + 1) = 1$ [48]. This condition is the same as that obtained for quadratic RS bent functions in Sect. 5.1.1. The two first classes described in that subsection correspond to the classes of bent idempotents given above. The third example in Sect. 5.1.1 gives a third general example here. For instance, for n not divisible by 3, we have the following bent idempotent:

$$\mathrm{Tr}_1^m(z^{2^m+1}) + \sum_{\substack{0 \le u,v,w \le m \\ u+v+w=m, 2u+v \in \{1,\ldots,m-1\}}} \frac{m!}{u!v!w!} \mathrm{Tr}_1^n(z^{2^{2u+v}+1}),$$

where the coefficients are taken modulo 2. Of course, what is written at Sect. 5.1.1 when n is a power of 2 is valid here. Note that more results, valid for more general values of n, can be found in [65].

- The Niho bent functions [31] recalled at Sect. 3.2

$$\mathrm{Tr}_1^m(az^{2^m+1}) + \mathrm{Tr}_1^n(bz^d)$$

are bent idempotents when the coefficients a and b equal 1. The extension of the second class by Leander et al. [45] gives also a bent idempotent.

- The generalized Dillon and Mesnager functions are potentially bent idempotents, under conditions involving Kloosterman sums:

  - For every m such that $K_m(1)$ is null, $g_1(x) = \mathrm{Tr}_1^n(x^{r(2^m-1)})$ is bent when $\gcd(r, 2^m + 1) = 1$.
  - For every m odd such that $K_m(1) = 4$, $g_2(x) = \mathrm{Tr}_1^n(x^{r(2^m-1)}) + \mathrm{Tr}_1^2(x^{\frac{2^n-1}{3}})$ is bent when $\gcd(r, 2^m + 1) = 1$.

  But the condition $K_m(1) = 0$ never happens as shown in [47, Theorem 2.2], and it can be checked by computer that the condition $K_m(1) = 4$ never happens as well for $5 \leq m \leq 20$.

- For $n = 2m = 6r, r \geq 1$, $\mathrm{Tr}_1^n(z^{1+2^r+2^{2r}}) + \mathrm{Tr}_1^{2r}(z^{1+2^{2r}+2^{4r}}) + \mathrm{Tr}_1^m(z^{1+2^t}) = \mathrm{Tr}_1^r((z + z^{2^{3r}})^{1+2^r+2^{2r}}) + \mathrm{Tr}_1^m(z^{1+2^t})$ is a bent idempotent [36].

5.3 Secondary Constructions of Rotation Symmetric and Idempotent Bent Functions

We precise the relationship between RS functions and the bivariate representation of idempotent functions; a proper relationship is between weak RS functions (invariant under circular permutation of indices by two positions) and weak idempotents (a natural notion that we introduce). This gives a way of constructing a new RS n-variable function where $n \equiv 2$ [mod 4], from two known semi-bent RS functions in m variables, by using the indirect sum (the definition of semi-bent functions is recalled below). It provides an infinite class of RS bent functions of algebraic degree 4 and an infinite class of bent idempotents of algebraic degree 4 as well. This section and the next one are a recall and an extension of results from [15].

5.3.1 Bivariate Representation of Idempotents

Most bent functions being known in bivariate form, it is useful to characterize the bivariate representation of idempotent and RS functions. The situation is easier when m is odd (which is the case of most known bent functions).

Given $w \in \mathbb{F}_{2^2} \setminus \mathbb{F}_2$, we have $w^2 = w + 1$, $w^4 = w$, and we can take (w, w^2) for basis of $\mathbb{F}_{2^n}$ over $\mathbb{F}_{2^m}$, since we have $\frac{w^2}{w} = w \notin \mathbb{F}_{2^m}$ for m odd. Any element of $\mathbb{F}_{2^n}$ can then be written in the form $xw + yw^2$, where $x, y \in \mathbb{F}_{2^m}$. Note that, given a normal basis $(\alpha, \ldots, \alpha^{2^{m-1}})$ of $\mathbb{F}_{2^m}$, a natural normal basis of $\mathbb{F}_{2^n}$ over $\mathbb{F}_2$ is

$$(\alpha w, (\alpha w)^2, (\alpha w)^4, (\alpha w)^8, \ldots, (\alpha w)^{2^{m-1}}, (\alpha w)^{2^m}, (\alpha w)^{2^{m+1}}, \ldots, (\alpha w)^{2^{n-1}}) =$$

$$(\alpha w, \alpha^2 w^2, \alpha^4 w, \alpha^8 w^2, \ldots, \alpha^{2^{m-1}} w, \alpha w^2, \alpha^2 w, \ldots, \alpha^{2^{m-1}} w^2). \tag{7}$$

Since $(xw + yw^2)^2 = y^2w + x^2w^2$, the shift $z \mapsto z^2$ corresponds to the mapping $(x, y) \mapsto (y^2, x^2)$. Given a function $f(x, y)$ in bivariate form, the related Boolean function over $\mathbb{F}_2^n$ obtained by decomposing the input $xw + yw^2$ over the basis (7) is then RS if and only if $f(x, y) = f(y^2, x^2)$. Note that applying this identity m times gives $f(x, y) = f(y, x)$ and applying it $m+1$ times gives $f(x, y) = f(x^2, y^2)$; the double condition "$f(x, y) = f(y, x)$ and $f(x, y) = f(x^2, y^2)$" is then necessary and is also sufficient.

Open Problem 18 : Handle the case m even.

More generally, let m and k be two co-prime integers and $n = mk$. Let α be a normal element of $\mathbb{F}_{2^m}$ over $\mathbb{F}_2$ and w a normal element of $\mathbb{F}_{2^k}$ over $\mathbb{F}_2$. We know that αw is a normal element of $\mathbb{F}_{2^n}$ over $\mathbb{F}_2$ (see [56, Proposition 5.2.3]). We have then the normal bases $(\alpha, \ldots, \alpha^{2^{m-1}})$ of $\mathbb{F}_{2^m}$ over $\mathbb{F}_2$ (which is in the same time a basis of $\mathbb{F}_{2^n}$ over $\mathbb{F}_{2^k}$), $(w, \ldots, w^{2^{k-1}})$ of $\mathbb{F}_{2^k}$ over $\mathbb{F}_2$ (also a basis of $\mathbb{F}_{2^n}$ over $\mathbb{F}_{2^m}$), and

$$(\alpha w, \alpha^2 w^2, \ldots, \alpha^{2^i \ (\mathrm{mod}\ m)} w^{2^i \ (\mathrm{mod}\ k)}, \ldots, \alpha^{2^{n-1} \ (\mathrm{mod}\ m)} w^{2^{n-1} \ (\mathrm{mod}\ k)})$$

of $\mathbb{F}_{2^n}$ over $\mathbb{F}_2$. Any element of $\mathbb{F}_{2^n}$ can be written in the form $\sum_{i=0}^{k-1} x_i w^{2^i}$, where $x_i \in \mathbb{F}_{2^m}$. Since $(\sum_{i=0}^{k-1} x_i w^{2^i})^2 = \sum_{i=0}^{k-1} x_i^2 w^{2^{i+1} \ (\mathrm{mod}\ k)}$, the univariate shift $z \mapsto z^2$ corresponds to the mapping

$$(x_0, \ldots, x_{k-1}) \mapsto \rho_k(x_0^2, \ldots, x_{k-1}^2),$$

where $\rho_k(x_0, \ldots, x_{k-1}) = (x_{k-1}, x_0, \ldots, x_{k-2})$ is the cyclic shift over $\mathbb{F}_{2^m}^k$.

Given a Boolean function $f(x_0, \ldots, x_{k-1})$ in k-variate form (where $x_i \in \mathbb{F}_{2^m}$), the related Boolean function over $\mathbb{F}_2^n$ obtained by decomposing $\sum_{i=0}^{k-1} x_i w^{2^i}$ over (7) is then RS if and only if

$$f(x_0, \ldots, x_{k-1}) = f(\rho_k(x_0^2, \ldots, x_{k-1}^2)).$$

Proposition 2 *Let m and k be two co-prime integers and $n = mk$. Let α be a normal element of $\mathbb{F}_{2^m}$ over $\mathbb{F}_2$ and w a normal element of $\mathbb{F}_{2^k}$ over $\mathbb{F}_2$. Then the n-variable Boolean idempotents are those polynomials $f(z)$ representing Boolean functions over $\mathbb{F}_{2^n}$ whose associate k-variate expressions, defined as $f(x_0, \ldots, x_{k-1}) = f(\sum_{i=0}^{k-1} x_i w^{2^i})$, satisfy $f(x_0, \ldots, x_{k-1}) = f(\rho_k(x_0^2, \ldots, x_{k-1}^2))$. In particular, if $k = 2$, the n-variable Boolean idempotents are those polynomials $f(z)$ representing Boolean functions over $\mathbb{F}_{2^n}$ whose associate bivariate expressions $f(x, y) = f(wx + w^2 y)$ satisfy $f(x, y) = f(y^2, x^2)$.*

Applying the identity m times gives $f(x_0, \ldots, x_{k-1}) = f(\rho_k^m(x_0, \ldots, x_{k-1}))$, and applying it k times gives $f(x_0, \ldots, x_{k-1}) = f(x_0^{2^k}, \ldots, x_{k-1}^{2^k})$. Since m and k are co-prime, there exist integers u and v such that $um + vk = 1$.

Then applying u times the identity $f(x_0, \ldots, x_{k-1}) = f(\rho_k^m(x_0, \ldots, x_{k-1}))$ and v times $f(x_0, \ldots, x_{k-1}) = f(x_0^{2^k}, \ldots, x_{k-1}^{2^k})$ gives $f(x_0, \ldots, x_{k-1}) = f(\rho_k(x_0, \ldots, x_{k-1}))$ and $f(x_0, \ldots, x_{k-1}) = f(x_0^2, \ldots, x_{k-1}^2)$. The double condition "$f(x_0, \ldots, x_{k-1}) = f(\rho_k(x_0, \ldots, x_{k-1}))$ and $f(x_0, \ldots, x_{k-1}) = f(x_0^2, \ldots, x_{k-1}^2)$" is then necessary, and it is also clearly sufficient.

Definition 5 Under the hypotheses of Proposition 2, we call any polynomial $f(z)$ whose k-variate expression satisfies $f(x_0, \ldots, x_{k-1}) = f(x_0^2, \ldots, x_{k-1}^2)$ a k-weak idempotent.

Note that the condition $f(x_0, \ldots, x_{k-1}) = f(x_0^2, \ldots, x_{k-1}^2)$ is equivalent to $f(x_0, \ldots, x_{k-1}) = f(x_0^{2^k}, \ldots, x_{k-1}^{2^k})$ since m and k are co-prime.

Proposition 3 *The set of n-variable idempotent functions is included in that of k-weak idempotents. An idempotent is a k-weak-idempotent invariant under the shift ρ_k.*

The corresponding definition at the bit level is obtained by decomposing the univariate representation over the normal basis (7) and the k-variate representation over the basis $(\alpha, \ldots, \alpha^{2^{m-1}})$:

Definition 6 Let m and k be two co-prime integers and $n = mk$. A Boolean function

$$f(x_{0,0}, y_{1,1}, \ldots, x_{n-1,n-1})$$

(where each first index is reduced modulo k and each second index is reduced modulo m) over $\mathbb{F}_2^n$ is k-weak RS if it is invariant under the cyclic shift by k positions.

For $n = 2m$, m odd, we can see that a function $f(x_0, y_1, x_2, y_3, \ldots, x_{n-2}, y_{n-1})$ (where each index is reduced modulo m; we skip the first index) over $\mathbb{F}_2^n$ is 2-weak RS if it is invariant under the transformation $\begin{cases} x_j \mapsto x_{j+1} \\ y_j \mapsto y_{j+1} \end{cases}$.

Such 2-weak RS function is RS if and only if in bivariate form, it is invariant under $(x, y) \mapsto (y, x)$.

Proposition 4 *A Boolean function $f(x_{0,0}, y_{1,1}, \ldots, x_{n-1,n-1})$ is RS if and only if it is m-weak RS and k-weak RS.*

We shall call the 2-weak idempotents (resp. the 2-weak RS functions) simply weak idempotents (resp. weak RS functions). An example of a weak RS function is the direct sum $f(x) + g(y)$ where f and g are RS m-variable functions; such function is RS when $f = g$.

Remark 4 All the functions derived from o-polynomials with coefficients equal to 1 are bent weak idempotents.

5.3.2 A Secondary Construction of RS and Idempotent Functions

We have seen that the direct sum allows constructing, for $n = 2m$, an n-variable weak idempotent from two m-variable idempotents. The indirect sum allows constructing, for $n = 2m$, an n-variable weak idempotent h from four m-variable idempotents f_1, f_2, g_1, g_2:

$$h(x, y) = f_1(x) + g_1(y) + (f_1 + f_2)(x)(g_1 + g_2)(y); \ x, y \in \mathbb{F}_{2^m}.$$

If $f_1 = g_1$ and $f_2 = g_2$, then we obtain the idempotent $h(x, y) = f_1(x) + f_1(y) + (f_1 + f_2)(x)(f_1 + f_2)(y)$. This gives also a secondary construction of an RS n-variable function from two RS m-variable functions ($n = 2m$, m odd). This function is bent if the two functions are semi-bent.

Definition 7 For odd m, a Boolean function on $\mathbb{F}_2^m$ is called semi-bent if its Walsh transform takes values 0 and $\pm 2^{\frac{m+1}{2}}$ only. See more on semi-bent functions in [19, 25, 42, 54].

Proposition 5 ([15]) *Let f_1 and f_2 be two m-variable RS semi-bent functions, m odd, and let $n = 2m$. If the Walsh supports of f_1 and f_2 are complementary, then $h(x_0, y_1, x_2, y_3, \ldots, x_{n-2}, y_{n-1}) = f_1(x_0, \ldots, x_{m-1}) + f_1(y_0, \ldots, y_{m-1}) + (f_1 + f_2)(x_0, \ldots, x_{m-1})(f_1 + f_2)(y_0, \ldots, y_{m-1})$ is bent RS.*

Indeed, the Walsh transform $\widehat{\chi}_h(a_0, b_1, a_2, b_3, \ldots, a_{n-2}, b_{n-1})$ of h is equal to $\frac{1}{2}\widehat{\chi}_{f_1}(a)\left[\widehat{\chi}_{f_1}(b) + \widehat{\chi}_{f_2}(b)\right] + \frac{1}{2}\widehat{\chi}_{f_2}(a)\left[\widehat{\chi}_{f_1}(b) - \widehat{\chi}_{f_2}(b)\right]$ (see [11]).

Note that, according to Parseval's relation, the Walsh supports of f_1 and f_2 have size 2^{m-1} and then can be complementary. Note also that the secondary construction of Proposition 5 is closely related to that of Theorem 9. It is well known that two m-variable functions f_1 and f_2 (m odd) are semi-bent with complementary Walsh supports if and only if the $(m + 1)$-variable function $f(x, x_{m+1}) = f_1(x) + x_{m+1}f_2(x); \ x \in \mathbb{F}_{2^m}, x_{m+1} \in \mathbb{F}_2$, is bent. Indeed, we have $W_f(a, a_{m+1}) = W_{f_1}(a) + (-1)^{a_{m+1}}W_{f_2}(a)$, implying that f is bent when f_1 and f_2 are semi-bent with complementary Walsh supports; and we have $W_{f_1}(a) = \frac{1}{2}\left(W_f(a, 0) + W_f(a, 1)\right)$ and $W_{f_2}(a) = \frac{1}{2}\left(W_f(a, 0) - W_f(a, 1)\right)$, implying that f_1 and f_2 are semi-bent with complementary Walsh supports when f is bent. But note that when f is RS, f_1 and f_2 are in general not RS, and when f_1 and f_2 are RS, f is in general not RS.

A case of application of the construction of Proposition 5 happens with the bent quadratic function involved in the definition of the Kerdock code (see Sect. 3.4):

$$f(x, x_{m+1}) = \sum_{i=1}^{\frac{m-1}{2}} \text{Tr}_1^m(x^{2^i+1}) + x_{m+1}\text{Tr}_1^m(x).$$

This function is bent and its semi-bent restrictions f_1 and f_2 to the hyperplanes of equations $x_{m+1} = 0$ and $x_{m+1} = 1$ are idempotent functions of x. But the resulting function h derived by the indirect sum is quadratic, because $f_1 + f_2$ is linear, and this reduces its interest. Another example, found in [15], of such a pair (f_1, f_2) will yield an infinite class of idempotent bent functions of algebraic degree 4:

Proposition 6 *For every odd m, the following m-variable idempotent functions $f_1(x) = tr(x) + tr(x^{2^{(m-1)/2}+1})$ and $f_2(x) = tr(x^3)$ are semi-bent functions with complementary Walsh supports.*

Proof We know (see [11, Theorem 8.23]) that a quadratic Boolean function f over $\mathbb{F}_{2^m}$ has for Walsh support the set of elements $a \in \mathbb{F}_{2^m}$ such that $tr(ax) + f(x)$ is constant on E_f where $E_f = \{x \in \mathbb{F}_{2^m} / \forall y \in \mathbb{F}_{2^m}, f(x + y) + f(x) + f(y) + f(0) = 0\}$ is the so-called linear kernel of f. We also know that function f is semi-bent, for m odd, if and only if E_f has dimension 1 (i.e., has size 2). Functions f_1 and f_2 have kernels of equations $x^{2^{(m-1)/2}} + x^{2^{(m+1)/2}} = 0$ and $x^2 + x^{2^{m-1}} = 0$, which are, respectively, equivalent to the equations $x + x^2 = 0$ and $x + x^4 = 0$. These two equations have the same set of solutions, equal to $\mathbb{F}_2$ (using that m is odd for the second one). Hence, both functions are semi-bent. The first function f_1 has then Walsh support $\{a / tr(a) = 0\}$, and the second one f_2 has Walsh support $\{a / tr(a) + tr(1) = 0\}$; these Walsh supports are complementary since $tr(1) = 1$. $\square$

Theorem 11 ([15]) *Let $n = 2m$, m odd. We define the m-variable idempotent functions $f_1(x) = \mathrm{Tr}_1^m(x) + \mathrm{Tr}_1^m(x^{2^{(m-1)/2}+1})$ and $f_2(x) = \mathrm{Tr}_1^m(x^3)$. Then $h(x, y) = f_1(x) + f_1(y) + (f_1 + f_2)(x)(f_1 + f_2)(y)$ is a bent idempotent with algebraic degree 4.*

Similarly, we define the RS functions $f_1^*(x) = \sum_{i=0}^{m-1}(x_i + x_i x_{(m-1)/2+i})$ and $f_2^*(x) = \sum_{i=0}^{m-1} x_i x_{1+i}$, where the subscripts are taken modulo m. Then function $h^*(x_0, y_1, x_2, y_3, \ldots, x_{n-2}, y_{n-1}) = f_1^*(x_0, \ldots, x_{m-1}) + f_1^*(y_0, \ldots, y_{m-1}) + (f_1^* + f_2^*)(x_0, \ldots, x_{m-1})(f_1^* + f_2^*)(y_0, \ldots, y_{m-1})$ is an RS bent function of algebraic degree 4.

Open Problem 19 : Construct classes of RS bent functions of all algebraic degrees between 5 and m.

6 A Transformation on Rotation Symmetric Bent Functions

We can observe a correspondence in Theorem 11 between the functions f_1, f_2, h from one hand side and the functions f_1^*, f_2^*, h^* for the other hand side. It is simpler to describe how f_1, f_2, h can be obtained from f_1^*, f_2^*, h^* rather than vice versa. Given an RS Boolean function f, we consider the function $f'(z)$ over the finite field of order 2^n, expressed in trace representation and obtained from the ANF of $f(x_0, \ldots, x_{n-1})$ by replacing x_i by z^{2^i}. Functions f_1, f_2, h are obtained from f_1^*, f_2^*, h^* by the transformation $f \mapsto f'$. Given an RS Boolean function f, function f' happens to be always a Boolean idempotent function (its idempotence is merely related to the fact that f is Boolean, and its binarity is related to the fact that f is RS).

Proposition 7 ([15]) *Let $f(x_0, \ldots, x_{n-1})$ be any Boolean RS function over $\mathbb{F}_2^n$, then*

$$f'(z) = f(z, z^2, \ldots, z^{2^{n-1}})$$

is a Boolean idempotent. In other words, if:

$$f(x_0, \ldots, x_{n-1}) = \sum_{u \in \mathbb{F}_2^n} a_u x^u, \ then:$$

$$f'(z) = \sum_{u \in \mathbb{F}_2^n} a_u z^{\sum_{i=0}^{n-1} u_i 2^i}$$

is a Boolean idempotent, and any idempotent Boolean function can be obtained this way.

Note that the trace representation of f' is directly deduced from the ANF of f, even for infinite classes of functions f, and has a very similar shape (note that this is not at all the case between an idempotent function and the related RS function obtained by decomposing the input over a normal basis); the question whether f and f' (or more coherently the RS function obtained from f' by decomposing the input over a normal basis) are the same function up to affine equivalence is then natural. We show now with examples that the two functions are in general affinely inequivalent.

Examples

- If f is the indicator of $\{(1, 0, 1, 0, \ldots, 1, 0), (0, 1, 0, 1, \ldots, 0, 1)\}$, then, as observed in [15], we have $f'(z) = z^{\frac{2^n-1}{3}}(1+z)^{\frac{2^n-1}{3}}\left((1+z)^{\frac{2^n-1}{3}} + z^{\frac{2^n-1}{3}}\right)$, which has Hamming weight $2^n - 2 - \frac{2^n-4}{3}$.
  If f is the indicator of $\{(0, \ldots, 0), (1, \ldots, 1)\}$, then f' is the indicator of $\mathbb{F}_2$.
  Hence, f and f' can be affinely equivalent or not, and two functions f and g can be affinely equivalent without that f' and g' be EA-equivalent.
- If $f(x) = \sum_{i=1}^{n}\left(\prod_{j \neq i} x_i\right)$, then f' is the inverse function $tr(z^{2^n-2})$.

6.1 Relationship Between the Bentness of f and f'

We study the relationship between the bentness of f and that of f': we check with infinite classes of RS functions that f can be bent when f' is not and that f' can be bent when f is not; we show that if f is quadratic, then it is bent if and only if f' is bent, and we study classes of bent RS non-quadratic functions f for which f' is bent.

6.1.1   Quadratic Functions

The characterizations recalled in Sects. 5.1.1 and 5.2.1 for the bentness of quadratic RS functions and bent idempotents, given, respectively, in [35] and [48], are the same. Then:

Theorem 12 *If f is a quadratic RS function, then f is bent if and only if f' is bent.*

6.1.2   An Infinite Class of Cubic Bent RS Functions f Such That f' Is Not Bent

Let

$$f_t(x) = \sum_{i=0}^{n-1}(x_i x_{t+i} x_{m+i} + x_i x_{t+i}) + \sum_{i=0}^{m-1} x_i x_{m+i}$$

over $\mathbb{F}_2^n$, where $n = 2m$ and $0 < t < m$ is such that $m/\gcd(m, t)$ is odd. Then we have recalled that f is bent and it is shown in [15] that

$$f_t'(z) = \operatorname{Tr}_1^n(z^{1+2^t+2^m}) + \operatorname{Tr}_1^n(z^{1+2^t}) + \operatorname{Tr}_1^m(z^{1+2^m})$$

is not bent.

6.1.3   An Infinite Class of Cubic Bent Idempotents f' Such That f Is Not Bent

Let

$$f'(z) = \operatorname{Tr}_1^m(x^{1+2^m}) + \operatorname{Tr}_1^n(x^d); \ d = (2^m - 1)/4 + 1$$

be the second Niho bent function given in [31]; then, as shown in [15]

$$f(x) = \sum_{i=0}^{n-1} x_i x_{1+i} x_{m+i} + \sum_{i=0}^{m-1} x_i x_{m+i}$$

can be written in the MM form where π is not a permutation, and f is then not bent.

6.1.4   Infinite Classes of Bent RS Functions f Such That f' Is Bent

A first example is given by Theorem 11. Let us give another example.

Let

$$f(x) = \sum_{i=0}^{n-1} x_i x_r x_{i+2r} + \sum_{i=0}^{2r-1} x_i x_{i+2r} x_{i+4r} + \sum_{i=0}^{m-1} x_i x_{i+m},$$

where $n = 2m = 6r$ with $r \geq 1$.

We know that f and $f'(z) = \mathrm{Tr}_1^n(z^{1+2^r+2^{2r}}) + \mathrm{Tr}_1^{2r}(z^{1+2^{2r}+2^{4r}}) + \mathrm{Tr}_1^m(z^{1+2^t}) = \mathrm{Tr}_1^r((z+z^{2^{3r}})^{1+2^r+2^{2r}}) + \mathrm{Tr}_1^{\nu(t)}(z^{1+2^t})$ are bent.

Our investigations suggest that searching RS bent functions f provides larger probability of success when we choose them such that f' is bent and vice versa.

The transformation $f \mapsto f'$ is however not an equivalence between RS bent functions; the bent functions are likely to be new under affine equivalence.

Open Question: what are the relationships between the cryptographic parameters of f' and those of f?

Conclusion
The research on bent functions continues to be very active. Much work has been done recently. The less recent results which are not recalled in this chapter can be found in [11], and results on bent vectorial functions can be found in [12, 13]. There are many connections with other domains of mathematics and computer science (designs, difference sets, Kloosterman sums, coding, cryptography, sequences, etc.) that we could not detail in this chapter. Important open problems remain (a few evoked in this chapter). Super-classes (partially-bent functions, plateaued functions, etc.), related classes (semi-bent functions), and subclasses (hyperbent functions) pose many problems not evoked here either (see [11]). A complete classification remains elusive.

Acknowledgments We thank Thomas Cusick, Guangpu Gao, and Sihem Mesnager for useful information.

References

1. L. Budaghyan, C. Carlet, CCZ-equivalence of single and multi output Boolean functions. AMS Contemp. Math. **518**, 43–54 (2010). Post-proceedings of the Conference Fq9
2. L. Budaghyan, C. Carlet, CCZ-equivalence of bent vectorial functions and related constructions. Des. Codes Cryptogr. **59**(1–3), 69–87 (2011). Post-proceedings of WCC 2009
3. L. Budaghyan, C. Carlet, A. Pott, New classes of almost bent and almost perfect nonlinear functions. IEEE Trans. Inf. Theory **52**(3), 1141–1152 (2006)
4. L. Budaghyan, C. Carlet, T. Helleseth, On bent functions associated to AB functions, in *Proceedings of IEEE Information Theory Workshop*, ITW'11, Paraty, October 2011

5. L. Budaghyan, C. Carlet, T. Helleseth, A. Kholosha, S. Mesnager, Further results on Niho bent functions. IEEE Trans. Inf. Theory **58**(11), 6979–6985 (2012)
6. A. Canteaut, P. Charpin, G. Kyureghyan, A new class of monomial bent functions. Finite Fields Appl. **14**(1), 221–241 (2008)
7. C. Carlet, Two new classes of bent functions, in *Proceedings of EUROCRYPT'93*. Lecture Notes in Computer Science, vol. 765 (1994), pp. 77–101
8. C. Carlet, A construction of bent functions, in *Finite Fields and Applications*. London Mathematical Society. Lecture Series, vol. 233 (Cambridge University Press, Cambridge, 1996), pp. 47–58
9. C. Carlet, On the secondary constructions of resilient and bent functions, in *Proceedings of the Workshop on Coding, Cryptography and Combinatorics 2003* (Birkhäuser, Basel, 2004), pp. 3–28
10. C. Carlet, On bent and highly nonlinear balanced/resilient functions and their algebraic immunities, in *Proceedings of AAECC 16*. Lecture Notes in Computer Science, vol. 3857 (2006), pp. 1–28
11. C. Carlet, Boolean functions for cryptography and error correcting codes. Chapter of the monography, in *Boolean Models and Methods in Mathematics, Computer Science, and Engineering*, ed. by Y. Crama, P. Hammer (Cambridge University Press, Cambridge, 2010), pp. 257–397. Preliminary version available at http://www-rocq.inria.fr/codes/Claude.Carlet/pubs.html
12. C. Carlet, Vectorial Boolean functions for cryptography. Chapter of the monography, in *Boolean Models and Methods in Mathematics, Computer Science, and Engineering*, ed. by Y. Crama, P. Hammer (Cambridge University Press, Cambridge, 2010), pp. 398–469. Preliminary version available at http://www-rocq.inria.fr/codes/Claude.Carlet/pubs.html
13. C. Carlet, C. Ding, Highly nonlinear mappings. J. Complex. **20**, 205–244 (2004). Special Issue "Complexity Issues in Coding and Cryptography", dedicated to Prof. Harald Niederreiter on the occasion of his 60th birthday
14. C. Carlet, P. Gaborit, Hyper-bent functions and cyclic codes. J. Combin. Theory Ser. A **113**(3), 466–482 (2006)
15. C. Carlet, G. Gao and W. Liu. A secondary construction and a transformation on rotation symmetric functions, and their action on bent and semi-bent functions. *Journal of Combinatorial Theory, Series A, vol.* 127(1), pp. 161–175 (2014).
16. C. Carlet, T. Helleseth, Sequences, Boolean functions, and cryptography, in *Handbook of Codes, Sequences and Their Applications*, ed. by S. Boztas (CRC Press, to appear)
17. C. Carlet, A. Klapper, Upper bounds on the numbers of resilient functions and of bent functions. This paper was meant to appear in an issue of Lecture Notes in Computer Sciences dedicated to Philippe Delsarte, Editor Jean-Jacques Quisquater. But this issue finally never appeared. A shorter version has appeared in the *Proceedings of the 23rd Symposium on Information Theory in the Benelux*, Louvain-La-Neuve, Belgium, 2002. The results are given in [11]
18. C. Carlet, S. Mesnager, On Dillon's class H of bent functions, Niho bent functions and o-polynomials. J. Combin. Theory Ser. A **118**, 2392–2410 (2011)
19. C. Carlet, S. Mesnager, On semi-bent Boolean functions. IEEE Trans. Inform. Theory **58**, 3287–3292 (2012)
20. C. Carlet, P. Charpin, V. Zinoviev, Codes, bent functions and permutations suitable for DES-like cryptosystems. Des. Codes Cryptogr. **15**(2), 125–156 (1998)
21. C. Carlet, L.E. Danielsen, M.G. Parker, P. Solé, Self dual bent functions. Int. J. Inf. Coding Theory **1**(4), 384–399 (2010). Special Issue, dedicated to Vera Pless
22. C. Carlet, F. Zhang, Y. Hu, Secondary constructions of bent functions and their enforcement. Adv. Math. Commun. **6**(3), 305–314 (2012)
23. P. Charpin, G. Gong, Hyperbent functions, Kloosterman sums and Dickson polynomials. IEEE Trans. Inf. Theory **54**(9), 4230–4238 (2008)
24. P. Charpin, G.M. Kyureghyan, Cubic monomial bent functions: a subclass of M. SIAM J. Discrete Math. **22**(2), 650–665 (2008)

25. P. Charpin, E. Pasalic, C. Tavernier, On bent and semi-bent quadratic Boolean functions. IEEE Trans. Inform. Theory **51**, 4286–4298 (2005)
26. T. Cusick, Y. Cheon, Affine equivalence of quartic homogeneous rotation symmetric Boolean functions. Inf. Sci. **259**, 192–211 (2014)
27. D.K. Dalai, S. Maitra, S. Sarkar, Results on rotation symmetric bent functions. Discrete Math. **309**, 2398–2409 (2009)
28. J. Dillon, Elementary Hadamard difference sets. Ph.D. dissertation, University of Maryland, 1974
29. J.F. Dillon, H. Dobbertin, New cyclic difference sets with Singer parameters. Finite Fields Appl. **10**(3), 342–389 (2004)
30. J.F. Dillon, G. McGuire, Near bent functions on a hyperplane. Finite Fields Appl. **14**(3), 715–720 (2008)
31. H. Dobbertin, G. Leander, A. Canteaut, C. Carlet, P. Felke, P. Gaborit, Construction of bent functions via Niho power functions. J. Combin. Theory Ser. A **113**, 779–798 (2006)
32. E. Filiol, C. Fontaine, Highly nonlinear balanced Boolean functions with a good correlation-immunity, in *Proceedings of EUROCRYPT'98*. Lecture Notes in Computer Science, vol. 1403 (1998), pp. 475–488
33. C. Fontaine, On some cosets of the first-order Reed-Muller code with high minimum weight. IEEE Trans. Inform. Theory **45**, 1237–1243 (1999)
34. S. Fu, L. Qu, C. Li, B. Sun, Balanced $2p$-variable rotation symmetric Boolean functions with maximum algebraic immunity. Appl. Math. Lett. **24**, 2093–2096 (2011)
35. G. Gao, X. Zhang, W. Liu, C. Carlet, Constructions of quadratic and cubic rotation symmetric bent functions. IEEE Trans. Inform. Theory **58**, 4908–4913 (2012)
36. C. Carlet, G. Gao and W. Liu. Results on Constructions of Rotation Symmetric Bent and Semi-bent Functions. To appear in the proceedings of Sequences and Their Applications - Seta 2014: 8th International Conference, Melbourne, Vic, Australia, November 24–28, LNCS 8865, 2014.
37. F. Gologlu, Almost bent and almost perfect nonlinear functions, exponential sums, geometries and sequences. Ph.D. dissertation, University of Magdeburg, 2009
38. P. Guillot, Completed GPS covers all bent functions. J. Combin. Theory Ser. A **93**, 242–260 (2001)
39. T. Helleseth, A. Kholosha, Private communication (2013)
40. T. Helleseth, A. Kholosha, S. Mesnager, Niho bent functions and Subiaco hyperovals. Contemp. Math. AMS **579**, 91–101 (2012). Proceedings of the 10-th International Conference on Finite Fields and Their Applications (Fq'10)
41. S. Kavut, S. Maitra, M.D. Yücel, Search for Boolean functions with excellent profiles in the rotation symmetric class. IEEE Trans. Inf. Theory **53**(5), 1743–1751 (2007)
42. K. Khoo, G. Gong, D. Stinson, A new characterization of semi-bent and bent functions on finite fields. Des. Codes Cryptogr. **38**, 279–295 (2006)
43. P. Langevin, G. Leander, Counting all bent functions in dimension eight 99270589265934370305785861242880. Des. Codes Cryptogr. **59**(1–3), 193–205 (2011)
44. G. Leander, Monomial bent functions, in *Proceedings of WCC 2006* (2005), pp. 462–470; and IEEE Trans. Inf. Theory **52**(2), 738–743 (2006)
45. G. Leander, A. Kholosha, Bent functions with 2^r Niho exponents. IEEE Trans. Inform. Theory **52**, 5529–5532 (2006)
46. N. Li, T. Helleseth, X. Tang, A. Kholosha, Several new classes of bent functions From Dillon exponents. IEEE Trans. Inf. Theory **59**(3), 1818–1831 (2013)
47. P. Lisoněk, M. Marko, On zeros of Kloosterman sums. Des. Codes Cryptogr. **59**, 223–230 (2011)
48. W. Ma, M. Lee, F. Zhang, A new class of bent functions. EICE Trans. Fundam. **E88-A**(7), 2039–2040 (2005)
49. F.J. MacWilliams, N.J. Sloane, *The Theory of Error-Correcting Codes* (North-Holland, Amsterdam, 1977)

50. S. Mesnager, A new family of hyper-bent Boolean functions in polynomial form, in *Proceedings of Twelfth International Conference on Cryptography and Coding (IMACC 2009)*. Lecture Notes in Computer Science, vol. 5921 (Springer, Heidelberg, 2009), pp. 402–417
51. S. Mesnager, Hyper-bent Boolean functions with multiple trace terms, in *Proceedings of International Workshop on the Arithmetic of Finite Fields (WAIFI 2010)*. Lecture Notes in Computer Science, vol. 6087 (2010), pp. 97–113
52. S. Mesnager, A new class of bent and hyper-bent Boolean functions in polynomial forms. Des. Codes Cryptogr. **59**(1–3), 265–279 (2011) (see also proceedings of WCC 2009)
53. S. Mesnager, Bent and hyper-bent functions in polynomial form and their link with some exponential sums and Dickson polynomials. IEEE Trans. Inf. Theory **57**(9), 5996–6009 (2011)
54. S. Mesnager, Semi-bent functions from Dillon and Niho exponents, Kloosterman sums, and Dickson polynomials. IEEE Trans. Inform. Theory **57**, 7443–7458 (2011)
55. S. Mesnager, J.P. Flori, Hyper-bent functions via Dillon-like exponents. IEEE Trans. Inf. Theory **59**(5), 3215–3232 (2013)
56. G.L. Mullen, D. Panario, *Handbook of Finite Fields*. Series: Discrete Mathematics and Its Applications (CRC Press, West Palm Beach, 2013)
57. S. Perlis, Normal bases of cyclic fields of prime-power degree. Duke Math J. **9**, 507–517 (1942)
58. J. Pieprzyk, C. Qu, Fast Hashing and rotation symmetric functions. J. Univ. Comput. Sci. **5**, 20–31 (1999)
59. O.S. Rothaus, On bent functions. J. Combin. Theory Ser. A **20**, 300–305 (1976)
60. R. Singh, B. Sarma, A. Saikia, Public key cryptography using permutation p-polynomials over finite fields. IACR Cryptology ePrint Archive 2009: 208 (2009)
61. P. Stănică, S. Maitra, Rotation symmetric Boolean functions-count and cryptographic properties. Discrete Appl. Math. **156**, 1567–1580 (2008)
62. P. Stănică, S. Maitra, J. Clark, Results on rotation symmetric bent and correlation immune Boolean functions, in *Proceedings of Fast Software Encryption 2004*. Lecture Notes in Computer Science, vol. 3017 (2004), pp. 161–177
63. N. Tokareva, On the number of bent functions from iterative constructions: lower bounds and hypotheses. Adv. Math. Commun. **5**(4), 609–621 (2011)
64. A.M. Youssef, G. Gong, Hyper-bent functions, in *Proceedings of EUROCRYPT 2001*. Lecture Notes in Computer Science, vol. 2045 (2001), pp. 406–419
65. N.Y. Yu, G. Gong, Construction of quadratic bent functions in polynomial forms. IEEE Trans. Inf. Theory **7**(52), 3291–3299 (2006)
66. F. Zhang, C. Carlet, Y. Hu, W. Zhang, New secondary constructions of bent functions. Preprint, 2013

On Semi-bent Functions and Related Plateaued Functions Over the Galois Field $\mathbb{F}_{2^n}$

Sihem Mesnager

Abstract Plateaued functions were introduced in 1999 by Zheng and Zhang as good candidates for designing cryptographic functions since they possess desirable various cryptographic characteristics. They are defined in terms of the Walsh–Hadamard spectrum. Plateaued functions bring together various nonlinear characteristics and include two important classes of Boolean functions defined in even dimension: the well-known bent functions and the semi-bent functions. Bent functions (including their constructions) have been extensively investigated for more than 35 years. Very recently, the study of semi-bent functions has attracted the attention of several researchers. Much progress in the design of such functions has been made. The chapter is devoted to certain plateaued functions. The focus is particularly on semi-bent functions defined over the Galois field $\mathbb{F}_{2^n}$ (n even). We review what is known in this framework and investigate constructions.

1 Introduction

The so-called plateaued functions in n variables (or r-plateaued functions) were introduced in 1999 by Zheng and Zhang in [54] for $0 < r < n$. They were first studied by these authors in [55, 56] and further by Carlet and Prouff in [7] as good candidates for designing cryptographic functions. The Walsh–Hadamard spectrum is a very important tool to define and design plateaued functions. An n-variable Boolean function is said to be r-plateaued if the values of its Walsh transform belong to the set $\{0, \pm 2^{\frac{n+r}{2}}\}$ for some fixed r, $0 \le r \le n$. Consequently, plateaued functions have low Hadamard transform, which provides protection against fast correlation attacks [33] and linear cryptanalysis [31]. It has been shown in [54] that plateaued functions are significant in cryptography as they possess desirable

S. Mesnager (✉)
Department of Mathematics, University of Paris VIII, LAGA (Laboratoire Analyse, Géometrie et Applications), UMR 7539, CNRS, Paris, France

University of Paris XIII, Sorbonne Paris Cité, 2 rue de la liberté, 93526 Saint-Denis Cedex, France
e-mail: smesnager@univ-paris8.fr

© Springer International Publishing Switzerland 2014
Ç.K. Koç (ed.), *Open Problems in Mathematics and Computational Science*,
DOI 10.1007/978-3-319-10683-0_11

various cryptographic characteristics such as high nonlinearity, resiliency, low additive autocorrelation, and high algebraic degree and satisfy propagation criteria. Plateaued functions bring together various nonlinear characteristics. They include three significant classes of Boolean functions: the well-known bent functions, the near-bent functions and the semi-bent functions. More precisely, the bent functions are exactly 0-plateaued functions, the near-bent (also called semi-bent in odd dimension) are 1-plateaued functions, and the semi-bent functions are 2-plateaued functions. 0-plateaued functions and 2-plateaued functions on $\mathbb{F}_{2^n}$ exist when n is even, while the 1-plateaued functions on $\mathbb{F}_{2^n}$ exist when n is odd.

For $r \in \{0, 1, 2\}$, r-plateaued functions have been actively studied and have attractive much attention due to their cryptographic, algebraic, and combinatorial properties.

In the mathematical field of combinatorics, bent functions (or 0-plateaued functions) are a special type of Boolean functions. Introduced and named in 1974 by Rothaus [46] in research not published until 1976, firstly studied by Dillon [14], bent functions are so called because they are as different as possible from all linear and affine functions (more precisely, they are at maximum Hamming distance from the set of all affine functions). They are extremal objects in combinatorics and Boolean function theory and have been studied for about 35 years (even more, under the name of difference sets in elementary Abelian 2-groups). The motivation for the study of these particular difference sets is mainly cryptographic, but bent functions play also a role in sequence theory, as difference sets, and especially in coding theory, as elements of Reed-Muller codes. Bent functions exist only with even number of inputs n and have 2-valued spectrum $\pm 2^{\frac{n}{2}}$. The definition of bent function has been extended in several ways, leading to different classes of generalized bent functions that share many of the useful properties of the original. A lot of research has been devoted to designing constructions of bent functions. The reader can refer to the book's chapter of Carlet [4] for general constructions of bent functions and to the following references [37, 41, 44] for a complete state of the art on bent functions defined over the Galois field $\mathbb{F}_{2^n}$, including the main constructions obtained until 2012.

Another special family of plateaued functions defined in even dimension is the set of semi-bent functions. The notion of *semi-bent function* has been introduced in 1994 by Chee et al. [11]. Nevertheless, these functions had been previously investigated in [2] under the name of three-valued almost optimal Boolean functions. Very recently, the development of the theory of semi-bent functions has increased. For very recent results on the treatment of semi-bent functions, we refer to [6,38–40,43]. The motivation for their study is firstly related to their use in cryptography (we recall that in the design of cryptographic functions, various characteristics need be considered simultaneously). Indeed, unlike bent functions, semi-bent functions can also be balanced and resilient. They also possess various desirable characteristics such as low autocorrelation, and a maximal nonlinearity among balanced plateaued functions, satisfy the propagation criteria, and have high algebraic degree. Secondly, besides their practical use in cryptography, they are also widely used in code division multiple access (CDMA) communication systems for sequence design (see, e.g., [17, 19–21, 23, 24, 45]). In this context, families of maximum-length sequences (maximum-length linear feedback shift-register sequences) having three-valued cross-correlation are used. Such sequences have received a lot of attention since the late 1960s and can be generated by a semi-bent function [10]. Up to 2011, the main constructions of semi-bent functions in even dimension are either quadratic functions [48] or derived from power polynomials $Tr_1^n(x^d)$ for a suitably chosen d (see [10]). Since then, several constructions of semi-bent have been proposed in the literature. The principal engine of this progress is the result of several important observations in connection with the construction of bent functions [5, 36, 42]. We shall describe this more precisely in Sect. 4.2.

The chapter is devoted to certain plateaued functions. Special attention is directed to semi-bent functions. We review what is known in this context and investigate new constructions. The chapter is organized as follows. In Sect. 2, we fix our main notation and recall the necessary background. Section 3 is devoted to r-plateaued functions. We recall some basic concepts concerning these functions. In Sects. 3.1–3.3, we treat special classes of r-plateaued functions and present an overview related to the notion of bent, near-bent, and semi-bent functions, respectively. Next, in Sect. 4, we focus on the class of semi-bent functions. We survey the constructions discovered recently. We first point out the relationship between the semi-bentness property of some type of functions and some exponential sums (involving Dickson polynomials). Secondly, we emphasize the link between semi-bent functions and some bent functions. Finally, we study the new connections between semi-bent functions and oval polynomials from projective finite geometry and investigate several constructions. Open problems related to semi-bent functions are given in Sect. 4.

2 Background

For any set E, $E^\star = E \setminus \{0\}$ and $\#E$ will denote the cardinality of E. For any positive integer k, $\mathbb{F}_{2^k}$ denotes the finite field of order 2^k.

Let n be a positive integer. A Boolean function f is a map from the vector space $\mathbb{F}_2^n$ of all binary vectors of length n to the finite field with two elements $\mathbb{F}_2$, i.e., $f : \mathbb{F}_2^n \to \mathbb{F}_2$. The *Hamming weight of a Boolean function* f on $\mathbb{F}_2^n$, denoted by $wt(f)$, is the size of the support of the function, i.e., the set $\{x \in \mathbb{F}_2^n / f(x) \neq 0\}$. The *Hamming distance* $d_H(f, g)$ between two functions f and g is the size of the set $\{x \in \mathbb{F}_2^n / f(x) \neq g(x)\}$. Thus it equals $w_H(f \oplus g)$.

In cryptography, the most usual representation of these functions is the *algebraic normal form* (ANF) :

$$f(x_1, \ldots, x_n) = \sum_{I \subseteq \{1, \ldots, n\}} a_I \left(\prod_{i \in I} x_i \right)$$

where the a_I's are in $\mathbb{F}_2$. The terms $\prod_{i \in I} x_i$ are called monomials. The *algebraic degree* of a Boolean function f equals the global degree of its (unique) ANF, that is, the maximum degree of those monomials whose coefficients are nonzero.

There exist several kinds of possible trace (univariate) representations of Boolean functions (see, e.g., [4, p. 266]) which are not necessary unique and use the identification between the vector space $\mathbb{F}_2^n$ and the field $\mathbb{F}_{2^n}$. A possible representation of Boolean functions using such an identification is to consider any Boolean function as a polynomial in one variable $x \in \mathbb{F}_{2^n}$ of the form $f(x) = \sum_{j=0}^{2^n-1} a_j x^j$ where the a_j's are elements of the field. This representation exists for every function from $\mathbb{F}_{2^n}$ to $\mathbb{F}_{2^n}$, and such a function f is Boolean if and only if a_0 and a_{2^n-1} belong to $\mathbb{F}_2$ and $a_{2j} = a_j^2$ for every $j \neq 0, 2^n - 1$, where $2j$ is taken modulo $2^n - 1$. This allows representing $f(x)$ in a (unique) trace expansion. Recall that for any positive integer k, and r dividing k, the trace function from $\mathbb{F}_{2^k}$ to $\mathbb{F}_{2^r}$, denoted by Tr_r^k, is the mapping defined as

$$Tr_r^k(x) := \sum_{i=0}^{\frac{k}{r}-1} x^{2^{ir}} = x + x^{2^r} + x^{2^{2r}} + \cdots + x^{2^{k-r}}.$$

In particular, we denote the *absolute trace* over $\mathbb{F}_2$ of an element $x \in \mathbb{F}_{2^n}$ by $Tr_1^n(x) = \sum_{i=0}^{n-1} x^{2^i}$.

A unique representation of a Boolean function over $\mathbb{F}_{2^n}$ by means of trace functions is of the form

$$f(x) = \sum_{j \in \Gamma_n} Tr_1^{o(j)}(a_j x^j) + \epsilon(1 + x^{2^n-1}) \tag{1}$$

called its *polynomial form*, where:

- Γ_n is the set of integers obtained by choosing one element in each cyclotomic class of 2 modulo $2^n - 1$ (the most usual choice for j is the smallest element in its cyclotomic class, called the coset leader of the class).
- $o(j)$ is the size of the cyclotomic coset of 2 modulo $2^n - 1$ containing j (recall that, the cyclotomic class of 2 modulo $2^n - 1$ denoted by $C(j)$ is defined as $C(j) := \{j, j2, j2^2, j2^3, \ldots, j2^{o(j)-1}\}$ where $o(j)$ is the smallest positive integer such that $j2^{o(j)} \equiv j \pmod{2^n - 1}$).
- $a_j \in \mathbb{F}_{2^{o(j)}}$.
- $\epsilon = wt(f)$ modulo 2 where $wt(f)$ is the *Hamming weight* of the image vector of f, that is, the cardinality of its support $supp(f) := \{x \in \mathbb{F}_{2^n} \mid f(x) = 1\}$.

Note that the expression of f given by (1) can also be written under a non-unique form $Tr_1^n(P(x))$ where $P(x)$ is a polynomial over $\mathbb{F}_{2^n}$.

The algebraic degree of f is then equal to the maximum 2-weight of an exponent j for which $a_j \neq 0$ if $\epsilon = 0$ and to n if $\epsilon = 1$. Recall that the 2-weight $w_2(j)$ of an integer j equals by definition the number of 1's in its binary expansion. In particular, affine functions are those of algebraic degree at most 1.

Quadratic functions are those of algebraic degree 2. They can be represented as follows: when n is even,

$$f(x) = \sum_{i=1}^{\frac{n}{2}-1} Tr_1^n(a_i x^{2^i+1}) + Tr_1^{\frac{n}{2}}(a_{\frac{n}{2}} x^{1+2^{\frac{n}{2}}})$$

where $a_i \in \mathbb{F}_{2^n}, \forall i, 0 \le i \le n/2$ and $a_{\frac{n}{2}} \in \mathbb{F}_{2^{n/2}}$.
When n is odd,

$$f(x) = \sum_{i=1}^{\frac{n-1}{2}} Tr_1^n(a_i x^{2^i+1}), a_i \in \mathbb{F}_{2^n}.$$

The rank of a quadratic function f is defined as follows:

$$\text{rank}(f) = n - \dim_{\mathbb{F}_2} \text{rad}(B_f)$$

where $\text{rad}(B_f) := \{x \in \mathbb{F}_{2^n} \mid B_f(x, y) = 0, \forall y \in \mathbb{F}_{2^n}\}$ with B_f the bilinear form defined as

$$B_f(x, y) := f(x + y) + f(x) + f(y).$$

Set $k_f := \dim_{\mathbb{F}_2} \text{rad}(B_f)$. Then 2 divides $(n - k_f)$. Any quadratic Boolean function on $\mathbb{F}_{2^n}$ has a rank $2t$ with $0 \le t \le \lfloor \frac{n}{2} \rfloor$ [29] and can be obtained as follows: set $\tilde{B}_f(x, y) := f(0) + f(x) + f(y) + f(x + y)$. Then the rank of f equals $2t$ if and only if the equation $\tilde{B}_f(x, y) = 0$ for any $y \in \mathbb{F}_{2^n}$ in x has exactly 2^{n-2t} solutions. The set $E_f := \{x \in \mathbb{F}_{2^n}, \mid \forall y \in \mathbb{F}_{2^n}, \tilde{B}_f(x, y) = 0\}$ is called the linear kernel of f.

Note that a significant result dealing with quadratic Boolean functions of rank $2t$ has been obtained by Helleseth and Kumar [21] (see Theorem 1).

The *bivariate representation* of Boolean functions is defined only when $n = 2m$ is even as follows: we identify $\mathbb{F}_2^n$ with $\mathbb{F}_{2^m} \times \mathbb{F}_{2^m}$, and we consider then the input to f as an ordered pair (x, y) of elements of $\mathbb{F}_{2^m}$. There exists a unique bivariate polynomial

$$\sum_{0 \le i, j \le 2^m - 1} a_{i,j} x^i y^j$$

over $\mathbb{F}_{2^m}$ such that f is the bivariate polynomial function over $\mathbb{F}_{2^m}$ associated to it. Then the algebraic degree of f equals

$$\max_{(i,j) \mid a_{i,j} \neq 0} (w_2(i) + w_2(j)),$$

and f being Boolean, its bivariate representation can be written in the form

$$f(x, y) = Tr_1^m(P(x, y))$$

where $P(x, y)$ is some polynomial in two variables over $\mathbb{F}_{2^m}$.

Now, let f be a Boolean function over $\mathbb{F}_{2^n}$ and $a \in \mathbb{F}_{2^n}$. The derivative of f with respect to a is defined as

$$D_{af}(x) = f(x) + f(x + a), \forall x \in \mathbb{F}_{2^n}.$$

For $(a, b) \in \mathbb{F}_{2^n} \times \mathbb{F}_{2^n}$, the second-order derivative of f with respect to (a, b) is defined as

$$D_b D_{af}(x) = f(x) + f(x + b) + f(x + a) + f(x + a + b), \forall x \in \mathbb{F}_{2^n}.$$

The notion of Walsh transform refers to a scalar product. When $\mathbb{F}_2^n$ is identified with the field $\mathbb{F}_{2^n}$ by an isomorphism between these two n-dimensional vector spaces over $\mathbb{F}_2$, it is convenient to choose the isomorphism such that the canonical scalar product "·" in $\mathbb{F}_2^n$ coincides with the canonical scalar product in $\mathbb{F}_{2^n}$, which is the trace of the product : $x \cdot y = Tr_1^n(xy)$ for $x, y \in \mathbb{F}_{2^n}$.

If f is a Boolean function defined on $\mathbb{F}_{2^n}$, then the Walsh–Hadamard transform of f is the discrete Fourier transform of the sign function $\chi_f := (-1)^f$ of f, whose value at $\omega \in \mathbb{F}_{2^n}$ is defined as follows:

$$\forall \omega \in \mathbb{F}_{2^n}, \quad \widehat{\chi_f}(\omega) = \sum_{x \in \mathbb{F}_{2^n}} (-1)^{f(x) + Tr_1^n(\omega x)}.$$

The Walsh transform satisfies the well-known Parseval's relation

$$\sum_{\omega \in \mathbb{F}_{2^n}} \widehat{\chi_f}^2(\omega) = 2^{2n}.$$

Note that not all values of the Walsh–Hadamard transform can have the same sign, except when the function is affine. This comes from the fact that we then have $\left(\sum_{\omega \in \mathbb{F}_{2^n}} \widehat{\chi_f}(\omega) \right)^2 = \sum_{\omega \in \mathbb{F}_{2^n}} \widehat{\chi_f}^2(\omega)$ which implies that all these values are null except one (see, for instance, [42]).

The Walsh–Hadamard transform is an important tool for research in cryptography. It plays an important role to characterize many cryptographic criteria for Boolean functions but also to define some significant cryptographic Boolean functions used in various type of symmetric cryptosystems.

Finally, the rank of quadratic Boolean functions is connected with the distribution of its Walsh–Hadamard transform values. The following result concerning the distribution of the Walsh transform of quadratic Boolean functions is due to Helleseth and Kumar.

Table 1 Walsh spectrum of quadratic function with rank $2t$

Value of $\widehat{\chi_f}(\omega)$, $\omega \in \mathbb{F}_{2^n}$	Number of occurrences
0	2^{n-2t}
2^{n-t}	$2^{2t-1} + 2^{t-1}$
-2^{n-t}	$2^{2t-1} - 2^{t-1}$

Theorem 1 ([21]) *Let f be a quadratic Boolean function on $\mathbb{F}_{2^n}$ with rank $2t$, $0 \leq t \leq \lfloor \frac{n}{2} \rfloor$. Then the distribution of its Walsh transform is given in Table 1.*

3 Plateaued Functions

Plateaued Boolean functions can be defined as follows.

Definition 1 A Boolean function f defined over $\mathbb{F}_{2^n}$ is said to be r-plateaued if the values of its Walsh transform $\widehat{\chi_f}$ are in $\{0, \pm 2^{\frac{n+r}{2}}\}$, for some fixed r, $r = 0, 1, \ldots, n$.

The r-plateaued functions exist only when $n - r$ is even; equivalently, if n and r have the same parity (which implies that 2 divides $n + r$). The value $\lambda := 2^{\frac{n+r}{2}}$ is usually called *the amplitude*.

Remark 1 Note that if f is an r-plateaued function on $\mathbb{F}_{2^n}$, then its Walsh transform $\widehat{\chi_f}$ can be expressed by $\widehat{\chi_f} = ((-1)^g + (-1)^h) 2^{\frac{n+r-2}{2}}$ for some Boolean g and h defined over $\mathbb{F}_{2^n}$.

Plateaued functions can be characterized by their second-order derivatives. More precisely:

Proposition 1 ([7]) *A Boolean function f on $\mathbb{F}_{2^n}$ is plateaued if and only if there exists λ (necessarily the amplitude of f) such that for every $x \in \mathbb{F}_{2^n}$*

$$\sum_{a,b \in \mathbb{F}_{2^n}} (-1)^{D_a D_b f(x)} = \lambda^2$$

where $D_a D_b f$ is the second-order derivative of f with respect to $(a, b) \in \mathbb{F}_{2^n}^2$.

A direct consequence of the previous proposition is that all the quadratic functions are plateaued. Several properties of plateaued functions have been studied. Concerning the degree of r-plateaued functions, it has been shown in [56] that for a given fixed n and r with $r > 0$, the maximum possible degree of r-plateaued on $\mathbb{F}_{2^n}$ is $\frac{n-r+2}{2}$ (while the maximum possible degree of 0-plateaued on $\mathbb{F}_{2^n}$ is $\frac{n}{2}$) and that this upper bound is sharp. Other properties of plateaued functions can be found in [2].

The existence of r-plateaued functions on $\mathbb{F}_{2^n}$ ($0 < r < n$) has been shown in [56]. However, there exist some results concerning the nonexistence of certain types of plateaued functions. More precisely, Xia et al. have proved in [52] that there are no homogeneous[1] 0-plateaued of degree $\frac{n}{2}$ when $n \geq 4$. This result on the nonexistence of homogeneous 0-plateaued functions has been extended on one hand by Meng et al. [34] for functions of degree $\frac{n}{2} - k$ ($0 \leq k \leq \frac{n}{2}$) and on the other hand by Hyun et al. [22] for 0-plateaued functions f (not necessarily homogeneous) of minimum degree (i.e., the lowest degree among the degrees of nonconstant terms in f $\frac{n}{2} - k$ ($0 \leq k \leq \frac{n}{2}$). Moreover, very recently, it has been proved in [22] the nonexistence of r-plateaued functions on $\mathbb{F}_{2^n}$ ($0 < r < n$) with certain degree for a given $n \geq N$ and r (where N is some integer depending on r). More precisely:

Proposition 2 ([22]) *For any nonnegative integer k, there exists an integer N such that for an integer $n \geq N$, there is no r-plateaued function ($0 < r < n$) over $\mathbb{F}_{2^n}$ of minimum degree $\frac{n-r+2}{2} - k$, where N is the smallest integer satisfying $\binom{\frac{N+r}{2}+k}{r+k} < 2^{\frac{N+r-2}{2}} - 1$.*

As a consequence, it has been shown in [22] that there is no homogeneous 1-plateaued function over $\mathbb{F}_{2^n}$ of degree $\frac{n+1}{2}$ when $n \geq 7$, and there is no homogeneous 2-plateaued function over $\mathbb{F}_{2^n}$ of degree $\frac{n}{2}$ when $n \geq 6$.

3.1 Plateaued Functions: The Special Class of 0-Plateaued Functions (Bent Functions)

Bent functions introduced in 1974 [14,46] are extremal objects in combinatorics and Boolean function theory. They are maximally nonlinear Boolean functions. Recall that the *nonlinearity* of a Boolean function f, denoted by $\mathrm{nl}(f)$, is defined as the minimum Hamming distance between f and all affine functions (i.e., of degree at most 1). It can be expressed by means of the Walsh transform as follows:

$$\mathrm{nl}(f) = 2^{n-1} - \frac{1}{2} \max_{b \in \mathbb{F}_{2^n}} \left| \widehat{\chi_f}(b) \right|.$$

Because of the well-known Parseval's relation $\sum_{b \in \mathbb{F}_{2^n}} \widehat{\chi_f}(b)^2 = 2^{2n}$, $\mathrm{nl}(f)$ is upper bounded by $2^{n-1} - 2^{n/2-1}$. This bound is tight for n even.

Definition 2 Let n be an even integer. A Boolean function on $\mathbb{F}_{2^n}$ is said to be bent if the upper bound $2^{n-1} - 2^{n/2-1}$ on its nonlinearity $\mathrm{nl}(f)$ is achieved with equality.

Bent functions on $\mathbb{F}_{2^n}$ exist then only when n is even. We have the following main characterization of the bentness for Boolean functions in terms of the Walsh transform:

---

[1]A Boolean function f is said to be homogeneous of degree r if $f(x) = \sum_{i=0}^{2^n-1} a_i x^i$ where $a_i = 0$ for $wt(i) \neq r$, where $wt(i)$ is the Hamming weight of i.

Table 2 Walsh spectrum of bent functions (0-plateaued) f with $f(0) = 0$

Value of $\widehat{\chi_f}(\omega)$, $\omega \in \mathbb{F}_{2^n}$	Number of occurrences
$2^{\frac{n}{2}}$	$2^{n-1} + 2^{\frac{n-2}{2}}$
$-2^{\frac{n}{2}}$	$2^{n-1} - 2^{\frac{n-2}{2}}$

Proposition 3 *Let n be an even integer. A Boolean function f is then bent if and only if its Walsh transform satisfies $\widehat{\chi_f}(a) = \pm 2^{\frac{n}{2}}$ for all $a \in \mathbb{F}_{2^n}$.*

Hence, the Walsh transform provides a basic characterization of bentness. However, for a given Boolean function f, the Walsh transform can definitely not be used in practice to test in an efficient way the bentness of f, especially if all its values are computed naively one at a time as exponential sums. Thanks to Parseval's identity, one can determine the number of occurrences of each value of the Walsh transform of a bent function (see Table 2).

Bent functions are not classified. A complete classification of these functions is elusive and looks hopeless. So it is important to design constructions in order to find as many of bent functions as possible. A good reference for general properties and general constructions of bent functions is the book's chapter of Carlet [4]. We refer to [37] and [41] for a survey and a general overview of the constructions discovered recently including the relationship between the bentness property of some type of bent functions and some exponential sums, namely, Kloosterman sums (involving Dickson polynomials). Finally, note that a nice construction of bent functions have been derived from plateaued functions in [8].

3.2 Plateaued Functions: The Special Class of 1-Plateaued Functions (Near-Bent Functions)

Near-bent functions (or 1-plateaued functions) on $\mathbb{F}_{2^n}$ exist only when n is odd. They are defined as follows.

Definition 3 Let n be an odd integer. A Boolean function on $\mathbb{F}_{2^n}$ is said to be near-bent if its Walsh transform satisfies $\widehat{\chi_f}(a) \in \{0, \pm 2^{\frac{n+1}{2}}\}$ for all $a \in \mathbb{F}_{2^n}$.

Note that a function from $\mathbb{F}_{2^n} \to \mathbb{F}_{2^n}$ is said to be almost bent if it has Walsh-Fourier spectrum $\{0, \pm 2^{\frac{n+1}{2}}\}$, that is, the same as a near-bent function. The difference between an almost bent function and a near-bent function is that almost bent functions map $\mathbb{F}_{2^n} \to \mathbb{F}_{2^n}$, whereas near-bent functions map $\mathbb{F}_{2^n} \to \mathbb{F}_2$. In this context, $f : \mathbb{F}_{2^n} \to \mathbb{F}_{2^n}$ is almost bent if and only if each of the Boolean functions $x \mapsto Tr_1^n(vf(x))$ is near-bent, for all $v \in \mathbb{F}_{2^n}^\star$.

Thanks to Parseval's identity, one can determine the number of occurrences of each value of the Walsh transform of a near-bent function (see Table 3).

Table 3 Walsh spectrum of near-bent functions (1-plateaued) f with $f(0) = 0$

Value of $\widehat{\chi_f}(\omega)$, $\omega \in \mathbb{F}_{2^n}$	Number of occurrences
0	2^{n-1}
$2^{\frac{n+1}{2}}$	$2^{n-2} + 2^{\frac{n-3}{2}}$
$-2^{\frac{n+1}{2}}$	$2^{n-1} - 2^{\frac{n-3}{2}}$

Again from Parseval's identity, it is straightforward to see that the support of the Walsh transform $\widehat{\chi_f}$ of a near-bent function f on $\mathbb{F}_{2^n}$ is of cardinality 2^{n-1} (i.e., $\#\mathrm{supp}(\widehat{\chi_f}) = 2^{n-1}$).

In the particular case of quadratic functions, there exists a criterion on the near-bentness involving the dimension of the linear kernel (see, e.g., [10]). More precisely, it is well known (see Sect. 8.5.2 in [4]) that a quadratic Boolean function f over $\mathbb{F}_{2^n}$ has for Walsh support the set of elements $\alpha \in \mathbb{F}_{2^n}$ such that $Tr_1^n(\alpha x) + f(x)$ is constant on E_f, where $E_f := \{x \in \mathbb{F}_{2^n}, \mid \forall y \in \mathbb{F}_{2^n}, f(x+y) + f(x) + f(y) + f(0) = 0\}$ is the linear kernel of f. It has been proved that f is near-bent over $\mathbb{F}_{2^n}$, if and only if E_f has dimension 1 (i.e., has size 2). Note that from Theorem 1, it is easy to see that quadratic Boolean function f is near-bent if and only if the rank of f is $n-1$, that is, $k_f = 1$.

Several constructions of quadratic near-bent functions have been obtained in the literature. We give a list of the known families of quadratic near-bent functions on $\mathbb{F}_{2^n}$, n odd:

- $f(x) = Tr_1^n(x^{2^i+1})$, $gcd(i,n) = 1$ [17].
- $f(x) = \sum_{i=1}^{\frac{n-1}{2}} Tr_1^n(x^{1+2^i})$ [1].
- $f(x) = \sum_{i=1}^{\lfloor \frac{n-1}{2} \rfloor} c_i Tr_1^n(x^{1+2^i})$, $c_i \in \mathbb{F}_2$ [10].
- $f(x) = Tr_1^n(x^{2^i+1} + x^{2^j+1} + x^{2^t+1})$, $1 \le i < j \le t \le \frac{n-1}{2}$, $i+j = t$, $gcd(n,i) = gcd(n,j) = gcd(n,i+j) = 1$ [10].
- $f(x) = \sum_{i=1}^{\frac{n-1}{2}} c_i Tr_1^n(x^{1+2^i})$, $c_i \in \mathbb{F}_2$, $gcd(x^n + 1, c(x)) = x + 1$ where $c(x) = \sum_{i=1}^{\frac{n-1}{2}} c_i(x^i + x^{n-i})$ [24].
- $f(x) = Tr_1^n(x^{2^i+1}) + Tr_1^n(x^{2^i+1})$, $gcd(n, i+j) = gcd(n, i-j)$ [24].
- $f(x) = \sum_{i=0}^{r} Tr_1^n(x^{1+2^{k+id}})$, $gcd(2k + rd, n) = 1$ [24].
- $f(x) = \sum_{i=1}^{\frac{q-1}{2}} Tr_1^n(x^{1+2^{pi}}) + Tr_1^n(x^{1+2^q})$, $n = pq$, $3 \not| p$, p odd, q odd, $gcd(p,q) = 1$ [16].

Because bent functions exist in even dimensions and near-bent functions exist in odd dimensions, the possibility exists of moving up and down between bent and near-bent functions. The four possibilities are discussed in [26]; see also some results in [2]. In [27], Leander and McGuire have considered the problem on going up from a near-bent function to a bent function and proposed constructions. In particular, it has been shown that two n-variable functions g and h (n odd) are near-bent with complementary Walsh supports (i.e., $\mathrm{supp}(\widehat{\chi_g}) \cap \mathrm{supp}(\widehat{\chi_h}) = \emptyset$) if and only if the $(n+1)$-variable function $x \mapsto f(x, x_{n+1}) = g(x) + x_{n+1}h(x)$; $x \in \mathbb{F}_2^n$, $x_{n+1} \in \mathbb{F}_2$ is bent. The restrictions to a $(2n)$-bent function to any hyperplan and to the complement of this hyperplan (view as $(2n - 1)$-Booleans functions) are near-bent. The problem of the construction of $(2n)$-bent functions from two $(2n-1)$-near-bent functions has also been considered by Wolfmann with a different point of view in [49]. Some progress on this question has been made very recently in [51] and [50]. In particular, Wolfmann [50] has introduced a way to construct new bent functions starting from a near-bent functions having a specific derivative or from a bent function such that the sum of the two components is a Boolean function of degree 1. Some open problems have been presented by Wolfmann [50] in the continuation of his interesting approach.

In 2005, Charpin et al. [10] have proved that some classes of near-bent functions can been derived via the composition with nonpermutation linear polynomials. In fact, the composition of any linear permutation polynomial P with a quadratic near-bent function gives rise again to a near-bent function $x \mapsto f(P(x))$. However, it is not necessary for P to be a permutation polynomial in order for $f \circ P$ to be near-bent. In fact, one may choose a linear mapping P from $\mathbb{F}_{2^n}$ to $\mathbb{F}_{2^n}$ which is still near-bent. Charpin et al. [10] have exhibited some nonpermutation linear polynomials that preserve the near-bentness property when composed with a quadratic near-bent function. For more details on the treatment of near-bent functions, we send the reader to [10].

Finally, very few secondary constructions of near-bent functions (i.e., constructions of new near-bent functions from two or several already known ones) have been proposed in the literature. The following statement shows that secondary constructions of near-bent functions can be derived under a condition involving the derivative functions.

Theorem 2 *Let n be an odd integer. Let f and g be two near-bent functions over $\mathbb{F}_{2^n}$. Assume that there exists an element a of $\mathbb{F}_{2^n}$ such that $D_{af} = D_a g$. Then the function $h = f + D_{af}(f + g)$ is a near-bent function on $\mathbb{F}_{2^n}$.*

Proof Let us compute the Walsh transform of h for every $\omega \in \mathbb{F}_{2^n}$. We have

$$\widehat{\chi_h}(\omega) = \sum_{x \in \mathbb{F}_{2^n}} \chi(h(x) + Tr_1^n(\omega x)) = \sum_{x \in \mathbb{F}_{2^n}} \chi(f(x) + D_{af}(x)(f+g)(x) + Tr_1^n(\omega x)).$$

Now, one can split the sum depending whether D_{af} is equal to 1 or not (recall that $D_{af}(x) = f(x) + f(x + a)$):

$$\widehat{\chi_h}(\omega) = \sum_{x \in \mathbb{F}_{2^n} | D_{af}=0} \chi(f(x) + Tr_1^n(\omega x)) + \sum_{x \in \mathbb{F}_{2^n} | D_{af}=1} \chi(g(x) + Tr_1^n(\omega x))$$

$$= \frac{1}{2}\left(\sum_{x \in \mathbb{F}_{2^n}} \chi(f(x) + Tr_1^n(\omega x)) + \sum_{x \in \mathbb{F}_{2^n}} \chi(f(x+a) + Tr_1^n(\omega x)) \right)$$

$$+ \frac{1}{2}\left(\sum_{x \in \mathbb{F}_{2^n}} \chi(g(x) + Tr_1^n(\omega x)) - \sum_{x \in \mathbb{F}_{2^n}} \chi(g(x+a) + Tr_1^n(\omega x)) \right).$$

Hence,

$$
\widehat{\chi_h}(\omega) = \frac{1}{2}\Big(\sum_{x\in\mathbb{F}_{2^n}} \chi(f(x) + Tr_1^n(\omega x)) + \sum_{x\in\mathbb{F}_{2^n}} \chi(f(x) + Tr_1^n(\omega(x+a))) \Big)
$$

$$
+\frac{1}{2}\Big(\sum_{x\in\mathbb{F}_{2^n}} \chi(g(x) + Tr_1^n(\omega x)) - \sum_{x\in\mathbb{F}_{2^n}} \chi(g(x) + Tr_1^n(\omega(x+a))) \Big)
$$

$$
= \frac{1}{2}\Big(\widehat{\chi_f}(\omega)(1 + \chi(Tr_1^n(\omega a))) \Big) + \frac{1}{2}\Big(\widehat{\chi_g}(\omega)(1 - \chi(Tr_1^n(\omega a))) \Big).
$$

Now, f and g being near bent, therefore if $Tr_1^n(\omega a) = 0$, then $\widehat{\chi_h}(\omega) = \widehat{\chi_f}(\omega) \in \{0, \pm 2^{\frac{n+1}{2}}\}$. And if $Tr_1^n(\omega a) = 1$, then $\widehat{\chi_h}(\omega) = \widehat{\chi_g}(\omega) \in \{0, \pm 2^{\frac{n+1}{2}}\}$, which completes the proof. $\square$

3.3 Plateaued Functions: The Special Class of 2-Plateaued Functions (Semi-Bent Functions)

Semi-bent functions (or 2-plateaued functions) on $\mathbb{F}_{2^n}$ exist only when n is even. So, in this section n denotes an even integer, and we set $m = \frac{n}{2}$. Semi-bent functions are defined as follows.

Definition 4 Let n be an even integer. A Boolean function on $\mathbb{F}_{2^n}$ is said to be semi-bent if its Walsh transform satisfies $\widehat{\chi_f}(a) \in \{0, \pm 2^{\frac{n+2}{2}}\}$ for all $a \in \mathbb{F}_{2^n}$.

Thanks to Parseval's identity, one can determine the number of occurrences of each value of the Walsh transform of a semi-bent function (see Table 4).

Using the relationship between the nonlinearity and the Walsh spectrum, it is immediate to see that the nonlinearity of a semi-bent function on $\mathbb{F}_{2^n}$ equals $2^{n-1} - 2^{\frac{n}{2}}$. In addition, the possible values of the Hamming weight of a semi-bent function are 2^{n-1}, $2^{n-1} - 2^m$ and $2^{n-1} + 2^m$.

Many recent progresses have been made on the treatment of semi-bent functions. In the next section, we focus on the constructions of such functions.

Table 4 Walsh spectrum of semi-bent functions (2-plateaued) f with $f(0) = 0$

Value of $\widehat{\chi_f}(\omega)$, $\omega \in \mathbb{F}_{2^n}$	Number of occurrences
0	$2^{n-1} + 2^{n-2}$
$2^{\frac{n+2}{2}}$	$2^{n-3} + 2^{\frac{n-4}{2}}$
$-2^{\frac{n+2}{2}}$	$2^{n-3} - 2^{\frac{n-4}{2}}$

4   Semi-Bent Functions (in Even Dimension): Constructions and Characterizations

In the following, we present a general overview of the main known constructions of semi-bent functions and investigate new constructions.

4.1   On Constructions of Quadratic Semi-Bent Functions

The first papers dealing with constructions of semi-bent functions have been dedicated to quadratic functions. In this particular case of functions, there exists a criterion on the semi-bentness involving the dimension of the linear kernel defined above (see, e.g., [10]). More precisely, it has been proved that f is semi-bent over $\mathbb{F}_{2^n}$, if and only if its linear kernel E_f (defined previously) has dimension 2. Note that from Theorem 1, it is easy to see that quadratic Boolean function is semi-bent if and only if the rank of f is $n-2$, that is, $k_f = 2$.

Several constructions of quadratic semi-bent functions have been obtained in the literature. We give a list of the known quadratic semi-bent functions on $\mathbb{F}_{2^n}$, $n = 2m$:

- $f(x) = \sum_{i=1}^{\lfloor \frac{n-1}{2} \rfloor} c_i Tr_1^n(x^{1+2^i})$, $c_i \in \mathbb{F}_2$, $gcd(\sum_{i=1}^{\frac{n}{2}-1} c_i(x^i + x^{n-i}), x^n + 1) = x^2 + 1$ [10].
- $f(x) = Tr_1^n(\alpha x^{2^i+1})$, $\alpha \in \mathbb{F}_{2^n}^\star$, i even, m odd [48].
- $f(x) = Tr_1^n(\alpha x^{2^i+1})$, m even, i odd, $\alpha \in \{x^3, x \in \mathbb{F}_{2^n}^\star\}$ where $\alpha \in \mathbb{F}_{2^n}^\star$ [48].
- $f(x) = Tr_1^n(\alpha x^{2^i+1})$, m odd, i odd, $gcd(m,i) = 1$, $\alpha \in \{x^3, x \in \mathbb{F}_{2^n}^\star\}$ where $\alpha \in \mathbb{F}_{2^n}^\star$ [48].
- $f(x) = Tr_1^n(x^{2^i+1} + x^{2^j+1})$, m odd, $1 \leq i < j < m$, $gcd(n, i+j) = gcd(n, j-i) = 1)$, $gcd(n, i+j) = gcd(n, j-i) = 2$ [48].
- $f(x) = \sum_{i=1}^{\frac{m-1}{2}} Tr_1^n(\beta x^{1+4^i})$, m odd, $\beta \in \mathbb{F}_4^\star$ [16].
- $f(x) = \sum_{i=1}^{\frac{m-1}{2}} c_i Tr_1^n(\beta x^{1+4^i})$, $c_i \in \mathbb{F}_2$, $\beta \in \mathbb{F}_4^\star$, m odd, $gcd(\sum_{i=1}^{\frac{m-1}{2}} c_i(x^i + x^{m-i}), x^m + 1) = x + 1$ [16].
- $f(x) = \sum_{i=1}^{k} Tr_1^n(\beta x^{1+4^{di}})$ $\beta \in \mathbb{F}_4^\star$, m odd, $d \geq 1$, $1 \leq k \leq \frac{m-1}{2}$, $gcd(k+1, m) = gcd(k, m) = gcd(d, m) = 1$ [16].
- $f(x) = Tr_1^n(\beta x^{1+4^i} + \beta x^{1+4^j})$ $\beta \in \mathbb{F}_4^\star$, m odd, $1 \leq i < j \leq \lfloor \frac{n}{4} \rfloor$, $gcd(i+j, m) = gcd(j-i, m) = 1$ [16].
- $f(x) = Tr_1^n(\beta x^{1+4^i} + x^{1+4^j} + x^{1+4^t})$, $\beta \in \mathbb{F}_4^\star$, m odd, $1 \leq i < j < t \leq \lfloor \frac{n}{4} \rfloor$, $i + j = t$, $gcd(i, m) = gcd(j, m) = gcd(j, t) = 1$ [16].
- $f(x) = Tr_1^n(\beta x^{1+4^i} + \beta x^{1+4^j} + \beta x^{1+4^t})$, $\beta \in \mathbb{F}_4^\star$, $1 \leq i < j < t \leq \lfloor \frac{n}{4} \rfloor$, $i + j = 2t$, $j - i = 3^h p$, $3 \nmid p$, $n = 3^k q$, $3 \nmid q$, $gcd(2t, m) = 1$, $h \geq k$ [16].
- $f(x) = Tr_1^n(\beta x^{1+4^i} + \beta x^{1+4^j} + \beta x^{1+4^t})$, $\beta \in \mathbb{F}_4^\star$, m odd, $1 \leq i, j, t \leq \lfloor \frac{n}{4} \rfloor$, $j - i = 2t$, $t \neq i$, $j + i = 3^u p$, $3 \nmid p$, $n = 3^v q$, $3 \nmid q$, $gcd(2t, m) = 1$, $u \geq v$ [16].

- $f(x) = Tr_1^n(\beta x^{1+4^i} + \beta x^{1+4^j} + \beta x^{1+4^t}), \beta \in \mathbb{F}_4^\star, 1 \le i, j, t \le \lfloor \frac{n}{4} \rfloor, j - i = 2t,$
  $t \ne i, j + i = 3^u p, 3 \nmid p, n = 3^v q, 3 \nmid q, gcd(2t, m) = 1, u \ge v$ [16].
- $f(x) = Tr_1^n(\beta x^{1+4^i} + \beta x^{1+4^j} + \beta x^{1+4^t} + \beta x^{1+4^s}), \beta \in \mathbb{F}_4^\star, 1 \le i, j, t, s \le \lfloor \frac{n}{4} \rfloor,$
  $i < j, t < s, i + j = t + s = r, t \ne i, gcd(r, m) = gcd(m, s - i) =$
  $gcd(m, s - j) = 1$ [16].

4.2 On Constructions of Semi-Bent Functions From Bent Functions

In the following subsections, we are dealing with the construction of semi-bent functions from bent functions. We shall present several such kinds of constructions. A natural problem arises is:

Problem 1 Find new primary constructions of bent functions from semi-bent functions.

4.2.1 Primary Constructions in Univariate Representation from Niho and Dillon Bent Functions

In 2011, many concrete constructions of semi-bent functions of maximum algebraic degree have been discovered. Indeed, in [38], the semi-bentness of several infinite families functions in polynomial form constructed via Dillon and Niho exponents has been studied in detail. From this study, explicit criteria in terms of Kloosterman sums for deciding whether a function expressed as a sum of trace functions is semi-bent or not have been derived. Kloosterman sums have been used as a very suitable tool to study the semi-bentness property of several functions in univariate representation. In particular, we have showed in [38] that the values 0 and 4 of Kloosterman sums defined on $\mathbb{F}_{2^m}$ give rise to semi-bent functions on $\mathbb{F}_{2^n}$. Below is the list of the known semi-bent functions constructed via the zero of Kloosterman sums:

- $f(x) = Tr_1^n(ax^{r(2^m-1)}) + Tr_1^n(cx^{(2^m-1)\frac{1}{2}+1}), K_m(a) = 0$ [38].
- $f(x) = Tr_1^n(ax^{r(2^m-1)}) + Tr_1^n(cx^{(2^m-1)\frac{1}{2}+1}) + Tr_1^n(x^{(2^m-1)\frac{1}{4}+1}), Tr_m^n(c) = 1, m$ odd, $K_m(a) = 0$ [38].
- $f(x) = Tr_1^n(ax^{r(2^m-1)}) + Tr_1^n(cx^{(2^m-1)\frac{1}{2}+1}) + Tr_1^n(x^{(2^m-1)3+1}), K_m(a) = 0$ $Tr_m^n(c) = 1$ [38].
- $f(x) = Tr_1^n(ax^{r(2^m-1)}) + Tr_1^n(cx^{(2^m-1)\frac{1}{2}+1}) + Tr_1^n(x^{(2^m-1)\frac{1}{6}+1}); Tr_m^n(c) = 1,$ $K_m(a) = 0, m$ even [38].
- $f(x) = Tr_1^n(ax^{r(2^m-1)}) + Tr_1^n(\alpha x^{2^m+1}) + Tr_1^n(\sum_{i=1}^{2^{v-1}-1} x^{(2^m-1)\frac{i}{2^v}+1}); gcd(v, m) = 1, \alpha \in \mathbb{F}_{2^n}, Tr_m^n(\alpha) = 1, K_m(a) = 0$ [38].

Below is the list of the known semi-bent functions constructed via the value four of Kloosterman sums:

- $f(x) = Tr_1^n(ax^{r(2^m-1)}) + Tr_1^2(bx^{\frac{2^n-1}{3}}) + Tr_1^n(cx^{(2^m-1)\frac{1}{2}+1})$; m odd, $K_m(a) = 4$ [38].

- $f(x) = Tr_1^n(ax^{3(2^m-1)}) + Tr_1^n(cx^{(2^m-1)\frac{1}{2}+1}) + Tr_1^2(bx^{\frac{2^n-1}{3}})$; m odd and $m \not\equiv 3$ (mod 6) $K_m(a) = 4$ [38].

- $f(x) = Tr_1^n(ax^{r(2^m-1)}) + Tr_1^2(bx^{\frac{2^n-1}{3}}) + Tr_1^n(cx^{(2^m-1)\frac{1}{2}+1}) + Tr_1^n(x^{(2^m-1)\frac{1}{4}+1})$, m odd, $K_m(a) = 4$ [38].

- $f(x) = Tr_1^n(ax^{r(2^m-1)}) + Tr_1^2(bx^{\frac{2^n-1}{3}}) + Tr_1^n(cx^{(2^m-1)\frac{1}{2}+1}) + Tr_1^n(x^{3(2^m-1)+1})$; $Tr_m^n(c) = 1$, m odd, $K_m(a) = 4$ [38].

- $f(x) = Tr_1^n(ax^{r(2^m-1)}) + Tr_1^n(\alpha x^{2^m+1}) + Tr_1^n(\sum_{i=1}^{2^{\nu-1}-1} x^{(2^m-1)\frac{i}{2^\nu}+1}) + Tr_1^2(bx^{\frac{2^n-1}{3}})$; $\gcd(\nu, m) = 1$, $\alpha \in \mathbb{F}_{2^n}$, $Tr_m^n(\alpha) = 1$, m odd, $K_m(a) = 4$ ([38]).

All the families of semi-bent functions presented above are of maximum algebraic degree m and then are suitable for use in symmetric cryptosystems.

The previous constructions can be generalized leading to general constructions of semi-bent functions via Dillon-like exponents and Niho exponents. First, recall that *Dillon-like exponents* are of the form $s(2^m - 1)$.

A positive integer s (always understood modulo $2^n - 1$) is said to be a *Niho exponent* and x^s a Niho power function, if the restriction of x^s to $\mathbb{F}_{2^m}$ is linear. One can show that the restriction of the power function $x \mapsto x^s$ to $\mathbb{F}_{2^m}$ is linear then $s = 2^j$ for some $j < n$. As we consider $Tr_1^n(x^d)$, without loss of generality, we can assume that s is in the normalized (unique) representation $s = (2^m - 1)d + 1$ with $1 \leq d \leq 2^m$.

The following statement is due to Carlet and the author [6]. An alternative direct proof has been proposed in [12].

Theorem 3 ([6, 12]) *Denote by Ω_n the set of Boolean functions f defined on $\mathbb{F}_{2^n}$ by $f(x) = \sum_{i \in \Gamma_{n,m}} Tr_1^{o(i)}(a_i x^i)$ where $\Gamma_{n,m}$ is the set of cyclotomic cosets $[i]$ such that $i \equiv 0$ (mod $2^m - 1$). Denote by Δ_n the set of Boolean functions f defined on $\mathbb{F}_{2^n}$ by $f(x) = \sum_{i \in \Lambda'_{n,m}} Tr_1^{o(i)}(a_i x^i)$ where $\Lambda'_{n,m}$ is the set of cyclotomic cosets $[i]$ such that $i \equiv 2^j$ (mod $2^m - 1$) for some j ($j < n$). Set*

$$\mathscr{D}_n := \{f \in \Omega_n \text{ such that } f \text{ is bent with } f(0) = 0\}$$

and set

$$\mathscr{N}_n := \{f \in \Delta_n \text{ such that } f \text{ is bent with } f(0) = 0\}.$$

Let $g \in \mathscr{D}_n$ and $h \in \mathscr{N}_n$. Then $g + h$ is semi-bent on $\mathbb{F}_{2^n}$.

Let us specify some infinite families of semi-bent functions in univariate form. Firstly, we give a list of infinite families containing bent functions defined on $\mathbb{F}_{2^n}$

belonging to the class $\mathscr{PS}_{ap}$; here, $K_m(a) := \sum_{x \in \mathbb{F}_{2^m}} \chi\big(Tr_1^m(ax + \frac{1}{x})\big)$ denotes the binary Kloosterman sums on $\mathbb{F}_{2^m}$ and $C_m(a, a) := \sum_{x \in \mathbb{F}_{2^m}} \chi\big(Tr_1^m(ax^3 + ax)\big)$ denotes the cubic sums on $\mathbb{F}_{2^m}$:

- $g_1(x) = Tr_1^n(ax^{r(2^m-1)})$; $\gcd(r, 2^m + 1) = 1$, $a \in \mathbb{F}_{2^m}^\star$ such that $K_m(a) = 0$ [9].
- $g_2(x) = Tr_1^n(ax^{r(2^m-1)}) + Tr_1^2(bx^{\frac{2^n-1}{3}})$; $\gcd(r, 2^m + 1) = 1$, $m > 3$ odd, $b \in \mathbb{F}_4^\star$, $a \in \mathbb{F}_{2^m}^\star$ such that $K_m(a) = 4$ [36].
- $g_3(x) = Tr_1^n(a\zeta^i x^{3(2^m-1)}) + Tr_1^2(\beta^j x^{\frac{2^n-1}{3}})$; m odd and $m \not\equiv 3 \pmod 6$, β is a primitive element of $\mathbb{F}_4$, ζ is a generator of the cyclic group U of $(2^m + 1)$-th of unity, $(i, j) \in \{0, 1, 2\}^2$, $a \in \mathbb{F}_{2^m}^\star$ such that $K_m(a) = 4$ and $Tr_1^m(a^{1/3}) = 0$ [35].
- $g_4(x) = Tr_1^n(a\zeta^i x^{3(2^m-1)}) + Tr_1^2(\beta^j x^{\frac{2^n-1}{3}})$; m odd and $m \not\equiv 3 \pmod 6$, β is a primitive element of $\mathbb{F}_4$, ζ is a generator of the cyclic group U of $(2^m + 1)$-th of unity, $i \in \{1, 2\}$, $j \in \{0, 1, 2\}$, $a \in \mathbb{F}_{2^m}^\star$ such that $K_m(a) + C_m(a, a) = 4$ and $Tr_1^m(a^{1/3}) = 1$ [35].
- $g_5(x) = \sum_{i=1}^{2^{m-1}-1} Tr_1^n\big(\beta x^{i(2^m-1)}\big)$; $\beta \in \mathbb{F}_{2^m} \setminus \mathbb{F}_2$ [18].
- $g_6(x) = \sum_{i=1}^{2^{m-2}-1} Tr_1^n\big(\beta x^{i(2^m-1)}\big)$; m odd and $\beta^{(2^m-4)^{-1}} \in \{x \in \mathbb{F}_{2^m}^\star; Tr_1^m(x) = 0\}$ [18].

Secondly, we give a list of known Niho bent functions in $\mathscr{N}_n$:

- $h_1(x) = Tr_1^m\big(a_1 x^{2^m+1}\big)$; $a_1 \in \mathbb{F}_{2^m}^\star$.
- $h_2(x) = Tr_1^n\big(a_1 x^{(2^m-1)\frac{1}{2}+1} + a_2 x^{(2^m-1)3+1}\big)$.
  $a_1 \in \mathbb{F}_{2^n}^\star$, $a_2^{2^m+1} = a_1 + a_1^{2^m} = \beta^5$ for some $\beta \in \mathbb{F}_{2^n}^\star$ [15];
- $h_3(x) = Tr_1^n\big(a_1 x^{(2^m-1)\frac{1}{2}+1} + a_2 x^{(2^m-1)\frac{1}{4}+1}\big)$.
  $a_1 \in \mathbb{F}_{2^n}^\star a_2^{2^m+1} = a_1 + a_1^{2^m}$, m odd [15].
- $h_4(x) = Tr_1^n\big(a_1 x^{(2^m-1)\frac{1}{2}+1} + a_2 x^{(2^m-1)\frac{1}{6}+1}\big)$; $a_1 \in \mathbb{F}_{2^n}^\star$ $a_2^{2^m+1} = a_1 + a_1^{2^m}$, m even [15].
- $h_5(x) = Tr_1^n\big(\alpha x^{2^m+1} + \sum_{i=1}^{2^{r-1}-1} x^{s_i}\big)$, $r > 1$ such that $\gcd(r, m) = 1$, $\alpha \in \mathbb{F}_{2^n}$ such that $\alpha + \alpha^{2^m} = 1$, $s_i = (2^m - 1)\frac{i}{2^r} \pmod{2^m + 1} + 1$, $i \in \{1, \ldots, 2^{r-1} - 1\}$ [25].

By Theorem 3, we recover the families in univariate form containing semi-bent functions derived previously by the author in [38].

A complete list of the known functions in $\mathscr{D}_n$ can be found in [44] with additional functions in [28] Now, note that $\mathscr{D}_n$ coincides with the set of Boolean functions $f : \mathbb{F}_{2^n} \to \mathbb{F}_2$ such that the restriction to $u\mathbb{F}_{2^m}^\star$ is constant for every $u \in U$ with $f(0) = 0$ while $\mathscr{L}_n$ coincides with the set of Boolean functions on $\mathbb{F}_{2^n}$ such that the restriction to $u\mathbb{F}_{2^m}^\star$ is linear for every $u \in U$ with $f(0) = 0$.

A stronger version of the previous statement has been proved in [6].

Theorem 4 ([6]) *Let* $n = 2m$ *with* $m > 2$. *Keeping the same notation as in Theorem 3. Set*

$$\mathscr{A}_n := \{f : \mathbb{F}_{2^n} \to \mathbb{F}_2 \text{ s.t the restriction to } u\mathbb{F}_{2^m}^\star \text{ is affine for every } u \in U\}.$$

Then a function f in $\mathscr{A}_n$ is semi-bent if and only if f can be written as the sum of a function in $\mathscr{D}_n$ and a function in $\mathscr{L}_n$.

Example 1 Identify the semi-bent Boolean function f over $\mathbb{F}_{64}$ of the form $f(x) = Tr_1^6(ax^{36}) + Tr_1^6(bx^{32}) + Tr_1^6(cx^{56})$. Set $f = g + h$ where $g : x \in \mathbb{F}_{64} \mapsto Tr_1^6(cx^{56})$ and $h : x \in \mathbb{F}_{64} \mapsto Tr_1^6(ax^{36}) + Tr_1^6(bx^{32})$. We have $36 \equiv 1 \pmod 7$, $36 \equiv 2^2 \pmod 7$ and $56 \equiv 0 \pmod 7$. So 36 and 32 are Niho exponents, while 56 is a Dillon exponent. According to the above result, f is semi-bent if and only if its Niho part (that is, the function h) is bent and its Dillon part (i.e., the function g) is bent. On one hand, the bentness of h depends only on the bentness of $x \mapsto Tr_1^6(ax^{36})$ (since $x \mapsto Tr_1^6(bx^{32})$ is linear). But $36 = 7 \times \frac{1}{2} + 1$ where $\frac{1}{2}$ is understood modulo 9. Thus, the function $x \mapsto Tr_1^6(ax^{36})$ is bent if and only if $Tr_3^6(a) = a + a^8 \neq 0$. Hence, h is bent if and only if $a + a^8 \neq 0$ ($a \in \mathbb{F}_{64}$). On the other hand, $g(x)$ is of the form $Tr_1^n(cx^{2^m-1})$ with $m = \frac{n}{2} = 3$ (the size of the cyclotomic class of 56 modulo $2^6 - 1 = 63$ is 6). Therefore, g is bent, if and only if $K_m(c^{2^m+1}) = K_3(c^9) = 0$ where K_m denotes the Kloosterman sums over $\mathbb{F}_{2^m}$. Let α be a primitive element of $\mathbb{F}_8$ such that $\alpha^3 + \alpha^2 + 1 = 0$. Then, it is easy to check that g is bent, if and only if $c^9 \in \{\alpha, \alpha^2, \alpha^4\}$, that is, $c^9 = \alpha^{2^j}$ for some j (since the Kloosterman sums is invariant under the Frobenius mapping). Finally, one can conclude that f is semi-bent on $\mathbb{F}_{64}$, if and only if $a + a^8 \neq 0$ and $c^9 = \alpha^{2^j}$ for some j where $\alpha \in \mathbb{F}_8$ such that $\alpha^3 + \alpha^2 + 1 = 0$.

Recall [14] that a *spread* is a collection $\{E_i, i = 1, \ldots, 2^m + 1\}$ of vector spaces of dimension $m = n/2$ such that $E_i \cap E_j = \{0\}$ for every i and j and $\bigcup_{i=1}^{2^m+1} E_i = \mathbb{F}_{2^n}$. The classical example of spread is $\{u\mathbb{F}_{2^m} ; u \in U\}$ where U is the multiplicative group $\{u \in \mathbb{F}_{2^n} ; u^{2^m+1} = 1\}$. Theorem 4 can be stated in more general setting as follows.

Theorem 5 ([6]) *Let $m \geq 2$ and $n = 2m$. Let $\{E_i, i = 1, \ldots, 2^m + 1\}$ be a spread in $\mathbb{F}_{2^n}$ and h a Boolean function whose restriction to every E_i is linear (possibly null). Let S be any subset of $\{1, \ldots, 2^m + 1\}$ and $g = \sum_{i \in S} 1_{E_i} \pmod 2$ where 1_{E_i} is the indicator of E_i. Then $g + h$ is semi-bent if and only if g and h are bent.*

Given a spread $(E_i)_{i=1,\ldots,2^m+1}$, the previous theorem provides a characterization of the semi-bentness for a function whose restriction to every E_i^* is affine (i.e., equal to the sum of a function whose restriction to every E_i is linear and of a function whose restriction to every E_i^* is constant).

Remark 2 One can modify the hypothesis of Theorem 5 by assuming that we have only a partial spread. There exists an example due for m even to Dillon [14] of a partial spread in $\mathbb{F}_{2^n} \approx \mathbb{F}_{2^m} \times \mathbb{F}_{2^m}$ which is not included in a spread: $E_\infty = \{0\} \times \{0\} \times \mathbb{F}_{2^{m-1}} \times \mathbb{F}_2$ and $E_a = \{(x, \epsilon, a^2 x + a Tr_1^{m-1}(ax) + a\epsilon, Tr_1^{m-1}(ax)); (x, \epsilon) \in \mathbb{F}_{2^{m-1}} \times \mathbb{F}_2\}$ for $a \in \mathbb{F}_{2^{m-1}}$ (the corresponding function g is quadratic bent). By modifying the hypothesis, we need then to add a condition on the E_i's, and we have only a sufficient condition for $g + h$ being semi-bent:

Let g be a bent function in the $\mathscr{PS}$ class, equal to the sum modulo 2 of the indicators of $l := 2^{m-1}$ or $2^{m-1} + 1$ pairwise "disjoint" vector spaces E_i having dimension m, and h a bent function which is linear on each E_i. Assume additionally that for every $c \in \mathbb{F}_{2^n}$ there exist at most 2 indices i such that $\forall e \in E_i$, $h(e) = Tr_1^n(ce)$. Then $g + h$ is semi-bent.

Problem 2 Find semi-bent functions obtained by applying the result of Remark 2.

Problem 3 Show that some semi-bent functions obtained above in [6] are not extendable to $(n + 2)$-variable bent functions (or deduce new bent functions from them).

4.2.2 Primary Constructions in Bivariate Representation from the Class $\mathscr{H}$ of Bent Functions

Semi-bent functions in bivariate representation have been derived from the class $\mathscr{H}$ of bent functions introduced by Carlet and the author in [5] and from the partial spread class $\mathscr{PS}_{ap}$ of bent functions introduced by Dillon [14]. Recall that functions of the class $\mathscr{PS}_{ap}$ are a subclass of the partial spread class $\mathscr{PS}$ defined as the set of all the sums (modulo 2) of the indicators of 2^{m-1} or $2^{m-1} + 1$ pairwise supplementary m-dimensional subspaces of $\mathbb{F}_{2^n}$. The elements of $\mathscr{PS}_{ap}$ can be defined in an explicit form as follows.

Definition 5 Let $n = 2m$ and let $\mathbb{F}_{2^n}$ be identified, as a vector space, with $\mathbb{F}_{2^m} \times \mathbb{F}_{2^m}$. The partial spread class $\mathscr{PS}_{ap}$ consists of all the functions f defined as follows: let g be a balanced Boolean function over $\mathbb{F}_{2^m}$ (i.e., $wt(g) = 2^{m-1}$) such that $g(0) = 0$ (but, in fact, this last condition is not necessary for f to be bent). Then f is defined from $\mathbb{F}_{2^m} \times \mathbb{F}_{2^m}$ to $\mathbb{F}_2$ as $f(x, y) = g(\frac{x}{y})$ (i.e., $g(xy^{2^m-2})$) with $\frac{x}{y} = 0$ if $y = 0$.

The functions from class $\mathscr{PS}_{ap}$ are those whose supports can be uniquely written as $\bigcup_{u \in S} u\mathbb{F}_{2^m}^\star$ where U is the set $\{u \in \mathbb{F}_{2^n}; u^{2^m+1} = 1\}$ and S is a subset of U of size 2^{m-1}. We shall also include in $\mathscr{PS}_{ap}$ the complements of these functions.

Now, functions of the class $\mathscr{H}$ are defined in bivariate form as follows.

Definition 6 ([5]) Functions h of the class $\mathscr{H}$ defined on $\mathbb{F}_{2^m} \times \mathbb{F}_{2^m}$ are of the form

$$h(x, y) = \begin{cases} Tr_1^m \left(x\psi \left(\frac{y}{x} \right) \right) & \text{if } x \neq 0 \\ Tr_1^m (\mu y) & \text{if } x = 0 \end{cases} \tag{2}$$

where $\psi : \mathbb{F}_{2^m} \to \mathbb{F}_{2^m}$ and $\mu \in \mathbb{F}_{2^m}$ and satisfying the following condition:

$$\forall \beta \in \mathbb{F}_{2^m}^\star, \text{ the function } z \mapsto G(z) + \beta z \text{ is 2-to-1 on } \mathbb{F}_{2^m}, \tag{3}$$

where G is defined as: $G(z) := \psi(z) + \mu z$.

The current list of examples of functions h from the class $\mathscr{H}$ is the following:

- $h(x, y) = Tr_1^m(x^{-5}y^6)$, m odd.
- $h(x, y) = Tr_1^m(x^{\frac{5}{6}} y^{\frac{1}{6}})$, m odd.

- $h(x, y) = Tr_1^m(x^{-3 \cdot (2^k+1)} y^{3 \cdot 2^k+4}), m = 2k - 1.$
- $h(x, y) = Tr_1^m(x^{-3 \cdot (2^{k-1}-1)} y^{3 \cdot 2^{k-1}-2}), m = 2k - 1.$
- $h(x, y) = Tr_1^m(x^{1-2^k-2^{2k}} y^{2^k+2^{2k}}), m = 4k - 1.$
- $h(x, y) = Tr_1^m(x^{2^{3k-1}-2^{2k}+2^k} y^{1-2^{3k-1}+2^{2k}-2^k}), m = 4k - 1.$
- $h(x, y) = Tr_1^m(x^{1-2^{2k+1}-2^{3k+1}} y^{2^{2k+1}+2^{3k+1}}), m = 4k + 1.$
- $h(x, y) = Tr_1^m(x^{2^{3k+1}-2^{2k+1}+2^k} y^{1-2^{3k+1}+2^{2k+1}-2^k}), m = 4k + 1.$
- $h(x, y) = Tr_1^m(x^{1-2^k} y^{2^k} + x^{-(2^k+1)} y^{2^k+2} + x^{-3 \cdot (2^k+1)} y^{3 \cdot 2^k+4}), m = 2k - 1.$
- $h(x, y) = Tr_1^m(y(y^{2^k+1} x^{-(2^k+1)} + y^3 x^{-3} + yx^{-1})^{2^{k-1}-1}), m = 2k - 1;$
- $h(x, y) = Tr_1^m(x^{\frac{5}{6}} y^{\frac{1}{6}} + x^{\frac{1}{2}} y^{\frac{1}{2}} + x^{\frac{1}{6}} y^{\frac{5}{6}}), m$ odd.
- $h(x, y) = Tr_1^m(x[D_{\frac{1}{5}}(\frac{y}{x})]^6), m$ odd, where $D_{\frac{1}{5}}$ is the Dickson polynomial of index $\frac{1}{5}$.

The following result provides constructions of semi-bent functions from the classes $\mathcal{H}$ and $\mathcal{PS}_{ap}$.

Theorem 6 ([6]) *The sum of a function defined on $\mathbb{F}_{2^m} \times \mathbb{F}_{2^m}$ from the class $\mathcal{PS}_{ap}$ and a function defined on $\mathbb{F}_{2^m} \times \mathbb{F}_{2^m}$ from the class $\mathcal{H}$ is semi-bent on $\mathbb{F}_{2^m} \times \mathbb{F}_{2^m}$.*

4.2.3   A Construction from Bent Functions via the Indirect Sum

In [3], Carlet has introduced a secondary construction (which means a construction of new functions from ones having the same properties) of bent functions. Later, such a construction was called as the *"indirect sum"* because it generalizes the well-known direct sum introduced by Dillon and Rothaus [14, 46]. The indirect sum is defined as follows.

Definition 7 ([3]) Let $n = r + s$ where r and s are positive integers. Let f_1, f_2 be Boolean functions defined on $\mathbb{F}_{2^r}$ and g_2, g_2 be two Boolean functions defined on $\mathbb{F}_{2^s}$. Define h as follows (i.e., h is the concatenation of the four functions f_1, $f_1 \oplus 1$, f_2, and $f_2 \oplus 1$, in an order controlled by $g_1(y)$ and $g_2(y)$):

$$\forall (x, y) \in \mathbb{F}_{2^r} \times \mathbb{F}_{2^s}, \quad h(x, y) = f_1(x) + g_1(y) + (f_1(x) + f_2(x))(g_1(y) + g_2(y)).$$

Using the indirect sum, we derive a general constructions of semi-bent functions from both bent and semi-bent functions.

Theorem 7 *Let $n = r + s$ with r and s two even integers. Let h be as in Definition 7. Assume that f_1 and f_2 are semi-bent on $\mathbb{F}_{2^r}$ and that g_1 and g_2 are bent on $\mathbb{F}_{2^s}$. Then h is semi-bent on $\mathbb{F}_{2^n}$.*

Proof Set $r = 2\rho$ and $s = 2\sigma$. Let's compute the Walsh transform of h for every $(a, b) \in \mathbb{F}_{2^r} \times \mathbb{F}_{2^s}$. We have

$$\widehat{\chi_h}(a, b) = \sum_{x \in \mathbb{F}_{2^r}} \sum_{y \in \mathbb{F}_{2^s}} \chi(f_1(x) + g_1(y) + (f_1(x) + f_2(x))(g_1(y) + g_2(y))$$

$$+ Tr_1^r(ax) + Tr_1^s(by)).$$

Now, one can split the sum depending whether $g_1(y) + g_2(y)$ is equal to 1 or not :

$$\widehat{\chi_h}(a, b) = \sum_{x \in \mathbb{F}_{2^r}} \sum_{y \in \mathbb{F}_{2^s} | g_1(y) + g_2(y) = 1} \chi(f_2(x) + g_1(y) + Tr_1^r(ax) + Tr_1^s(by))$$

$$+ \sum_{y \in \mathbb{F}_{2^s} | g_1(y) + g_2(y) = 0} \chi(f_1(x) + g_1(y) + Tr_1^r(ax) + Tr_1^s(by)).$$

Now, note that the indicator of the set $\{y \in \mathbb{F}_{2^s} \mid g_1(y) + g_2(y) = 1\}$ can be written as $\frac{1 - \chi(g_1(y) + g_2(y))}{2}$. Similarly, one can write the indicator of the set $\{y \in \mathbb{F}_{2^s} \mid g_1(y) + g_2(y) = 0\}$ as $\frac{1 + \chi(g_1(y) + g_2(y))}{2}$. Hence,

$$\widehat{\chi_h}(a, b) = \widehat{\chi_{f_1}}(a) \left(\frac{\widehat{\chi_{g_1}}(b) + \widehat{\chi_{g_2}}(b)}{2} \right) + \widehat{\chi_{f_2}}(a) \left(\frac{\widehat{\chi_{g_1}}(b) - \widehat{\chi_{g_2}}(b)}{2} \right).$$

Now, if g_1 and g_2 are bent, then

$$\left(\frac{\widehat{\chi_{g_1}}(b) - \widehat{\chi_{g_2}}(b)}{2} \right) \left(\frac{\widehat{\chi_{g_1}}(b) + \widehat{\chi_{g_2}}(b)}{2} \right) = \frac{1}{4} \left((\widehat{\chi_{g_1}}(b))^2 - (\widehat{\chi_{g_2}}(b))^2 \right) = 0.$$

and thus only the two following situations can occur

$$\frac{\widehat{\chi_{g_1}}(b) - \widehat{\chi_{g_2}}(b)}{2} = 0 \text{ and } \frac{\widehat{\chi_{g_1}}(b) + \widehat{\chi_{g_2}}(b)}{2} = \pm 2^\sigma$$

or

$$\frac{\widehat{\chi_{g_1}}(b) - \widehat{\chi_{g_2}}(b)}{2} = \pm 2^\sigma \text{ and } \frac{\widehat{\chi_{g_1}}(b) + \widehat{\chi_{g_2}}(b)}{2} = 0.$$

Now f_1 and f_2 being semi-bent : $\widehat{\chi_{f_1}}(a) \in \{0, \pm 2^{\rho+1}\}$ and $\widehat{\chi_{f_2}}(a) \in \{0, \pm 2^{\rho+1}\}$. Therefore $\widehat{\chi_h}(a, b) \in \{0, \pm 2^{\rho+\sigma+1}\}$ proving that h is semi-bent.   □

Remark 3 Obviously, the roles of f_1 and f_2 can be exchanged with those of g_1 and g_2. This means that one can exchange the property of bentness and semi-bentness in Theorem 7.

4.2.4   A Simple Construction of Semi-Bent Functions from Bent Functions by Field Extension

Another kind of construction of semi-bent functions from bent functions is given by the simple following statement. When we identify $\mathbb{F}_{2^n}$ with the vector space $\mathbb{F}_2^n$, it corresponds to a simple construction of an $(n+2)$-variable semi-bent function from an n-variable bent function.

Proposition 4 ([12]) *Let n be an even positive integer. Let f be a Boolean function over $\mathbb{F}_{2^{n+2}} \simeq \mathbb{F}_{2^n} \times \mathbb{F}_4$. For $\delta \in \mathbb{F}_4$, we define a Boolean function f_δ over $\mathbb{F}_{2^n} \times \mathbb{F}_4$ by*

$$f_\delta(y,z) = f(y) + Tr_1^2(\delta z), \forall y \in \mathbb{F}_{2^n}, z \in \mathbb{F}_4.$$

If f is bent over $\mathbb{F}_{2^n}$ then f_δ is semi-bent over $\mathbb{F}_{2^{n+2}}$.

4.2.5   Construction of Semi-Bent Functions from Bent Functions by Considering the Derivative Functions

Recall that the derivative of a Boolean function f on $\mathbb{F}_{2^n}$ with respect $a \in \mathbb{F}_{2^n}$ is defined by $D_{af}(x) = f(x) + f(x+a)$. The following construction of semi-bent functions from bent functions under a strong condition on the derivatives functions has been shown in [48].

Theorem 8 ([48]) *Let n be an even positive integer. Let f and g be two bent functions over $\mathbb{F}_{2^n}$. Assume that there exists $a \in \mathbb{F}_{2^n}$ such that $D_{af}(x) = D_a g(x) + 1$ for all $x \in \mathbb{F}_{2^n}$. Then the function $h = f + g + D_{af} + D_a(fg)$ is semi-bent over $\mathbb{F}_{2^n}$.*

A possible construction of semi-bent functions by applying Theorem 8 is provided by the following statement.

Proposition 5 *Let f be a bent function defined over $\mathbb{F}_{2^n}$ (with n even). Define a Boolean function g by $g(x) = f(x+a) + Tr_1^n(bx), \forall x \in \mathbb{F}_{2^n}$ where a and b are elements of $\mathbb{F}_{2^n}$ such that $Tr_1^n(ab) = 1$. Then the function $h = f + g + D_{af} + D_a(fg)$ is semi-bent over $\mathbb{F}_{2^n}$.*

Proof The bentness is invariant under the addition of linear functions. Thus g is also bent. Moreover, one has $D_a g(x) = g(x) + g(x+a) = f(x+a) + Tr_1^n(bx) + f(x) + Tr_1^n(bx) + Tr_1^n(ab) = D_{af}(x) + Tr_1^n(ab) = D_{af}(x) + 1$. The proposition follows from Theorem 8.   □

Notice that quadratics semi-bent functions can be easily derived from Proposition 5.

Problem 4 Find other examples of constructions of non-quadratic semi-bent functions h starting from two bent functions f and g satisfying $D_{af}(x) = D_{ag}(x) + 1$ for some $a \in \mathbb{F}_{2^n}$.

4.3 A General Construction of Semi-Bent Functions Based on Maiorana–McFarland's Construction

Recall that the Maiorana–McFarland's constructions are the best known primary constructions of bent functions [14,32]. The *Maiorana–McFarland class* is the set of all the Boolean functions on $\mathbb{F}_{2^m} \times \mathbb{F}_{2^m}$ of the form $f(x, y) = x \cdot \pi(y) + g(y)$; $x, y \in \mathbb{F}_{2^m}$ where "$\cdot$" denotes an inner product in $\mathbb{F}_{2^m}$, π is any permutation on $\mathbb{F}_{2^m}$, and g is any Boolean function on $\mathbb{F}_{2^m}$. Any such function is bent (the bijectivity of π is a necessary and sufficient condition for f being bent). By computing the Walsh transform, it is easy to see that if π is a 2-to-1 mapping from $\mathbb{F}_{2^m}$ to on $\mathbb{F}_{2^m}$, then f is semi-bent on $\mathbb{F}_{2^m} \times \mathbb{F}_{2^m}$. Consequently, the reader notices that using the Maiorana–McFarland method, any permutation leads to the construction of bent functions and any mapping 2-to-1 leads to the construction of semi-bent functions.

The following statement provides an example of construction of semi-bent functions via the Maiorana–McFarland method.

Proposition 6 *Let r be a positive integer. Set $m = 2r - 1$. Let g be any Boolean function over $\mathbb{F}_{2^m}$. Define over $\mathbb{F}_{2^m} \times \mathbb{F}_{2^m}$ a Boolean function by $f(x, y) = Tr_1^m(xy^{2^r+2} + xy) + g(y)$, $\forall(x, y) \in \mathbb{F}_{2^m} \times \mathbb{F}_{2^m}$. Then f is semi-bent.*

Proof We have to prove that f is semi-bent, that is, its Walsh transform takes only the values 0, 2^{m+1} and -2^{m+1}. Compute the Walsh transform of f. For every $(a, b) \in \mathbb{F}_{2^m} \times \mathbb{F}_{2^m}$, we have:

$$\widehat{\chi_f}(a, b) = \sum_{x \in \mathbb{F}_{2^m}} \sum_{y \in \mathbb{F}_{2^m}} (-1)^{Tr_1^m(xy^{2^r+2}+xy)+g(y)+Tr_1^m(ax)+Tr_1^m(by)}$$

$$= \sum_{y \in \mathbb{F}_{2^m}} (-1)^{g(y)+Tr_1^m(by)} \sum_{x \in \mathbb{F}_{2^m}} (-1)^{Tr_1^m(xy^{2^r+2}+xy))+Tr_1^m(ax)}$$

$$= \sum_{y \in \mathbb{F}_{2^m}} (-1)^{g(y)+Tr_1^m(by)} \sum_{x \in \mathbb{F}_{2^m}} (-1)^{Tr_1^m((y^{2^r+2}+y)x)}$$

$$= 2^m \sum_{y \in \mathbb{F}_{2^m} \mid y^{2^r+2}+y=a} (-1)^{g(y)+Tr_1^m(by)}.$$

Now, according to Cusick and Dobbertin [13], the equation $y^{2^r+2} + y = a$ has 0 or 2 solutions in $\mathbb{F}_{2^m}$. The mapping $y \in \mathbb{F}_{2^m} \mapsto y^{2^r+2} + y + a$ is 2-to-1 for every $a \in \mathbb{F}_{2^m}$. Therefore,

$$\widehat{\chi_f}(a, b) \in \{0, \pm 2^{m+1}\}$$

which completes the proof.   □

4.4 A Construction from APN Functions

Let us recall the definition of *almost perfect nonlinear* (APN) functions.

Definition 8 Let F be a mapping from $\mathbb{F}_{2^m}$ to itself (m a positive integer). The function f is said to be APN if, $max_{a \in \mathbb{F}_{2^m}^{\star}} max_{b \in \mathbb{F}_{2^m}} \#\{x \in \mathbb{F}_{2^m} \mid F(x+a)+F(x) = b\} = 2$.

APN functions are important research objects in cryptography and coding theory. Given an APN function, one can derive a construction of semi-bent function in the sprit of Maiorana–McFarland's method.

Proposition 7 *Let m be a positive integer. Let $F : \mathbb{F}_{2^m} \to \mathbb{F}_{2^m}$ be an APN function, g a Boolean function over $\mathbb{F}_{2^m}$ and $\alpha \in \mathbb{F}_{2^m}^{\star}$. Denote by $D_\alpha F$ the derivative function of F with respect to α defined by $D_\alpha F(x) = F(x + \alpha) + F(x), \forall x \in \mathbb{F}_{2^m}$. Define over $\mathbb{F}_{2^m} \times \mathbb{F}_{2^m}$ a Boolean function by $f(x, y) = Tr_1^m(xD_\alpha F(y)) + g(y), \forall (x, y) \in \mathbb{F}_{2^m} \times \mathbb{F}_{2^m}$. Then f is semi-bent.*

Proof Let us compute the Walsh transform of f. For every $(a, b) \in \mathbb{F}_{2^m} \times \mathbb{F}_{2^m}$, we have

$$\widehat{\chi_f}(a, b) = \sum_{x \in \mathbb{F}_{2^m}} \sum_{y \in \mathbb{F}_{2^m}} (-1)^{Tr_1^m(xD_\alpha F(y))+g(y)+Tr_1^m(ax)+Tr_1^m(by)}$$

$$= \sum_{y \in \mathbb{F}_{2^m}} (-1)^{g(y)+Tr_1^m(by)} \sum_{x \in \mathbb{F}_{2^m}} (-1)^{Tr_1^m(x(D_\alpha F(y)+a))}$$

$$= 2^m \sum_{y \in \mathbb{F}_{2^m} \mid D_\alpha F(y)=a} (-1)^{g(y)+Tr_1^m(by)}.$$

Now, since F is APN, the mapping $y \in \mathbb{F}_{2^m} \mapsto D_\alpha F(y)$ is 2-to-1 for every $\alpha \in \mathbb{F}_{2^m}^{\star}$. Hence, $\widehat{\chi_f}(a, b) \in \{0, \pm 2^{m+1}\}$ which completes the proof. $\square$

4.5 Several Constructions from Hyperovals and Oval Polynomials

Let $PG_2(2^n)$ be the two-dimensional projective space over $\mathbb{F}_{2^n}$. The one-dimensional subspaces of $\mathbb{F}_{2^n}^3$ are then the points, and the two-dimensional subspaces of $\mathbb{F}_{2^n}^3$ are called the lines. A hyperoval in $PG_2(2^n)$ can be defined as follows.

Definition 9 (Hyperoval) A hyperoval in $PG_2(2^n)$ is a set of $2^n + 2$ points; no three of them are collinear (i.e., lie in a line[2]).

---

[2] We say a point $p = (x_0, \ldots, x_n)$ is on a line $L[y_0, \ldots, y_n]$ if and only if $x_0 y_0 + x_1 y_1 + \cdots x_n y_n = 0$.

A particular type of polynomials on $\mathbb{F}_{2^n}$ give rise to hyperovals in $PG_2(2^n)$. More precisely:

Definition 10 *An oval polynomial on $\mathbb{F}_{2^n}$ is a polynomial G on $\mathbb{F}_{2^n}$ such that the set of points $\{(1, t, G(t)), t \in \mathbb{F}_2^n\} \cup \{(0, 0, 1), (0, 1, 0)\}$ (denoted by $D(G)$) forms a hyperoval of $PG_2(2^n)$ (for short, an o-polynomial).*

There is a close connection between the hyperovals and the o-polynomials since a hyperoval of $PG_2(2^n)$ can be represented by $D(G)$ where G is an o-polynomial on $\mathbb{F}_{2^n}$. In fact, there exists a necessary and sufficient condition for a mapping over $\mathbb{F}_{2^n}$ to give a hyperoval of $PG_2(2^n)$. This leads to a reformulation of the definition of an o-polynomial given as follows.

Definition 11 *A permutation polynomial G over $\mathbb{F}_{2^n}$ is an o-polynomial if, for every $\gamma \in \mathbb{F}_{2^n}$, the function*

$$z \in \mathbb{F}_{2^n} \mapsto \begin{cases} \frac{G(z+\gamma)+G(\gamma)}{z} & \text{if } z \neq 0 \\ 0 & \text{if } z = 0 \end{cases}$$

is a permutation of $\mathbb{F}_{2^n}$.

Note that if G is an *o*-polynomial over $\mathbb{F}_{2^n}$ then, $z \in \mathbb{F}_{2^n} \mapsto G(z) + \alpha z$ is 2-to-1 for every $\alpha \in \mathbb{F}_{2^n}^\star$.

The current list, up to equivalence, of the known o-polynomials on $\mathbb{F}_{2^m}$ is given in [5].

A simple construction of semi-bent functions from hyperovals of $PG_2(2^m)$ with $m > 2$ is given by the following statement.

Theorem 9 *Let k be a positive integer such that $2 \leq k \leq 2^m - 2$. Let $D(k) := \{(1, t, t^k), t \in \mathbb{F}_{2^m}\} \cup \{(0, 0, 1), (0, 1, 0)\}$ $(m > 2)$ be a hyperoval of $PG_2(2^m)$ and g be a Boolean function on $\mathbb{F}_{2^m}$. Then the function f defined over $\mathbb{F}_{2^m} \times \mathbb{F}_{2^m}$ by $f(x, y) = Tr_1^m(xy^k + xy) + g(y)$ is semi-bent.*

Proof We have to prove that f is semi-bent, that is, its Walsh transform takes only the values 0, 2^{m+1} and -2^{m+1}. Compute the Walsh transform of f. For every $(a, b) \in \mathbb{F}_{2^m} \times \mathbb{F}_{2^m}$, we have:

$$\widehat{\chi_f}(a, b) = \sum_{x \in \mathbb{F}_{2^m}} \sum_{y \in \mathbb{F}_{2^m}} \chi\left(Tr_1^m(xy^k + xy) + g(y) + Tr_1^m(ax) + Tr_1^m(by)\right)$$

$$= \sum_{y \in \mathbb{F}_{2^m}} \chi\left(g(y) + Tr_1^m(by)\right) \sum_{x \in \mathbb{F}_{2^m}} \chi\left(Tr_1^m(xy^k + xy) + Tr_1^m(ax)\right)$$

$$= \sum_{y \in \mathbb{F}_{2^m}} \chi\left(g(y) + Tr_1^m(by)\right) \sum_{x \in \mathbb{F}_{2^m}} \chi\left(Tr_1^m((y^k + y + a)x)\right)$$

$$= 2^m \sum_{y \in \mathbb{F}_{2^m} | y^k + y = a} \chi\left(g(y) + Tr_1^m(by)\right).$$

Now, since $D(k)$ is a hyperoval of $PG_2(2^m)$ then according to Maschietti [30], the equation $y^k + y + a = 0$ has either zero or two distinct solutions in $\mathbb{F}_{2^m}$ for every $a \in \mathbb{F}_{2^m}$ ($m > 2$). Therefore, $\widehat{\chi_f}(a, b) \in \{0, \pm 2^{m+1}\}$ which completes the proof.

$\square$

An application of Theorem 9 is given by the next proposition.

Proposition 8 *Let m be a positive odd integer with $m > 2$. Let g be a Boolean function on $\mathbb{F}_{2^m}$. Then the function f defined over $\mathbb{F}_{2^m} \times \mathbb{F}_{2^m}$ by $f(x, y) = Tr_1^m(xy^6 + xy) + g(y)$ is semi-bent.*

Proof According to Theorem 9, f is semi-bent if $D(6) := \{(1, t, t^6), t \in \mathbb{F}_{2^m}\} \cup \{(0, 0, 1), (0, 1, 0)\}$ ($m > 2$) is a hyperoval of $PG_2(2^m)$. According to Segre and Bartocci [47], for m odd with $m > 3$, $D(6)$ is a hyperoval of $PG_2(2^m)$. It remains to check the case $m = 3$. According to Maschietti [30], it suffices to prove that the equation $y^6 + y = a$ has either zero solution or two distinct solutions in $\mathbb{F}_{2^m}$, for every $a \in \mathbb{F}_{2^m}$. The result is trivial for $a = 0$. Now, let $a \in \mathbb{F}_{2^m}^\star$. Using the fact that $y^7 = 1$ for $y \neq 0$, it is easy to see that the number of solutions of the equation $y^6 + y = a$ in $\mathbb{F}_{2^m}$ is equal to the number of solutions of $y^2 + ay + 1 = 0$ in $\mathbb{F}_{2^m}^\star$, which equals 2 (since if $y^2 + ay + 1 = 0$ has two identical solutions implies that $a = 0$, which contradicts the hypothesis). $\square$

In the following, we show how one can construct several infinite classes of semi-bent functions from o-polynomials. The first result in this direction was given in [6] which is closely related to the construction of semi-bent functions in bivariate representation from the class $\mathcal{H}$ of bent functions and the class of partial spreads $\mathcal{PS}_{ap}$ given by Theorem 6.

Theorem 10 ([6]) *Let G be an o-polynomial on $\mathbb{F}_{2^m}$, and g be Boolean function on $\mathbb{F}_{2^m}$ such that $g(0) = 0$ and $wt(g) = 2^{m-1}$ (i.e., g is balanced on $\mathbb{F}_{2^m}$). Let $\mu \in \mathbb{F}_{2^m}$. Define over $\mathbb{F}_{2^m} \times \mathbb{F}_{2^m}$ the Boolean function f by*

$$f(x, y) = Tr_1^m(\mu y + xG(yx^{2^m-2})) + g(yx^{2^m-2}), \quad (x, y) \in \mathbb{F}_{2^m} \times \mathbb{F}_{2^m}.$$

Then f is semi-bent.

Very recently, several more constructions of semi-bent functions have been derived from o-polynomials [40]. An important point is that the notion of oval polynomial over $\mathbb{F}_{2^m}$ appears to be suitable to build 2-to-1 mappings on $\mathbb{F}_{2^m}$. Such a property has been used to built infinite classes of semi-bent functions.

Theorem 11 ([40]) *Let α be a primitive element of $\mathbb{F}_{2^m}$ and j a positive integer in the range $[0, 2^m - 2]$. Let G be an o-polynomial on $\mathbb{F}_{2^m}$ and g a Boolean function on $\mathbb{F}_{2^m}$. Define over $\mathbb{F}_{2^m} \times \mathbb{F}_{2^m}$ a Boolean function f by*

$$f(x, y) = Tr_1^m(xG(y) + \alpha^j xy) + g(y), \quad (x, y) \in \mathbb{F}_{2^m} \times \mathbb{F}_{2^m}.$$

Then f is semi-bent.

Problem 5 Find other permutations G than oval polynomials having the property that $y \mapsto G(y) + \alpha^j y$ is 2-to-1 (which is the key in the proof of Theorem 11).

In the following, we emphasize the following observation.

Proposition 9 ([40]) *Any semi-bent function of Theorem 11 is the sum of two bent functions in the class of Maiorana–McFarland.*

Remark 4 Note that if we take at random two bent functions, even in the class of Maiorana–McFarland, their sum would not be probably semi-bent in most cases (the reader should notice that semi-bent functions of Theorem 10 can also be decomposed in the sum of two bent functions).

Problem 6 Find new constructions of semi-bent functions using permutations other than oval polynomials.

Another construction of semi-bent function in bivariate representation has been derived by the author in [40].

Theorem 12 ([40]) *Let m be a positive integer. Assume $m = 2m_1 + 1$ odd. Let G be an o-polynomial on $\mathbb{F}_{2^m}$ and g be a Boolean function on $\mathbb{F}_{2^m}$. Define a Boolean function f in bivariate representation as*

$$f(x, y) = Tr_1^m \left(x G^{2^{m_1+1}+1}(y) + xy G^{2^{m_1+1}}(y) + x G^3(y) + xy G^2(y) \right)$$

$$+ Tr_1^m \left((xy^{2^{m_1+1}} + xy^2 + x)G(y) + xy^{2^{m_1+1}+1} + xy + xy^3 \right)$$

$$+ g(y), (x, y) \in \mathbb{F}_{2^m} \times \mathbb{F}_{2^m}.$$

Then f is semi-bent on $\mathbb{F}_{2^m} \times \mathbb{F}_{2^m}$.

Now, Theorems 11 and 12 can be generalized since other semi-bent functions of a more general form can be obtained from o-polynomials.

Theorem 13 ([40]) *Let π_1 and π_2 be two permutations of $\mathbb{F}_{2^m}$ whose composition $\pi_1 \circ \pi_2^{-1}$ is an o-polynomial on $\mathbb{F}_{2^m}$. Let g be a Boolean function over $\mathbb{F}_{2^m}$. Let f be the Boolean function defined on $\mathbb{F}_{2^m} \times \mathbb{F}_{2^m}$ by*

$$(x, y) \in \mathbb{F}_{2^m} \times \mathbb{F}_{2^m}, \quad f(x, y) = Tr_1^m(x(\pi_1(y) + \pi_2(y))) + g(y).$$

Then f is semi-bent.

A first consequence of the previous theorem is the following statement which provides another primary construction of semi-bent functions.

Theorem 14 ([40]) *Let m be an odd positive integer. Define the Boolean function f on $\mathbb{F}_{2^m} \times \mathbb{F}_{2^m}$ as*

$$(x, y) \in \mathbb{F}_{2^m} \times \mathbb{F}_{2^m}, \quad f(x, y) = Tr_1^m \left(y^6 x + y^5 x + y^3 x + yx \right) + g(y)$$

where g is any Boolean function over $\mathbb{F}_{2^m}$. Then f is semi-bent.

A generalization of Theorem 12 is given by the following statement.

Theorem 15 ([40]) *Let π be a permutation of $\mathbb{F}_{2^m}$. Let α be a primitive element of $\mathbb{F}_{2^m}$ and j a nonnegative integer. Let G be an o-polynomial and g a Boolean function over $\mathbb{F}_{2^m}$. Define*

$$\forall (x, y) \in \mathbb{F}_{2^m} \times \mathbb{F}_{2^m}, \quad f(x, y) = Tr_1^m(\pi(G(y) + \alpha^{jy})x) + g(y).$$

Then f is semi-bent.

Let $L(x) = \sum_{s=0}^{m-1} \alpha_s x^{2^s}$ and $l(x) = \sum_{s=0}^{m-1} \alpha_s x^s$ be two polynomial over $\mathbb{F}_{2^m}$. $L(x)$ and $l(x)$ are the 2-associate of each other. More specifically, $l(x)$ is the conventional 2-associate of $L(x)$ and $L(x)$ is the linearized 2-associate of $l(x)$. It is well known that L is a linear permutation polynomial, if and only if, the determinant of the matrix $(\alpha_{i-j}^{2^i})_{0 \le i, j \le m-1}$ is not zero.

A possible construction of semi-bent functions involving linearized polynomials and oval polynomials is given by the following statement.

Proposition 10 *Let $L(x)$ and $l(x)$ two polynomials on $\mathbb{F}_{2^m}$ defined as above. Assume that $l(x)$ is co-prime with $x^m - 1$. Let $a \in \mathbb{F}_{2^m}$ such that $Tr_1^m(a) = 0$ and δ be a non zero elements of $\mathbb{F}_{2^m}$. Let G be an o-polynomial on $\mathbb{F}_{2^m}$ and g any Boolean function on $\mathbb{F}_{2^m}$. Then the function f defined on $\mathbb{F}_{2^m} \times \mathbb{F}_{2^m}$ as*

$$f(x, y) = Tr_1^m \left(a x Tr_1^m(G(y) + \delta y) + x L(G(y) + \delta y) \right) + g(y)$$

is semi-bent.

Proof The proposition follows from Theorem 15 and Corollary 3.6 in [53]. □

In [5], we have introduced the notion of *o-equivalence* between two oval polynomials.

Definition 12 ([5]) Two functions G and G' are o-equivalent if one can be obtained from the other by a sequence of the following list of transformations:

1. $G \mapsto G'$ where $G' : z \in \mathbb{F}_{2^m} \mapsto G'(z) := G(\lambda z + \mu)$ with $\lambda \in \mathbb{F}_{2^m}^\star$ and $\mu \in \mathbb{F}_{2^m}$,
2. $G \mapsto G'$ where $G' : z \in \mathbb{F}_{2^m} \mapsto G'(z) := \lambda G(z) + \mu$ with $\lambda \in \mathbb{F}_{2^m}^\star$ and $\mu \in \mathbb{F}_{2^m}$,
3. $G \mapsto G'$ where $G' : z \in \mathbb{F}_{2^m} \mapsto G'(z) := z G(z^{2^m-2})$ (with $G(0) = 0$),
4. $G \mapsto G'$ where $G' : z \in \mathbb{F}_{2^m} \mapsto G'(z) := G(z^{2^j})^{2^{m-j}}$ where $j \in \mathbb{N}$,
5. $G \mapsto G'$ where $G' : z \in \mathbb{F}_{2^m} \mapsto G'(z) := G^{-1}(z)$.

Recall the notion of extended affine equivalence between two Boolean functions.

Definition 13 Two Boolean functions f and f' defined on $\mathbb{F}_{2^n}$ are called extended affine equivalent (EA-equivalent) if $f' = f \circ \phi + \ell$ where the mapping ϕ is an affine automorphism on $\mathbb{F}_{2^n}$ and ℓ is an affine Boolean function (affine functions are those whose algebraic degree is at most 1).

A discussion about the *EA-equivalence* between two semi-bent Boolean functions constructed from o-equivalent ovals polynomials can be found in [40].

4.6   Secondary Constructions of Semi-Bent Functions

In general, "secondary constructions" means constructions of new functions from ones having the same properties. Only few secondary constructions of semi-bent functions have been considered in the literature. An example of a secondary construction of semi-bent functions based on a strong condition on the derivative functions has been given in [48].

Theorem 16 ([48]) *Let f and g be two semi-bent functions over $\mathbb{F}_{2^n}$ (with n even). Assume that there exists an element a in $\mathbb{F}_{2^n}$ such that $D_a f = D_a g$. Then the function $h = f + D_a f (f + g)$ is semi-bent on $\mathbb{F}_{2^n}$.*

The reader notices that Theorem 7 shows that the indirect sum could be used to construct semi-bent functions from both bent and semi-bent functions. The construction derived from Theorem 7 can be therefore viewed as a secondary-like construction of semi-bent functions.

Problem 7 Find new secondary constructions of semi-bent functions, that is, constructions of new semi-bent functions from two or several already known ones.

Conclusion
The research activity on bent functions has lasted over 35 years and remains intensive. However, very recently, many advances have been made subsequently on super classes of bent functions (plateaued functions, etc.) and related classes of bent functions (semi-bent functions, etc.). In particular many new connections in the framework of semi-bent functions with other domains of mathematics and computer science (Dickson polynomial, Kloosterman sums, spreads, oval polynomial, finite geometry, coding, cryptography, sequences, etc.) have been exhibited. The research in this framework is relatively new (comparatively to the context of bent functions) and is becoming very active. Despite recent progress, much remains to do. In particular, although many concrete constructions of semi-bent functions have been discovered, the general structure of semi-bent functions is still unclear.

Acknowledgements The author wishes to thank Claude Carlet for his careful reading and interesting comments.

References

1. S. Boztas, P.V. Kumar, Binary sequences with Gold-like correlation but larger linear span. IEEE Trans. Inf. Theory **40**(2), 532–537 (1994)
2. A. Canteaut, C. Carlet, P. Charpin, C. Fontaine, On cryptographic properties of the cosets of R(1,m). IEEE Trans. Inf. Theory **47**, 1494–1513 (2001)
3. C. Carlet, On the secondary constructions of resilient and bent functions, in *Proceedings of the Workshop on Coding, Cryptography and Combinatorics 2003* (Birkhäuser, Basel, 2004), pp. 3–28
4. C. Carlet, Boolean functions for cryptography and error correcting codes, in Chapter of the monography *Boolean Models and Methods in Mathematics, Computer Science, and Engineering*, ed. by Y. Crama, P.L. Hammer (Cambridge University Press, Cambridge, 2010), pp. 257–397
5. C. Carlet, S. Mesnager, On Dillon's class H of bent functions, niho bent functions and O-polynomials. J. Comb. Theory Ser. A **118**(8), 2392–2410 (2011)
6. C. Carlet, S. Mesnager, On Semi-bent Boolean Functions. IEEE Trans. Inf. Theory **58**(5), 3287–3292 (2012)
7. C. Carlet, E. Prouff, On plateaued functions and their constructions, in *Proceedings of Fast Software Encryption (FSE)*. Lecture Notes in Computer Science, vol. 2887 (2003), pp. 54–73
8. A. Cesmelioglu, W. Meidl, A construction of bent functions from plateaued functions. Des. Codes Cryptogr. **66**(1–3), 231–242 (2013)
9. P. Charpin, G. Gong, Hyperbent functions, Kloosterman sums and Dickson polynomials. IEEE Trans. Inform. Theory **54**(9), 4230–4238 (2008)
10. P. Charpin, E. Pasalic, C. Tavernier, On bent and semi-bent quadratic Boolean functions. IEEE Trans. Inf. Theory **51**(12), 4286–4298 (2005)
11. S. Chee, S. Lee, K. Kim, Semi-bent functions, in *Advances in Cryptology-ASIACRYPT94. Proceedings of 4th International Conference on the Theory and Applications of Cryptology*, Wollongong, ed. by J. Pieprzyk, R. Safavi-Naini. Lecture Notes in Computer Science, vol. 917 (1994), pp. 107–118
12. G. Cohen, S. Mesnager, On constructions of semi-bent functions from bent functions. Journal Contemporary Mathematics 625, Discrete Geometry and Algebraic Combinatorics, Americain Mathematical Society, 141–154 (2014)
13. T.W. Cusick, H. Dobbertin, Some new three-valued crosscorrelation functions for binary m-sequences. IEEE Trans. Inf. Theory **42**(4), 1238–1240 (1996)
14. J. Dillon, Elementary Hadamard difference sets, Ph.D. dissertation, University of Maryland, 1974
15. H. Dobbertin, G. Leander, A. Canteaut, C. Carlet, P. Felke, P. Gaborit, Construction of bent functions via Niho Power Functions. J. Comb. Theory Ser. A **113**, 779–798 (2006)
16. D. Dong, L. Qu, S. Fu, C. Li, New constructions of semi-bent functions in polynomial forms. Math. Comput. Model. **57**, 1139–1147 (2013)
17. R. Gold, Maximal recursive sequences with 3-valued recursive crosscorrelation functions. IEEE Trans. Inform. Theory **14** (1), 154–156 (1968)
18. F. Gologlu, Almost bent and almost perfect nonlinear functions, exponential sums, geometries and sequences, Ph.D. dissertation, University of Magdeburg, 2009
19. T. Helleseth, Some results about the cross-correlation function between two maximal linear sequences. Discrete. Math. **16**, 209–232 (1976)
20. T. Helleseth, Correlation of m-sequences and related topics, in *Proceedings of SETAO98, Discrete Mathematics and Theoretical Computer Science*, ed. by C. Ding, T. Helleseth, H. Niederreiter (Springer, London, 1999), pp. 49–66
21. T. Helleseth, P.V. Kumar, Sequences with low correlation, in *Handbook of Coding Theory, Part 3: Applications*, chap. 21, ed. by V.S. Pless, W.C. Huffman, R.A. Brualdi (Elsevier, Amsterdam, 1998), pp. 1765–1853

22. J.Y. Hyun, H. Lee, Y. Lee, Nonexistence of certain types of plateaued functions. Discrete Appl. Math. **161**(16–17), 2745–2748 (2013)
23. K. Khoo, G. Gong, D.R. Stinson, A new family of Gold-like sequences, in *Proceedings IEEE International Symposium on Information Theory*, Lausanne (2002)
24. K. Khoo, G. Gong, D.R. Stinson, A new characterization of semi-bent and bent functions on finite fields. J. Design Codes Cryptogr. **38**(2), 279–295 (2006)
25. G. Leander, A. Kholosha, Bent functions with 2^r Niho exponents. IEEE Trans. Inf. Theory **52**(12), 5529–5532 (2006)
26. G. Leander, G. McGuire, Spectra of functions, subspaces of matrices, and going up versus going down, in *International Conference on Applied Algebra, Algebraic Algorithms and Error-Correcting Codes (AAECC)*. Lecture Notes in Computer Science, vol. 4851 (Springer, Berlin, 2007), pp. 51–66
27. G. Leander, G. McGuire, Construction of bent functions from near-bent functions. J. Comb. Theory Ser. A **116**, 960–970 (2009)
28. N. Li, T. Helleseth, X. Tang, A. Kholosha, Several new classes of bent functions from Dillon exponents. IEEE Trans. Inf. Theory **59**(3), 1818–1831 (2013)
29. F.J. MacWilliams, N.J. Sloane, *The Theory of Error-Correcting Codes* (North-Holland, Amsterdam, 1977)
30. A. Maschietti, Difference sets and hyperovals. J. Design Codes Cryptogr. **14**(1), 89–98 (1998)
31. M. Matsui, Linear cryptanalysis method for DES cipher, in *Proceedings of EUROCRYPT'93*. Lecture Notes in Computer Science, vol. 765 (1994), pp. 386–397
32. R.L. McFarland, A family of noncyclic difference sets. J. Comb. Theory Ser. A **15**, 1–10 (1973)
33. W. Meier, O. Staffelbach, Fast correlation attacks on stream ciphers, in *Advances in Cryptology, EUROCRYPT'88*. Lecture Notes in Computer Science, vol. 330 (1988), 301–314
34. Q. Meng, H. Zhang, M. Yang, J. Cui, On the degree of homogeneous bent functions. Discrete Appl. Math. **155**(5), 665–669 (2007)
35. S. Mesnager, A new family of hyper-bent boolean functions in polynomial form, in *Proceedings of Twelfth International Conference on Cryptography and Coding, IMACC 2009*. Lecture Notes in Computer Science, vol. 5921 (Springer, Heidelberg, 2009), pp. 402–417
36. S. Mesnager, A new class of bent and hyper-bent Boolean functions in polynomial forms. J. Design Codes Cryptogr. **59**(1–3), 265–279 (2011)
37. S. Mesnager, Bent and hyper-bent functions in polynomial form and their link with some exponential sums and Dickson polynomials. IEEE Trans. Inf. Theory **57**(9), 5996–6009 (2011)
38. S. Mesnager, Semi-bent functions from Dillon and Niho exponents, Kloosterman sums and Dickson polynomials. IEEE Trans. Inf. Theory **57**(11), 7443–7458 (2011)
39. S. Mesnager, Semi-bent functions with multiple trace terms and hyperelliptic curves, in *Proceeding of International Conference on Cryptology and Information Security in Latin America (IACR), Latincrypt 2012*. Lecture Notes in Computer Science, vol. 7533 (Springer, Berlin, 2012), pp. 18–36
40. S. Mesnager, Semi-bent functions from oval polynomials, in *Proceedings of Fourteenth International Conference on Cryptography and Coding*, Oxford, IMACC 2013. Lecture Notes in Computer Science, vol. 8308 (Springer, Heidelberg, 2013), pp. 1–15
41. S. Mesnager, Contributions on boolean functions for symmetric cryptography and error correcting codes, Habilitation to Direct Research in Mathematics (HdR thesis), December 2012
42. S. Mesnager, Bent functions from Spreads. Journal of the American Mathematical Society (AMS), Contemporary Mathematics 632. to appear.
43. S. Mesnager, G. Cohen, On the link of some semi-bent functions with Kloosterman sums, in *Proceedings of International Workshop on Coding and Cryptology, IWCC 2011*. Lecture Notes in Computer Science, vol. 6639 (Springer, Berlin, 2011), pp. 263–272
44. S. Mesnager, J.P. Flori, Hyper-bent functions via Dillon-like exponents. IEEE Trans. Inf. Theory **59**(5), 3215–3232 (2013)

45. Y. Niho, Multi-valued cross-correlation functions between two maximal linear recursive sequences, Ph.D. dissertation, University of Sothern California, Los Angeles, 1972
46. O.S. Rothaus, On "bent" functions. J. Comb. Theory Ser. A **20**, 300–305 (1976)
47. B. Segre, U. Bartocci, Ovali ed altre curve nei piani di Galois di caratteristica due. Acta Arith. **18**(1), 423–449 (1971)
48. G. Sun, C. Wu, Construction of semi-bent Boolean functions in even number of variables. Chin. J. Electron. **18**(2), 231–237 (2009)
49. J. Wolfmann, Cyclic code aspects of bent functions, in *Finite Fields Theory and Applications, Contemporary Mathematics Series of the AMS*, vol. 518 (American Mathematical Society, Providence, 2010), pp. 363–384
50. J. Wolfmann, Special bent and near-bent functions. Adv. Math. Commun. **8**(1), 21–33 (2014)
51. J. Wolfmann, Bent and near-bent functions (2013). arxiv.org/abs/1308.6373
52. T. Xia, J. Seberry, J. Pieprzyk, C. Charnes, Homogeneous bent functions of degree n in $2n$ variables do not exist for $n > 3$. Discrete Appl. Math. **142**(1–3), 127–132 (2004)
53. P. Yuan, C. Ding, Permutation polynomials over finite fields from a powerful lemma. Finite Fields Appl. **17**, 560–574 (2011)
54. Y. Zheng, X.M. Zhang, Plateaued functions, in *Advances in Cryptology ICICS 1999*. Lecture Notes in Computer Science, vol. 1726 (Springer, Berlin, 1999), 284–300
55. Y. Zheng, X.M. Zhang, Relationships between bent functions and complementary plateaued functions. Lecture Notes in Computer Science, vol. 1787 (1999), pp. 60–75
56. Y. Zheng, X.M. Zhang, On plateaued functions. IEEE Trans. Inform. Theory **47**(3), 1215–1223 (2001)

True Random Number Generators

Mario Stipčević and Çetin Kaya Koç

Abstract Random numbers are needed in many areas: cryptography, Monte Carlo computation and simulation, industrial testing and labeling, hazard games, gambling, etc. Our assumption has been that random numbers cannot be computed; because digital computers operate deterministically, they cannot produce random numbers. Instead, random numbers are best obtained using physical (true) random number generators (TRNG), which operate by measuring a well-controlled and specially prepared physical process. Randomness of a TRNG can be precisely, scientifically characterized and measured. Especially valuable are the information-theoretic provable random number generators (RNGs), which, at the state of the art, seem to be possible only by exploiting randomness inherent to certain quantum systems. On the other hand, current industry standards dictate the use of RNGs based on free-running oscillators (FRO) whose randomness is derived from electronic noise present in logic circuits and which cannot be strictly proven as uniformly random, but offer easier technological realization. The FRO approach is currently used in 3rd- and 4th-generation FPGA and ASIC hardware, unsuitable for realization of quantum RNGs. In this chapter we compare weak and strong aspects of the two approaches. Finally, we discuss several examples where use of a true RNG is critical and show how it can significantly improve security of cryptographic systems, and discuss industrial and research challenges that prevent widespread use of TRNGs.

M. Stipčević (✉)
Centre of Excellence for Advanced Materials and Sensors,
Rudjer Bošković Institute, Zagreb, Croatia

University of California Santa Barbara, Santa Barbara, CA 93106, USA
e-mail: mstipcev@physics.ucsb.edu; stipcevi@gmail.com

Ç.K. Koç
University of California Santa Barbara, Santa Barbara, CA 93106, USA
e-mail: koc@cs.ucsb.edu

© Springer International Publishing Switzerland 2014
Ç.K. Koç (ed.), *Open Problems in Mathematics and Computational Science*,
DOI 10.1007/978-3-319-10683-0_12

1   Introduction

True random numbers and physical nondeterministic random number generators (RNGs) seem to be of an ever-increasing importance. Random numbers are essential in cryptography (mathematical, stochastic, and quantum), Monte Carlo calculations, numerical simulations, statistical research, randomized algorithms, lotteries, etc. Today, true random numbers are most critically required in cryptography and its numerous applications to our everyday life: mobile communications, e-mail access, online payments, cashless payments, ATMs, e-banking, Internet trade, point of sale, prepaid cards, wireless keys, general cybersecurity, distributed power grid security (SCADA), etc.

Without loss of generality in the rest of this chapter, we will assume that generators produce random bits.

In applications where provability is essential, randomness sources (if involved) must also be provably random; otherwise, the whole chain of proofs collapses. In cryptography, where due to Kerckhoffs' principle all parts of protocols are publicly known except some secret (the key or other information) known only to the sender and the recipient, it is clear that the secret must not be calculable by an eavesdropper, i.e., it must be random. For example, the well-known BB84 quantum key distribution protocol [5] (described in Sect. 3.4) would be completely insecure if only an eavesdropper could calculate (or predict) either Alice's random numbers or Bob's random numbers or both. From analysis of the secret key rate presented therein, it is obvious that any predictability of random numbers by the eavesdropper would leak relevant information to him, thus diminishing the effective key rate. It is intriguing [79] that in the case that the eavesdropper could calculate the numbers exactly, the cryptographic potential of the BB84 protocol would be zero. Indeed one of the recent successful attacks on quantum cryptography exploits the possibility to control local quantum RNGs by exploiting a design flaw of two commercial quantum cryptographic systems and one practical scientific system. This example, discussed below, shows that the local RNGs assumed in BB84 are essential for its security and may not be exempt from the security proof.

Lotteries are yet another serious business where random numbers are essential. Due to the large sum of money involved (estimated six billion USD annually only online and only in the USA [36]), some countries have set explicit requirements for RNGs for use in online gambling and lottery machines and have set certificate issuing authorities. For example, the Lotteries and Gaming Authority (LGA) of Malta has prescribed a list of requirements for RNGs, stipulated in the Remote Gaming Regulations Act [45]. An RNG that does not conform to this act may not be legally used for gambling business. These rules have been put forward in order to ensure fair game by providers and to prevent possibility that gamers manipulate the system by foreseeing outcomes.

RNGs have been an occupation of scientists and inventors for a long time. Whole branches of mathematics have been invented out of a need to understand random numbers and ways to obtain them. At the dawn of the modern computing era, John von Neumann was one of the first to note that deterministic Turing computers are not able to produce true random numbers, as he put it in his well-known statement that "Anyone who considers arithmetical methods of producing random digits is, of course, in a state of sin."

RNGs are one of the hottest topics of research in recent years. There have been about 83 patents per year in the last decade, 1418 in total since 1970, and countless scientific articles published regarding true RNGs. Still, a sharp discrepancy between the number of publications and very modest number of products (only four quantum RNGs and a handful of Zener noise-based mostly phased-out RNGs) that ever made it to the market [33,34,57,60] clearly indicates immaturity of most of the art. In our view, the main problems are lack of randomness proofs and poor reproducibility of the majority of solutions presented so far. The search for true randomness continues.

2 Pseudorandom Number Generators

Historically, there have been two approaches to random number generation: algorithmic (pseudorandom) and by a physical process (nondeterministic).

Pseudorandom number generators (PRNG) are well known in the art and we are not going to address them here in great detail. Surveys and individual examples of PRNGs can be found elsewhere [32, 40, 48, 49, 92]. In a nutshell, a PRNG is nothing more than a mathematical formula, which produces a deterministic, periodic sequence of numbers, which is completely determined by the initial state called the seed. By definition such generators are not provably random. In practice, PRNGs feature perfect balance between 0s and 1s (zero bias) but also strong long-range correlations, which undermine cryptographic strength and can show up as unexpected errors in Monte Carlo calculations and modeling.

While most modern PRNGs pass all known statistical tests, there are myths about some PRNGs being much better than the others. The truth is that every PRNG shows its weakness in some particular application. Indeed PRNGs are often found to be the cause of erroneous stochastic simulations and calculations [11,12,15,21,29,40,45, 55, 58, 70, 87]. As for cryptographic purposes, all major families of PRNGs have been cryptanalyzed so far [40, 61, 74], and use of PRNG where an RNG should be used will therefore present a big security risk for the protocol in question. We will revisit this point in more detail in Sect. 6.

In any case, due to strict determinism of PRNG algorithms, no PRNG is random by any reasonable definition of randomness. Let us illustrate this by a fictitious anecdote. Alice wanted to impress Bob, by a particular version of Mersenne Twister PRNG [49] for which she claimed that it produces true random numbers, by asking him to test them. Bob agreed but asked a minimum of 1 Giga bytes of random data to be sent to him via e-mail. Alice produced the huge file but her mailing program refused to send such a big file. Cutting a file into small pieces and sending multiple e-mails, etc. was an option but too big a nuisance for both of them. Finally, Bob received from Alice a 1 kilobyte e-mail containing the following short notice: "Dear

Bob, Please find attached a program in C++. Compile it, use the following seed: 12345678 and stop the program after producing 1 Giga bytes of data. That is what I wanted to send you." Instead of reproducing the file and running on his computer very time-consuming tests, Bob shortly answered: "Dear Alice, if you think that 1 Giga bytes of truly random data can, under any circumstances, can be compressed without loss to just 1,000 bytes, than I have nothing more to say to you!"

Advantages of PRNGs are their low cost, ease of implementation, and user-friendliness, especially in a CPU-available environment such as a PC computer, but one has to be cautious when it comes to use of PR numbers for simulations, cryptography, and in fact any use.

3 True Random Number Generators

Due to Kerckhoffs' principle, the definition of a RNG suitable for cryptography must include that even if every detail is known about the generator (schematic, algorithms, etc.), it still must produce totally unpredictable bits. In contrast to PRNGs, physical (true, hardware) RNGs extract randomness from physical processes that behave in a fundamentally nondeterministic way which makes them better candidates for true random number generation. A physical RNG is a piece of hardware separate from the computer, usually connected to it via USB or PCI bus. Importing random numbers into a user program is complicated and requires original drivers. Prices range from 1k USD to 30k USD for bit production rates from 4 to 150 megabits per second [33, 34, 60]. Examples of physical processes used to generate randomness include: Johnson's noise [54], Zener noise [77], radioactive decay [22, 26], photon path splitting at the two-way beam splitter, photon arrival times, etc. [9, 13, 22, 23, 26, 35, 38, 63, 77, 80, 88–90, 93]. Unlike the PRNGs, physical random number generators suffer from uneven probabilities of zeros and ones, that is, bias (b), defined as the difference of probabilities of 1s and 0s:

$$b = \frac{p(1) - p(0)}{2} \tag{1}$$

and short-range correlations which are best captured by serial autocorrelation coefficients (a_k), defined, for example, in [40]:

$$a_k = \frac{\sum_{i=1}^{N-k} (b_i - \bar{b})(b_{i+k} - \bar{b})}{\sum_{i=1}^{N-k} (b_i - \bar{b})^2} \tag{2}$$

where $\{b_1, b_2, \ldots, b_N\}$ is an N bits long random string and k is the lag or "order" of the coefficient. Both b and a_k are normalized such that they can take on values in the interval $[-1, 1]$ and that an ideal RNG exhibits $b = 0$ and $a_k = 0$. True RNGs are generally constructed such that the correlation among bits is small—which is, namely, the idea of randomness. In some cases the physical system that is measured

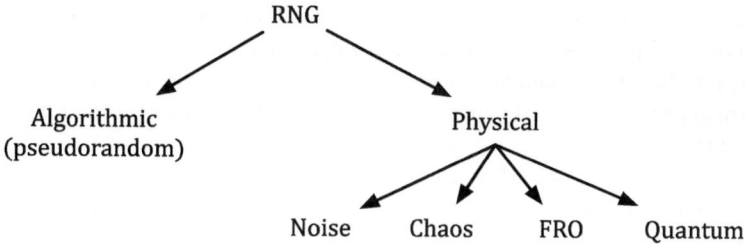

Fig. 1 Classification of random number generators

is being "reset" to an initial condition after production of each bit in order to reduce autocorrelation. Therefore in most cases only a few lowest-order autocorrelation coefficients are significant, ideally only the first one, which is named autocorrelation and denoted by a.

There are very many constructions or true RNGs and research is still getting impetus, but in our view one can roughly classify the present art into four families:

- Noise-based RNGs
- Free-running oscillator RNGs
- Chaos RNGs
- Quantum RNGs

The tree of RNGs is illustrated in Fig. 1. Mathematical, pseudorandom generators can also be divided into several categories depending on the type of algorithm used.

Note that our definition of a true RNG is not to be confused with a pseudorandom number generator implemented in CMOS logic or similar hardware; such a generator is still a PRNG, since it is just a hardware implementation of a mathematical method. Next, we are going to address each of the above families in some detail.

3.1 Noise-Based RNGs

Johnson's effect [54] creates random voltage on terminals of any resistive material which is held at a temperature higher than absolute zero. Johnson's noise is due to random thermal motion of the quantized electric charge (i.e., carriers). However, long-range carrier correlations in conductors cause correlations in movements of electric charges, and, therefore, the resulting voltage is not completely random [4].

Zener noise (in semiconductor Zener diodes) is caused by tunneling of carriers through quantum barriers of ideally constant height and width. If current is sufficiently low, individual "jumps" of carriers through barriers will be seen as voltage peaks across the diode, forming a pink noise of perfect randomness. An interesting property of this kind of noise is that at sufficiently high inverse voltage, the diode exhibits high internal avalanche gain. Such a gain mechanism leads to large amplitude of the noise and is highly insensitive to electromagnetic radiation from the environment. However, the Zener effect is never found well isolated in physical devices from other effects nor is the quantum barrier constant. Most of the aforementioned processes in resistors and Zener diodes have some memory effect. This means that an instantaneous voltage across the device depends on voltages in the (near) past and this in turn leads to a correlation among random numbers extracted therefrom.

Other popular sources of noise include: inverse base-emitter breakdown in bipolar transistors, laser phase noise [30], chaos noise [44], etc. The biggest problem with all kinds of noises is that randomness of noise sources cannot be well characterized, measured, or even controlled during fabrication of the device. Furthermore, some noise mechanisms (notably Johnson's noise) produce rather tiny voltages that need to be strongly amplified before conversion to digital form. The strong amplification introduces further deviations from randomness due to the limited amplifier bandwidth and gain nonlinearity. Also, fast electrical switching of binary logic used in the RNG circuitry produces strong electromagnetic interference so that multiple nearby RNGs (especially if on chip) tend to mutually synchronize causing the dramatic drop of overall entropy. On top of that, highly sensitive amplifiers allow easy manipulation of noise-based RNGs by external electromagnetic fields which can be exploited for cryptographic attacks.

The general idea of noise-based true RNG is the following. The random analog voltage is sampled periodically and compared to a certain predefined threshold: if higher, then "1" is generated; otherwise, "0" is generated (Fig. 2). It is obvious that the threshold can be set so that the probabilities of 1s and 0s are roughly the same. However, fine-tuning of the threshold poses an insurmountable time-consuming problem and can never be done properly. For example, if tuning of bias to value of 0.1 requires 10 s, then tuning to 10 times the lower value (0.01) would take 100 times longer (the required timescales as square of improvement ratio). And then there is a problem of stability: even the smallest drift of the mean value (e.g., due to temperature or supply voltage change) will create a noticeable bias. Provability of any noise RNG is complicated and eventually made impossible for three reasons:

1. Provability of randomness of the exploited noise source
2. Effect of the sampling/digitizing procedure
3. Eventual use of deterministic post-processing

Going from this basic circuit, researchers have proposed many circuits whose aim is to improve the randomness, notably the bias.

Fig. 2 Noise-based RNG. Noise is fed to a level comparator whose output is either 0 or 1 depending on whether its positive input is below or above the threshold value VBIAS. Upon Request, a fresh new random bit will sit on the Output

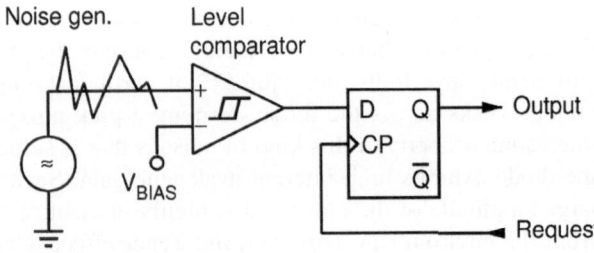

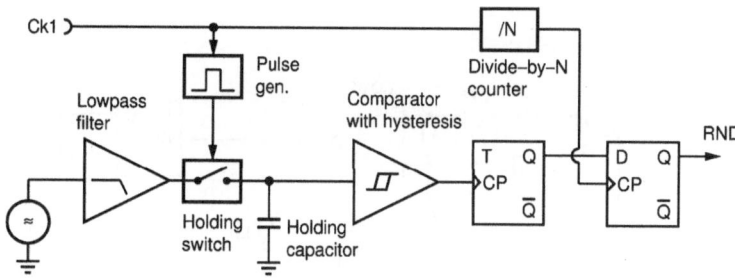

Fig. 3 A zero-bias noise-based RNG by Bagini and Bucci. The biased output produced by an imperfect threshold principle is divided by 2 by a T flip-flop. The output of the T flip-flop spends exactly 50 % of the time in state 1 and is sampled periodically by a pulse generator. The idea is that when sampled (by the D flip-flop), it will yield either 0 or 1 with perfectly equal probabilities. However, in practice non-negligible deviation from perfect bias will occur and correlations will exist

First, the most obvious improvement would be to somehow de-bias the raw noise stream in hardware without the need for any adjustment of the threshold voltage. An interesting solution to that has been discovered by Vincent [93], generalized by Chevalier and Menard [9], and independently rediscovered later by Bagini and Bucci [2] and Stipcevic [77].

In the Bagini–Bucci generator [2] shown in Fig. 3, the analog voltage from the free-running noise source is periodically sampled at frequency f_{Ck1} and compared to a threshold value at the comparator. Whenever the comparator produced logical "1" the T-type flip-flop (TFF) changes its state. If the sampled process is random and stationary, because of time symmetry of this process, the output of the TFF spends half of the time in the low state and the other half in the high state. There are a couple of problems with that design. First, the holding capacitor acts as a memory that remembers the previous analog voltage. Due to finite impedances in the circuit when charged with the next voltage level, the voltage will be to some extent dependent on the previous one, thus creating the autocorrelation. The second problem is that if the TFF is interrogated at too high a rate, it will tend to give the same answer several times in a row, thus producing positively autocorrelated output, even when the basic random process is truly random! The only way to circumvent this problem is to use a bit sampling frequency f_{Ck2} much lower than the noise sampling frequency f_{Ck1}, for example, $f_{Ck2} = f_{Ck1}/N$, thus arriving at an asymptotically random sequence of bits in the limit of $N \to \infty$.

In the variation of this principle named "time summation of a random signal" [76, 77] shown in Fig. 4, time-wise random pulses at the output of the comparator COMP are counted by a modulo 2 counter (TFF) whose output gets sampled upon a request sent over the Request input. The results are similar to the Bagini–Bucci circuit except that bits can be generated faster because both the low-pass filter and sampling circuits are not needed. Also, it features a naturally incorporated automated zero-bias loop consisting of the comparator COMP, a low-pass filter with

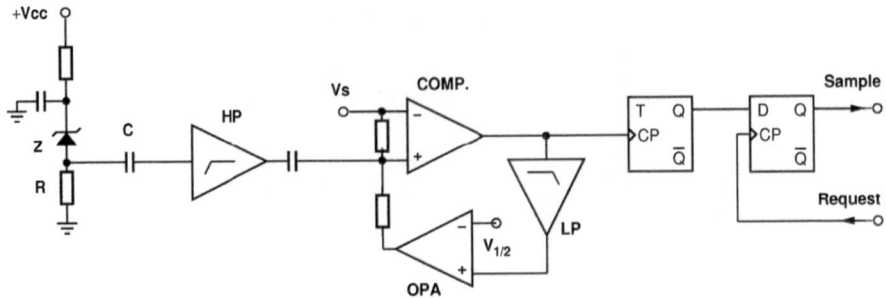

Fig. 4 A zero-bias noise-based RNG by Stipcevic. Time-wise random events appearing at COMP are summed at the input of the toggle flip-flop (TFF), and when the sum becomes bigger than the predefined time interval bit sampling period T, random output is equal to the number of random pulses in that interval mod 2. This is similar to a Bagini–Bucci generator except that there is no need for a low-pass filter and a sampling circuit. There is no requirement that bits be sampled periodically. On top of that, there is an automated zero-bias loop

a time constant much bigger than the bit sampling rate, and an amplifier OPA. The loop sets the threshold for the comparator in such a way that a comparator spends half of the time in state "1" which is important to minimize autocorrelation. The TFF then takes care of complete canceling of the bias. In case of periodic bit sampling, again, correlation among bits will be nonvanishing even if the pulses are completely random unless the ratio between mean frequencies of random pulses at the sampling frequency (N) goes to infinity. In practice however N only needs to be sufficiently large to keep correlations at the desired level.

The bad side of this "sampling" principle, illustrated in Figs. 3 and 4, is that it required N random events to produce one random bit (low efficiency). The good side is that by letting N be large enough, one can obtain any desired level of randomness quality, at least theoretically. Practically however, small imperfections in logical circuits, such as uneven high-to-low and low-to-high transitions, will ultimately limit the achievable randomness. Regarding provability of randomness of this principle, technical imperfections of the individual components, unclear theory of operation of the "noise source," and overall complexity of the circuit make it impossible to arrive to a credible proof of randomness.

The next example of noise class of RNGs is the Intel RNG [37] implemented in a limited series of computer processors (Fig. 5). It uses amplified thermal noise of a resistor to disturb a voltage-controlled oscillator, thus arriving to a "slow" random pulse generator which is used to sample a "high-speed" periodic oscillator. This fast-slow dichotomy is similar to the above described sampling RNGs and is known not to generate theoretically perfect randomness unless the ratio of fast to slow does tend to infinity. A particular peculiarity of this construction is that a voltage-controlled oscillator (a steady oscillator has zero entropy) is disturbed by a noisy voltage, thus very probably yielding a lesser entropy than available from the noise source. The important property of such a construction is that its frequency cannot surpass a certain limit, thus guaranteeing a high enough ratio between the aforementioned

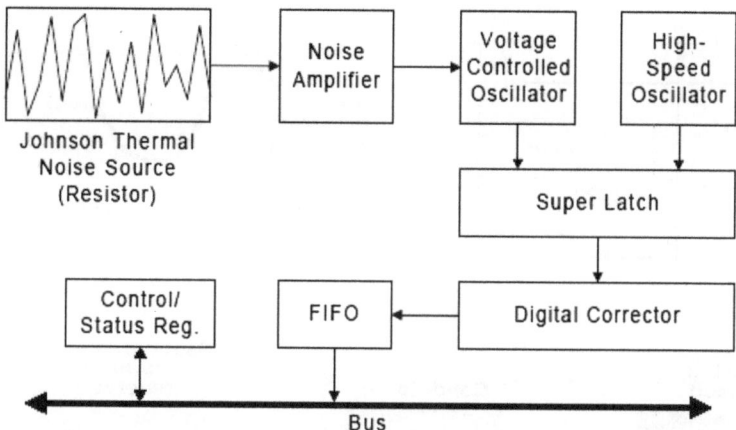

Fig. 5 A Johnson noise-based RNG by Intel. A high-speed periodic digital oscillator is sampled at approximately random times defined by a Johnson noise signal. Time-wise random events appearing at COMP are summed at the input of the toggle flip-flop (TFF)

high and low frequencies. It is therefore clear that the bits generated at the latch flip-flop (Super Latch) are not very random and require post-processing which consists of a modified (and patented) von Neumann method of efficiency 1/4 [83].

Yet another Intel RNG appeared in 2011 after "10 years of research" which is apparently extremely simple [83] (Fig. 6a). The idea is to obtain a circuit that does not have any (apparent) analog parts and is therefore compatible with logic chips. The circuit consists of two Yin-Yang connected inverters and two "oddly connected transistors." The authors explain that this circuit has two stable states: 0 and 1. If everything is perfectly symmetric, when transistors are driven high, the output will end up in either low or high state. The authors further explain that even though ideally the output value should be random, even the smallest difference in speed or strength of inverters would lead to high imbalance between zeros and ones (we would add: and possibly to complete lockup). Therefore Intel has put an additional current-injecting mechanism that makes inverters controllable enough so that they can be made "equal." The quality of random numbers must be very low, because Intel uses 2-stage post-processing in order to remove bias and correlations (Fig. 6b). The first stage is an unspecified randomness corrector after which "raw" bits become "high-quality random seeds." The second stage is a PRNG seeded by these high-quality seeds. It remains unclear why high-quality true random numbers would be passing through a PRNG, but there might be only two reasons. Either these hardware numbers are not very good and must be further processed by the PRNG or Intel must comply with FIPS PUB-140 [19] which explicitly does not endorse any true RNG for cryptographic purposes and in this way numbers technically exit from a PRNG.

All the above examples utilize electronic noise: a resource which is becoming less and less available because manufacturers of electronics components and chips make every possible effort to make it ever smaller. Therefore researchers have turned to sources capable of producing fluctuating voltage similar to electronics noise but

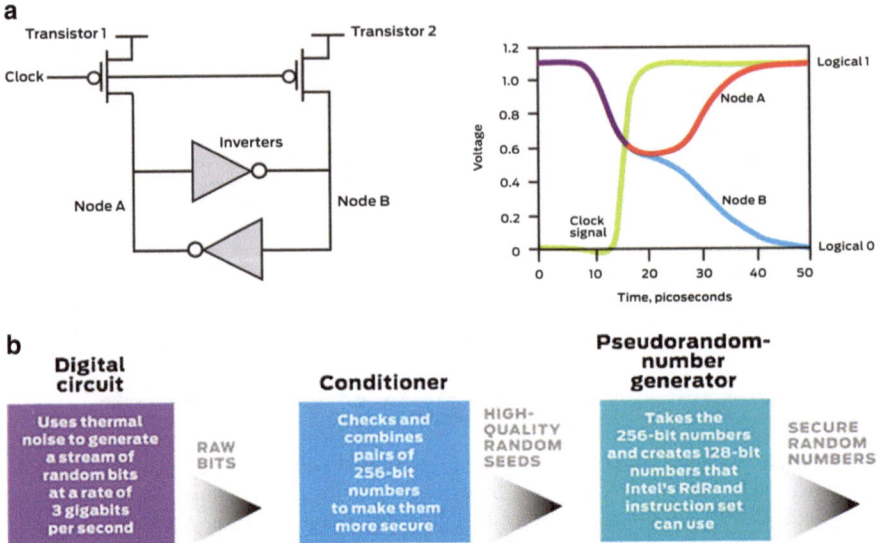

Fig. 6 Intel's "quantum" random number generator. (**a**) Basic digital RNG circuit. Upon each pulse, output stabilizes in random binary state. This is in fact yet another noise-driven generator based on a specially prepared trimmed RS-type flip-flop whose both set and reset inputs are tied together and driven at the same time. The specialty of this flip-flop is that its inner gates may be current-trimmed in such a way as to make it possible that the output may stabilize in either low or high state. (Normally it would be locked to either state or produce high bias because of the smallest asymmetry of its internal gates.) (**b**) The post-processing scheme

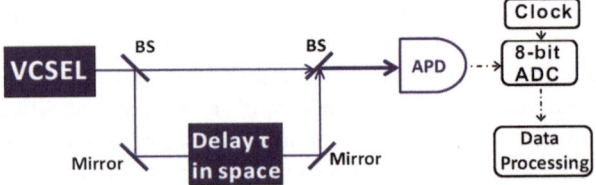

Fig. 7 Laser phase noise-based true RNG. Intensity of noise is determined by fundamental uncertainty of phase, while its whiteness, that is, a Gaussian distribution of instantaneous amplitude, is due to the central limit theorem

whose origin is more fundamental and therefore has less sensitivity to technological advances. For example, very fast noise can be obtained by lasers. Lasers exhibit very fast fluctuations which can be detected by fast PIN or avalanche photodiodes (APDs), thus producing wide-band electrical noise.

One such example is the phase noise of a single laser (Fig. 7) invented by the CREAM group [30].

This is an example of a white noise-based generator where a Gaussian-shaped distribution of analog electrical amplitudes has been obtained by optical rather than electrical means (e.g., such as discussed in the Bagini–Bucci generator [2] and some others described above). The noise source, shown in Fig. 7, is realized by use of a single mode VCSEL laser where the signal and its delayed copy have been brought to interference on an APD detector via a Michelson–Morley type interferometer, a system also known as "homodyne" detection. The electrical noise produced at a very high gain-bandwidth photodiode is the result of phase jitter of the laser. The noise voltage produced by the APD is then digitized by a fast (40 MHz) analog to digital converter (ADC) with 8-bit resolution and the numbers so obtained are further processed to obtain random bits at 20 Mbit/s. Authors show that if the delay is much longer than the laser coherence time of 1.6 ns, then the phase jitter is dominated by quantum effects which are separate from any construction detail and depend only on the laws of physics. In that regime, adding sufficient jitter leads to near-perfect Gaussian distribution via the central limit theorem, similar to the principle utilized in [77]. The authors further measure the autocorrelation function of the analog noise and show that after about 10 ns, all correlations die off. To be on the safe side, sampling of noise is made every 25 ns, and after further simple post-processing, one obtained 20 Mbit/s of random data that passed all relevant statistical tests (mentioned in Chap. 4). The similar phase self-interference principle is exploited in [59]. The advantage of the quantum phase noise over the electronic noise is that its amplitude is determined by fundamental laws and is therefore (in the ideal case) independent of technological details of the laser. In our finding though, the authors here were not considering two important points. First, the time delay introduces a "rolling" memory effect that necessarily leads to autocorrelation of the noise voltage generated by the APD, and, therefore, the bits obtained therefrom would not be random even if the phase jitter itself is random. Second, the bit generating algorithm, which most critically includes digitization of an analog quantum-random effect, is only approximate and good care has to be exercised in order to keep randomness at the desired level at all times. Even so, this is one of the very rare noise-based generators which are characterized by clean sequence of in-principle provable and well-understood physical and algorithmic processes.

More examples of noise-based true random number generators can be found in the scientific literature and in the free-access worldwide patent database Espacenet [20].

For all noise-based generators, some kind of post-processing is required. In some cases a simple *ad hoc* post-processing such as XORing several subsequent bits or von Neumann [94] de-biasing may be good enough. But if the raw bits exhibit strong correlations, simple procedures may not be sufficient to eliminate correlations among bits which can even be enhanced by simple de-biasing procedures or changed from short-range ones to long-range ones. A better approach is found in complex, often offline post-processing which however brings in its own problems (see Sect. 3.4).

There is a strong tendency among researchers to name noise-based RNGs "quantum RNG" because noise is ultimately caused by small particles governed by the laws of quantum mechanics. But noise is also a collective effect, a summation of many individual motions, and therefore its quantum property is "blurred" by a collective behavior which is somewhere between quantum and classical worlds.

Furthermore, motion of particles which generate noise (e.g., electrons in a wire) is usually intercorrelated by action of forces among them to such extent that the noise may not be completely random [4]. Note that an autocorrelation of the order of a percent may not be important when motion of electrons is considered but if so generated random numbers with the serial autocorrelation of the same order (0.01) are used in numerical simulations, the results may be completely wrong. Finally noise cannot be "restarted" in order to interrupt correlations between successive measurement/bit production.

In conclusion, a decent proof of randomness for present noise-based RNGs seems impossible because the underlying physical processes are not well isolated and do not rely on obvious or scientifically provable randomness.

3.2 Chaos RNG

Probably the most objectionable principle for physical generation of random numbers is to obtain them from repeated measurements of a physical system in chaos. The philosophical problem here is that chaos assumes the existence of an underlying order in what is seemingly random. So why would someone knowingly make use of a nonrandom system in order to generate random numbers? We are not aware of anyone so far asking or answering this question. In our opinion, authors often resort to this type of generators because of three reasons:

1. Conceptual mixing of chaos and randomness
2. (Mis)belief that hard-to-describe systems necessarily behave in random fashion
3. Robustness of certain chaotic systems to produce macroscopic levels of "noise" easily utilizable to generate random numbers essentially via noise RNG methods (as described in Sect. 3.1)

At present state-of-the-art most convenient chaotic systems for fast generation of random numbers are optical, electrical, or opto-electrical, although mechanical constructions have also been demonstrated, for example, in [52]. In this section we present several typical designs.

Lasers can be brought to chaotic fluctuation of output power by many different mechanisms. Well known are chaotic constructions involving distributed feedback lasers [44]. One very simple but extremely fast self-feedback chaotic laser system

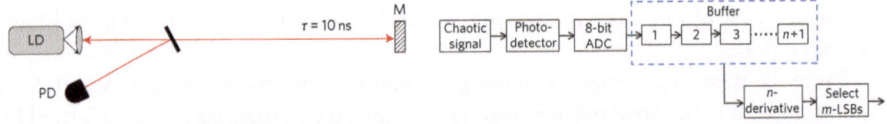

Fig. 8 The chaotically behaved intensity of a self-feedback laser is read by a photodiode (PD) whose amplitude is sampled by a fast ADC and further processed by performing a high-order differentiation, to yield a world record bit production speed of 300 Gigabit/s

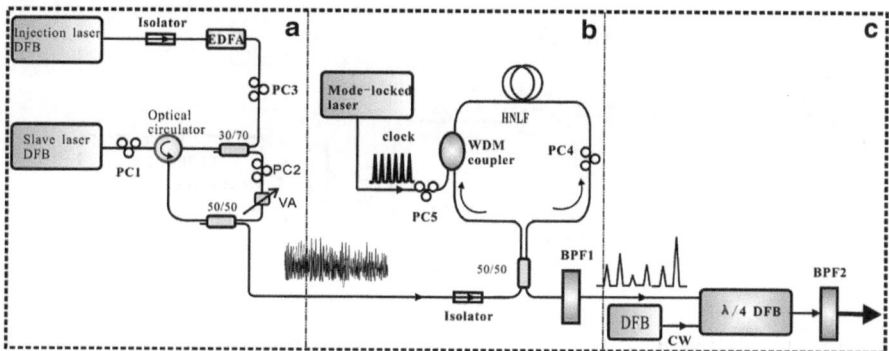

Fig. 9 All-optical laser consisting of (**a**) ultra-wide-band chaotic laser (UWB), (**b**) all-optical sampler, and (**c**) all-optical comparator

[38] is shown in Fig. 8. Again, the light of chaotically fluctuating amplitude is detected by a fast photodiode (PD) whose amplitude is sampled by a fast 8-bit ADC and further processed by performing a high-order differentiation, to yield a world record bit production speed of 300 Gigabit/s.

Lasers offer means for realizing very fast chaotic systems and are frequently used for random number generation. Due to the possibility to build tiny lasers, resonators, and various passive and active optical elements on a chip, such generators can be completely integrated and can feature a low power consumption.

A RNG shown in Fig. 9 [44] consists of an ultra-wide-band (UWB) chaotic laser (a), amplitude sampler (b), and comparator (c). Its principle of operation is a copy-paste of the Bagini–Bucci noise generator described earlier (Fig. 3) with the difference that instead of electrical noise here the light intensity of a chaotic laser is used as a source of randomness. The interesting distinguishing characteristic of this RNG is that it is "all optical," meaning that all signals and signal processing are done at the optical level, even the output numbers are in fact digital levels of light intensity: low light intensity signifies "0," while high intensity signifies "1." This is interesting for use in all-optical processing chips, and furthermore, if so needed, the output can be easily converted into an electrical signal by use of a fast photodiode and a suitable amplifier.

The UWB chaotic laser is made of two distributed feedback lasers, "master" and "slave" (Fig. 9a) with master disturbing the feedback loop of the slave in such a way as to enhance its bandwidth in a chaotic regime [97]. The output intensity is extracted from the feedback loop by means of a beam splitter and sampled by an optical sampler at a constant sampling frequency determined by the mode-locked laser (Fig. 9b). Each sampled value of light intensity is then compared to a threshold value by means of an all-optical comparator (Fig. 9c) resulting in either high output intensity ("1") or low intensity ("0"). The random bits are produced at the pace of the mode-locked laser.

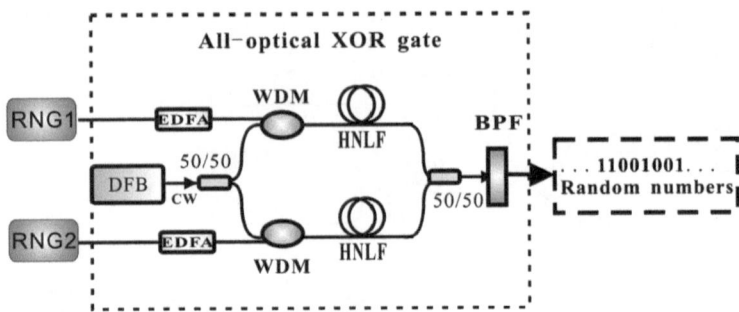

Fig. 10 All-optical XORing of two independent RNGs reduces bias and correlations among bits

Bits so obtained are biased and somewhat autocorrelated. Since they are produced at periodic times, the authors resort to a convenient bias and correlations-reducing procedure by XORing simultaneous output bits of two identical, independent RNGs, as shown in Fig. 10. The resulting random bits pass relevant statistical tests [44]. The chaotic behavior of the master-slave UWB laser has been theoretically modeled and the bandwidth of the model shown to agree with experimental data [98], in an attempt to support the claim of randomness of the above RNG. However, modeling or proving the shape and width of the noise spectrum of a source proves nothing about its randomness.

In a body of research related to chaotic RNGs, some authors claim to use system(s) in chaos without actually providing any direct evidence that the system in use for random number generation is indeed in chaos [85], some are able to demonstrate chaotic behavior, for example, by studying ballistic maps or Lyapunov exponents [62], and some even go so far as to model the chaotic behavior of the system and confirm it experimentally [44, 97, 98]. But whichever the case, chaotic RNGs have a theoretic base common to those PRNGs which operate by simulating a deterministic chaotic system, for example, and therefore in the long run became short-breathed in producing new entropy, inevitably ending in producing not more than a small fraction of 1 bit of entropy per each new generated random bit.

A general objection to the very idea of the generation of random numbers by chaos is that chaotic behavior is defined as a specific type of solution of the differential equation which, supplemented by initial conditions, describes the system. Because any such equation and data contain only a limited (small) amount of information, once that much information is extracted from the system by measurements there is no new information that can be extracted from it, and consequently all further measurements contain (asymptotically) zero new information. In particular it means that a chaotic system, in theory, can only produce a limited set of random bits and that all the rest must be perfectly or near perfectly correlated to that set. Having said that, we understand that a realistic chaotic system never behaves exactly as it would by obeying the "equation of motion" that models it because of random quantum or statistical effects which randomize the system's phase-space trajectory all the time. However, these additional effects are not the

Fig. 11 Schematic diagram of fast (*left*) and slow (*right*) FROs. Oscillation frequency is determined by internal delays and stray capacitances

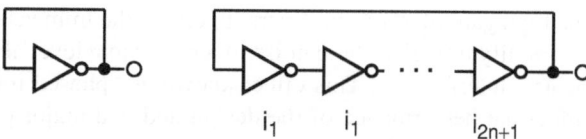

basis for a chaotic RNG (and therefore not accounted for in its definition) and also are usually too tiny or ineffective to make any significant difference in a system whose behavior is mostly determined by a macroscopically observable chaos. On the other hand, fundamental quantum randomness alone can be harnessed for the production of provable random numbers, as we will discuss in Sect. 3.4.

3.3 Free-Running Oscillator RNGs

When the output of a logical inverter circuit is fed to its input, the circuit turns to an oscillator, a so-called free-running oscillator (FRO) (Fig. 11).

An inverting gate is in practice a very high-gain inverting amplifier. Connecting its output to the input creates the Zeno paradox: if the output is in logical HIGH state, then the input will be as well and the NOT action will drive the output to go LOW. Once the output goes LOW, the NOT action will drive it to HIGH and so forth. Theoretical Boolean logic analysis will yield that the output is undetermined, but in practice due to the finite propagation delay of the NOT element, the circuit will oscillate. The peculiarity of this oscillation is that it appears in a circuit with negative feedback (180° phase shift), while in electronics theory negative feedback leads to "stabilization" rather than to oscillation. The reason for that is that by analyzing logical states we assumed infinite gain. However, since in practice gain is never infinite, it may happen that the circuit locks (stabilizes) into some voltage state between zero and one without any or with very small amplitude oscillations which are not capable of driving further logic circuits. To help oscillations, one may intentionally add some reactance in the feedback loop so as to produce phase shift different from ±180°. The same function may be provided with stray reactances. In that case, the Barkhausen criterion may be satisfied for some high-frequency pole and oscillations will appear. Due to the complex mechanism of free oscillations, their frequency is typically quite sensitive to variation of power supply voltage and temperature but these changes are slow compared to the oscillation frequency. On the other hand, the electronic noise present at the input adds to the signal fed back from the output and after being strongly amplified causes very fast, random jitter of frequency and phase of oscillations. In that sense, FRO RNG can be regarded as a special case of a noise-based generator. Since the noise of each such circuit is individual, it is reasonable to assume that the multiple oscillators even when on the same chip have different frequencies and that their mutual phases walk off randomly in time. But when multiple such oscillators are close to each other (e.g., on a single chip), they tend to synchronize through electromagnetic interaction facilitated by the high gain of FRO amplifiers. In effect, the immense gain of NOT gates required to amplify tiny electronic noise to a noticeable level also helps to pick up any other nearby interference. This effect known as "phased interlock" [54] may adversely affect the performance of the design and is a major problem inherent with FROs. Interlocked rings have waveforms that share (nearly) the same phase and this will lead to (near) pseudorandom operation. The same effect of high gain makes FROs vulnerable to attacks with external electromagnetic radiation.

The basic principle of random number generation with FROs is that output of a fast FRO (which can be either logical 0 or logical 1) is sampled by a slow FRO. This is an equivalent to abrupt stopping of a quickly turning wheel of fortune. Because the wheel spins so "fast," it appears stopped at a "random" position. In case of two FROs, it is important that the relative phase jitter, between the fast and the slow FRO, is both random and large enough. Clearly, if there is no relative phase jitter the output will provide repetitive binary patterns. If the jitter is random but small, deviation from the repetitive pattern will be small as well leading to near pseudorandom behavior. If FROs synchronize or at least partially synchronize, a pattern with stochastic excursion (noise) would appear. Apart from that, another very important problem with FRO RNGs is that the output amplitude of an FRO depends on the details of the stray reactances and delays in the circuit. As explained above, for a particular circuit it may well happen that the output amplitude of an FRO is too small to drive the logic circuitry or that the FRO locks in some state and stops oscillating. Schmitt action at the input of the first inverter (Fig. 11) can help minimize this problem but at the expense of lowering the oscillation frequency and complicating the fabrication.

In spite of all these problems, current security standards [65] practically dictate the use of RNGs based on free FROs. The NIST standard FIPS140-2 [19] says: "There are no FIPS Approved nondeterministic random number generators." Consequently, the FRO approach currently is used in 3rd- and 4th-generation FPGA, CPLD, and ASIC hardware for various cryptographic purposes. One real-life example that illustrates well the combinatorial cuisine typically needed to obtain a decent RNG is the entropy source for PadLock "quantum" RNG implemented in VIA C3 processors [88–91]. It consists of four FROs, 3 fast (450–810 MHz) and

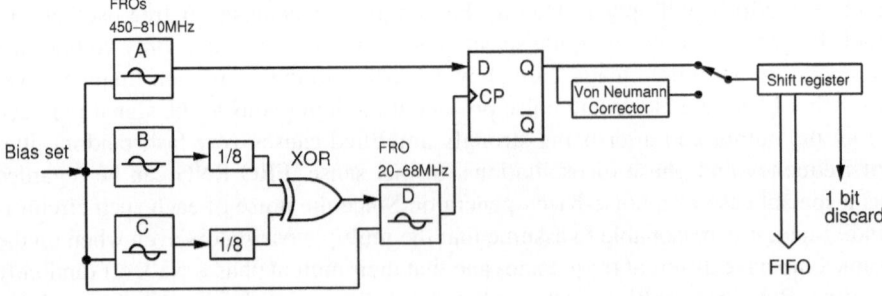

Fig. 12 VIA C3 PadLock random number generator samples fast FRO (A) by slow FRO (D)

1 slow (20–68 MHz). Wide tolerance on the frequencies already shows problems that we mentioned before: it is very hard to control the parameters of FROs during fabrication. In this topology fast FRO (A) is sampled by a slow FRO (D) as discovered in the patent application [78]. At least one of the two FROs must be of good randomness, and since it is easier to achieve with the slower one, VIA went for that option. The slow generator is made of FROs B, C, and D. First, B and C are slowed down by 1/8 dividers and their XORed outputs are used to disturb slow FRO D (which is the only one featuring digital input). Resulting bits appear at the output Q of the D-type flip-flop in synchronization with pulses from the FRO D. Optionally, the output is filtered through a von Neumann corrector [94] which cuts the bit production rate roughly by a factor of 4 (see description in Chap. 4). Looking at this schematic, it is clear that it is impossible to arrive to a proof of its randomness. According to VIA [88], the analog bias voltage injected into this otherwise digital circuitry "may (or may not!) improve the statistical characteristics of the random bits." The bottom line is that the random numbers are still of low quality and in order to pass tests must be corrected (Sect. 3) by a full-blown secure hash algorithm SHA-1 which is hardwired into the logic circuits on the same chip [88].

Because the digital logic chip infrastructure is unsuitable for realization of a quantum RNG (Sect. 3.4), an FRO approach seems to be a reasonable viable alternative. However, a caveat with FROs is that the semiconductor industry is making an enormous effort to make the electronics noise as small as possible and it generally goes down with newer versions of a chip. Consequently the effect of jitter can be very small and cause the FRO-based RNG to operate in nearly PRNG regime. Therefore the implementation details of an FRO-based RNG most often have to be tailored for each specific type or generation, and the technology of a programmable/reconfigurable or ASIC chip and its uniformity of operation cannot be guaranteed from batch to batch.

A partial solution to the above-mentioned problems has been recently found in a novel synergistic combination of a linear feedback shift register (LFSR) [27] and FRO, called the Fibonacci ring oscillator (FIRO) and Galois ring oscillator (GARO) [16]. The idea is to have a seeded LFSR-like PRNG which is realized as a clocked FRO. Such true random number generators do receive an initial state (seed), but although the seed sets the initial state, two identical generators with identical seed would diverge in time as they are under the influence of (at least partially) individual noises. Figure 13 shows the schematic of GARO and FIRO.

Still, even with this interesting and innovative principle, the problem is cross-platform non-portability of the design and the requirement of sufficiently large noise for the scheme to work in a reasonably random (far from pseudorandom) regime. Furthermore, the authors warn that the design must be done most carefully in order to minimize interlocking with the system clock and other logic circuits in the chip, including nearby FROs. Therefore they experimented with spatial placement of FROs in the chip. They also conclude that randomness of either of the two generator families by itself is not perfect and could be "enhanced" by XORing two independent generators, most favorably one GARO and one FIRO.

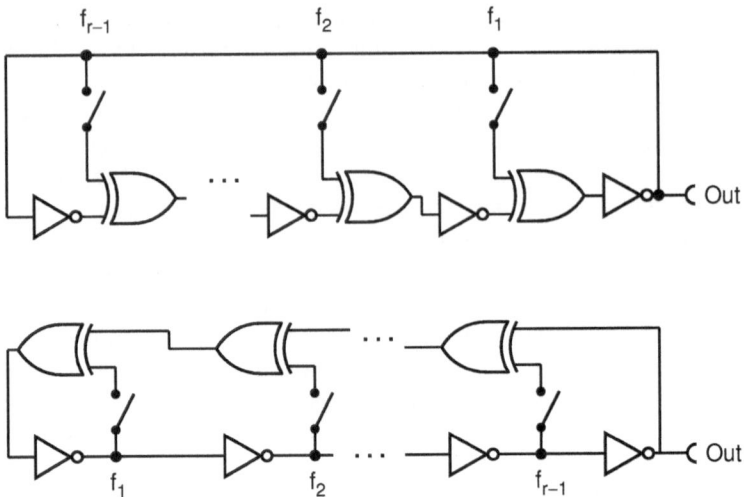

Fig. 13 Galois ring oscillator (*up*) and Fibonacci ring oscillator (*down*). Number of stages defines order (r), while switches f_i define coefficients of the feedback polynomial

More examples on FRO pre- and post-processing gymnastics, including XORing multiple generators, combinations with LFSRs, etc., can be found in [81]. The complexity of post-processing procedures required to pass the statistical tests with FRO-based RNGs is often such that any randomness proof is impossible, but even more interestingly the authors almost never seem to be aware that a proof is needed. A rare exemption in that respect is the work of Sunar et al. [82] where a theoretical model of an FRO-based RNG has been presented, analyzed, and proven but later criticized in [101] as nonrealistic. We however found the whole proof unsatisfactory because it is based on McNeill's model of FRO which simply postulates that free oscillations occur as a nonstationary random process without actually linking the postulate to reality, for example, by means of laws of physics. An excellent further reading and summary of problems and cuisine used to minimize them is found in [101]. Further reading on FRO-based RNGs is given in [81].

In conclusion, FRO-based RNGs are low-cost, low-entropy solutions whose only good side is the fact that they can be easily implemented in conventional programmable or reconfigurable logical chips which are used in various cybersecurity solutions, but they do not offer either very good or provable randomness.

3.4 Quantum RNGs

What is a quantum random number generator? Since we live in a world governed by the laws of quantum physics, any true random number generator (e.g., a rolling dice, or a flipping coin) may be named "quantum." However, we want to reserve this name for only those generators which utilize a single intrinsically random quantum effect (realized as close as possible to its theoretical idealization) measured over and over again in order to produce random bits in such a way that between any two sets of measurements used to deduce random bits, the system is reset to the same initial conditions. It may seem strange that such a physical setup (generator) is even possible, namely, that starting from exactly the same initial conditions and measured in exactly the same way, it gives different results, but quantum physics allows it. In this section we describe and explain multiple examples described in scientific articles and patents.

It turns out that some things in nature come in the smallest amounts known as quanta. For example, the electron carries the smallest quantity of charge, e. Similarly, there is the smallest quantity of information, called *qubit*. A single quantum of light (photon) can be used as a carrier of one qubit, but there are many other examples and they are not limited only to elementary particles. Qubit can be thought of as a linear combination of two bit values: 0 and 1. When a certain type of measurement is performed on a qubit, it will "project" to either pure 0 or pure 1 state in the basis in which the measurement has been carried out. Very often photons are used in QRNGs because they are easy to create, manipulate, and detect. To illustrate this let us consider circularly polarized photon entering a polarizing beam splitter (PBS) (Fig. 14). The PBS decomposes polarization of incident light and sends the linear horizontal component to one output port and the linear vertical component to the other port.

Thus, a circularly polarized photon has equal content of both linear polarizations, but since it cannot be split in half, it has exactly 50 % chance to exit either port. If now we label one of the ports as "0" and the other as "1," we immediately get a theoretically perfect RNG whose randomness is guaranteed by the laws of quantum physics. Note that the system being "measured" is always the same yet it always gives a new random result. This is completely different from chaotic and noisy generators where in order to get a different result systems must change.

Quantum RNGs based on this (or other principles) can be made pretty good, and the imperfections of any type (multiphoton emission, non-perfect circular polarization, beam splitter port axis misalignment, detector dead time, afterpulsing and memory effects, etc.) can be measured independently of the bit generation

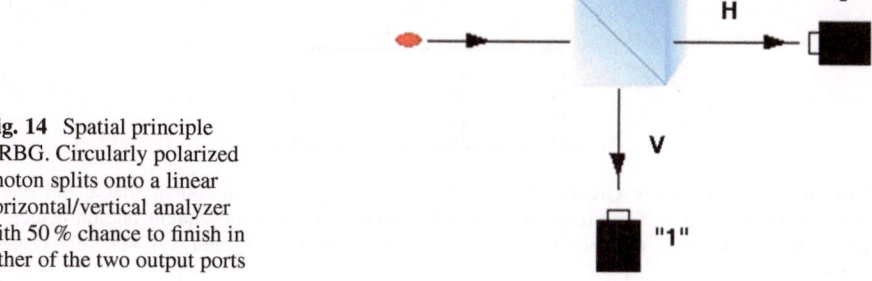

Fig. 14 Spatial principle QRBG. Circularly polarized photon splits onto a linear horizontal/vertical analyzer with 50 % chance to finish in either of the two output ports

process so their effect on random numbers can be estimated with precision and dealt with in post-processing (see Sect. 3.5). This method is a basis for a commercial generator [33].

The main problem in practical realization of the beam splitting RNG is that it requires two detectors. Their initial differences and subsequent walk-off with time due to aging and/or temperature effects will have an immediate impact on the quality of random numbers. For example, if the photon detection efficiencies of detectors are not perfectly equal, or if the beam splitter is not perfectly 50/50 %, then the probability of ones will not be equal to the probability of zeros. This problem can be minimized by use of a beam splitting scheme which utilizes only one photon detector [75] shown in Fig. 15, but still the beam splitting ratio must be precisely adjusted mechanically. Leftover problems arise from detector dead time and afterpulsing leading to correlations which are impossible to eliminate completely but can be reduced below any desired level by targeted post-processing.

The beam splitter RNG is an example of a "spatial principle" in which the value of the random bit, 0 or 1, is determined by the place at which the photon ends up. A complementary "temporal principle" uses time information of random photon emission, for example, in direct atomic (or quantum dot) relaxation, from well-saturated lasers, etc.

A simple time interval method shown in Fig. 16, which is particularly immune to hardware imperfections, has been proposed in [80]. It uses time rather than

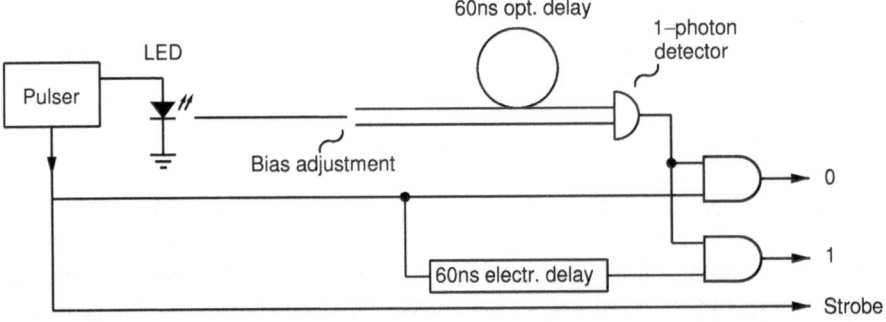

Fig. 15 Optical quantum random number generator based on beam splitting which makes use of only one photon detector in order to avoid bias fluctuation with aging and initial tolerances

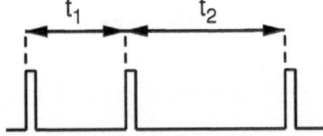

Fig. 16 Timing principle QRBG. Photons from a single photon Poissonian source fall onto a single photon detector. Time intervals t_1 and t_2 spanned by three subsequent photon detections are compared: if $t_1 > t_2$ then produce "0," if $t_2 > t_1$ then produce "1," and if $t_1 = t_2$ then produce nothing (skip)

Light Source: Photomultiplier   Amplifier   Window Comparator
LED, LD,...   tube

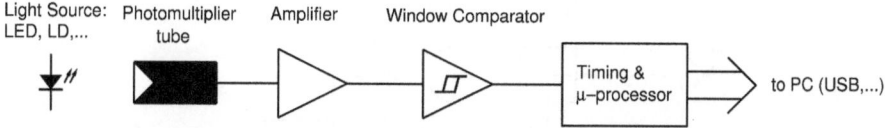

Fig. 17 A general processing scheme of the temporal principle QRBG. Time-random photons fall onto the single photon detector consisting of a photomultiplier, an amplifier, and a comparator, such that each detected photon generates one logical pulse. Pulses are then processed according to the desired bit extraction principle and transmitted to a computer

space information contained in a random event generator (REG). In [80] photon emission and detection processes are used for the first time instead of much slower (and more dangerous!) radioactive decay [22, 26]. The bit production principle is as follows. Time intervals t_1 and t_2 spanned by three subsequent photon detections are compared: if $t_1 > t_2$ then produce "0," if $t_2 > t_1$ then produce "1," and if $t_1 = t_2$ then produce nothing (skip).

The schematic of the physical setup is shown in Fig. 17. Because only one photon detector is used, both bias and correlations are suppressed to almost undetectable levels, yet there is nothing to be adjusted (unlike with the beam splitting principle).

The problem with this method is how the time intervals are measured. The crucial improvement made in [80] is the notion that clock measuring time intervals (t_i) must be started in synchronization with beginning of each interval; otherwise, the method would produce correlated bits even if fed by perfectly random events. This was not understood in previous works and patents [22] which consequently must have yielded correlated output, but this was not detected at the time because the clock frequency ($\approx 10\,\text{MHz}$) was much higher than the source mean frequency ($\approx 10\,\text{kHz}$) in which case correlations are small. It can be shown that this method not only performs well at low ratio between clock and RPG frequencies but that it also cancels out almost all imperfections: intensity change of the source, efficiency change, dead time, and afterpulsing of the photon detector. It is also highly immune to actual distribution of random interval times, as long as events are independent of each other. Furthermore, random bit production is self-clocked so if either source or detector dies, there will be no bits at the output. This generator was the first one found to pass all known tests including "usual" t-statistical tests [47, 67, 68, 96] and some undisclosed algorithmic randomness tests [31].

A mixture of beam splitter and temporal principle is described in [35]. An unpolarized photon stream from the light source (LED) is passed through a polarizer reaching a polarizing beam splitter (NPBS), much the same as in the aforementioned beam splitter RNG (Fig. 14). With careful adjustment of the relative angle between polarizer and NPBS axis (ideally 45°), detectors D1 and D2 should produce random, mutually uncorrelated pulses of equal frequency (however adjustment of the polarizer angle is an insurmountable task, as explained for RNG in Fig. 3). While pulses from D1 reset (input R) the RS-type flip-flop setting the output to LOW state, D2 sets (input S) the flip-flop to HIGH state. The output of the said flip-flop is sampled at periodic times in order to generate random bits (Fig. 18).

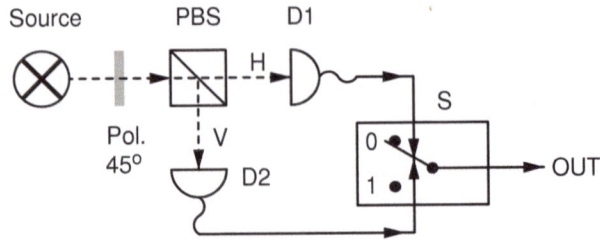

Fig. 18 Optical quantum random number generator based on beam splitting and periodic sampling principles

Fig. 19 Optical quantum random number generator based on near-exponential statistics of photon time detection. The main detector imperfections, the dead time and pileup, have been found to work in favor of smaller bias and serial autocorrelation which have been found to be as small as $2 \cdot 10^{-5}$ without post-processing

Being a combination of beam splitting and sampling principles, this construction inherits the worst of both:

1. Bias is unstable (sensitive to temperature variations) and only mechanically adjustable.
2. Correlations due to the finite sampling period as discussed in noise generators, and all that even if a perfectly random source of photons is assumed.

A commercial QRNG of Fürst et al. [23, 60] utilizing only the temporal principle is shown in Fig. 19. The data-taking schematic is equivalent to the general scheme given in Fig. 17 with the light source being a low-intensity operated LED weakly coupled to a photomultiplier tube. Low coupling ensured low photon sampling rate on the order of 10^{-8} which suppresses any eventual photon correlations far beyond the detectable level. The bit extraction method is implemented in an FPGA reconfigurable chip and is as follows. The number of detected photons is counted in intervals of a constant time yielding a Poissonian statistics. An even number of events within an interval is interpreted as "1" and odd as "0." The authors note that due to the nonsymmetric shape of the Poissonian distribution, the probability of ones is not equal to the probability of zeros. However, due to the two imperfections in the photon detector (nonzero dead time and dependence of dead time with the detection frequency), the resulting distribution is not Poissonian but more bell shaped, thus favorably leading to a bias that quickly tends to zero as the counting interval length rises. The authors show and compare experimental and theoretical results for modeled bias; however they do not model or prove anything about correlations. Instead, correlations are simply evaluated from generated bits using a linear autocorrelation coefficient. Theoretically, bias tends to zero as detection frequency goes to infinity. Empirically, the preferred operating condition is close to as high as possible a detection frequency but a bit smaller due to rising problems in the photon detector. But in the same limit, it is to be expected that fluctuating bias produces an increasing level of complex short-range correlations among bits—which however has not been mathematically modeled and/or brought into connection with the imperfections of the setup. The problem with this approach is that it fails to describe a theoretical model of an RNG that gives perfect random numbers based on a (nearly) ideally random quantum effect (e.g., low-intensity emission from LED) and assuming ideal apparatus. Consequently it fails to clearly prove randomness and to model deviation from perfect randomness introduced by implementation-related imperfections. Nevertheless, this generator has a practical value because it apparently passes all relevant statistical tests. It is however to be understood that an acceptable randomness proof cannot be obtained by passing any number of randomness tests (as will be discussed in Sect. 3.5).

Yet another commercial quantum RNG which utilizes photon arrival time information has been presented by Picoquant [57, 95]. Here the complete chain of reasoning required for a convincing randomness proof has been at least attempted and, according to the authors, successfully established. As in the previous example, a random event source of the type shown in Fig. 17 is made utilizing essentially the same technique as in [23] (LED + photomultiplier tube). The specific difference of this construction with respect to previously described ones which use high-speed photon detection and produce ≤ 1 bit per detection [23, 35, 75, 80] is that the random detections are made at a relatively low mean frequency of 12.5 MHz, thus operating in a regime far from dead time and pileup effects producing a highly precise exponential distribution of time intervals (Fig. 20 left). The time intervals

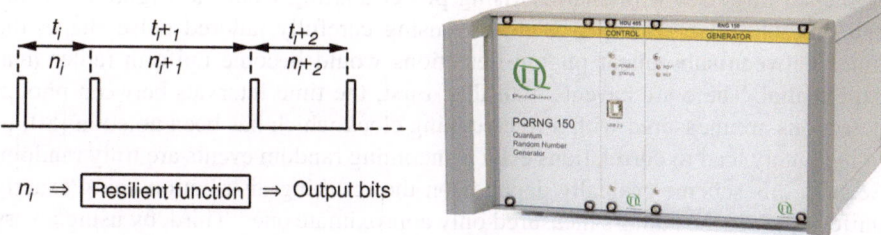

Fig. 20 Optical quantum random number generator based on highly precise exponential distribution of photon detection times: schematic (*left*) product photo (*right*). The times between subsequent random events are measured by a very precise timing hardware resulting in integer numbers that represent the time. These numbers are then used to extract much more than 1 bit per detected photon resulting in 150 Mbit/s overall average bit production rate obtained after postprocessing with resilient functions (see Sect. 3.5)

$t_1, t_2, t_3, \ldots$ are measured by a nanosecond precision, and quasi-exponentially distributed integer numbers so obtained are used to generate on average ≈ 14 random bits per each detected photon yielding ≈ 160 million raw random bits per second. The imperfections both in the extraction method and in hardware (timers, detectors, light source) are modeled resulting in a convincing lower bound on the average per-bit entropy of the raw bits. The average entropy is then improved by compression of the raw stream by resilient functions (see Sect. 3.5) to the level theoretically indistinguishable from true randomness even for bit strings of unrealistic length. The weakest link, in our opinion, is this last post-processing step because it is not clearly proven that resilient functions are effective against the specific type of imperfections present in raw bits, that is, that bounds on post-processed bits hold. However, raw bits are already very close to randomness, and further post-processing by resilient functions clearly improves the pass rate of statistical tests indicating that the resulting bits are very close to true randomness. Indeed, post-processed bits at an average speed of 150 Mbit/s pass all relevant statistical tests as well as some undisclosed statistical and algorithmic tests performed by the University of Twente research group [31]. Still, caution is maybe in order when resilient functions are used because some researchers [81] point out that resilient functions appear to be limited in their ability to eliminate the effects of active adversaries on the output bits. A similar principle but by digitization of an analog quantum amplitude is described in [1].

An example of a very fast (110 Mbit/s) generator of similar construction and philosophy as the previous one has been presented in [99, 100, Figure 21]. In the first article, faint continuous light (from an LED) shines upon a photon detector which produces random events (detections) quite similar to the general system shown in Fig. 17. Times between subsequent events are measured with a high resolution clock in order to obtain integer numbers that approximately follow exponential distribution. These numbers presented in binary form do not yield random bits because they have been drawn from a highly nonuniform distribution (namely, exponential cut-off near zero at the dead time). In order to obtain more uniformly distributed numbers, in the subsequent article the light from the source (LED) is shaped in pulses with sharply rising power starting from the beginning to the end of each pulse. The idea is that by using carefully tailored pulse shape, the times between subsequent photon detections would become uniform rather than exponential. There are caveats with this. First, the time intervals between photon detections are measured with a free-running clock which has been noted in [80] to immediately lead to correlations even if incoming random events are truly random. Second, this scheme critically depends on the resulting distribution being exactly uniform, while the authors measured only approximate ones. Third, by using a very high-speed clock, the authors try to "squeeze out" as many random bits as they can from a single photon event (≈ 20 bits per detected photon) which generally leads to great amplification of hardware imperfections, thus leading to pretty bad raw random bits, as indeed was found. Fourth, the approximate results relating to the variable pulse power are both fundamental (i.e., pulse power should tend to infinity at the end of pulse proportional to $1/(t - t_0)$ where t_0 is the pulse length)

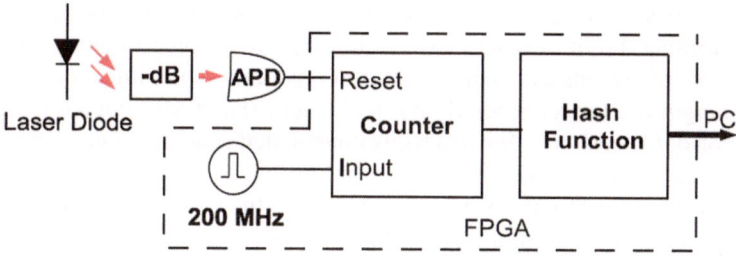

Fig. 21 Optical quantum random number generator based on near-uniform photon arrival times from a specially shaped optical pulses

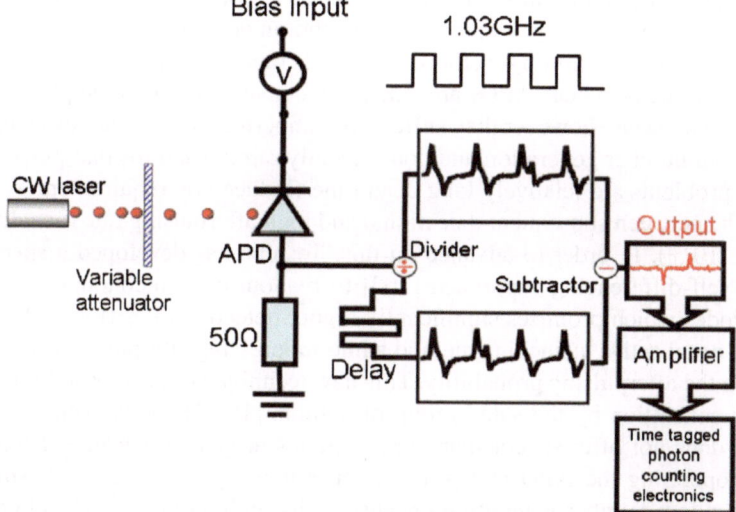

Fig. 22 Optical quantum random number generator based on periodically gated, "self-differencing" approach operated avalanche photodiode (APD). The power of DFB cw laser (1,550 nm) is adjusted (by means of variable attenuator) such that the strength of the electromagnetic field falling on the surface of the APD causes roughly 0.004 avalanche detections per gate, resulting in 4.01 MHz of random bits

and practical (pulse shape is achieved by an analog, only partially precise circuit, thus not allowing us to properly conduct proof of randomness. The authors also note that this circuit produces strong electrical disturbances in nearby circuitry which, in our view, makes it unsuitable for miniaturization to a chip level. And finally, the theoretical basis for exponential time-arrival distribution is drawn out of a steady field assumption, whereas here the strength of the light electromagnetic field is wildly varying so even theoretical grounds for this generator are not clean (Figs. 21 and 22).

This generator belongs to a broad niche of RNG constructs whose general philosophy is to produce partially random data and then filter it through a pseudorandom hash function (such as SHA-256 used in this example) in the hope of improving the randomness (see Sect. 3.5). We believe this is a very problematic approach and here is our reasoning. Proof of randomness in this case relies on estimating the entropy of the source of raw bits and on the process of randomness amplification by hashing. The hashing procedure is generally not foolproof [3] and does not allow just blind application of the hash function to a badly constructed generator. Let us imagine, for example, that raw RNG source produces some sequences more often than the others (which indeed is the case if it is nonrandom). Then the hash of these sequences (the hash function being deterministic) would also produce some sequences more often than the others meaning that even the "corrected" bits would not be random. A nice confirmation of this comes from this very example: even after hashing, the produced bits are not completely random and fail some statistical tests.

We saw that photon emission and photon detection techniques are often used in quantum random number generators. The photon detection rate of current single photon detectors is a limiting factor in achievable random bit product rate especially for semiconductor APDs. APDs are small and convenient for single photon detection on a chip scale; however they suffer from imperfections that are especially bad for random number generation and consequently rarely used for that purpose. The biggest problems are relatively long dead time (induced by requirement to quench avalanche between subsequent detections) and high afterpulsing rate (usually in the range 1–10%). In order to advance on this, Toshiba has developed a special, so-called "self-differencing" approach [102] to readout of semiconductor avalanche photodiodes which promises significantly higher detection rates (lower dead time) than the usual active quenching method while suppressing afterpulses by effectively squaring the afterpulsing probability. This new technique has been used for random number generation by the same group of authors [18]. Namely, even though this method does not offer spectacular improvements in general because it inherently prefers operating the APD at low detection efficiency, it is very well suited for use in random number generation because of its high gating speed and complete irrelevance of the photon detection efficiency for that application.

A distributed feedback (DFB) laser in continuous wave (cw) mode is used as the light source. The power of the DFB laser (1,550 nm) is adjusted (by means of variable attenuator) such that the strength of the electromagnetic field reaching the surface of the APD causes roughly 0.004 avalanche detections per gate. When detection occurs, a new random bit is generated and its value is "0" if it occurred on an even gate or "1" if on an odd gate. Taking into account the detection efficiency of 0.004, this method yields 4.01 MHz of random bits. This bit generating process is intrinsically biasless (probabilities of zeros and ones are equal) but (what is not noted by the authors of this article) there is an intrinsic negative autocorrelation which rises with detection efficiency. Namely, in the limiting case of efficiency 1 (one detection per gate), there would always be a "1" after "0" and vice versa, thus producing a completely deterministic sequence $01010101\cdots$ which has autocorrelation equal to -1. Even though the authors claim that this method of generating random numbers could, in principle, be extended to much higher rates by using a higher laser power and a detection rate of up to 100 MHz (efficiency of 0.100), it is clear that at that point the autocorrelation would amount to approximately -0.1 and the bits would not pass any randomness test.

There are numerous other variations of space and time principles that can be found in the scientific and patent literature.

In conclusion, the most distinctive characteristic of a quantum approach to random number generation is that, at least in principle, it makes it possible to establish a simple relation between the randomness of numbers, the exploited physical process, and the implementation imperfections, thus offering a possibility for scientific proof of randomness. Careful practical realizations come sufficiently close to theoretical idealization and allow for an independent assessment of implementation imperfections, the effects of which can, if required, be dealt with by information-theoretic post-processing (see Sect. 3.5). On top of that, quantum random detection processes exist that are inherently highly insensitive to electromagnetic radiation (e.g., avalanche amplification in semiconductor photodiodes), thus offering immunity to side-channel manipulation by external fields. Because of all said, quantum RNGs are the best choice for true random number generation for cryptography and other applications which critically require true random numbers. The most significant drawback of the present solutions is that they make use of bulky physical objects and therefore cannot be miniaturized to the chip level using present technologies. Furthermore, due to the frequent use of photon detectors, QRNGs are typically very expensive and much slower than software PRNGs. Fortunately, the nascent science and technology of optical chips offers a promising avenue for fast, miniature, and affordable quantum RNGs, and significant advances can be expected in this exciting field in the near future.

3.5 Post-processing

True random number generators can never be made perfect and therefore some post-processing is usually required. There exist a plethora of post-processing algorithms whose purpose is to eliminate imperfections present in "raw" random numbers produced by physical generators. A good review of post-processing methods is given in [81]. Here we will only categorize and shortly describe the main principles.

The general idea of post-processing (Fig. 23) is to sacrifice a certain percentage of bits in order to arrive at a smaller but more random set. There are basically four techniques:

1. Ad hoc simple correctors
2. Whitening with cryptographic hash functions

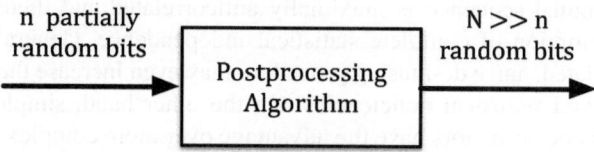

Fig. 23 General schematic of random number post-processing

3. Extractor algorithms [71, 72]
4. Resilient functions [10, 42, 43, 69]

Although there is a "gray zone" of what part of random number production belongs to bit extraction method and which to post-processing, the bit extraction is usually a first and very simple step which converts physical measurement of an analog or digital signal into the "raw" digital random binary number (such as digitizing analog noise via a threshold comparator shown in Fig. 2), whereas post-processing is a more complex process designed to reduce or completely remove imperfections that are necessarily present after the first step. While bit extraction is always made in hardware, post-processing algorithms are usually so complicated that they can only be executed by a computer (or a microcontroller or FPGA) although the most valuable post-processing techniques are those simple enough to be suitable for direct implementation in hardware.

Generally, post-processing takes a lot of resources and blurs. In our opinion, a good true RNG should be post-processing-free or use minimal ad hoc post-processing. Most popular post-processing techniques can be categorized into four families as described in the following.

3.5.1 Ad Hoc Simple Correctors

Ad hoc corrector examples are XORing two or more neighboring bits from the same RNG [72], omitting bits (decimator), feeding an LFSR with imperfect random numbers [84], Latin square bit reshuffling [47], von Neumann [94] and Peres [56] de-biasing, XORing two or more RNGs that work in parallel [14, 44], etc.

It is important to note that ad hoc, naïve processing can lead to unexpected problems. For example, it is usually considered a good idea to apply a von Neumann de-biasing scheme [94] in order to completely remove any bias from the sequence of bits. The scheme works as follows. The biased bit sequence is cut into a sequence of non-overlapping pairs of bits. Pairs 11 and 00 are discarded, 01 is converted to "0," and 10 is converted into "1." While it is tempting to think that the probability of occurrence of 10 is equal to the probability of 01 (and therefore the resulting sequence has no bias), it is often overlooked that this is true only if the bits are completely independent (no correlations). The following extreme example illustrates how miserably von Neumann's procedure can fail. Let us consider the sequence: $101010101010\cdots$. It obviously readily has no bias. After application of von Neumann de-biasing, the sequence reads: $111111\cdots$ which is a maximally biased and maximally correlated sequence. The reason for this unexpected result is that the original sequence is maximally anticorrelated and therefore quite far from the assumption of complete statistical independence. Generally, if the raw string is correlated, naïve de-biasing procedure may even increase the bias or create other unexpected statistical deficiencies. On the other hand, simple and easy-to-understand ad hoc correctors have the advantage over more complex procedures, in that they are easier to include in a randomness proof.

3.5.2 Cryptographic One-Way (Hash) Functions

A one-way hash function is a mathematical function whose domain is a whole set of integer numbers and whose output is a binary number of exactly N bits, where N usually is in the range from 128 to 512. Hash functions are characterized by two requirements:

- Given an output value, there is no faster way to find a corresponding input than by random guessing (i.e., a hash function is "one way").
- The probability of two different inputs yielding the same output is less than or equal to $1/2N$.

One of the most popular post-processing techniques is "whitening" of output of a TRNG by means of a cryptographic hash function, such as MD5, SHA-1, SHA-2, SHA-256, and SHA-512. Many authors believe that a bad RNG that does not pass statistical tests, and runs through a "cryptographic" hash compression procedure would magically become very good, without actually demonstrating any theoretical understanding on why this should be the case. Indeed a very interesting example given in [99] demonstrates that hashing a bad generator can fail to enhance randomness enough to pass statistical tests.

From a performance perspective, implementing a hash function in hardware chips is pretty resource demanding, so in most cases hashing is done on a computer; two exemptions to this rule are the aforementioned Intel's RNG in Fig. 5 and VIA C3 in Fig. 12, which make use of SHA-1 hardwired right next to the RNG on the same chip. Regarding the provability of randomness of the hashed output, even though interesting results on privacy (and randomness) amplification have been theoretically exercised for Wegman's Universal Hash Function(s) [7], in the case of real-life, black-box hash functions (which probably contain unknown statistical or security weaknesses), it is hard to perform a convincing proof of randomness. For example, a hash function may contain statistical problems like some output strings being more probable than others which would then be inherited by the output bits even if the function is fed by perfect random numbers. On top of that, hash functions are usually used at the end of the post-processing leaving a bitter aftertaste in the mouth that physically generated random numbers actually exit out of a deterministic, complex, black-box piece of software which has not been specifically designed for the purpose.

3.5.3 Extractor Functions

A more scrutinized approach to randomness healing is offered by the young theory of extractors [71]. A randomness extractor is an algorithm that converts a long weakly random sequence into a shorter sequence with almost perfect randomness. For some randomness sources, provable extractors exist but no single randomness extractor currently exists that has been proven to work when applied blindly to any type of a high-entropy source. The problem with extractor algorithms is that they require a memory buffer and a lot of CPU which complicates the hardware and slows down the overall output bit rate.

Extractor functions for post-processing of true random number generators were proposed by Barak et al. [3]. The initial purpose was to achieve designs robust against changes in the physical generators due to, for example, aging, temperature changes, or attacks. Extractor functions are stateless functions with quantifiable properties originally developed as a tool for complexity theory. The aforementioned group of authors has developed a mathematical model to capture an adversary's influence on the randomness source and give an explicit construction based on universal hash functions which is proven for its output properties even if nonlocal correlations exist in the input source.

More on the theory and practice of extractors can be found in [72].

3.5.4 Resilient Functions

Yet another approach to enhancing randomness by filtering through some deterministic process is the use of resilient functions that were introduced by Sunar et al. [82] as the post-processing step for an FRO RNG design. The idea is, according to the authors, to "filter out any deterministic bits" from the raw string in the environment where some bits may be under the control of an attacker and that bits are then considered "deterministic." The authors of [82] study the degree of resilience of the procedure against active adversaries (therefrom comes the name of these functions). In short, an (n, m, k)-resilient function is a function $f : F^n \to F_m$ such that every possible output m-tuple is equally likely to occur when the values of k input bits are fixed and the remaining $n - k$ bits are each chosen at random. The elements of F are binary values 0 and 1. The important distinguishing characteristic of resilient functions is that they have been constructed specifically to nullify the attacks on a (certain percentage of) random bits—a point of high importance in cryptographic applications of random numbers (see Sects. 4–6).

More on the theory and practice resilient functions can be found in [10, 42, 43, 69, 82].

4 Randomness Evaluation (Testing)

The most important notion about statistical testing is the following: if a generator passes all known statistical tests, this does not prove that it is random—it only means that it passes all currently known randomness tests. Tomorrow it can fail some new test or it already fails in the way known only to its constructors.

Most randomness tests check one or more statistical properties of long sequences of random numbers, for example, bias, serial autocorrelation, etc. Some compilations of tests are more oriented toward problems in PRNGs (e.g., DIEHARD [47]), some more to true RNGs (e.g., ENT [96]), while some are of general nature (e.g., Universal Test [50], NIST STS [68]). The unfortunate fact is that there is an infinite number of statistical properties which truly random numbers must satisfy. Tests themselves are not perfect: some contain errors discovered later [44, 65] or constants of questionable precision obtained by simulation using "trusted" RNGs such as combination of white noise and "black noise" [47].

Running a comprehensive set of tests takes many CPU hours: to test 1E9 bits with NIST STS, it takes about 6 h on the fastest single core CPU, while to produce that many bits with a commercial QRNG, it takes between 7 and 250 s.

Randomness tests are very time consuming—it takes much shorter time to generate numbers than to test them. Nevertheless, randomness testing is important for constructors of RNGs. Therefore in some cases where one can reasonably expect only certain type of imperfections (especially for quantum RNGs), one will tend to use only special tests sensitive to these particular imperfections in order to arrive at more efficient testing.

5   Random Numbers in Quantum Cryptography

Quantum cryptography is a protocol of public agreement of a symmetric cryptographic key, meaning if two parties A and B possess a small common secret key, then using this protocol they will be able to establish a common secret key of any length. This cryptographic function is also known as "secret key growing." The ultimate goal of establishing a long secret key is to use it as a one-time pad and thus obtain transfer of data in absolute secrecy. There are several mathematically identical QC protocols. The first one, named after its creators as BB84, appeared in 1984 and was experimentally realized in 1991 [5].

In the BB84 scenario, Alice and Bob are connected via two different channels: the quantum channel (usually well-shielded optic fiber) capable of conducting single photons of light and an unsecured "classical" channel such as a telephone line, radio link, or the Internet.

Here is the simplified schematic of how the protocol works: Alice can prepare photons in different polarization states. In order to establish a secret key, Alice sends to Bob a sequence of random numbers encoded in photon polarizations as follows: "1" is equiprobably encoded either as linear vertical (LV) or left circular (LC) polarization, while "0" is equiprobably encoded either as linear horizontal (LH) or right circular (RC). Bob has two polarization analyzers: one which can correctly measure linear polarizations (L) and the other which can correctly measure circular polarizations (C). Alice chooses one polarization at random, prepares the photon, and sends it to Bob. Bob chooses one of the two analyzers at random and measures with it the photon received from Alice. If, by chance, Bob has chosen the right polarizer, he will receive 0 or 1 as sent by Alice. If Bob has chosen the wrong polarizer, he will receive 0 or 1 with equal probability regardless of what Alice has sent. So after receiving a photon from Alice, Bob announces (over an authenticated but not secret public channel) which polarizer he has just used (L or C). Note that this says nothing to a potential eavesdropper about the value of the bit Bob has got. Alice responds with "Keep it" or "Trash it." So bit by bit the two of them are building their secret key. The laws of quantum mechanics prevent qubits from being faithfully copied, so an eavesdropper can obtain only limited information about Alice's and Bob's string and furthermore eavesdropping can be detected by Alice and Bob.

It is straightforward to see that the whole protocol would be completely insecure if only the eavesdropper could calculate (or predict) either Alice's random numbers or Bob's random numbers or both. From the analysis of the secret key rate presented in [6], it is obvious that any predictability of random numbers by the eavesdropper would leak relevant information to him, thus diminishing the effective key rate. It is intriguing (and obvious) that in the case that the eavesdropper could calculate the numbers exactly, the cryptographic potential of the BB84 protocol would be zero. This example shows that the local RNGs assumed in BB84 are essential for its security and should not be taken for granted.

Apart from what has been described above, the BB84 protocol has two more subprotocols. Namely, due to the quantum incoherence, losses in the quantum channel or eavesdropping Alice and Bob will not have the exact same strings of bits after the first phase, although the two strings will have a lot of common information. Therefore the second subprotocol, the "Data Reconciliation," is used to equalize the two strings, albeit at a cost of leaking some small information to an eavesdropper. Fortunately, Alice and Bob can calculate a lower limit of their mutual information after the two initial phases and then perform the privacy amplification phase in order to arrive to a shorter but much more private key. These two subprotocols require further random numbers.

The protocol BB84 is considered information theoretically proven [28, 73] meaning that an attacker simply has not enough information to calculate the plaintext even given infinite computing resources. This is in strong contrast with the widely used "deterministic cryptography" where an attacker has enough information to calculate the key except that it would probably require insurmountably large computation resources and/or time. The caveat with QC is that the security proof holds only against the family of attacks considered in the proof. Unfortunately, with time, it became evident that unexpected attacks on QC which utilize various quantum effects are feasible which makes QC much less "untouchable."

For example, in 2007 an MIT group presented an attack that gave Eve as much as 100 % of information about the key albeit at an expense of elevated BER [39], but the attack was reassuringly classified as "simulation only" because it assumed that Eve has a specific information about Bob's receiver that she apparently could not get.

As with any other cryptographic procedure, some problems in real-world implementation of the protocol, especially of the quantum channel and real photon detectors, could be used to weaken the cryptographic security of the protocol and open pathways for attacks.

A beautiful demonstration of serious weakening and even 100 % breaking of the key without any notice to legitimate parties has been made by Makarov et al. in 2010 [24, 46]. The demonstration has been made on the commercial QC systems from

Swiss company ID Quantique, based in Geneva, Switzerland, and one by MagiQ Technologies, based in Boston, Massachusetts. Improvements that would make QC resistant to those attacks are possible and have been proposed [46], but the lesson learned from this is that even protocols whose theoretical base is proven secure in some scenario are not to be automatically assumed immune to all practical attacks. The attack was made possible because the authors have found a way to manipulate RNG at the receiving station by exploiting weaknesses of single photon detectors. To make things even worse, this strategy made the previously mentioned MIT attack truly viable (not anymore just a simulation). This is yet another example of the importance of (local) RNGs to the security of a cryptographic scheme.

In conclusion, the quantum cryptographic protocol BB84 requires that both Alice and Bob possess their private (local) provable RNGs. This is a highly critical requirement. Note that a public server of random numbers cannot substitute for local generators because the random numbers would have to be delivered to Alice and Bob in perfect secrecy in the first place, and the server would have to be trusted.

6 Random Numbers in Statistical Cryptography

Statistical cryptography was invented by Ueli Maurer in 1991. The so-called SKAPD protocol [51] resembles quantum cryptography and likewise consists of three subprotocols. In fact the last two subprotocols (the Data Reconciliation and the Privacy Amplification) are the same as in QC. However, the first subprotocol, named "Advantage Distillation" (AD), is completely different and it does not involve the quantum channel which potentially makes it much more practical. Instead, it requires something called "binary channel with noise" which is theoretically a classical communication channel complemented with a provable RNG.

The condition for successful key agreement is that prior to the AD protocol, the common information shared by Alice and Bob is greater than the common information shared by either Alice and Eve or Bob and Eve.

The practical problem with SKAPD is that it contains an unspoken "zeroth phase" in which Alice and Bob obtain their partially correlated initial strings of bits which satisfy the above condition. There is no known plausible way to make the zeroth phase possible although some scenarios have been proposed (scanning the surface of the Moon, listening to noise from faraway galaxies, taking big chunks of Internet data, etc.).

7 Random Numbers in Deterministic Cryptography

What we call here "deterministic cryptography" in this chapter is what is widely known as just "cryptography." Some authors use the name "mathematical cryptography." It is the contemporary cryptography based on the difficulty of computing discrete logarithms in Galois groups and elliptic curve groups, and also the factorization of composite numbers into primes. It also needs and uses random data; an excellent short survey is given in [8]. Since all such security protocols are by definition deterministic and therefore reversible, the only true security resource is that nondeterministic part: a key or one-time data which is supposed to be "random." The quality and provability of randomness are therefore crucial for the security of the whole system.

In fact deterministic cryptography is the only version in wide use, and most cryptographers are not aware of or do not care about the existence of either quantum cryptography or statistical cryptography because apparently they are not yet practical or sufficiently trusted. Therefore it is important to explore what makes contemporary commercial-grade protocols secure and what could be done to get the maximum security out of them. Our hypothesis is that if a protocol requires random numbers, then use of a TRNG maximizes its security. Without ambition to make a strict proof or to give a comprehensive review, here let us have a look at several examples supporting this hypothesis:

1. The Diffie–Hellman key establishment protocol [17] enables the same functionality as the above-mentioned BB84 and SKAPD protocols and is used, for example, in the "https" protocol in order to establish a session key. The protocol requires both parties (Alice and Bob) to generate private random data and after some operations send them to each other. A more resistant version of DH requires further random data used for digital signatures. A vulnerability of the PRNG built in an early version of the Netscape Internet browser led to a complete compromise of the subsequent cryptographic protocol. An example is the attack on Netscape's 40-bit RC4-40 [64] challenge data and encryption keys, which was able to break the https protocol in a minute or so, as described in [25]. The authors of this article stipulate that 128-bit version RC4-128 would not be much harder to break either if seeding is done in a similar fashion.

2. The RSA public key protocol relies on the generation of public and private keys separately by Alice and Bob. In order to create a private/public pair of keys, it is necessary to generate two unique, large prime numbers. Already calculating prime number candidates involves random numbers. After that, candidates need to be tested for primality using the Miller–Rabin algorithm which requires random numbers as the bases in order to properly test for primality. Additional one-time random numbers may be used in the process of actual communication. Where high-entropy physical random bits are not available or are time expensive (like on a typical PC computer), there is a tendency to "expand" a short random string to a long one by pseudorandom methods. This approach can create serious cryptographic weaknesses because an attacker must guess a much smaller number of bits than he or she would in the case of use of truly random numbers.

3. Similarly, research into a cryptographic attack on the partially pseudorandom number generator of an AES-based commercial cryptographic system is described in [66].

To conclude, in deterministic cryptography, random numbers are the only part of the protocol which is different from point to point, and furthermore their true randomness is sometimes a prerogative for correct calculations. Therefore, even though most deterministic cryptography primitives are not secure, using true random numbers ensures the highest achievable security with these methods.

8   Open Problems and Outlook

In this survey chapter we attempt to show the importance of random numbers for the strength of cryptographic protocols not only for quantum and stochastic cryptographies where random numbers are an essential part of the data exchanged between communicating parties but also for contemporary deterministic cryptography where the unpredictability and maximal entropy of the random numbers used therein maximize the overall cryptographic strength.

True random number generators seem to be in modest use, even though some companies make a good profit from them [33]. From the available data, it seems that TRNGs are mainly sold to online gambling companies, state security agencies, and the product labeling and testing industry. At the time of writing this survey, the main problems preventing more widespread use of true random number generators in general are:

1. the lack of generator designs whose proofs of randomness would be at the same time correct, convincing, and demonstrated to be resilient to expected imperfections in hardware
2. the (widespread) lack of understanding that pseudorandom numbers cannot be used as a substitute for true random numbers in so many applications, notably cryptography, both classical and quantum, computer security, Monte Carlo simulations, lotteries, testing of products and their functionality, and many more
3. the high price of true random number generators
4. the lack of support of true random number generators in various popular software that requires random numbers which makes them hard to use

9   Additional Comments and References

A distinctive difference between PRNG and TRNG is the provability of the latter. While mathematical proof of randomness is impossible, for TRNGs we do not rely solely on mathematics but also on sets of physical postulates which lie outside of mathematics. Indeed, the only provable feature of a PRNG is that it is not random because all numbers produced thereby can be calculated from a single initial number: the seed. On the other hand, TRNGs seem to be inevitably plagued with "small imperfections" in hardware which turn into measurable deviations from randomness which calls for post-processing. But complex post-processing blurs or weakens our belief in the eventual randomness proof. Furthermore, the practice of withholding information on the operating of a TRNG as well as a scientific proof of its randomness seems to be almost a rule when it comes to commercial TRNGs. Manufacturers justify this by the need to protect their intellectual property and technology. While such justification is fine when it regards common products (e.g., a dishwasher), it is exactly what ruins the purpose of a TRNG because without a clear insight into the technology and randomness proof of a TRNG, one falls back to the unprovability situation of a PRNG. On the other side of the coin, in scientific publications proofs of randomness are offered very rarely too, probably because the proof is the hard part of the research while it does not seem to be required by the editors of scientific journals. Most researchers therefore fall back to the minimum-action strategy: make a TRNG, obtain at least one random number sequence that passes the chosen set of randomness tests, and publish. However, without a detailed investigation of the sensitivity of the extracted randomness on small variations in hardware and without a randomness proof, a scientific design cannot proceed toward a product. In our view this situation has been improving and will continue to improve very slowly over time, thus ensuring a longevity and freshness of the research on TRNGs.

Even though there is a large collection of publications which document the fact that PRNGs may fail their purpose as RNGs [11, 12, 15, 19, 21, 29, 36, 40, 45, 55, 58, 70, 78, 87], we see that PRNGs are still in much more widespread use than TRNGs even in most critical applications. Among the reasons is the fact that PRNGs are so much more convenient, simpler, and cheaper to use than TRNGs, and also there is a ubiquitous lack of understanding of what randomness is and what it isn't, supported by the nonexistence of a widely accepted definition of randomness [40]. Clearly, further research on that subject is needed.

All commercial TRNGs whose speed is at least 1 Mbit/s are bulky and the price is in the range $5–25k, which is more expensive than most of the software that would use a TRNG. It therefore generally does not pay a software manufacturer to make its product much more expensive by requiring a third-party TRNG for generating random numbers. In extreme rare cases though, a software has been married to a selected TRNG: for example, Mathematica and Quantis (by a third party) [53].

Commercial TRNGs typically come with drivers that support transfer of random numbers to programming languages such as Pascal or C++ on selected operating systems, e.g., [33], using a product specific subroutine or program library function. This is probably the maximum that a manufacturer can reasonably do to support its product. On the other hand, most commercial or free software that uses or needs random numbers does not come with support for any TRNG. This means that having a precompiled software, there is practically no way to connect it to any TRNG. The only viable solution to include a TRNG in a software package would be to write it from scratch and include in it specific function calls associated with the specific chosen TRNG. Since there is no industry standard for access to a TRNG from within a computer program (unlike, e.g., to access printers or other common peripherals), one could support only a specific TRNG per programming effort. In our view it is clear that as long as there is no standardized way to access TRNGs, or, better yet, until TRNGs are physically integrated in computers and are accessible in major programming languages, their popularity will remain minimal.

While it is clear that true randomness cannot be generated by deterministic operations and that therefore it must rely on physical phenomena, the problem of generating good enough randomness and the provability of randomness remain the main open problems with physical RNGs. New directions in the development of physical RNGs will probably concentrate on self-calibrating [41, 86] or no-calibration devices [80] with fundamentally random quantum phenomena as a source of randomness.

References

1. C. Abellán, W. Amaya, M. Jofre, M. Curty, A. Acín, J. Capmany, V. Pruneri, M.W. Mitchell, Ultra-fast quantum randomness generation by accelerated phase diffusion in a pulsed laser diode. Opt. Express **22**, 1645–1654 (2014)
2. V. Bagini, M. Bucci. A design of reliable true random number generator for cryptographic applications, in *Cryptographic Hardware and Embedded Systems (CHES)*, ed. by Ç.K. Koç, C. Paar (Springer, Berlin, 2002), pp. 204–218
3. B. Barak, R. Shaltiel, E. Tromer, True random number generators secure in a changing environment, in *Cryptographic Hardware and Embedded Systems (CHES)*, ed. by C.D. Walter, Ç.K. Koç, C. Paar (Springer, Berlin, 2003), pp. 166–180
4. C.W.J. Beenakker, M. Büttiker, Suppression of shot noise in metallic diffusive conductors. Phys. Rev. B **46**, 1889–1892 (1992)
5. C.H. Bennett, G. Brassard, Quantum cryptography: public key distribution and coin tossing, in *Proceedings of the IEEE International Conference on Computers, Systems, and Signal Processing*, 10–12 Dec 1984, pp. 175–179
6. C.H. Bennett, F. Bessette, G. Brassard, L. Salvail, J. Smolin, Experimental quantum cryptography. J. Cryptol. **5**(1), 3–28 (1992)
7. C.H. Bennett, T.J. Watson, G. Brassard, C. Crepeau, U.M. Maurer, Generalized privacy amplification. IEEE Trans. Inf. Theory **41**(6), 1915–1923 (1995)
8. D.J. Bernstein, J. Buchmann, E. Dahmen (eds.), *Post-Quantum Cryptography* (Springer, Heidelberg, 2009)
9. P. Chevalier et al., Random number generator. U.S. Patent Number 3,790,768, 5 February 1974
10. B. Chor, O. Goldreich, J. Hasted, J. Freidmann, S. Rudich, R. Smolensky. The bit extraction problem or t-resilient functions, in *26th Annual Symposium on Foundations of Computer Science (FOCS)* (IEEE, New York, 1985), pp. 396–407
11. T. Click, A. Liu, G. Kaminski, Quality of random number generators significantly affects results of Monte Carlo simulations for organic and biological systems. J. Comput. Chem. **32**, 513–524 (2011)
12. P.D. Coddington. Tests of random number generators using Ising model simulations. Int. J. Mod. Phys. C **7**, 295–303 (1996)
13. Cryptography Research. Evaluation summary: VIA C3 Nehemiah random number generator (2003), http://www.via.com.tw/en/downloads/whitepapers/initiatives/padlock/evaluation_summary_padlock_rng.pdf
14. R.B. Davies, Exclusive OR (XOR) and Hardware Random Number Generators, http://www.robertnz.net/pdf/xor2.pdf. February 28, 2002
15. A. De Matteis, S. Pagnutti, Long-range correlations in linear and non-linear random number generators. Parallel Comput. **14**(2), 207–210 (1990)

16. M. Dichtl, J.D. Golic, High-speed true random number generation with logic gates only, in *Cryptographic Hardware and Embedded Systems (CHES)*, ed. by P. Paillier, I. Verbauwhede (Springer, Berlin, 2007), pp. 45–62
17. W. Diffie, M.E. Hellman. New directions in cryptography. IEEE Trans. Inf. Theory **22**, 644–654 (1976)
18. J.F. Dynes, Z.L. Yuan, A.W. Sharpe, A.J. Shields, A high speed, postprocessing free, quantum random number generator. Appl. Phys. Lett. **93**, 031109 (2008)
19. R.J. Easter, C. French, Annex C: approved random number generators for FIPS PUB 140-2, in *Security Requirements for Cryptographic Modules*, NIST, February 2012
20. ESPACENET, European Patent Office, http://www.espacenet.com
21. A.M. Ferrenberg, D.P. Landau, Y. J. Wong, Monte Carlo simulations: hidden errors from 'good' random number generators. Phys. Rev. Lett. **69**, 3382–3384 (1992)
22. A. Figotin et al., Random number generator based on the spontaneous alpha-decay. U.S. Patent Number 6,745,217, 1 June 2004
23. M. Fürst, H. Weier, S. Nauerth, D.G. Marangon, C. Kurtsiefer, H. Weinfurter, High speed optical quantum random number generation. Opt. Exp. **18**, 13029–13037 (2010)
24. I. Gerhardt, Q. Liu, A. Lamas-Linares, J. Skaar, C. Kurtsiefer, V. Makarov, Perfect eavesdropping on a quantum cryptography system, 18 March 2012. arXiv:1011.0105v1 [quant-ph]
25. I. Goldberg, D. Wagner, Randomness in the Netscape browser. Dr. Dobb's Journal, January 1996
26. L. Gollub, Vorrichtung zur gewinnung von zufallszahlen. Germany Patent Number DE19743856A1, 8 April 1999
27. M. Goresky, A. Klapper, *Algebraic Shift Register Sequences* (Cambridge University Press, Cambridge, 2012)
28. D. Gottesman, H.-K. Lo, N. Lutkenhaus, J. Preskill, Security of quantum key distribution with imperfect devices. Quantum Inf. Comput. **4**, 325–360 (2004)
29. P. Grassberger, On correlations in "good" random number generators. Phys. Lett. A **181**, 43–46 (1993)
30. H. Guo, W. Tang, Y. Liu, W. Wei, Truly random number generation based on measurement of phase noise of a laser. Phys. Rev. E **81**, 051137 (2010)
31. R. Heinen, Private communication. University of Twente, Twente, Netherlands
32. P. Hellekalek, Good random number generators are (not so) easy to find. Math. Comput. Simulat. **46**, 485–505 (1998)
33. IdQuantique, Quantis: True random number generator exploiting quantum physics (2012), http://www.idquantique.com/random-number-generators/products/products-overview.html
34. Institut Ruder Bošković. QRBG 121 (2012), http://qrbg.irb.hr
35. T. Jennewein, U. Achleitner, G. Weihs, H. Weinfurter, A. Zeilinger, A fast and compact quantum random number generator. Rev. Sci. Instrum. **71**, 1675–1680 (2000)
36. P. Jonsson, Boom in Internet gambling ahead? US policy reversal clears the way, http://tinyurl.com/86b9aaz, 26 December 2011
37. B. Jun, P. Kocher, The Intel random number generator. Cryptography Research Inc., White Paper Prepared for Intel Corporation, 22 April 1999
38. I. Kanter, Y. Aviad, I. Reidler, E. Cohen, M. Rosenbluh, An optical ultrafast random bit generator. Nat. Photon. **4**(1), 58–61 (2010)
39. T. Kim, I.S. Wersborg, F.N.C. Wong, J.H. Shapiro, Complete physical simulation of the entangling-probe attack on the Bennett-Brassard 1984 protocol. Phys. Rev. A **75**, 042327 (2007)
40. D.E. Knuth, High speed single photon detection in the near infrared, in *The Art of Computer Programming*, vol. 2, 3rd edn. (Addison Wesley, Reading, 1997)
41. O. Kwon, Quantum random number generator using photon-number path entanglement. Appl. Opt. **48**, 1774–1778 (2009)
42. P. Lacharme, Post processing functions for a biased physical random number generator, in *Fast Software Encryption (FSE)* (2008), pp. 334–342

43. P. Lacharme, Analysis and construction of correctors. IEEE Trans. Inf. Theory **55**(10), 4742–4748 (2009)
44. X. Li, A.B. Cohen, T.E. Murphy, R. Roy, Scalable parallel physical random number generator based on a superluminescent LED. Opt. Lett. **36**, 1020–1022 (2011)
45. Lotteries and Gaming Authority. Remote gaming regulations, Legal notice 176 of 2004, 110 of 2006, 2760 and 426 of 2007, and 90 of 2011. Malta, 2011
46. L. Lydersen, V. Makarov, J. Skaar, Secure gated detection scheme for quantum cryptography, 29 Jan 2011. arXiv:1101.5698 [quant-ph]
47. G. Marsaglia, DIEHARD Battery of Stringent Randomness Tests (1995), http://stat.fsu.edu/~geo/diehard.html
48. G. Marsaglia, W.W. Tsang, The ziggurat method for generating random variables. J. Stat. Softw. **5**(8), 1–7 (2000). http://www.jstatsoft.org/v05/i08
49. M. Matsumoto, T. Nishimura, Mersenne twister: a 623-dimensionally equidistributed uniform pseudo-random number generator. ACM Trans. Model. Comput. Simulat. **8**, 3–30 (1998). http://www.math.sci.hiroshima-u.ac.jp/~m-mat/MT/emt.html
50. U.M. Maurer, A universal statistical test for random bit generators. J. Cryptol. **5**(2), 89–105 (1992)
51. U. Maurer, Secret key agreement by public discussion from common information. IEEE Trans. Inf. Theory **39**, 733–742 (1993)
52. T. McNichol, Totally random. Wired **11**(8) (2003). http://www.wired.com/wired/archive/11.08/random.html.
53. J.A. Miszczak, Generating and using truly random quantum states in Mathematica, 19 Oct 2011. arXiv:1102.4598v2 [quant-ph]
54. H. Nyquist, Thermal agitation of electric charge in conductors. Phys. Rev. **32**, 110–113 (1928)
55. G. Parisi, F. Rapuano, Effects of the random number generator on computer simulations. Phys. Lett. B **157**, 301–302 (1985)
56. Y. Peres, Iterating von Neumann's procedure for extracting random bits. Ann. Stat. **20**, 590–597 (1992)
57. PicoQuant, PQRNG 150 (2012), http://www.picoquant.com/products/pqrng150/pqrng150.htm
58. A. Proykova, How to improve a random number generator. Comput. Phys. Commun. **124**, 125–131 (2000)
59. B. Qi, Y.-M. Chi, H.-K. Lo, L. Qian, High speed quantum random number generation by measuring phase noise of single mode laser. Opt. Lett. **35**, 312–314 (2010)
60. qutools GmbH. quRNG (2012), http://www.qutools.com/products/quRNG/
61. J.A. Reeds, N.J.A. Sloane, Shift-register synthesis (Modulo m). SIAM J. Comput. **14**, 505–513 (1985)
62. I. Reidler, Y. Aviad, M. Rosenbluh, I. Kanter, Ultra high-speed random number generation based on a chaotic semiconductor laser. Phys. Rev. Lett. **103**(2), 024102 (2009)
63. T. Ritter, Random Number Machines: A Literature Survey, http://www.ciphersbyritter.com/RES/RNGMACH.HTM, 4 Dec 2002
64. R.L. Rivest, The RC4 encryption algorithm. RSA Data Security Inc., March 1992
65. F. Rodriguez-Henriquez, N.A. Saqib, A. Diaz-Perez, Ç.K. Koç, *Cryptographic Algorithms on Reconfigurable Hardware* (Springer, Berlin, 2007)
66. C.B. Roellgen, Visualisation of potential weakness of existing cipher engine implementations in commercial on-the-fly disk encryption software. Global IP Telecommunications, Ltd. & PMC Ciphers, Inc., 15 Aug 2008
67. A. Ruhkin, Statistical testing of randomness: Old and new procedures, in *Randomness Through Computation*, ed. by H. Zenil (World Scientific, Singapore, 2011)
68. A. Ruhkin et al., *A Statistical Test Suite for Random and Pseudorandom Number Generators for Cryptographic Applications*. NIST Special Publication 800-22rev1a, April 2010
69. D. Schellekens, B. Preneel, I. Verbauwhede, *FPGA Vendor Agnostic True Random Number Generator* (2006), http://citeseerx.ist.psu.edu/viewdoc/summary?doi=10.1.1.86.5319

70. F. Schmid, N.B. Wilding, Errors in Monte Carlo simulations using shift register random number generators. Int. J. Mod. Phys. **6**, 781–787 (1995)
71. R. Shaltiel, Recent developments in explicit constructions of extractors. Bull. EATCS **77**, 67–95 (2002)
72. R. Shaltiel, How to get more mileage from randomness extractors. Random Struct. Algorithm **33**, 157–186 (2008)
73. P. Shor, J. Preskill, Simple proof of security of the BB84 quantum key distribution protocol. Phys. Rev. Lett. **85**, 441–444 (2000)
74. A. Sidorenko, B. Schoenmakers, State recovery attacks on pseudorandom generators, in *Western European Workshop on Research in Cryptology* (Springer, Berlin, 2005), pp. 53–63
75. A. Stefanov, N. Gisin, O. Guinnard, L. Guinnard, H. Zbinden, Optical quantum random number generator. J. Mod. Opt. **47**, 595–598 (2000)
76. M. Stipčević, Apparatus and method for generating true random bits based on time integration of an electronic noise source. WIPO Patent Number WO03040854, 17 October 2001
77. M. Stipčević, Fast nondeterministic random bit generator based on weakly correlated physical events. Rev. Sci. Instrum. **75**, 4442–4449 (2004)
78. M. Stipčević, Quantum random bit generator. WIPO Patent Number WO2005106645 (A2), 30 April 2004
79. M. Stipčević, Preventing detector blinding attack and other random number generator attacks on quantum cryptography by use of an explicit random number generator, (2014). arXiv:1403.0143v3 [quant-ph]
80. M. Stipčević, B.M. Rogina, Quantum random number generator based on photonic emission in semiconductors. Rev. Sci. Instrum. **78**, 1–7 (2007)
81. B. Sunar, True random number generators for cryptography, in *Cryptographic Engineering*, ed. by Ç.K. Koç (Springer, Berlin, 2009), pp. 55–73
82. B. Sunar, W.J. Martin, D.R. Stinson, A provably secure true random number generator with built-in tolerance to active attacks. IEEE Trans. Comput. **56**(1), 109–119 (2007)
83. G. Taylor, G. Cox, Behind Intel's new random-number generator. IEEE Spectrum, http://spectrum.ieee.org/computing/hardware/behind-intels-new-randomnumber-generator, 24 Aug 2011
84. T.E. Tkacik, A hardware random number generator, in *Cryptographic Hardware and Embedded Systems (CHES)*, ed. by B.S. Kaliski Jr., Ç.K. Koç, C. Paar (Springer, Berlin, 2002), pp. 450–453
85. A. Uchida et al., Fast physical random bit generation with chaotic semiconductor lasers. Nat. Photon. **2**, 728–732 (2008)
86. G. Vallone, D. Marangon, M. Tomasin, P. Villoresi, Self-calibrating quantum random number generator based on the uncertainty principle, 30 Jan 2014. arXiv:1401.7917 [quant-ph]
87. I. Vattulainen, T. Ala-Nissila, K. Kankaala, Physical tests for random numbers in simulations. Phys. Rev. Lett. **73**, 2513–2516 (1994)
88. VIA Inc. Via security application note (2005), www.via.com.tw/en/downloads/whitepapers/initiatives/padlock/security_application_note.pdf
89. VIA Inc. AES encryption (2012), http://www.via.com.tw/en/initiatives/padlock/hardware.jsp
90. VIA Inc. Random number generation (2012), http://www.via.com.tw/en/initiatives/padlock/hardware.jsp
91. VIA Inc. Via padlock security engine (2012), http://www.via.com.tw/en/initiatives/padlock/hardware.jsp
92. J. Viega, Practical random number generation in software, in *Proceedings of 19th Annual Computer Security Applications Conference* (2003), pp. 129–140
93. C.H. Vincent, The generation of truly random binary numbers. J. Phys. E: Sci. Instrum. **3**, 594–598 (1970)
94. J. von Neumann, Various techniques for use in connection with random digits. John von Neumann Collect. Works **5**, 768–770 (1963)
95. M. Wahl, M. Leifgen, M. Berlin, T. Roehlicke, H.J. Rahn, O. Benson, An ultrafast quantum random number generator with provably bounded output bias based on photon arrival time measurements. Appl. Phys. Lett. **98**, 171105 (2011)

96. J. Walker, Ent: A pseudorandom number sequence test program, http://www.fourmilab.ch/random/.
97. A.B. Wang, Y.C. Wang, H.C. He, Enhancing the bandwidth of the optical chaotic signal generated by a semiconductor laser with optical feedback. IEEE Photon. Technol. Lett. **20**, 1633–1635 (2008)
98. A.B. Wang, Y.C. Wang, J.F. Wang, Route to broadband chaos in a chaotic laser diode subject to optical injection. Opt. Lett. **34**, 1144–1146 (2009)
99. M.A. Wayne, P.G. Kwiat, Low-bias high-speed quantum random number generator via shaped optical pulses. Opt. Exp. **18**, 9351–9357 (2010)
100. M.A. Wayne, E.R. Jeffrey, G.M. Akselrod, P.G. Kwiat, Photon arrival time quantum random number generation. J. Mod. Opt. **56**, 516–522 (2009)
101. S.-K. Yoo, D. Karakoyunlu, B. Birand, B. Sunar, Improving the robustness of ring oscillator TRNGs. ACM Trans. Reconfigur. Technol. Syst. **3**(2), 9:1–30 (2010)
102. Z.L. Yuan, B.E. Kardynal, A.W. Sharpe, A.J. Shields, High speed single photon detection in the near infrared. Appl. Phys. Lett. **91**, 041114 (2007)

How to Sign Paper Contracts? Conjectures and Evidence Related to Equitable and Efficient Collaborative Task Scheduling

Eric Brier, David Naccache, and Li-yao Xia

Abstract This chapter explores ways of performing a kind of commutative task by N parties, of which a particular scenario of contract signing is a canonical example. Tasks are defined as *commutative* if the order in which parties perform them can be freely changed without affecting the final result. It is easy to see that arbitrary N-party commutative tasks cannot be completed in less than $N - 1$ basic time units.

We conjecture that arbitrary N-party commutative tasks cannot be performed in $N - 1$ time units by exchanging less than $4N - 6$ messages and provide computational evidence in favor of this conjecture. We also explore the most equitable commutative task protocols.

1 Introduction

This chapter explores ways of performing commutative tasks by N parties denoted $\mathcal{A}_0, \ldots, \mathcal{A}_{N-1}$. Tasks are defined as *commutative* if the order in which parties perform them can be freely changed without affecting the final result. Furthermore, another requirement is that this result is distributed among the N parties in the end.

A typical formulation, used throughout this work, is the material signature of a contract by N parties. As the contract signing protocol ends, each party obtains a printed contract bearing the N signatures of all other parties. Empty contracts can be printed by all parties. Each contract must transit through all parties to eventually bear all the required signatures.

This problem is not only of theoretical interest. Cryptography conceals the meaning of information but not its existence. In many cases network monitoring allows to infer useful information from the message flow. This attack is called *traffic analysis*. A well-known way to defeat traffic analysis consists in continuously

E. Brier (✉)
Ingenico, 1, rue Claude Chappe, BP 346, 07503 Guilherand-Granges, France
e-mail: eric.brier@ingenico.com

D. Naccache • L.-y. Xia
École normale supérieure, Département d'informatique, 45, rue d'Ulm, 75230, Paris Cedex 05, France
e-mail: david.naccache@ens.fr; li-yao.xia@ens.fr

© Springer International Publishing Switzerland 2014
Ç.K. Koç (ed.), *Open Problems in Mathematics and Computational Science*,
DOI 10.1007/978-3-319-10683-0_13

padding the communication channel with dummy packets to simulate constant bandwidth occupation.

Ferguson and Schneier [1] states that "...*it is very hard to hide information about the size or timing of messages. The known solutions require Alice to send a continuous stream of messages at the maximum bandwidth she will ever use... This might be acceptable for military applications, but it is not for most civilian applications...*"

We also refer the reader to [2] who mentions that: "...*In practice this problem has been known for a very long time, and countermeasures are routinely used in modern link encryptors, by making sure that they always send information between sender and receiver, inserting dummy information if necessary [3]. By doing so, they seek to obscure the difference between actual communication and non-communication. Unfortunately, the approach taken by link encryptors to "keep the channel full" is infeasible on the Internet, due to the requirement that the communication infrastructure serves the needs of multiple parties...*"

It is hence useful to look for *economical* ways in which parties can exchange information without revealing their activity. Here envelopes represent constant-size encrypted data containers.[1] We show how to exchange containers between N parties in a way that ascertains that $\forall i \neq j$, party $\mathcal{A}_i$ can send a message to $\mathcal{A}_j$ in $N - 1$ elementary time units, provided that the container's capacity has not been exceeded.

Situation N parties want to sign a contract. Signatories consider the contract valid when each party possesses a copy of the contract bearing all N genuine signatures (which can only be affixed by their respective owners). Parties are unable to meet physically, so they have to employ a postal service.

Firstly, a total of N copies with no signature must be printed, any party can print some of these empty contracts (printing doesn't have to be done by one unique party, quite the opposite in fact).

Secondly, these copies must be sent among the parties. If at some point, party $\mathcal{A}$ wants to send k contracts to party $\mathcal{B}$, $\mathcal{A}$ can put these in one single envelope and pay a postal fee for the envelope independently of its contents.

We can assume that whenever a party receives a contract it has not yet signed, the party signs it immediately. The problem consists in finding a contract signing *protocol* such that each contract has gone through every party at least once and such that at the end, the N contracts are distributed among the N parties.

We denote by $\mathfrak{P}$ such a protocol.

---

[1] $\mathcal{A}_i$ gets a container, decrypts it, and examines its contents. $\mathcal{A}_i$ extracts any messages sent to him and erases these messages from the container. $\mathcal{A}_i$ potentially inserts into the container new messages for other parties and re-encrypts the container for the next receiving party *without changing the container's size.*

The notation $\mathcal{A}_i \overset{k}{\leadsto} \mathcal{A}_j$ will mean "$\mathcal{A}_i$ *signs k contracts and sends them to $\mathcal{A}_j$*". We study protocols according to the following three natural criteria:

Cost The cost of a protocol $\mathfrak{P}$ is the cumulated postal fee paid by all parties. We also make the assumption that this fee also does not depend on the sender and receiver, so we can consider that cost proportional to the number of envelopes sent globally (hereafter \$1/envelope).

A first natural goal consists in *minimizing the postage fees* $\mathrm{Cost}(\mathfrak{P}, N)$. We prove that $\min_{\mathfrak{P}} \mathrm{Cost}(\mathfrak{P}, N) = 2N - 2$.

The cheapest protocols are referred to simply as *cheap* protocols.

Time In this work, we assume that transmitting an envelope takes 1 day while neglecting the administrative delay to have the contract signed once it has been received.

It is easy to see that the contract signing task cannot be completed in less than $N - 1$ days. We call protocols that run in $N - 1$ days *fast protocols*. If N days are allowed, reaching the \$$(2N - 2)$ cost's lower bound is simple (e.g., protocol $\mathfrak{P}_{\mathrm{seq}}$ in Sect. 2). Hence, we will focus our attention on the costs of fast protocols. We show how to construct some fast protocols that cost \$$(4N - 6)$ and conjecture that this cost is optimal:

Conjecture 1 For all N the cheapest fast protocol costs \$$(4N - 6)$.

We checked this conjecture for $N \leq 8$ by exploiting problem symmetries and by using backtracking.

Equitableness It is interesting to find protocols in which postage costs are distributed between parties *as evenly as possible*.

We observed that for $6 \leq N \leq 8$, there exist fast protocols in which $N - 6$ parties pay \$4 and 6 parties pay \$3.

We do not know how to construct such optimally equitable protocols otherwise than by computerized search. We call such protocols *equitable*.

Even though current evidence that equitable protocols exist for all N is very limited, heuristics (more details are given in Sect. 10) suggest that all fast protocols are inherently inequitable in the following sense:

Conjecture 2 In every fast protocol for N parties, the most burdened party must pay \$$\Omega(N)$.

Convention In "*xxxx-protocol*" the xxxx will stand for any combination of the letters F,C,E,M meaning: fast, cheap, equitable and minimal.

2 Straightforward Non-fast Protocols

A trivial sequential protocol is the following:

The sequential protocol $\mathfrak{P}_{seq}$	
Day	Event
0	$\mathcal{A}_0$ prints N empty contracts
$i = 0, \ldots, N - 2$	$\mathcal{A}_i \overset{N}{\rightsquigarrow} \mathcal{A}_{i+1}$
$N - 1$	For $j = 0, \ldots, N - 2$:
	$\mathcal{A}_N - 1 \overset{1}{\rightsquigarrow} \mathcal{A}_j$

Note that:

- $\mathfrak{P}_{seq}$ is not fast because $\mathfrak{P}_{seq}$ validates the contracts on day N, assuming that indexing days starts from 0.
- $\mathfrak{P}_{seq}$ is cost optimal, i.e., $\text{Cost}(\mathfrak{P}_{seq}, N) = 2N - 2$.
- $\mathfrak{P}_{seq}$ is inequitable because $\mathcal{A}_{N-1}$ pays $\$(N - 1)$ while all other parties pay \$1.

3 Graphical Representation

A protocol is entirely defined by the path followed by each contract, i.e., the sequence of $\mathcal{A}_i$s that the contracts transit through each day (one row in Fig. 1).

For such a matrix to reflect a valid protocol, each $\mathcal{A}_i$ must appear at least once in each row and once in the last column.

We will use a very convenient graphical representation to illustrate protocols (e.g., Fig. 1). The graph of a protocol for N parties and D days is a bidimensional graph with $N \times (D + 1)$ vertices.

Vertex (d, i) represents $\mathcal{A}_i$ on day d.

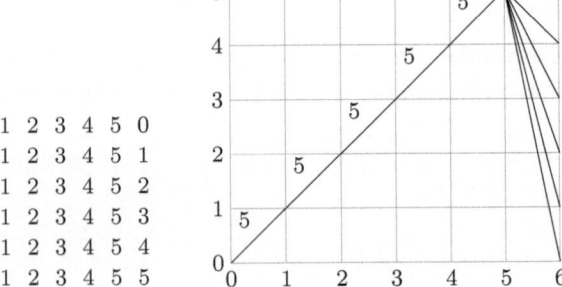

Fig. 1 The matrix and the graph of $\mathfrak{P}_{seq}$

```
0 1 2 3 4 5 0
0 1 2 3 4 5 1
0 1 2 3 4 5 2
0 1 2 3 4 5 3
0 1 2 3 4 5 4
0 1 2 3 4 5 5
```

Fig. 2 A graph may correspond to several different protocols

$$
\begin{array}{cccc}
0 & 1 & 2 & 0 \\
1 & 0 & 2 & 1 \\
0 & 1 & 2 & 2 \\
\end{array}
\qquad
\begin{array}{cccc}
0 & 1 & 2 & 0 \\
0 & 1 & 2 & 1 \\
1 & 0 & 2 & 2 \\
\end{array}
\qquad
\begin{array}{cccc}
1 & 0 & 2 & 0 \\
0 & 1 & 2 & 1 \\
0 & 1 & 2 & 2 \\
\end{array}
$$

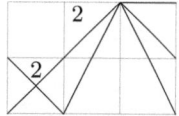

An edge is drawn between (d, i) and $(d + 1, j)$ if $\mathcal{A}_i$ sends an envelope to $\mathcal{A}_j$ on day d. Edges may be labeled with the number of contracts in the corresponding envelope.

Note that such graphs may not uniquely characterize a protocol (see Fig. 2).

4 Fast Protocols

It is easy to see that it takes at least $N - 1$ days to complete the contract signing process and that there is a very simple solution for doing so:

The circular protocol $\mathfrak{P}_{\mathrm{cir}}$	
Day	Event
0	Each party prints one empty contract
$i = 0, \ldots, (N - 1)$	For $j = 0, \ldots, N - 1$: $\mathcal{A}_j \overset{1}{\rightsquigarrow} \mathcal{A}_{j+1 \bmod N}$

- $\mathfrak{P}_{\mathrm{cir}}$ is fast because $\mathfrak{P}_{\mathrm{cir}}$ validates the contracts on day $N - 1$, assuming that indexing days starts from 0.
- $\mathfrak{P}_{\mathrm{cir}}$ is far from being cost optimal, i.e., $\mathrm{Cost}(\mathfrak{P}_{\mathrm{cir}}, N) = N(N - 1)$.
- $\mathfrak{P}_{\mathrm{cir}}$ is equitable because each party pays \$(N - 1)$.

We observe that $\mathfrak{P}_{\mathrm{cir}}$ outperforms $\mathfrak{P}_{\mathrm{seq}}$ by one day, but this (small) improvement comes at the rather high price of a quadratic increase in postage costs.

It is hence natural to ask if linear-cost fast protocols exist and, more generally, find out what the cost $\mathrm{CFP}(N)$ of the cheapest fast protocol is.

5 A Linear Protocol

The following protocol was designed following the *intuition* that to reduce costs, contracts must follow very similar routes. The obstruction to this is that each contract must carefully avoid one participant, namely, the party at which this contract's route will end. We hence design two parallel routes with one contract jumping from one route to the other, at each step.

The linear protocol $\mathfrak{P}_{\text{lin}}$	
Day	Event
0	$\mathcal{A}_0$ prints $N - 1$ empty contracts
	$\mathcal{A}_1$ prints one empty contract
$i = 0, \ldots, N - 3$	$\triangleright \mathcal{A}_i$ has $N - i - 1$ contracts
	$\mathcal{A}_i \overset{1}{\leadsto} \mathcal{A}_{i+2}$
	$\mathcal{A}_i \overset{N-i-2}{\leadsto} \mathcal{A}_{i+1}$
	$\triangleright \mathcal{A}_{i+1}$ has $i + 1$ contracts
	$\mathcal{A}_{i+1} \overset{i+1}{\leadsto} \mathcal{A}_{i+2}$
$N - 2$	$\mathcal{A}_{N-2} \overset{1}{\leadsto} \mathcal{A}_{N-1}$
	For $j = 0, \ldots, N - 2$:
	$\mathcal{A}_{N-1} \overset{1}{\leadsto} \mathcal{A}_j^{\text{a}}$

[a] Each $\mathcal{A}_j$ gets from $\mathcal{A}_{N-1}$ the contract unsigned by $\mathcal{A}_j$

Fig. 3 The matrix and the graph of $\mathfrak{P}_{\text{lin}}$

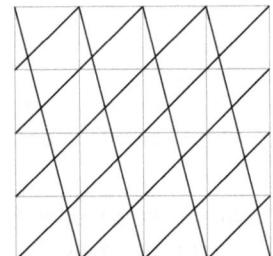

```
1 2 3 4 5 0
0 2 3 4 5 1
0 1 3 4 5 2
0 1 2 4 5 3
0 1 2 3 5 4
0 1 2 3 4 5
```

Fig. 4 The matrix and the graph of $\mathfrak{P}_{\text{cir}}$

```
1 2 3 4 0
2 3 4 0 1
3 4 0 1 2
4 0 1 2 3
0 1 2 3 4
```

$\text{Cost}(\mathfrak{P}_{\text{lin}}, N) = 4N - 6$. The cost vector of $\mathfrak{P}_{\text{lin}}$ (fees paid by $\{\mathcal{A}_0, \ldots, \mathcal{A}_{N-1}\}$) is

$$(2, \underbrace{3, 3, \ldots, 3, 3}_{N-3 \text{ times}}, 2, N - 1)$$

As mentioned previously, we conjecture $\$(4N - 6)$ to be optimal, i.e., $\text{CFP}(N) = 4N - 6$. We thus call $\$(4N - 6)$ protocols *cheap protocols*. The matrix and graph of a circular (resp. linear) protocol are given in Fig. 3 (resp. 4).

6   Counting Protocols

We denote by

$\mathbb{S}_N$:   the set $\{0, \ldots, N-1\}$
$\mathfrak{S}_N$:   the set of $N!$ permutations of $\mathbb{S}_N$

6.1   Observations

Label each contract by the index of the party that will eventually own this contract; the sequence of parties that each contract n goes through in $N-1$ days must be a permutation of the set $\{\mathcal{A}_0, \ldots, \mathcal{A}_{N-1}\}$. As such we can identify a fast protocol with an ordered set of N permutations,[2] in which the nth permutation ends with n.

Also note that for any fast protocol, on day $N-1$ there will always be N envelopes sent, one to each party.

6.2   Number of Protocols

We have tried to enumerate fast protocols and look for some pattern in their numbers.

As pointed out *supra*, a fast protocol can be bijectively mapped to an ordered set of N permutations of $\mathbb{S}_N$, $\mathfrak{P} = (\mathfrak{P}_0, \ldots, \mathfrak{P}_{N-1})$ where $\mathfrak{P}_n(N-1) = n$.

Using $\mathfrak{P} = \mathfrak{P}_{\text{lin}}$ in Fig. 3 as an example, the nth row $\mathfrak{P}_n$ is the cycle $\gamma(n, \ldots, N) = (0, \ldots, n-1, n+1, n+2, \ldots, N, n)$.

For $n = 0, \ldots, N-1$, consider the nth row without its last coordinate : $(\mathfrak{P}_n(0), \ldots, \mathfrak{P}_n(N-2))$ is a permutation of $\mathbb{S}_N \setminus \{n\} \simeq \mathbb{S}_{N-1}$.

The last coordinate that was removed must be equal to the row index. Consequently, fast protocols can be *bijectively* mapped onto sets of N permutations of $\mathbb{S}_{N-1}$. There are therefore $((N-1)!)^N$ fast protocols. Using that identification, we denote the set of fast protocols by $(\mathfrak{S}_{N-1})^N$.

6.2.1   Using Symmetry

There is a lot of symmetry in this problem, that we exploited to examine a (somewhat) lesser number of protocols.

---

[2]Of $\mathbb{S}_N$.

The *relabeling* of $\mathfrak{P}$ by a permutation $\sigma \in \mathfrak{S}_N$ is the protocol obtained by renaming each party $\mathcal{A}_n$ as $\mathcal{A}_{\sigma(n)}$:

$$\sigma(\mathfrak{P}) = (\sigma \circ \mathfrak{P}_{\sigma^{-1}(0)}, \sigma \circ \mathfrak{P}_{\sigma^{-1}(1)}, \ldots, \sigma \circ \mathfrak{P}_{\sigma^{-1}(N-1)}).$$

Notice that the change of index is such that $(\sigma(\mathfrak{P}))_n(n) = n$.

6.2.2 Protocol Isomorphism

Two protocols $\mathfrak{P}, \mathfrak{P}'$ are *truly isomorphic*,[3] if $\mathfrak{P}$ can be transformed into $\mathfrak{P}'$ by relabeling. We denote this relation by $\mathfrak{P} \equiv \mathfrak{P}'$.

$$\mathfrak{P} \equiv \mathfrak{P}' \overset{\text{def}}{\iff} \exists \sigma \; \mathfrak{P}' = \sigma(\mathfrak{P})$$

The number of number of fast protocols up to true isomorphism $\mathrm{NFP}^{\mathsf{T}}(N)$ as a function of N is currently unknown for $N > 6$.

A naïve algorithm for deciding if $\mathfrak{P} \equiv \mathfrak{P}'$ requires $O(N^2 \cdot N!)$ time. We will now show that the protocol isomorphism decision problem[4] can be solved in $O(N^3)$ time.

An interesting relabeling is $\sigma_{\mathrm{Id}} = \mathfrak{P}_n^{-1}$ for some $n \in \mathbb{S}_N$. σ_{Id} satisfies:

$$(\sigma_{\mathrm{Id}}(\mathfrak{P}))_{N-1} = \mathrm{Id}$$

that is to say that the last $((N-1)$th) row of the relabeled matrix is the identity permutation.

And this equality holds if and only if $\sigma_{\mathrm{Id}} = \mathfrak{P}_n^{-1}$ for some n.

In the lexicographical order on permutations $\pi = (\pi(0), \ldots, \pi(N-1))$ seen as words of length N, Id is the smallest of all permutations.

Hence, when looking at protocols, which are ordered sets of N permutations $(\mathfrak{Q}_n)_{n=0,\ldots,N-1}$ as the concatenation $(\mathfrak{Q}_{N-1}, \ldots, \mathfrak{Q}_0)$ (this is a relation on $N \times N$ matrices observed as words of length N^2), we notice that the set $I_{\mathfrak{P}} = \{\mathfrak{P}_n^{-1}(\mathfrak{P}) \mid n = 0, \ldots, N-1\}$ contains the lexicographically smallest protocols which are isomorphic to $\mathfrak{P}$: it is exactly the set of protocols isomorphic to $\mathfrak{P}$ such that the last row of their matrix is Id.

Note that $I_{\mathfrak{P}}$ does not always have cardinality N, e.g., in $\mathfrak{P}_{\mathrm{cir}}$ illustrated in Fig. 4, $I_{\mathfrak{P}_{\mathrm{cir}}} = \{\mathfrak{P}_{\mathrm{cir}}\}$ is a singleton.

Furthermore, the family of sets $I_{\mathfrak{P}}$ defines a partition of the set of matrices whose last row is the identity permutation. Each $I_{\mathfrak{P}}$ has size at most N. And only one protocol per set is minimal in its true isomorphism class.

---

[3]Or *isomorphic* when there is no ambiguity as is the case until the other relation is presented.
[4]That is, given $\mathfrak{P}, \mathfrak{P}' \in \mathfrak{S}_{N-1}^N$, decide if $\mathfrak{P} \equiv \mathfrak{P}'$.

Fig. 5 Two isomorphic protocols $\mathfrak{P}$ (*above*) and $\mathfrak{P}' = \sigma(\mathfrak{P})$ (*below*), where $\sigma = \mathfrak{P}_0^{-1} = (4\,1\,0\,3\,2)$. $\mathfrak{P}'$ is the lexicographically smallest protocol isomorphic to $\mathfrak{P}$

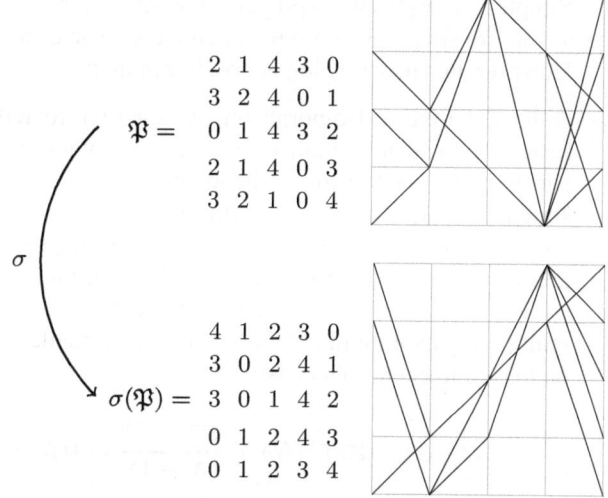

$$\mathfrak{P} = \begin{matrix} 2 & 1 & 4 & 3 & 0 \\ 3 & 2 & 4 & 0 & 1 \\ 0 & 1 & 4 & 3 & 2 \\ 2 & 1 & 4 & 0 & 3 \\ 3 & 2 & 1 & 0 & 4 \end{matrix}$$

σ

$$\sigma(\mathfrak{P}) = \begin{matrix} 4 & 1 & 2 & 3 & 0 \\ 3 & 0 & 2 & 4 & 1 \\ 3 & 0 & 1 & 4 & 2 \\ 0 & 1 & 2 & 4 & 3 \\ 0 & 1 & 2 & 3 & 4 \end{matrix}$$

Hence a lower bound of the number of fast and minimal protocols is:

$$\mathrm{NFP}^{\mathrm{T}}(N) \geq \frac{((N-1)!)^{N-1}}{N}$$

This is also a rough estimate of the actual number $\mathrm{NFP}^{\mathrm{T}}(N)$, if we assume that for most protocols, informally, $|S_{\mathfrak{P}}| \sim N$ (approximately equal). This approximation means that protocols are rarely their own relabeling by a nontrivial permutation; this indicates that the roles played by all parties are generally asymmetrical in some sense. Although limited, the current evidence shows a rather accurate lower bound.

For $N = 5$, we found that there are $\mathrm{NFP}^{\mathrm{T}}(5) = 66,360$ different protocols up to isomorphism, which is pretty close to $\frac{(4!)^4}{5} = 66,355.2$.

For $N = 6$ we get $\mathrm{NFP}^{\mathrm{T}}(6) = 4,147,236,820 \simeq \frac{(5!)^5}{6} = 4,147,200,000$.

By examining only the N permutations that constitute $\mathfrak{P}$, it is possible to determine in $O(N^3)$ time the smallest protocol isomorphic to $\mathfrak{P}$ (e.g., Fig. 5).

It is then a matter of checking equality between those single minimal representatives to decide if two protocols are isomorphic.

All in all, this process claims $O(N^3)$ time.

6.2.3 Simple Isomorphism

Another equivalence relation can be defined by only considering the last row of every matrix, and the corresponding relabeling. Two protocols $\mathfrak{P}$ and $\mathfrak{P}'$ are *simply isomorphic* if their respective relabelings by the permutation found in the last row are equal:

$$(\mathfrak{P}_{N-1})(\mathfrak{P}) = (\mathfrak{P}'_{N-1})(\mathfrak{P}').$$

Simple isomorphism is a proper subrelation of true isomorphism: two isomorphic protocols are simply isomorphic, but the converse is not always true.

However, this new relation is peculiar in that:

- compared to true isomorphism, it is easier to tell whether a matrix is the lexicographical minimum of its simple isomorphism class, as it amounts to checking that the last row is the identity,
- simple isomorphism classes partition the set of protocols evenly into sets of size $(N - 1)!$, as every class can be bijectively mapped to $\mathfrak{S}_{N-1}$ by associating an arbitrary permutation to its last row. We can find the number of protocols with a given cost by only enumerating protocols up to simple isomorphism, and then multiplying their number by $(N - 1)!$. In particular, the number of fast protocols up to simple isomorphism is:

$$\mathrm{NFP}^{\mathrm{S}}(N) = \frac{\mathrm{NFP}(N)}{(N - 1)!} = ((N - 1)!)^{N-1}.$$

True isomorphism does not define an even partition, in fact the number of matrices isomorphic to $\mathfrak{P}$ is $(N - 1)! \cdot |I_{\mathfrak{P}}|$, which is not a constant.

6.2.4 Backtracking

We have designed a backtracking algorithm to enumerate all fast protocols whose costs are bounded by a certain value.

The algorithm consists in incrementally completing a partial protocol in every possible way while keeping track of a lower bound on the cost, and backtracking as soon as the upper limit is reached (e.g., when the lower bound exceeds $4N - 6$).

As pointed out earlier, to enumerate matrices up to (true or simple) isomorphism, we can consider only matrices whose last row is the identity permutation, as all minimal representatives of isomorphism classes are to be found among these.

The number of such matrices is $((N-1)!)^{N-1}$; compared with the original $((N-1)!)^N$, this saves the effort of one iterative layer over a set of permutations.

In the case of true isomorphism, when a complete protocol is obtained, we check if it is lexicographically minimal in its isomorphism class, in which case it can be processed or stored for further examination.

To prune even more possibilities, we can further exploit the fact that the protocols we are looking for need to be lexicographically minimal. For example, instead of checking minimality once the protocol has been completed, it is possible to relabel the partial protocol to see that any completion of it will not be minimal. Unfortunately, in our attempt, the resulting overhead outweighted the pruning. We

Table 1 Number of protocols per N and per cost up to simple isomorphism. Note that there are no \$13 protocols. Here TOTAL is equal to $((N-1)!)^{N-1}$

$\mathrm{NFP}^S(N)$ ↘	$N = 4$	$N = 5$	$N = 6$	$N = 7$	$N = 8$
Cost = \$10	32				
Cost = \$11	80				
Cost = \$12	104				
Cost = \$14		305			
Cost = \$15		2,080			
Cost = \$16		9,590			
Cost = \$17		31,500			
Cost = \$18		76,105	3,960		
Cost = \$19		105,900	49,236		
Cost = \$20		106,296	414,612		
Cost = \$21			2,601,276		
Cost = \$22			13,618,017	59,703	
Cost = \$23			59,672,844	1,305,388	
Cost = \$24			221,523,600	16,320,507	
Cost = \$25			686,256,012	158,145,372	
Cost = \$26			1,792,257,378	1,268,548,841	1,078,176
Cost = \$27			3,770,289,744	8,844,900,603	37,965,696
Cost = \$28			6,119,608,548	54,834,944,423	694,507,192
Cost = \$29			7,281,092,136	305,436,177,578	Unknown
Cost = \$30			4,935,812,637	Unknown	Unknown
Cost > \$30				Unknown	Unknown
Total	$3!^3$	$4!^4$	$5!^5$	$6!^6$	$7!^7$

assume this is due to the small values of N we could examine, and that this modification results in a faster algorithm for greater protocols.

By exhaustively examining all protocols whose last row is Id, we could enumerate all fast protocols for $N \le 6$ (Tables 1 and 2).

And using backtracking as described above, we enumerated some of the cheapest protocols for $N = 7, 8$ while checking[5] that protocols cheaper than \$(4N - 6)$ do not exist.

Table 1 also provides the number of \$c$ protocols for $4N - 6 \le c \le N(N - 1)$.

---

[5] Our Ocaml code is available at http://www.eleves.ens.fr/home/xia/posting.

Table 2 Number of protocols per N and per cost up to true isomorphism. Note that there are no $13 protocols

NFP$^T(N)$ ↘	$N = 4$	$N = 5$	$N = 6$	$N = 7$	$N = 8$
Cost = $10	9				
Cost = $11	10				
Cost = $12	104				
Cost = $14		61			
Cost = $15		416			
Cost = $16		1,918			
Cost = $17		6,300			
Cost = $18		15,221	663		
Cost = $19		21,180	8,206		
Cost = $20		21,264	69,138		
Cost = $21			433,554		
Cost = $22			2,269,917	8,529	
Cost = $23			9,945,474	186,484	
Cost = $24			36,922,032	2,331,501	
Cost = $25			114,376,002	22,592,196	
Cost = $26			298,714,009	181,221,263	134,772
Cost = $27			628,381,792	1,263,557,229	4,745,712
Cost = $28			1,019,946,014	7,833,563,489	86,813,703
Cost = $29			1,213,515,356	43,633,739,654	Unknown
Cost = $30			822,654,663	Unknown	Unknown
Cost > $30				Unknown	Unknown
Total	123	66,360	4,147,236,820	Unknown	Unknown

7 Equitableness

In $\mathfrak{P}_{\text{lin}}$, all parties but one pay a fixed fee, and one party pays a fee that increases with N. This is not an equitable protocol. We hence looked for the most equitable cheap protocol.

We measure equitableness using the Theil index:

$$T_N(\mathfrak{P}) = \frac{1}{N} \sum_{n=1}^{N} \frac{m_n}{\tilde{m}} \log\left(\frac{m_n}{\tilde{m}}\right)$$

where m_n is the fee paid by $\mathcal{A}_n$ and

$$\tilde{m} = \frac{1}{N} \sum_{n=1}^{N} m_n$$

is the average fee.

Table 3 Protocol enumeration up to simple isomorphism

N	$\text{NCP}^S(N)$	$\text{NCEP}^S(N)$	$\text{ATICP}^S(N)$	$\text{TICEP}^S(N)$
2	1	1	0	0
3	4	4	0	0
4	32	32	0.020136	0.020136
5	305	40	0.037728	0.011069
6	3,960	24	0.057973	0
7	59,703	84	0.077496	0.005786
8	1,078,176	216	0.094730	0.008475

NCP number of cheap protocols
NCEP number of cheap and equitable protocols
ATICP average Theil index of cheap protocols
TICEP Theil index of cheap and equitable protocols

Table 4 Protocol enumeration up to true isomorphism

N	$\text{NCP}^T(N)$	$\text{NCEP}^T(N)$	$\text{ATICP}^T(N)$	$\text{TICEP}^T(N)$
2	1	1	0	0
3	2	2	0	0
4	9	9	0.020136	0.020136
5	61	8	0.037728	0.011069
6	663	5	0.057825	0
7	8,529	12	0.077496	0.005786
8	134,772	27	0.094730	0.008475

NCP number of cheap protocols
NCEP number of cheap and equitable protocols
ATICP average Theil index of cheap protocols
TICEP Theil index of cheap and equitable protocols

A smaller $T_N(\mathfrak{P})$ value expresses a more equitable protocol.

We computed the average Theil index of (fast and) cheap protocols, and we also enumerated those that are equitable with results in Tables 3 and 4.

For $N = 7$, the average Theil index computed over all minimal representatives of protocol isomorphism classes is $\simeq 0.077$, whereas the minimum index is $\simeq 0.0058$, reached by the 12 FCEM protocols given in the appendix. We also illustrate in Fig. 6 one of the 27 FCEM protocols for $N = 8$, found by automated search.

7.1 Symbol Insertion Experiments

It is natural to wonder if FCE protocols can be constructed from smaller ones. To get a hint, we took all 27 eight-party FCEM protocols $\mathfrak{P}_1, \ldots, \mathfrak{P}_{27}$ and performed the following exploration:

for $i = 1 \rightarrow 27$ **do**

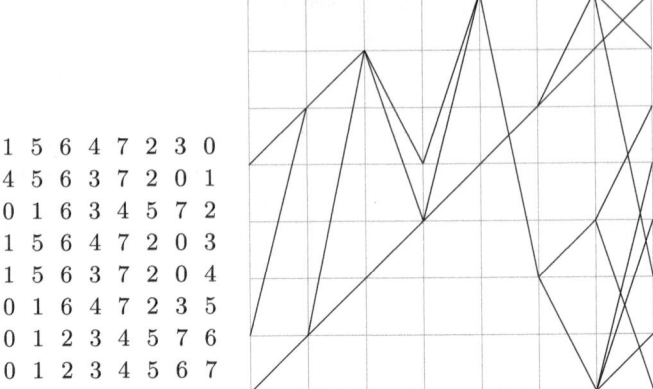

```
1  5  6  4  7  2  3  0
4  5  6  3  7  2  0  1
0  1  6  3  4  5  7  2
1  5  6  4  7  2  0  3
1  5  6  3  7  2  0  4
0  1  6  4  7  2  3  5
0  1  2  3  4  5  7  6
0  1  2  3  4  5  6  7
```

Fig. 6 Equitable protocol, $N = 8$ of cost vector $(4, 3, 3, 4, 3, 3, 3, 3)$ (example)

```
5  7  2  3  6  1  4  0

0̸  7  2  3  4  5  6  1          7  2  3  4  5  6  1
5  7  0̸  3  6  1  4  2          5  7  3  6  1  4  2
0̸  1  2  7  4  5  6  3          1  2  7  4  5  6  3
5  7  0̸  3  6  1  2  4   ⟶      5  7  3  6  1  2  4
0̸  7  2  3  6  1  4  5          7  2  3  6  1  4  5
0̸  1  2  7  4  5  3  6          1  2  7  4  5  3  6
0̸  1  2  3  4  5  6  7          1  2  3  4  5  6  7
```

Fig. 7 Symbol deletion experiment (example). The deleted symbol is 0

for $\ell = 1 \rightarrow 8$ **do**

    $M \leftarrow$ the matrix of $\mathfrak{P}_i$ where row ℓ was suppressed.

    $M' \leftarrow M$ where all occurrences of ℓ were suppressed.

    check if the protocol corresponding to M' is an FCE-protocol.

  end for

end for

Indeed, the above algorithm detected 168 different ways to build (non necessarily minimal) eight-party FCE protocols by inserting new symbols into 7 seven-party FCEM protocols. The process is illustrated in Fig. 7.

The experiment was repeated *mutatis mutandis* by eliminating all possible combinations of two rows (and their corresponding pairs of symbols). There were 136 ways to obtain eight-party FCE protocols using symbol insertions into six-party FCEM protocols. Only two protocols out of the five equitable six-party protocols enabled these insertions, and 17 out of the 27 eight-party FCEM protocols could be reached that way.

Results are available online.[6]

---

[6]http://www.eleves.ens.fr/home/xia/posting.

We doubt that this process would allow to infer a general process for constructing $(N+1)$ party FCE protocols by extending N-party FCE protocols for the following reason: for $N = 6, 7, 8$ all FCE protocols have four active parties on day $N-2$, never 2 or 3 nor 5.

The exhaustive list of matrices for $N = 6, 7, 8$ hints that we cannot do better than four parties on day $N-2$. If there was an algorithm allowing to build FCE protocols from smaller ones, this algorithm would have to add active parties on day $N-2$, and it would be unexpected for it not to work for six, seven, or eight parties.

We regard this as evidence that the algorithmic construction of FCEM protocols is a nontrivial problem.

This approach can be used to find a way of generating cheap protocols rather than equitable protocols. Indeed, there is a pattern we have discovered, though without the use of this approach, as explained in Sect. 9.1.

8 Lower Bounds

8.1 General Case

With no conditions on the protocol's duration D, we show that

$$\min_{\mathfrak{P}} \text{Cost}(\mathfrak{P}, N) = 2N - 2$$

as achieved by $\mathfrak{P}_{\text{seq}}$.

It should be noted that in general having some party hold a contract for several consecutive days without sending it away is an allowed "move," and is of course free of charge.

Only with the now unassumed constraint of validation in $N-1$ days, it becomes necessary to have all contracts circulating in envelopes every day.

The same applies as well to the fact that a contract can transit through one same party multiple times.

Proof The proof is done by induction on the number of parties N.

When $N = 1$, it is clear that $\text{Cost}(\mathfrak{P}, 1) = 0$.

Assume that for every N-party protocol $\mathfrak{P}'$, $\text{Cost}(\mathfrak{P}') \geq 2N - 2$. Let us prove that for every $(N+1)$-party protocol, $\text{Cost}(\mathfrak{P}) \geq 2N$.

Let $\mathfrak{P}$ be a \$$c$ $(N+1)$-party protocol.

By conveniently removing one party from $\mathfrak{P}$, we will create an N-party protocol $\mathfrak{P}'$ that costs at most \$$(c-2)$.

$$c - 2 \geq \text{Cost}(\mathfrak{P}')$$

Fig. 8 Transformation on
day 0 (first transformation)

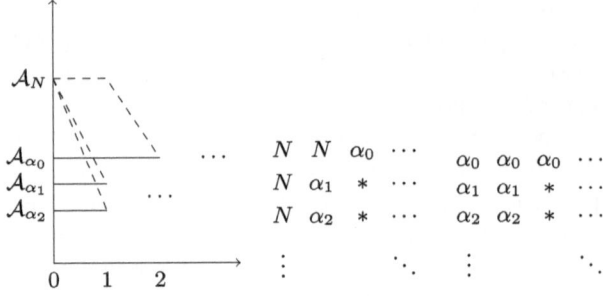

Then, using the inductive hypothesis for $\mathfrak{P}'$,

$$\mathrm{Cost}(\mathfrak{P}') \geq 2N - 2$$

will conclude the inductive step.

At least one party is to print an empty copy of the contract, which will be sent using one envelope. Without loss of generality, we can assume that $\mathcal{A}_N$ is one of those who print contracts, that is the party we will want to remove from this protocol.

8.1.1 A First Protocol Transformation

Instead of having $\mathcal{A}_N$ print contracts (on day 0) and send them to $\mathcal{A}_{\alpha_0}, \mathcal{A}_{\alpha_1}, \ldots$ (not necessarily on day 0), we will have $\mathcal{A}_{\alpha_0}, \mathcal{A}_{\alpha_1}, \ldots$ print these contracts. Because $\mathcal{A}_N$ is assumed to print at least one contract, at least one less envelope will be used (Fig. 8).

We now have to consider the points in time at which $\mathcal{A}_N$ receives some contracts.

This must happen at least once, as every party must receive a final copy of the contract at some point.

8.1.2 A Second Protocol Transformation

The following transformation removes another envelope from the process.

– If $\mathcal{A}_N$ receives only one envelope containing only the contract that $\mathcal{A}_N$ is to own, then we can just remove this envelope from the protocol.
– Otherwise, $\mathcal{A}_N$ receives some contracts which are to be signed by $\mathcal{A}_N$. Then these contracts need to be rerouted away (not necessarily on the same day), excluding the contract that is ultimately bound to reach $\mathcal{A}_N$ that we will just remove.

   Denote by $\mathcal{A}_{\beta_0}, \mathcal{A}_{\beta_1}, \ldots$ the parties those contracts will be sent to next. There must be at least one of them, $\mathcal{A}_{\beta_0}$. Since $\mathcal{A}_N$ is to be removed, we can change the

Fig. 9 Transformation when $\mathcal{A}_N$ receives envelopes (second transformation)

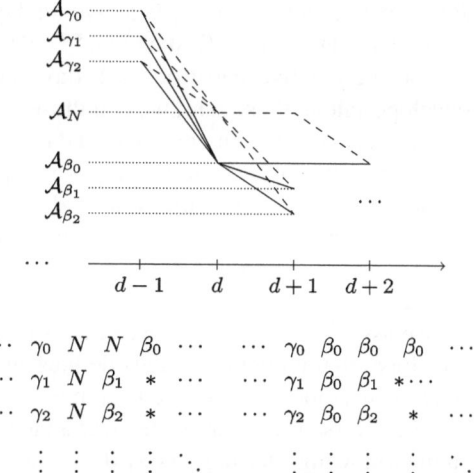

destination of the contracts to $\mathcal{A}_{\beta_0}$ instead of $\mathcal{A}_N$, and one less envelope will be used as $\mathcal{A}_{\beta_0}$ does not need to send a contract to himself (Fig. 9).

With the above two transformations, we can obtain an N-party protocol instead of an $N + 1$ one while removing at least one envelope with each transformation. Therefore the resulting protocol costs at most $\$(c - 2)$.

We can conclude that for all N-party protocols, $\texttt{Cost}(\mathfrak{P}) \geq 2N - 2$.

8.2 Fast Protocols

Although still unsatisfactory, a lower bound $\texttt{CFP}(N) \geq 3N - 5 + \log_2(N)$ can be proven.

We first prove that $\texttt{CFP}(N) \geq 3N - 4$.

Proof Assume that $\texttt{CFP}(N) \leq 3N - 5$ for some N.

Since $\texttt{CFP}(2) = 2$, we can assume that $N \geq 3$.

Let $\mathfrak{P}$ be a $\$\texttt{CFP}(N)$ protocol, i.e., a protocol using less than $3N - 5$ envelopes.

We know that on day $N - 2$, exactly N envelopes are sent. Hence between days 0 and $N - 3$, strictly less than $2(N - 2)$ envelopes would be sent.

On at least one day $\leq N - 3$, only one envelope is sent; therefore all contracts go through one same party, and on the last day the contract that this party receives would have gone through it twice, which is impossible.

The N contracts must follow N different paths between days 0 and $N - 2$, as the final destination of each contract is the only party it hasn't gone through during days 0 to $N - 2$. Moreover, we can bound the number of different available paths when using $3N - 4 + q$ envelopes by 2^{q+1}.

Proof We say that party $\mathcal{A}_n$ is *active* on day d in protocol $\mathfrak{P}$ if $\mathcal{A}_n$ has at least one contract on day d, i.e., $\exists k$ such that $\mathfrak{P}_k(d) = n$.

For every active party on each day between days 1 and $N - 3$, choose one envelope among those sent; we call those chosen envelopes *default* envelopes. Also choose only one default envelope on day 0.

The number of default envelopes is equal to the cumulated number of active parties in days 1 to $N - 3$, plus one on day 0. That is at least $2N - 5$ as a consequence of the previous proof. There are also N envelopes sent on day $N - 2$.

Therefore there are at most $q + 1$ non-default envelopes between days 0 and $N - 2$.

We associate any path between days 0 and $N - 2$ with the set L of non-default envelopes that it contains.[7] This defines an injection into the set of subsets of non-default envelopes, whose size is at most 2^{q+1}.

The reverse procedure to recover a path Δ from its associated subset L consists in the following, starting on day 0:

- If no envelope sent on day 0 is in L, then path Δ starts with the default envelope.
- Otherwise there should be a unique such envelope, and this is the first envelope in the path.

The reason why we chose only one default envelope on day 0 is that we don't know yet where the path begins from. This default envelope allows to set a default starting party at the same time.

Once the first envelope in Δ is found, $\Delta(0)$ and $\Delta(1)$ are known.

We carry on by induction. On each day $d = 1, \ldots, N - 2$, assume that $\Delta(d)$ is known; there is at most one envelope in L which was sent on day d. If there is none, then $\Delta(d + 1)$ is the recipient of the default envelope sent by $\Delta(d)$.

A conflict in this procedure, where there are several envelopes in L among those sent on day d, means that L is not associated with any path.

This procedure shows that there are at most 2^{q+1} paths.

Since there must be at least N paths, $q \geq \log_2(N) - 1$.

In conclusion,

$$\text{CFP}(N) \geq 3N - 5 + \log_2(N).$$

---

[7]The reader is referred to Fig. 10 for a clarifying example.

An edge $(d, \oslash) - (d + 1, \odot)$ means that $\mathcal{A}_\alpha$ sends an envelope to $\mathcal{A}_\beta$ on day d.

The path $(1, 5, 6, 4, 7, 2, 3)$ is associated to the set $\{(0, 1) - (1, 5); (2, 6) - (3, 4); (5, 2) - (6, 3)\}$ (the final destination of the corresponding contract is $\mathcal{A}_0$).

Note that all available paths on this graph are not necessarily taken by a contract, e.g., $(4, 5, 6, 4, 7, 2, 3)$ is associated to $\{(0, 4) - (1, 5); (5, 2) - (6, 3)\}$.

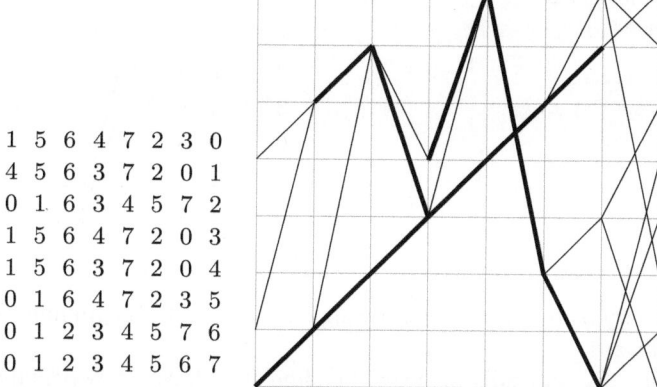

```
1 5 6 4 7 2 3 0
4 5 6 3 7 2 0 1
0 1 6 3 4 5 7 2
1 5 6 4 7 2 0 3
1 5 6 3 7 2 0 4
0 1 6 4 7 2 3 5
0 1 2 3 4 5 7 6
0 1 2 3 4 5 6 7
```

Fig. 10 Protocol from Fig. 6 where edges representing default envelopes are drawn in *thick lines*

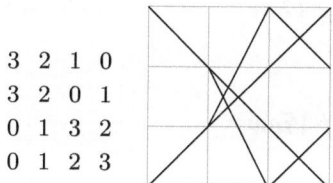

```
3 2 1 0
3 2 0 1
0 1 3 2
0 1 2 3
```

Fig. 11 On day 1, only two envelopes are sent

9 Leads

Looking at the proof of the previous lower bound, it is natural to wonder whether we can improve on the lower bound of two envelopes/day.

This is however the best we can do as illustrated in Fig. 11.

This example can be generalized to all $N \geq 4$.

Other examples were found where only two envelopes were sent on a day other than $N - 3$.

9.1 A Wider Class of Protocols

Given one \$($4N - 6$) protocol $\mathfrak{P}$ for $N \geq 2$, such that on day 0 only two parties print empty contract copies, we can build an $(N + 1)$-party protocol verifying the same property.

This construction can produce the $\mathfrak{P}_{lin}$ protocol and many more cheap protocols which do not comply with the above property, starting with one smaller-size protocol.

Fig. 12 Extension of the
protocol of Fig. 11

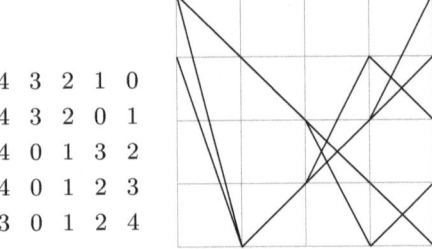

$$
\begin{array}{ccccc}
4 & 3 & 2 & 1 & 0 \\
4 & 3 & 2 & 0 & 1 \\
4 & 0 & 1 & 3 & 2 \\
4 & 0 & 1 & 2 & 3 \\
3 & 0 & 1 & 2 & 4
\end{array}
$$

Extend every path in protocol $\mathfrak{P}$ by appending N at the beginning (N paths are defined that way).

Choose $n \in \mathbb{S}_N$, consider the path Δ ending at n, draw a new path that begins at n, ends at N, and follows Δ in between.

This method hints that the number of $\$(4N - 6)$ protocols grows at least a fast as a factorial. Example: Fig. 12.

9.2   Another Point of View

This problem of contract signature can be equivalently formulated as a variant of the traveling salesman problem. This alternative point of view may provide more insight into the problem, although to this date it is still unclear in what way precisely.

There are N salesmen who are located at different cities at the beginning, and each of them wishes to visit all N cities while minimizing transport fees. To that end, they can cooperate by carpooling, such that if k salesmen are at the same city at the same time and if their current destinations are the same, then the overall cost of transport is the same as for 1 salesman.

This formulation amounts to looking at signing protocols in reversed time. Moreover, this point of view reveals several possible generalizations. Indeed, since in this chapter we consider that any party can send mail to any other party, this relationship can be described by a complete graph; it is then natural to wonder how the problem changes with more general classes of graphs: not necessarily complete, with weights indicating different fees or routing delays associated to different sources and destinations, etc.

However a notion such as equitableness for the parties becomes meaningless, as focus is shifted from parties/cities to contracts/salesmen, instead we can then define equitableness for the salesmen.

10 Open Questions and Further Research

Besides proving (or refuting) Conjectures 1 and 2, we encourage readers to research the following open problems:

Algorithmic Construction of FCE Protocols If equitable protocols exist for all N, design an efficient strategy for constructing FCE protocols. Here *"efficient"* means constructing a protocol in $O(N^t)$ for some fixed t.

It seems there cannot be more than four "active" parties on day $N - 2$, whatever the protocol is. If that is true then, and because there must be N envelopes sent on this last day, one of the parties is going to pay at least $\$N/4$. The best case is $\$N/4$ for every four parties.

Assume on the contrary that equitable protocols exist for all N, or even a weaker form of equitableness where the individual cost is bounded by a constant C. On day $N - 2$, N envelopes must be sent. Since every party pays less than $\$C$, there are at least N/C active parties on day $N - 2$, N/C^2 on day $N - 3$, and so on. There would be a treelike structure at the end of the corresponding graph, which means a lot of active parties—whereas the idea behind the current minimal cost protocols is quite the opposite. When N is large enough, this takes a lot of envelopes to set up; in fact we believe that it takes too many and that there won't be enough on the first days. But we are unsure and maybe such a tree would end up being possible.

Being able to look for equitable protocols for $N = 9$ would be a first step. Finding out if equitable protocols still exist for $N = 12$ or 13 would provide very strong evidence in favor of or against the existence of equitable protocols for all N.

Finding a Protocol Matching (or Best Matching) a Given Cost Vector Given a cost vector:

$$c = \left\{ c_0, c_1, \ldots, c_{N-2}, 4N - 6 - \sum_{i=0}^{N-2} c_i \right\}$$

identify the protocol $\mathfrak{P}_s$ that deviates as little as possible from c.

What Are the Possible Cost Vectors? Let $\mathbb{P}_N = \{\mathfrak{P}_1, \ldots, \mathfrak{P}_{\mathrm{NCP}(N)}\}$ be all N-party FC protocols. Let s_i denote the cost vector of $\mathfrak{P}_i \in \mathbb{P}_N$ with elements sorted by increasing order.[8] How many different s_is are there? What can be said about their frequencies?

Nonconstant Postage Fees We assumed that the cost of an envelope is independent of the number of contracts sent. What happens for a general cost function $f(k)$, for instance, $f(k) = ak + b$ or $f(k) = a\lceil k/b \rceil$?

---

[8]That is, renumbering the parties by increasing workload.

Continuous Flow Communication This chapter dealt with a latency of $N - 1$ days. What happens if $N - 1$ shifted protocols are started simultaneously so that $N - 1$ protocols are always run in parallel? Here the most equitable setting would be $N - 6$ parties paying $\$(4N - 10)$ and six parties paying $\$(4N - 9)$ but is this achievable?[9] If so, how regular can the spending rate of each party be? (i.e., avoid sudden "spending bursts").

If Conjecture 2 Is True Does relaxing the fast protocol hypothesis enable equitableness?

Acknowledgements The authors thank Oğuzhan Külekci for interesting discussions and useful remarks notably concerning the variant of the traveling salesmen problem.

---

[9]For example, if we launch seven shifted instances of Fig. 6 we get a very uneven split of cost where $\mathcal{A}_0$ and $\mathcal{A}_3$ pay \$4 every day (i.e., a total of \$28 each) whereas the other six parties pay \$3 every day (i.e., a total of \$21 each).

Appendix: FCEM Protocols for $N = 7$

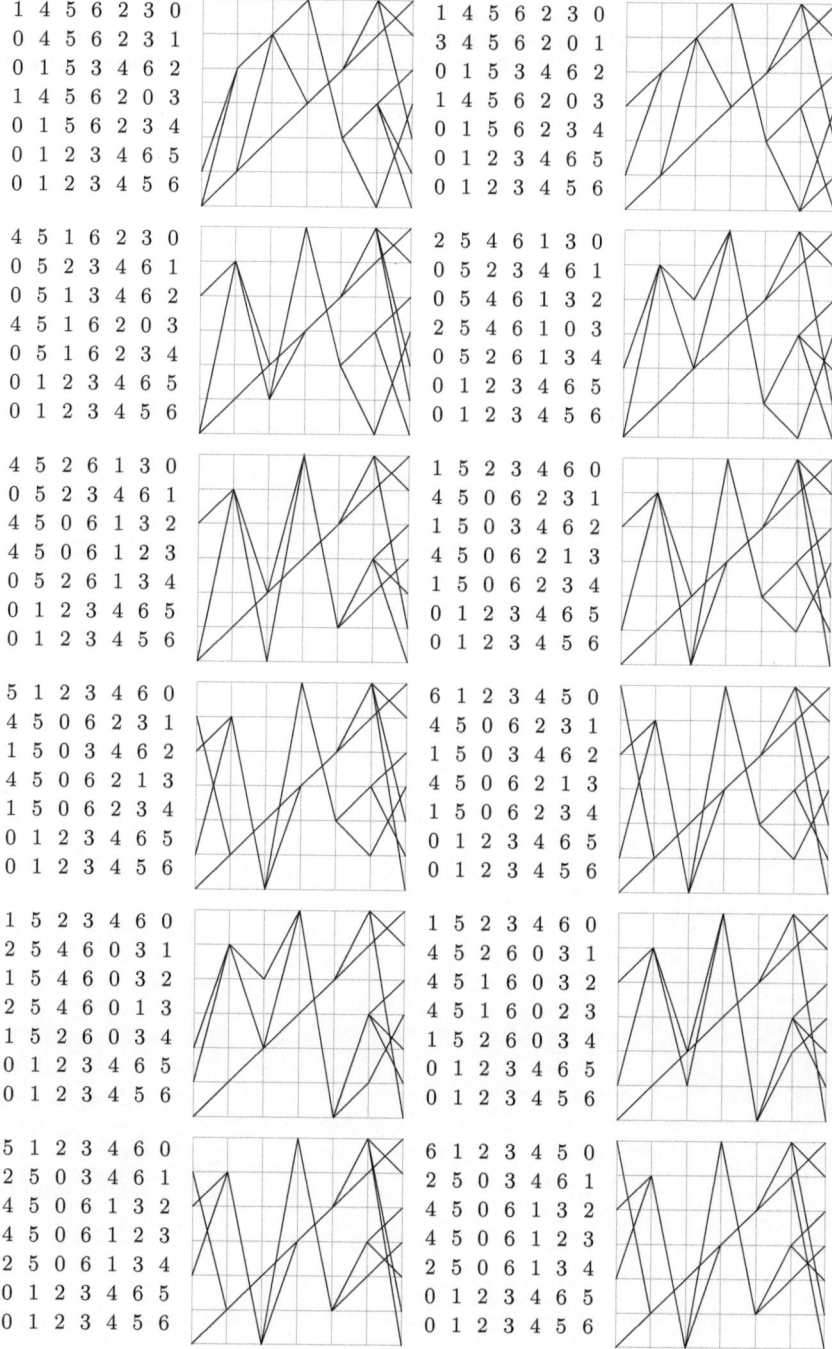

```
1 4 5 6 2 3 0        1 4 5 6 2 3 0
0 4 5 6 2 3 1        3 4 5 6 2 0 1
0 1 5 3 4 6 2        0 1 5 3 4 6 2
1 4 5 6 2 0 3        1 4 5 6 2 0 3
0 1 5 6 2 3 4        0 1 5 6 2 3 4
0 1 2 3 4 6 5        0 1 2 3 4 6 5
0 1 2 3 4 5 6        0 1 2 3 4 5 6

4 5 1 6 2 3 0        2 5 4 6 1 3 0
0 5 2 3 4 6 1        0 5 2 3 4 6 1
0 5 1 3 4 6 2        0 5 4 6 1 3 2
4 5 1 6 2 0 3        2 5 4 6 1 0 3
0 5 1 6 2 3 4        0 5 2 6 1 3 4
0 1 2 3 4 6 5        0 1 2 3 4 6 5
0 1 2 3 4 5 6        0 1 2 3 4 5 6

4 5 2 6 1 3 0        1 5 2 3 4 6 0
0 5 2 3 4 6 1        4 5 0 6 2 3 1
4 5 0 6 1 3 2        1 5 0 3 4 6 2
4 5 0 6 1 2 3        4 5 0 6 2 1 3
0 5 2 6 1 3 4        1 5 0 6 2 3 4
0 1 2 3 4 6 5        0 1 2 3 4 6 5
0 1 2 3 4 5 6        0 1 2 3 4 5 6

5 1 2 3 4 6 0        6 1 2 3 4 5 0
4 5 0 6 2 3 1        4 5 0 6 2 3 1
1 5 0 3 4 6 2        1 5 0 3 4 6 2
4 5 0 6 2 1 3        4 5 0 6 2 1 3
1 5 0 6 2 3 4        1 5 0 6 2 3 4
0 1 2 3 4 6 5        0 1 2 3 4 6 5
0 1 2 3 4 5 6        0 1 2 3 4 5 6

1 5 2 3 4 6 0        1 5 2 3 4 6 0
2 5 4 6 0 3 1        4 5 2 6 0 3 1
1 5 4 6 0 3 2        4 5 1 6 0 3 2
2 5 4 6 0 1 3        4 5 1 6 0 2 3
1 5 2 6 0 3 4        1 5 2 6 0 3 4
0 1 2 3 4 6 5        0 1 2 3 4 6 5
0 1 2 3 4 5 6        0 1 2 3 4 5 6

5 1 2 3 4 6 0        6 1 2 3 4 5 0
2 5 0 3 4 6 1        2 5 0 3 4 6 1
4 5 0 6 1 3 2        4 5 0 6 1 3 2
4 5 0 6 1 2 3        4 5 0 6 1 2 3
2 5 0 6 1 3 4        2 5 0 6 1 3 4
0 1 2 3 4 6 5        0 1 2 3 4 6 5
0 1 2 3 4 5 6        0 1 2 3 4 5 6
```

References

1. N. Ferguson, B. Schneier, *Practical Cryptography* (Wiley, Indianapolis, 2003)
2. K. McCurley, Language modeling and encryption on packet switched networks, in *Advances in Cryptology - Eurocrypt 2006*. Lecture Notes in Computer Science, vol. 4004 (Springer, Berlin, 2006), pp. 359–372
3. V. Voydoc, S. Kent, Security mechanisms in high-level network protocols. ACM Comput. Surv. **15**, 135–171 (1983)

Theoretical Parallel Computing Models for GPU Computing

Koji Nakano

Abstract The latest GPUs are designed for general purpose computing and attract the attention of many application developers. The main purpose of this chapter is to introduce theoretical parallel computing models, the Discrete Memory Machine (DMM) and the Unified Memory Machine (UMM), that capture the essence of CUDA-enabled GPUs. These models have three parameters: the number p of threads and the width w of the memory and the memory access latency l. As examples of parallel algorithms on these theoretical models, we show fundamental algorithms for computing the sum and the prefix-sums of n numbers. We first show that the sum of n numbers can be computed in $O(\frac{n}{w} + \frac{nl}{p} + l \log n)$ time units on the DMM and the UMM. We then go on to show that $\Omega(\frac{n}{w} + \frac{nl}{p} + l \log n)$ time units are necessary to compute the sum. We also present a simple parallel algorithm for computing the prefix-sums that runs in $O(\frac{n \log n}{w} + \frac{nl \log n}{p} + l \log n)$ time units on the DMM and the UMM. Clearly, this algorithm is not optimal. We present an optimal parallel algorithm that computes the prefix-sums of n numbers in $O(\frac{n}{w} + \frac{nl}{p} + l \log n)$ time units on the DMM and the UMM. We also show several experimental results on GeForce Titan GPU.

1 Introduction

Research into parallel algorithms has a long history of more than 40 years. Sequential algorithms have been developed mostly on the random access machine (RAM) [1]. In contrast, since there are a variety of connection methods and patterns between processors and memories, many parallel computing models have been presented and many parallel algorithmic techniques have been shown on them. The most well-studied parallel computing model is the parallel random access machine (PRAM) [5,7,30], which consists of processors and a shared memory. Each processor on the PRAM can access any address of the shared memory in a time unit. The PRAM is a good parallel computing model in the sense that parallelism of each

K. Nakano (✉)
Hiroshima University, Higashi-Hiroshima 739-8527, Japan
e-mail: nakano@cs.hiroshima-u.ac.jp

© Springer International Publishing Switzerland 2014
Ç.K. Koç (ed.), *Open Problems in Mathematics and Computational Science*,
DOI 10.1007/978-3-319-10683-0_14

problem can be revealed by the performance of parallel algorithms on the PRAM. However, since the PRAM requires a shared memory that can be accessed by all processors at the same time, it is not feasible.

The graphics processing unit (GPU) is a specialized circuit designed to accelerate computation for building and manipulating images [10, 11, 17, 31]. Latest GPUs are designed for general purpose computing and can perform computation in applications traditionally handled by the CPU. Hence, GPUs have recently attracted the attention of many application developers [10, 26, 27]. NVIDIA provides a parallel computing architecture called *CUDA* (compute unified device architecture) [29], the computing engine for NVIDIA GPUs. CUDA gives developers access to the virtual instruction set and memory of the parallel computational elements in NVIDIA GPUs. In many cases, GPUs are more efficient than multicore processors [18], since they have hundreds of processor cores and very high memory bandwidth.

CUDA uses two types of memories in the NVIDIA GPUs: *the global memory* and *the shared memory* [29]. The global memory is implemented in off-chip DRAMs with large capacity, say, 1.5–6 GB, but its access latency is very high. The shared memory is an extremely fast on-chip memory with lower capacity, say, 16–48 KB. The efficient usage of the global memory and the shared memory is a key for CUDA developers to accelerate applications using GPUs. In particular, we need to consider *the coalescing* of the global memory access and *the bank conflicts* of the shared memory access [17, 18, 28]. To maximize the bandwidth between the GPU and the DRAM chips, the consecutive addresses of the global memory must be accessed at the same time. Thus, threads of CUDA should perform coalesced access when they access to the global memory. The address space of the shared memory is mapped into several physical memory banks. If two or more threads access to the same memory banks at the same time, the access requests are processed sequentially. Hence, to maximize the memory access performance, threads should access to distinct memory banks to avoid the bank conflicts of the memory access.

Memory machine models, *the Discrete Memory Machine (DMM)* and *the Unified Memory Machine (UMM)* [22], are parallel computing models which reflect the essential features of the shared memory and the global memory of NVIDIA GPUs. The outline of the architectures of the DMM and the UMM are illustrated in Fig. 1. In both architectures, *a sea of threads (Ts)* are connected to *the memory banks (MBs)* through *the memory management unit (MMU)*. Each thread is a random access machine (RAM) [1], which can execute fundamental operations in a time unit. We do not discuss the architecture of the sea of threads in this chapter, but we can imagine that it consists of a set of multicore processors which can execute many threads in parallel. Threads are executed in SIMD [4] fashion, and the processors run on the same program and work on the different data.

MBs constitute a single address space of the memory. A single address space of the memory is mapped to the MBs in an interleaved way such that a word of data of address i is stored in the (i mod w)th bank, where w is the number of MBs. The main difference of the two architectures is the connection of the address line between the MMU and the MBs, which can transfer an address value. In the DMM, the address lines connect the MBs and the MMU separately, while a single address

Fig. 1 The architectures of the DMM and the UMM

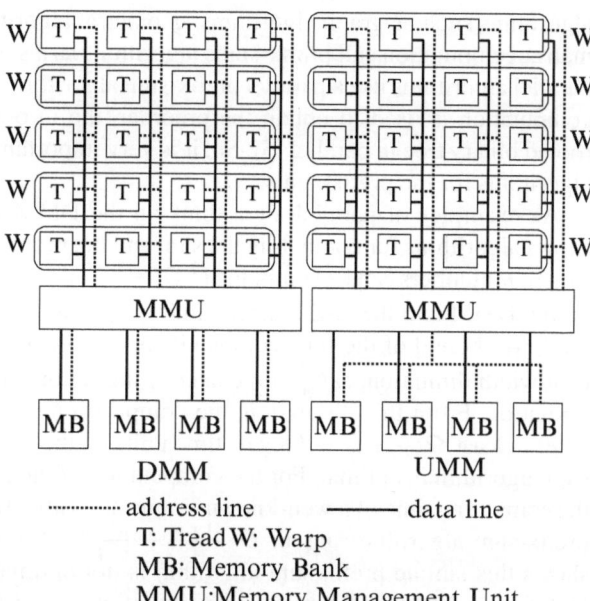

address line .... ———— data line

T: Tread W: Warp
MB: Memory Bank
MMU:Memory Management Unit

line from the MMU is connected to the MBs in the UMM. Hence, in the UMM, the same address value is broadcast to every MB, and the same address of the MBs can be accessed in each time unit. On the other hand, different addresses of the MBs can be accessed in the DMM. Since the memory access of the UMM is more restricted than that of the DMM, the UMM is less powerful than the DMM.

The performance of algorithms of the PRAM is usually evaluated using two parameters: the size n of the input and the number p of processors. For example, it is well known that the sum of n numbers can be computed in $O(\frac{n}{p} + \log n)$ time on the PRAM [5]. On the other hand, four parameters, the size n of the input, the number p of threads, the width w, and the latency l of the memory are used when the performance of algorithms on the DMM and on the UMM is evaluated. The width w is the number of memory banks and the latency l is the number of time units to complete the memory access. Hence, the performance of algorithms on the DMM and the UMM is evaluated as a function of n (the size of a problem), p (the number of threads), w (the width of a memory), and l (the latency of a memory). In NVIDIA GPUs, the width w of global and shared memory is 16 or 32. Also, the latency l of the global memory is several hundreds of clock cycles. In CUDA, a grid can have at most 65,535 blocks with at most 1,024 threads each [29]. Thus, the number p of threads can be 65 million.

Suppose that an array a of n numbers is given. The prefix-sums of a is the array of size n such that the ith ($0 \le i \le n - 1$) element is $a[0] + a[1] + \cdots + a[i]$. Clearly, a sequential algorithm can compute the prefix-sums by executing $a[i+1] \leftarrow a[i + 1] + a[i]$ for all i ($0 \le i \le n - 2$). The computation of the prefix-sums of an array is one of the most important algorithmic procedures. Many sequential algorithms such as graph algorithms, geometric algorithms, image processing, and matrix computation call prefix-sums algorithms as a subroutine. In particular, many parallel algorithms use a parallel prefix-sum algorithm. For example, the prefix-sum computation is used to obtain the preorder, the in-order, and the post-order of a rooted binary tree in parallel [5]. So, it is very important to develop efficient parallel algorithms for the prefix-sums.

As examples of parallel algorithms on the DMM and the UMM, this chapter shows algorithms for computing the sum and the prefix-sums. We first show that the sum of n numbers can be computed in $O(\frac{n}{w} + \frac{nl}{p} + l \log n)$ time units using p threads on the DMM and the UMM with width w and latency l. We then go on to discuss the lower bound of the time complexity and show three lower bounds, $\Omega(\frac{n}{w})$-time bandwidth limitation, $\Omega(\frac{nl}{p})$-time latency limitation, and $\Omega(l \log n)$-time reduction limitation. From this discussion, the computation of the sum and the prefix-sums takes at least $\Omega(\frac{n}{w} + \frac{nl}{p} + l \log n)$ time units on the DMM and the UMM. Thus, the sum algorithm is optimal. For the computation of the prefix-sums, we first evaluate the computing time of a well-known simple algorithm [8,30]. We show that a simple prefix-sum algorithm runs in $O(\frac{n \log n}{w} + \frac{nl \log n}{p} + l \log n)$ time. Hence, this fact shows this simple prefix-sum algorithm is not optimal, and it has an overhead of factor $\log n$ both for the bandwidth limitation $\frac{n}{w}$ and for the latency limitation $\frac{nl}{p}$. We show an optimal parallel algorithm that computes the prefix-sums of n numbers in $O(\frac{n}{w} + \frac{nl}{p} + l \log n)$ time units on the DMM and the UMM.

This chapter is organized as follows. Section 2 introduces the Discrete Memory Machine (DMM) and the Unified Memory Machine (UMM) [22]. In Sect. 3, we evaluate the computing time of the contiguous memory access to the memory of the DMM and the UMM. The contiguous memory access is a key ingredient of parallel algorithm development on the DMM and the UMM. Using the contiguous access, we show that the sum of n numbers can be computed in $O(\frac{n}{w} + \frac{nl}{p} + l \log n)$ time units in Sect. 4. We then go on to discuss the lower bound of the time complexity and show three lower bounds, $\Omega(\frac{n}{w})$-time bandwidth limitation, $\Omega(\frac{nl}{p})$-time latency limitation, and $\Omega(l \log n)$-time reduction limitation in Sect. 5. Section 6 shows a simple prefix-sum algorithm, which runs in $O(\frac{n \log n}{w} + \frac{nl \log n}{p} + l \log n)$ time units. In Sect. 7, we show an optimal parallel prefix-sum algorithm running in $O(\frac{n}{w} + \frac{nl}{p} + l \log n)$ time units. Section 8 presents several implementation and experimental results of the sum and the prefix-sum algorithms using GeForce Titan GPU. In Sect. 9, we briefly show several published results on memory machine models. The final section offers conclusion of this chapter.

2 Memory Machine Models: DMM and UMM

The main purpose of this section is to define the Discrete Memory Machine (DMM) and the Unified Memory Machine (UMM) [22].

Fig. 2 Banks and address groups for $w = 4$

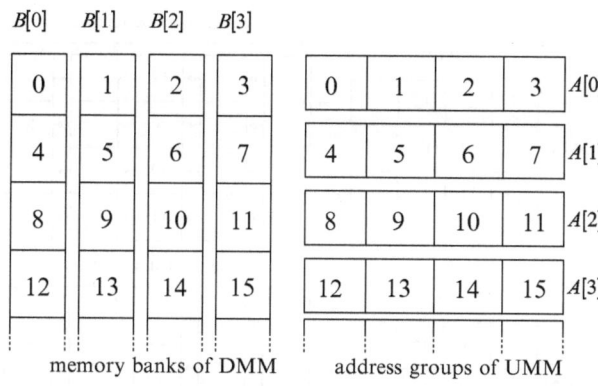

memory banks of DMM   address groups of UMM

We first define *the Discrete Memory Machine (DMM)* of width w and latency l. Let $m[i]$ $(i \geq 0)$ denote a memory cell of address i in the memory. Let $B[j] = \{m[j], m[j+w], m[j+2w], m[j+3w], \ldots\}$ $(0 \leq j \leq w-1)$ denote *the jth bank* of the memory as illustrated in the Fig. 2. Clearly, a memory cell $m[i]$ is in the $(i \bmod w)$th memory bank. We assume that memory cells in different banks can be accessed in a time unit, but no two memory cells in the same bank can be accessed in a time unit. Also, we assume that l time units are necessary to complete an access request and continuous requests are processed in a pipeline fashion through the MMU. Thus, it takes $k + l - 1$ time units to complete k access requests to a particular bank.

We assume that p threads are partitioned into $\frac{p}{w}$ groups of w threads called *warps*. More specifically, p threads are partitioned into $\frac{p}{w}$ warps $W(0), W(1), \ldots, W(\frac{p}{w}-1)$ such that $W(i) = \{T(i \cdot w), T(i \cdot w + 1), \ldots, T((i+1) \cdot w - 1)\}$ $(0 \leq i \leq \frac{p}{w} - 1)$. Warps are dispatched for memory access in turn, and w threads in a warp try to access the memory at the same time. In other words, $W(0), W(1), \ldots, W(\frac{p}{w}-1)$ are dispatched in a round-robin manner if at least one thread in a warp requests memory access. If no thread in a warp needs memory access, such warp is not activated for memory access and is skipped. When $W(i)$ is activated, w thread in $W(i)$ sends memory access requests, one request per thread, to the memory. We also assume that a thread cannot send a new memory access request until the previous memory access request is completed. Hence, if a thread sends a memory access request, it must wait l time units to send a new memory access request.

For the reader's benefit, let us evaluate the time for memory access using Fig. 3 on the DMM for $p = 8$, $w = 4$, and $l = 5$. In the figure, $p = 8$ threads are partitioned into $\frac{p}{w} = 2$ warps $W(0) = \{T(0), T(1), T(2), T(3)\}$ and $W(1) = \{T(4), T(5), T(6), T(7)\}$. As illustrated in the figure, four threads in $W(0)$ try to access $m[7], m[5], m[15]$, and $m[0]$, and those in $W(1)$ try to access $m[10], m[11], m[12]$, and $m[9]$. The time for the memory access is evaluated under the assumption that memory access is processed by imaginary l pipeline stages with w registers each as illustrated in the figure. Each pipeline register in the first stage receives memory access request from threads in an activated warp. Each ith $(0 \leq i \leq w - 1)$ pipeline register receives the request to the ith memory bank. In

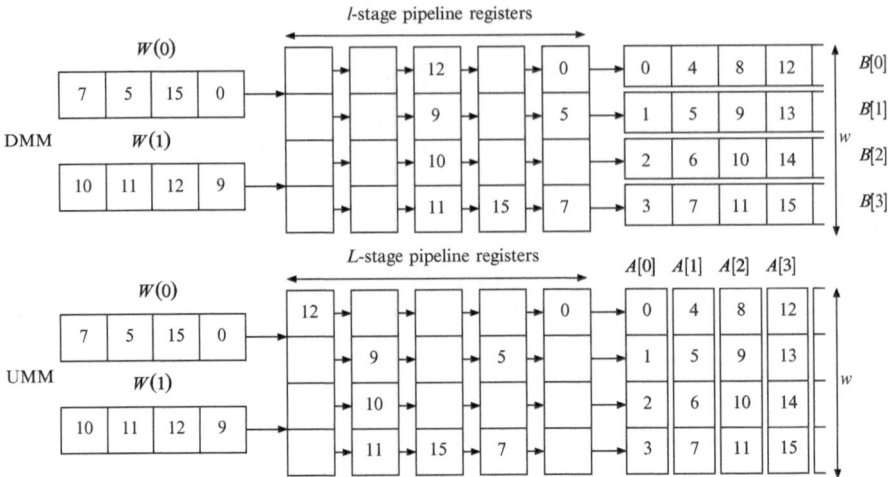

Fig. 3 An example of memory access

each time unit, a memory request in a pipeline register is moved to the next one. We assume that the memory access completes when the request reaches the last pipeline register.

Note that the architecture of pipeline registers illustrated in Fig. 3 are imaginary, and it is used only for evaluating the computing time. The actual architecture should involve a multistage interconnection network [6, 14] or sorting network [2, 3] to route memory access requests.

Let us evaluate the time for memory access on the DMM. First, the access request for $m[7], m[5], m[0]$ is sent to the first pipeline stage. Since $m[7]$ and $m[15]$ are in the same bank $B[3]$, their memory requests cannot be sent to the first stage at the same time. Next, the $m[15]$ is sent to the first stage. After that, memory access requests for $m[10], m[11], m[12], m[9]$ are sent at the same time, because they are in different memory banks. Finally, after $l - 1 = 4$ time units, these memory requests are processed. Hence, the DMM takes $2 + 1 + 4 = 7$ time units to complete the memory access.

We next define *the Unified Memory Machine (UMM)* of width w as follows. Let $A[j] = \{m[j \cdot w], m[j \cdot w + 1], \ldots, m[(j + 1) \cdot w - 1]\}$ denote the jth address group as illustrated in Fig. 2. We assume that memory cells in the same address group are processed at the same time. However, if they are in the different groups, one time unit is necessary for each of the groups. Also, similarly to the DMM, p threads are partitioned into warps and each warp access to the memory in turn.

Again, let us evaluate the time for memory access using Fig. 3 on the UMM for $p = 8, w = 4$, and $l = 5$. The memory access requests by $W(0)$ are in three address groups. Thus, three time units are necessary to send them to the first stage of pipeline registers. Next, two time units are necessary to send memory access requests by $W(1)$, because they are in two address groups. After that, it takes $l - 1 = 4$ time units to process the memory access requests. Hence, totally $3 + 2 + 4 = 9$ time units are necessary to complete all memory access.

3   Contiguous Memory Access

The main purpose of this section is to review the contiguous memory access on the DMM and the UMM shown in [22]. Suppose that an array a of size n ($\geq p$) is given. We use p threads to access all of n memory cells in a such that each thread accesses to $\frac{n}{p}$ memory cells. Note that "accessing" can be "reading from" or "writing in." Let $a[i]$ ($0 \leq i \leq n - 1$)denote the ith memory cells in a. When $n \geq p$, the contiguous access can be performed as follows:

[Contiguous memory access]
for $t \leftarrow 0$ to $\frac{n}{p} - 1$ do
  for $i \leftarrow 0$ to $p - 1$ do in parallel
    $T(i)$ accesses $a[p \cdot t + i]$

Let us evaluate the computing time. For each t ($0 \leq t \leq \frac{n}{p} - 1$), p threads access p memory cells $a[pt], a[pt + 1], \ldots, a[p(t + 1) - 1]$. This memory access is performed by $\frac{p}{w}$ warps in turn. More specifically, first, w threads in $W(0)$ access $a[pt], a[pt + 1], \ldots, a[pt + w - 1]$. After that, p threads in $W(1)$ access $a[pt + w], a[pt+w+1], \ldots, a[pt+2w-1]$, and the same operation is repeatedly performed. In general, p threads in $W(j)$ ($0 \leq j \leq \frac{p}{w} - 1$) accesses to $a[pt + jw], a[pt + jw + 1], \ldots, a[pt + (j + 1)w - 1]$. Since w memory cells accessed by a warp are in the different bank, the access can be completed in l time units on the DMM. Also, these w memory cells are in the same address group, and thus, the access can be completed in l time units on the UMM.

Recall that the memory access is processed in pipeline fashion such that w threads in each $W(j)$ send w memory access requests in one time unit. Let us consider two cases:

Case 1: $\frac{p}{w} \leq l$ If this is the case, $\frac{p}{w}$ warps send memory access requests in turn, and the first memory access requests by the first warp $W(0)$ are completed in l time units as illustrated in Fig. 4. After that, w threads in $W(0)$ can send next memory access requests immediately, and they can be completed in l time units. This is repeated $\frac{n}{p}$ times. After all memory access requests by $W(0)$ are completed, it takes $\frac{p}{w} - 1$ time units to complete the last memory access requests by $W(\frac{p}{w} - 1)$. Hence, the contiguous access can be done in $\frac{nl}{p} + \frac{p}{w} - 1$ time units.

Case 2: $\frac{p}{w} > l$ When the memory access requests by w threads in $W(0)$ are completed, they cannot send next memory requests immediately. They must wait until w threads in $W(\frac{p}{w} - 1)$ send the memory access requests. Hence, all warps send memory access request continuously in turn as illustrated in Fig. 4. Since each of $\frac{p}{w}$ warps sends memory access requests $\frac{n}{p}$ times, it takes $\frac{p}{w} \cdot \frac{n}{p} = \frac{n}{w}$ time units to send all memory access requests. After that, the last memory access

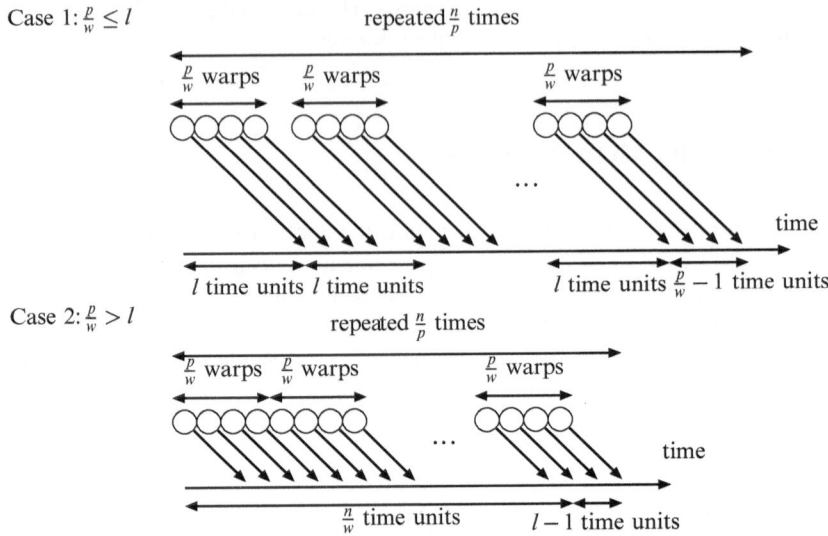

Fig. 4 The time chart of the contiguous memory access

requests by the last warp are completed in $l - 1$ time units. Hence, the contiguous access can be done in $\frac{n}{w} + l - 1$ time units.

With cases 1 and 2 combined, the prefix-sums can be computed in $O(\frac{n}{w} + \frac{nl}{p} + l)$ time units. Further, if $n < p$ then the contiguous memory access can be simply done using n threads out of the p threads. If this is the case, the memory access can be done by $O(\frac{n}{w} + l)$ time units. Therefore, we have,

Lemma 1 *The contiguous access to an array of size n can be done in $O(\frac{n}{w} + \frac{nl}{p} + l)$ time using p threads on the UMM and the DMM with width w and latency l.*

4   An Optimal Parallel Algorithm for Computing the Sum

The main purpose of this section is to show an optimal parallel algorithm for computing the sum on the memory machine models.

Let a be an array of $n = 2^m$ numbers. Let us show an algorithm to compute the sum $a[0] + a[1] + \cdots + a[n - 1]$. The algorithm uses a well-known parallel computing technique which repeatedly computes the sums of pairs. We implement this technique to perform contiguous memory access. The details are spelled out as follows:

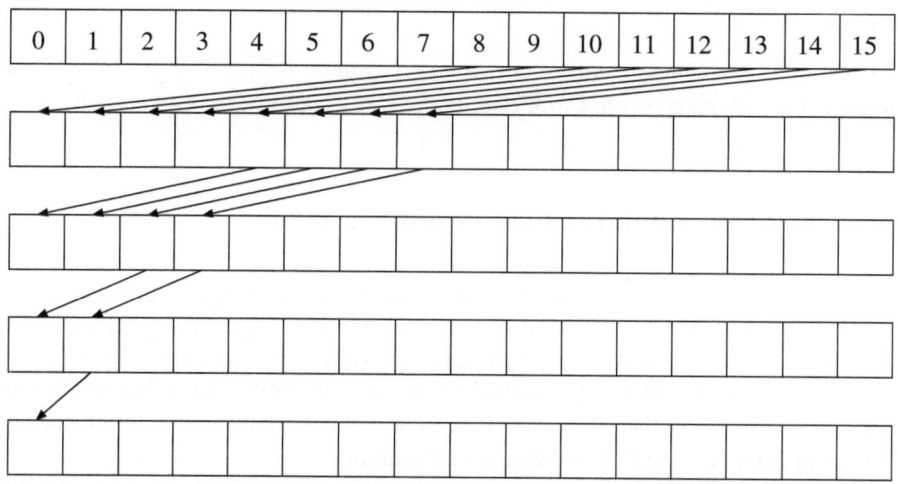

| 0 | 1 | 2 | 3 | 4 | 5 | 6 | 7 | 8 | 9 | 10 | 11 | 12 | 13 | 14 | 15 |

Fig. 5 Illustrating the summing algorithm for $n = 16$ numbers

[Optimal algorithm for computing the sum]

for $t \leftarrow m - 1$ downto 0 do

  for $i \leftarrow 0$ to $2^t - 1$ do in parallel

    $a[i] \leftarrow a[i] + a[i + 2^t]$

Figure 5 illustrates how the sums of pairs are computed. From the figure, the reader should have no difficulty to confirm that this algorithm computes the sum correctly.

We assume that p threads to compute the sum. For each t $(0 \leq t \leq m - 1)$, 2^t operations "$a[i] \leftarrow a[i] + a[i + 2^t]$" are performed. These operations involve the following memory access operations:

- reading from $a[0], a[1], \ldots, a[2^t - 1]$,
- reading from $a[2^t], a[2^t + 1], \ldots, a[2 \cdot 2^t - 1]$, and
- writing in $a[0], a[1], \ldots, a[2^t - 1]$.

Since these memory access operations are contiguous, they can be done in $O(\frac{2^t}{w} + \frac{2^t l}{p} + l)$ time using p threads both on the DMM and on the UMM with width w and latency l from Lemma 1. Thus, the total computing time is

$$\sum_{t=0}^{m-1} O(\frac{2^t}{w} + \frac{2^t l}{p} + l) = O(\frac{2^m}{w} + \frac{2^m l}{p} + lm)$$

$$= O(\frac{n}{w} + \frac{nl}{p} + l \log n)$$

and we have

Theorem 1 *The sum of n numbers can be computed in* $O(\frac{n}{w} + \frac{nl}{p} + l \log n)$ *time units using p threads on the DMM and on the UMM with width w and latency l.*

5 The Lower Bound of the Computing Time

Let us discuss the lower bound of the time necessary to compute the sum on the DMM and the UMM to show that our parallel summing algorithm for Theorem 1 is optimal. Since the sum is the last value of the prefix-sums, this lower bound discussion for the sum can be applied to that for the prefix-sums. We will show three lower bounds of the sum, $\Omega(\frac{n}{w})$-time bandwidth limitation, $\Omega(\frac{nl}{p})$-time latency limitation, and $\Omega(l \log n)$-time reduction limitation.

Since the width of the memory is w, at most w numbers in the memory can be read in a time unit. Clearly, all of the n numbers must be read to compute the sum. Hence, $\Omega(\frac{n}{w})$ time units are necessary to compute the sum. We call the $\Omega(\frac{n}{w})$-time lower bound *the bandwidth limitation*.

Since the memory access takes latency l, a thread can send at most $\frac{t}{l}$ memory read requests in t time units. Thus, p threads can send at most $\frac{pt}{l}$ total memory requests in t time units. Since at least n numbers in the memory must be read to compute the sum, $\frac{pt}{l} \geq n$ must be satisfied. Thus, at least $t = \Omega(\frac{nl}{p})$ time units are necessary. We call the $\Omega(\frac{nl}{p})$-time lower bound *the latency limitation*.

Next, we will show *the reduction limitation*, the $\Omega(l \log n)$-time lower bound. The formal proof is more complicated than those for the bandwidth limitation and the latency limitation.

Imagine that each of n input numbers stored in the shared memory (or the global memory) is a token and each thread is a box. Whenever two tokens are placed in a box, they are merged into one immediately. We can move tokens to boxes and each box can accept at most one token in l time units. Suppose that we have n tokens outside boxes. We will prove that it takes at least $l \log n$ time units to merge them into one token. For this purpose, we will prove that if we have n' tokens at some time, we must have at least $\frac{n'}{2}$ tokens l time units later. Suppose that we have n' tokens such that k of them are in k boxes and the remaining $n' - k$ tokens are out of boxes. If $k \leq n' - k$, then we can move k tokens to k boxes and can merge k pairs of tokens in l time units. After that, $n' - k$ tokens remain. If $k > n' - k$, then we can merge $n' - k$ pairs of tokens and we have k tokens after l time units. Hence, after l time units, we have at least $\max(n' - k, k) \geq \frac{n'}{2}$ tokens. Thus, we must have at least $\frac{n'}{2}$ tokens l time units later. In other words, it is not possible to reduce the number of tokens by less than half. Hence, in t time units, we have at least $\frac{n}{2^{\frac{t}{l}}}$ tokens. Since $\frac{n}{2^{\frac{t}{l}}} \leq 1$ must be satisfied, it takes at least $t \geq l \log n$ time units to merge n tokens into one. It should be clear that reading a number by a thread from the shared memory (or the global memory) corresponds to a token movement to a box. Therefore, it takes at least $\Omega(l \log n)$ time units to compute the sum of n numbers.

From the discussion above, we have

Theorem 2 *Both the DMM and the UMM with p threads, width w, and latency l take at least $\Omega(\frac{n}{w} + \frac{nl}{p} + l \log n)$ time units to compute the sum of n numbers.*

From Theorem 2, the parallel algorithm for computing the sum shown for Theorem 1 is optimal.

6 A Simple Prefix-Sum Algorithm

We assume that an array a with $n = 2^m$ numbers is given. Let us start with a well-known simple prefix-sum algorithm for array a [8,9] and show it is not optimal. The simple prefix-sum algorithm is written as follows:

[A simple prefix-sum algorithm]
for $t \leftarrow 0$ to $m - 1$ do
 for $i \leftarrow 2^t$ to $n - 1$ do in parallel
  $a[i] \leftarrow a[i] + a[i - 2^t]$

Figure 6 illustrates how the prefix-sums are computed.

We assume that p threads are available and evaluate the computing time of the simple prefix-sum algorithm. The following three memory access operations are performed for each t $(0 \leq t \leq m - 1)$ by:

- reading from $a[0], a[1], \ldots, a[n - 2^t - 1]$,

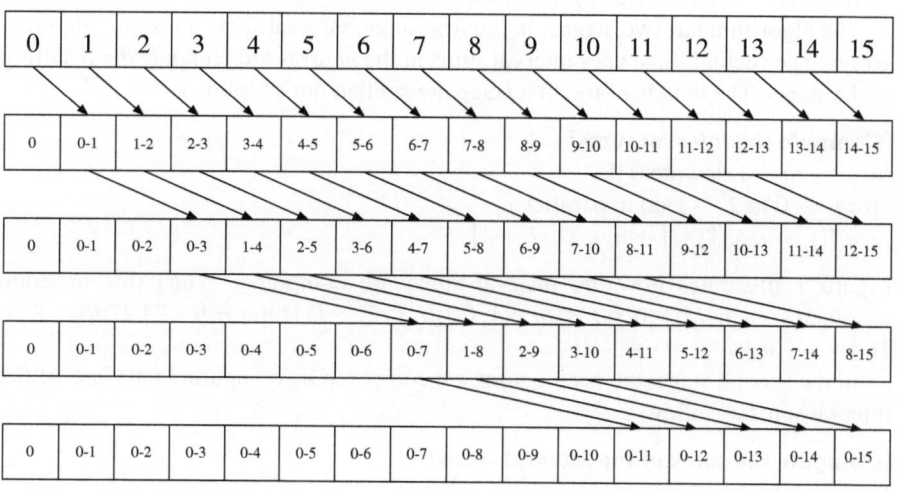

Fig. 6 Illustrating the simple prefix-sum algorithm for $n = 16$ numbers

- reading from $a[2^t], a[2^t + 1], \ldots, a[n - 1]$, and
- writing in $a[2^t], a[2^t + 1], \ldots, a[n - 1]$.

Each of the three operations can be done by contiguous memory access for $n - 2^t$ elements. Hence, the computing time of each t is $O(\frac{n-2^t}{w} + \frac{(n-2^t)l}{p} + l)$ from Lemma 1. The total computing time is

$$\sum_{t=0}^{m-1} O(\frac{n - 2^t}{w} + \frac{(n - 2^t)l}{p} + l) = O(\frac{n \log n}{w} + \frac{nl \log n}{p} + l \log n).$$

Thus, we have

Lemma 2 *The simple prefix-sum algorithm runs $O(\frac{n \log n}{w} + \frac{nl \log n}{p} + l \log n)$ time units using p threads on the DMM and on the UMM with width w and latency l.*

If the computing time of Lemma 2 matches the lower bound shown in Theorem 2, the prefix-sum algorithm is optimal. However, it does not match the lower bound. In the following section, we will show an optimal prefix-sum algorithm whose running time matches the lower bound.

7 An Optimal Prefix-Sum Algorithm

This section shows an optimal algorithm for the prefix-sums running $O(\frac{n}{w} + \frac{nl}{p} + l \log n)$ time units. We use m arrays $a_0, a_1, \ldots a_{m-1}$ as work space. Each a_t ($0 \leq t \leq m - 1$) can store 2^t numbers. Thus, the total size of the m arrays is no more than $2^0 + 2^1 + \cdots + 2^{m-1} - 1 = 2^m - 1 = n - 1$. We assume that the input of n numbers are stored in array a_m of size n.

The algorithm has two stages. In the first stage, interval sums are stored in the m arrays. The second stage uses interval sums in the m arrays to compute the resulting prefix-sums. The details of the first stage are spelled out as follows.

[Compute the interval sums]
for $t \leftarrow m - 1$ downto 0 do
  for $i \leftarrow 0$ to $2^t - 1$ do in parallel
    $a_t[i] \leftarrow a_{t+1}[2 \cdot i] + a_{t+1}[2 \cdot i + 1]$

Figure 7 illustrates how the interval sums are computed. When this program terminates, each $a_t[i]$ ($0 \leq t \leq m - 1, 0 \leq i \leq 2^t - 1$) stores $a_t[i \cdot \frac{n}{2^t}] + a_t[i \cdot \frac{n}{2^t} + 1] + \cdots + a_t[(i + 1) \cdot \frac{n}{2^t} - 1]$.

In the second stage, the prefix-sums are computed by computing the sums of the interval sums as follows:

[Compute the sums of the interval sums]
for $t \leftarrow 0$ to $m - 1$ do
  for $i \leftarrow 0$ to $2^t - 1$ do in parallel

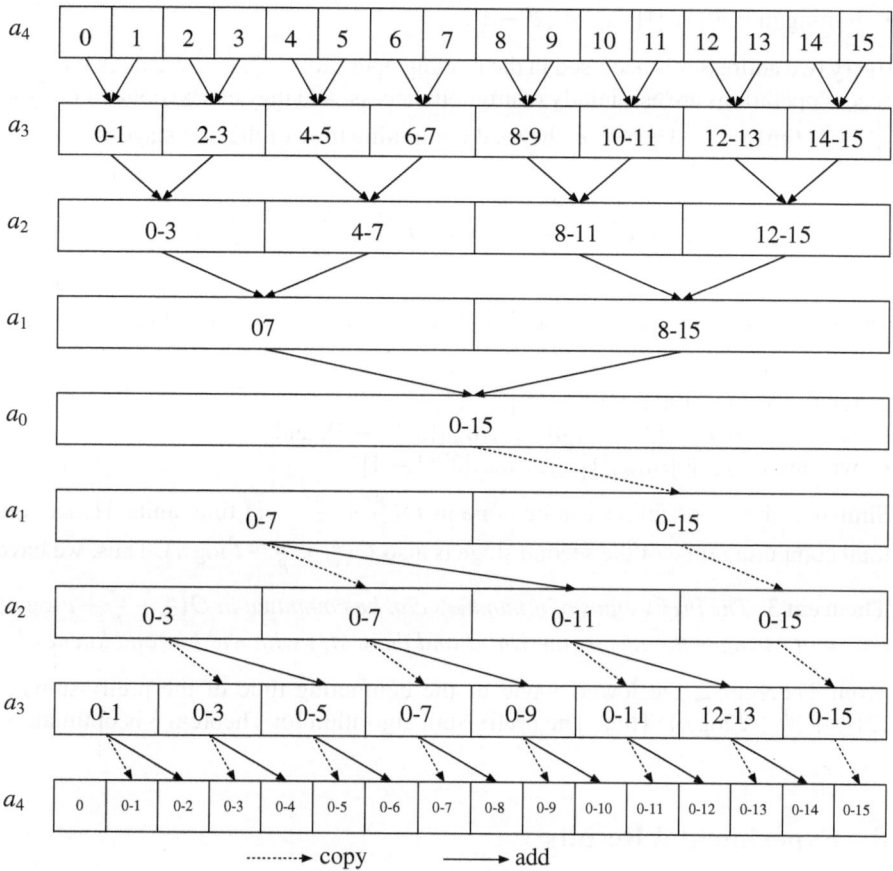

Fig. 7 Illustrating the optimal prefix-sum algorithm for $n = 16$ numbers

$$a_{t+1}[2 \cdot i + 1] \leftarrow a_t[i]$$
for $i \leftarrow 0$ to $2^t - 2$ do in parallel
$$a_{t+1}[2 \cdot i + 2] \leftarrow a_{t+1}[2 \cdot i + 2] + a_t[i]$$

Figure 7 shows how the prefix-sums are computed. In the figure, "$a_{t+1}[2 \cdot i + 1] \leftarrow a_t[i]$" and "$a_{t+1}[2 \cdot i + 2] \leftarrow a_{t+1}[2 \cdot i + 2] + a_t[i]$" correspond to "copy" and "add," respectively.

When this algorithm terminates, each $a_m[i]$ ($0 \le i \le 2^t - 1$) stores the prefix-sum $a_m[0] + a_m[1] + \cdots + a_m[i]$. We assume that p threads are available and evaluate the computing time. The first stage involves the following memory access operations for each t ($0 \le t \le m - 1$):

- reading from $a_{t+1}[0], a_{t+1}[2], \ldots, a_{t+1}[2^{t+1} - 2]$,
- reading from $a_{t+1}[1], a_{t+1}[3], \ldots, a_{t+1}[2^{t+1} - 1]$, and

- writing in $a_t[0], a_t[1], \ldots, a_t[2^t - 1]$.

Every two addresses is accessed in the reading operations. Thus, these three memory access operations are essentially contiguous access, and they can be done in $O(\frac{2^t}{w} + \frac{2^t l}{p} + l)$ time units. Therefore, the total computing time of the first stage is

$$\sum_{t=1}^{m-1} O(\frac{2^t}{w} + \frac{2^t l}{p} + l) = O(\frac{n}{w} + \frac{nl}{p} + l \log n).$$

The second stage consists of the following memory access operations for each t $(0 \le t \le m - 1)$:

- reading from $a_t[0], a_t[1], \ldots, a_t[2^t - 1]$,
- reading from $a_{t+1}[2], a_{t+1}[4], \ldots, a_{t+1}[2^{t+1} - 2]$, and
- writing in $a_{t+1}[0], a_{t+1}[1], \ldots, a_{t+1}[2^{t+1} - 1]$.

Similarly, these operations can be done in $O(\frac{2^t}{w} + \frac{2^t l}{p} + l)$ time units. Hence, the total computing time of the second stage is also $O(\frac{n}{w} + \frac{nl}{p} + l \log n)$. Thus, we have

Theorem 3 *The prefix-sums of n numbers can be computed in* $O(\frac{n}{w} + \frac{nl}{p} + l \log n)$ *time units using p threads on the DMM and the UMM with width w and latency l.*

From Theorem 2, the lower bound of the computing time of the prefix-sums is $\Omega(\frac{n}{w} + \frac{nl}{p} + l \log n)$. Thus, The prefix-sum algorithm for Theorem 3 is optimal.

8 Experimental Results

This section is devoted to show experimental results. We have implemented an algorithm for computing the sum (Theorem 1) on the shared memory and the global memory on GeForce Titan and evaluated the performance. We have also implemented the simple prefix-sum algorithm (Lemma 2) and the optimal prefix-sum algorithm (Theorem 3) on GeForce Titan. GeForce Titan has 2688 processor cores in 14 streaming multiprocessors.

Figure 8 shows the computing time of the three algorithms on GeForce Titan. The computing time is evaluated for 4,096 32-bit (float) numbers for the shared memory. We have used $\frac{n}{2}$ threads of a CUDA block for n 32-bit (float) numbers, when $n \le 2,048$ and 1,024 threads for $n = 4,096$, because a CUDA block can have up to 1,024 threads. Since the capacity of the shared memory is up to 48 KB, we can implement these algorithms up to 4,096 32-bit numbers. The running times of the sum algorithm and that of the simple prefix-sum algorithm are almost the same for small n, because the latency overhead $l \log n$ is dominant in the computing time. On the other hand, the optimal prefix-sum algorithm takes much more computing time, because its latency overhead is more than $2l \log n$. However, the bandwidth

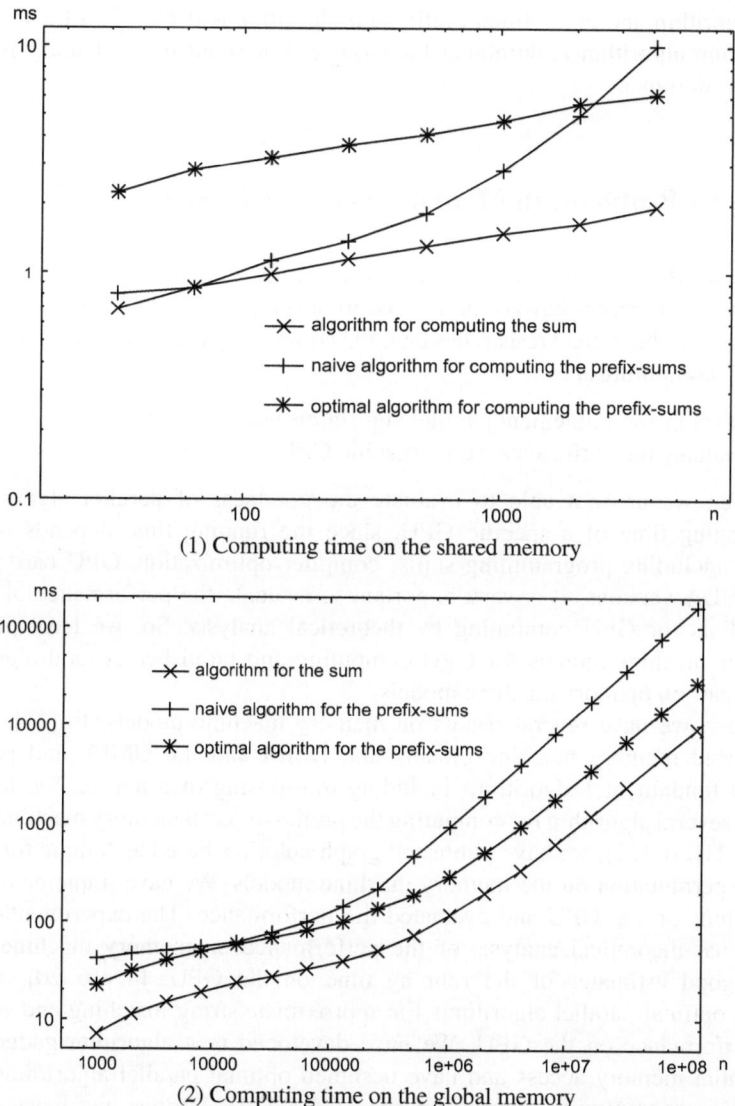

Fig. 8 The computing time of the sum and the prefix-sums on the shared memory and the global memory on GeForce Titan

overhead $\frac{n \log n}{w}$ of the naive algorithm is dominant when n is large. Hence, the running time of the naive algorithm is larger than the others for large n.

Figure 8 also shows the computing time for 1K-128M ($2^{10} - 2^{27}$) 32-bit (float) numbers on the global memory. We have used $\frac{n}{2}$ threads in multiple CUDA blocks for n 32-bit (float) numbers. We can see that the bandwidth overhead $O(\frac{n}{w})$ for the sum algorithm and the optimal prefix-sum algorithm and $O(\frac{n \log n}{w})$ for the simple prefix-sum algorithm is dominant for large n. For small n, the latency overhead $l \log n$ is dominant.

9   Open Problems in Memory Machine Models

Although GPUs have recently attracted the attention of many application developers and many researchers have been devoted to develop efficient algorithms on GPUs, there are few theoretical researches on GPU computing. Most of the research results on GPU computing are:

- to develop and implement parallel algorithms using CUDA, and
- to evaluate the performance on a specific GPU.

However, we are not able to evaluate the goodness of parallel algorithms by the running time of a specific GPU, since the running time depends on many factors including programming skills, compiler optimization, GPU card models, and CUDA versions. It is very important to evaluate the performance of parallel algorithms for GPU computing by theoretical analysis. So, we have published memory machine models for GPU computing and published several algorithmic techniques on memory machine models.

So far, we have several results on memory machine models. In [22], we first introduced memory machine models, the DMM and the UMM, and presented several fundamental algorithms including transposing of a matrix. We have presented several algorithm for computing the prefix-sums on memory machine models in [19, 21]. In [12], we have published graph coloring-based technique for optimal offline permutation on the memory machine models. We have implemented these algorithms on the GPU and evaluated the performance. The experimental results show that theoretical analysis of the performance on memory machine models gives good estimates of the running time on the GPU. In [16, 20], we have shown optimal parallel algorithms for approximate string matching and evaluated the performance on the GPU. We have developed new algorithm gadget called sequential memory access and have designed optimal parallel algorithms for the dynamic programming running on the UMM [24]. Further, we have shown a new generic algorithmic technique called random address shift on the DMM to reduce the memory access congestion to memory banks [25]. Quite recently, we have introduced Hierarchical Memory Machine (HMM), which consists of multiple DMMs and a single UMM [13, 23]. The HMM features the hierarchical structure of CUDA-enabled GPUs. Figure 9 illustrates the architecture of HMM. Each DMM corresponds to a streaming multiprocessors of CUDA-enabled GPUs. Also, all DMMs combined correspond to the UMM. Thus, theoretical analysis of algorithms on the HMM is more realistic for GPU computing.

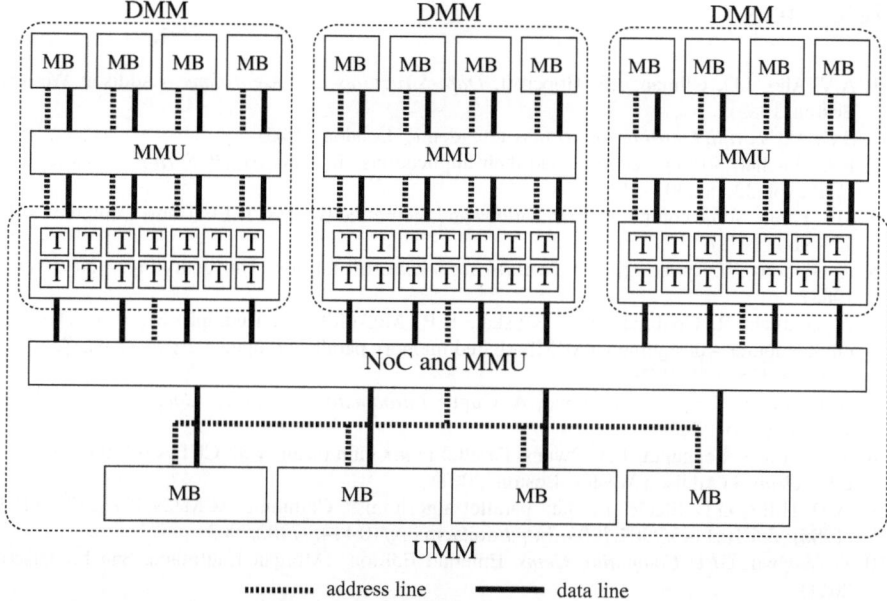

Fig. 9 The Hierarchical Memory Machine (HMM)

We still have a lot of things to do in the area of theoretical research for GPU computing. Many researchers have developed many parallel algorithmic techniques on traditional parallel computing models [5, 7, 15, 30] such as PRAMs. However, in many cases, direct implementation of the memory machine models is not efficient. We need to consider the memory access characteristics of the memory machine models when we design parallel algorithms on them. There are a lot of open problems to develop algorithmic techniques for graph problems, geometric problems, and optimization problems, among others.

Conclusion

This chapter introduces theoretical parallel computing models, the Discrete Memory Machine (DMM) and the Unified Memory Machine (UMM), that capture the essence of CUDA-enabled GPUs. We have shown that the sum and the prefix-sums can be computed in $O(\frac{n}{w} + \frac{nl}{p} + l \log n)$ time units on the DMM and the UMM. We have also shown that $\Omega(\frac{n}{w} + \frac{nl}{p} + l \log n)$ time units are necessary to compute the sum. We also show several experimental results on a CUDA-enabled GPU.

References

1. A.V. Aho, J.D. Ullman, J.E. Hopcroft, *Data Structures and Algorithms* (Addison Wesley, Boston, 1983)
2. S.G. Akl, *Parallel Sorting Algorithms* (Academic, London, 1985)
3. K.E. Batcher, Sorting networks and their applications, in *Proc. AFIPS Spring Joint Comput. Conf.*, vol. 32, pp. 307–314, 1968
4. M.J. Flynn, Some computer organizations and their effectiveness. IEEE Trans. Comput. **C-21**, 948–960 (1972)
5. A. Gibbons, W. Rytter, *Efficient Parallel Algorithms* (Cambridge University Press, New York, 1988)
6. A. Gottlieb, R. Grishman, C.P. Kruskal, K.P., McAuliffe, L. Rudolph, M. Snir, The nyu ultracomputer – designing an MIMD shared memory parallel computer. IEEE Trans. Comput. **C-32**(2), 175–189 (1983)
7. A. Grama, G. Karypis, V. Kumar, A. Gupta, *Introduction to Parallel Computing* (Addison Wesley, Boston, 2003)
8. M. Harris, S. Sengupta, J.D. Owens, Parallel prefix sum (scan) with CUDA (Chapter 39), in *GPU Gems 3* (Addison Wesley, Boston, 2007)
9. W.D. Hillis, G.L. Steele Jr., Data parallel algorithms. Commun. ACM **29**(12), 1170–1183 (1986). doi:10.1145/7902.7903. http://doi.acm.org/10.1145/7902.7903
10. W.W. Hwu, *GPU Computing Gems*, Emerald Edition (Morgan Kaufmann, San Francisco, 2011)
11. Y. Ito, K. Ogawa, K. Nakano, Fast ellipse detection algorithm using Hough transform on the GPU, in *Proc. of International Conference on Networking and Computing*, pp. 313–319, 2011
12. A. Kasagi, K. Nakano, Y. Ito, Offline permutation algorithms on the discrete memory machine with performance evaluation on the GPU. IEICE Trans. Inf. Syst. **Vol. E96-D**(12), 2617–2625 (2013)
13. A. Kasagi, K. Nakano, Y. Ito, An optimal offline permutation algorithm on the hierarchical memory machine, with the GPU implementation, in *Proc. of International Conference on Parallel Processing*, pp. 1–10, 2013
14. D.H. Lawrie, Access and alignment of data in an array processor. IEEE Trans. Comput. **C-24**(12), 1145–1155 (1975)
15. F.T. Leighton, *Introduction to Parallel Algorithms and Architectures: Arrays, Trees, Hypercubes* (Morgan Kaufmann, San Francisco, 1991)
16. D. Man, K. Nakano, Y. Ito, The approximate string matching on the hierarchical memory machine, with performance evaluation, in *Proc. of International Symposium on Embedded Multicore/Many-core System-on-Chip*, pp. 79–84, 2013
17. D. Man, K. Uda, Y. Ito, K. Nakano, A GPU implementation of computing Euclidean distance map with efficient memory access, in *Proc. of International Conference on Networking and Computing*, pp. 68–76, 2011
18. D. Man, K. Uda, H. Ueyama, Y. Ito, K. Nakano, Implementations of a parallel algorithm for computing euclidean distance map in multicore processors and GPUs. Int. J. Netw. Comput. **1**(2), 260–276 (2011)
19. K. Nakano, Asynchronous memory machine models with barrier synchronization, in *Proc. of International Conference on Networking and Computing*, pp. 58–67, 2012
20. K. Nakano, Efficient implementations of the approximate string matching on the memory machine models, in *Proc. of International Conference on Networking and Computing*, pp. 233–239, 2012
21. K. Nakano, An optimal parallel prefix-sums algorithm on the memory machine models for GPUs, in *Proc. of International Conference on Algorithms and Architectures for Parallel Processing (ICA3PP)*. Lecture Notes in Computer Science, vol. 7439 (Springer, Berlin, 2012), pp. 99–113

22. K. Nakano, Simple memory machine models for GPUs, in *Proc. of International Parallel and Distributed Processing Symposium Workshops*, pp. 788–797, 2012
23. K. Nakano, The hierarchical memory machine model for GPUs, in *Proc. of International Parallel and Distributed Processing Symposium Workshops*, pp. 591–600, 2013
24. K. Nakano, Sequential memory access on the unified memory machine with application to the dynamic programming, in *Proc. of International Symposium on Computing and Networking*, pp. 85–94, 2013
25. K. Nakano, S. Matsumae, Y. Ito, The random address shift to reduce the memory access congestion on the discrete memory machine, in *Proc. of International Symposium on Computing and Networking*, pp. 95–103, 2013
26. K. Nishida, Y. Ito, K. Nakano, Accelerating the dynamic programming for the matrix chain product on the GPU, in *Proc. of International Conference on Networking and Computing*, pp. 320–326, 2011
27. K. Nishida, Y. Ito, K. Nakano, Accelerating the dynamic programming for the optimal poygon triangulation on the GPU, in *Proc. of International Conference on Algorithms and Architectures for Parallel Processing (ICA3PP)*. Lecture Notes in Computer Science, vol. 7439 (Springer, Berlin, 2012), pp. 1–15
28. NVIDIA Corporation, NVIDIA CUDA C best practice guide version 3.1 (2010)
29. NVIDIA Corporation, NVIDIA CUDA C programming guide version 5.0 (2012)
30. M.J. Quinn, *Parallel Computing: Theory and Practice* (McGraw-Hill, New York, 1994)
31. A. Uchida, Y. Ito, K. Nakano, Fast and accurate template matching using pixel rearrangement on the GPU, in *Proc. of International Conference on Networking and Computing*, pp. 153–159, 2011

Membrane Computing: Basics and Frontiers

Gheorghe Păun

Abstract Membrane computing is a branch of natural computing inspired by the structure and the functioning of the living cell, as well as by the cooperation of cells in tissues, colonies of cells, and neural nets. This chapter briefly introduces the basic notions and (types of) results of this research area, also discussing open problems and research topics. Several central classes of computing models (called P systems) are considered: cell-like P systems with symbol objects processed by means of multiset rewriting rules, symport/antiport P systems, P systems with active membranes, spiking neural P systems, and numerical P systems.

1 Introduction

Membrane computing is a branch of natural computing initiated in [22] and having as its main goal to abstract computing models from the architecture and the functioning of living cells, considered alone or as parts of higher order structures, such as tissues, organs (brain included), and colonies of cells (e.g., of bacteria). Several classes of computing devices, called *P systems*, were introduced in this framework. Their basic features/ingredients are the following: (1) *the membrane structure*, of a cell-like (hierarchical, described by a tree) or a tissue-like (described by an arbitrary graph) type, defining compartments (also called regions), where (2) *multisets of objects* (i.e., sets with multiplicities associated with their elements) evolve according to given (3) *evolution rules* inspired from the biochemistry of the cell; the objects and the rules are placed in the compartments; the functioning of the model is *distributed*, as imposed by the compartments defined by membranes; and *parallel* evolution rules are applied simultaneously in all regions, to all objects which can evolve.

We will enter immediately into some details. What is important here is to understand membrane computing as a framework for devising computing models (where "computing" is understood in the Turing sense, of an input–output well-defined process, which provides a result after halting) which are distributed and parallel

---

G. Păun (✉)

Institute of Mathematics of the Romanian Academy, PO Box 1-764, 014700 București, Romania

e-mail: george.paun@imar.ro

© Springer International Publishing Switzerland 2014

Ç.K. Koç (ed.), *Open Problems in Mathematics and Computational Science*,

DOI 10.1007/978-3-319-10683-0_15

and rather different from the many computing devices known in (theoretical) computer science, dealing with a data structure which is not very common to computer science, but it is fundamental for biology, the multiset. We interpret the chemicals which swim in water in the compartments of a cell, from ions to large macromolecules, as atomic objects (in the etymological sense), identified by symbols in a given alphabet; their multiplicity in a compartment matters, but there is no ordering, and no positional information (like in the strings usually processed in computer science by automata, grammars, rewriting machineries). We ignore at this stage chemicals bound on membranes or on the cytoskeleton (as well as the structure of chemicals), although also such details were taken into consideration in certain variants of P systems. In turn, the rules by which the objects evolve are also inspired by the functioning of the cell; the most investigated are the multiset rewriting rules, similar to the biochemical reactions, but also many other types of rules were considered: abstraction of biological operations such as symport and antiport, membrane division, creation, separation, etc. In neural-like P systems, one uses specific operations, for instance, spiking rules.

At this level, membrane computing is interested in understanding the computing processes taking place in cells, in order to learn something possibly useful for computer science, in the same way as many (actually, most: the only exception is DNA computing, which comes with a different goal, that of using DNA, and other ingredients from biology, as a support for computations) areas of natural computing look to biology in order to improve the use/the efficiency of existing computers. The domain developed very much in this direction—but also well developed are the applications, especially in biology and biomedicine, in ecology, linguistics, computer science, economics, approximate optimization, etc. We will discuss here only some theoretical issues, and we refer the reader to [2, 3, 30] for details about applications.

When introducing a new computability model, the basic theoretical questions concern the *computing power* and the *computing efficiency*, in both cases comparing the new model with standard models and classifications in computer science: the power of Turing machines and of their restrictions and the classes in computational complexity. The equivalence in power with Turing machines is desired from two points of view: according to Turing–Church thesis, this is the maximal power an algorithmic model can achieve, and, moreover, the equivalence with Turing machine also means *programmability* (the existence of universal computing devices in the sense of universal Turing machines). In turn, the efficiency question is expected to have answers indicating a speedup when passing from Turing machines to the new model, if possible, indicating (even only theoretical) ways to solve classically intractable problems (typically, **NP**-complete problems) in a feasible time (typically, polynomial).

Membrane computing provides encouraging answers to both these questions. Most of the classes of P systems are Turing complete, even when using ingredients of a reduced complexity—a small number of membranes, rules of simple forms, and ways of controlling the use of rules directly inspired from biology, while certain classes of P systems are also efficient; they can solve **NP**-complete (even

PSPACE-complete) problems in a polynomial (often, linear) time; this speedup is obtained by means of a space-time trade-off, with the (exponential) space obtained during the computation, in a linear time, by means of bioinspired operations, such as membrane division, membrane creation, string replication, etc.

For the fifteen years since its beginning, membrane computing is quite well developed at all levels (theory, applications, software), and its bibliography is very large. The reader is advised to look for details, at various levels of developments of this research area, in [25] and in the handbook [30]. A comprehensive source of information (with many papers, PhD theses, and pre-proceedings volumes available for downloading) can be found on the P systems website from [36]. In general, we refer the reader to these bibliographical sources, so that in what follows we only specify a few references.

2 Cell-Like P Systems

We introduce now, in some detail, the basic (the first introduced and the most investigated) class of P systems, the *cell-like P systems* processing multisets of symbol objects. We discuss first the basic ingredients: membrane structure, multisets, and multiset processing rules.

2.1 Membrane Structure

The starting point is the (eukaryotic) cell and its compartmentalization by means of membranes, hierarchically arranged. We represent such a (spatial) structure in the way suggested in Fig. 1. Please notice the intuitive terminology used, the way

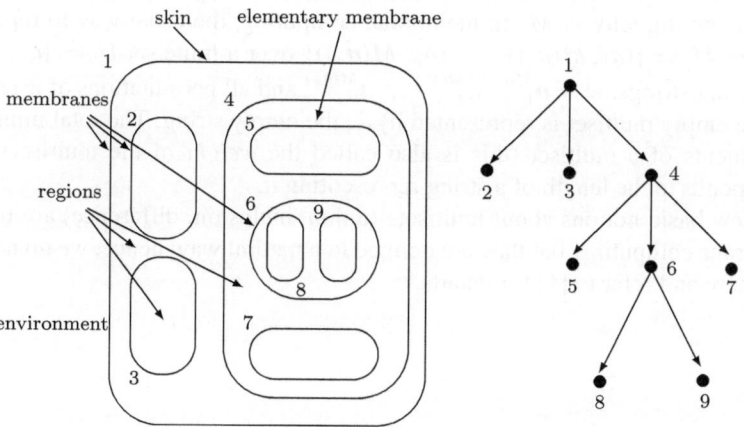

Fig. 1 A membrane structure and its tree representation

of defining compartments ("protected reactors," where specific biochemistry takes place), and the one-to-one correspondence between membranes and compartments. The membranes are usually labeled; thus, we can identify by a label both the membrane and its associated region.

A hierarchical structure of membranes can be represented by a rooted unordered tree, with labeled nodes; the tree which describes the membrane structure in Fig. 1 is given in the right hand side of the same figure. The root of the tree is associated with the skin membrane, and the leaves are associated with the elementary membranes. The tree representation directly suggests a formal way to represent a membrane structure as a string of labeled matching parentheses. For instance, a string corresponding to the structure from Fig. 1 is the following:

$$[_1 \, [_2 \,]_2 \, [_3 \,]_3 \, [_4 \, [_5 \,]_5 \, [_6 \, [_8 \,]_8 \, [_9 \,]_9 \,]_6 \, [_7 \,]_7 \,]_4 \,]_1$$

Of course, several strings can represent the same membrane structure, because the tree is not ordered; hence, membranes placed at the same level (and hence the corresponding parentheses) can interchange their place.

2.2 Multisets

In the compartments of a cell, there are various chemicals swimming in water (at this stage, we ignore the chemicals, mainly proteins, bound on membranes, but there are classes of P systems taking them into account). Therefore, the natural data structure to use in this framework is the *multiset*, the set with multiplicities associated with its elements.

Formally, a *multiset* over a given set U is a mapping $M : U \longrightarrow \mathbf{N}$, where $\mathbf{N}$ is the set of nonnegative integers. For $a \in U$, $M(a)$ is the *multiplicity of a in M*. If the set U is finite, $U = \{a_1, \ldots, a_n\}$, then the multiset M can be explicitly given in the form $\{(a_1, M(a_1)), \ldots, (a_n, M(a_n))\}$, thus specifying for each element of U its multiplicity in M. In membrane computing, the usual way to represent a multiset $M = \{(a_1, M(a_1)), \ldots, (a_n, M(a_n))\}$ over a finite set $U = \{a_1, \ldots, a_n\}$ is by using strings: $w = a_1^{M(a_1)} a_2^{M(a_2)} \ldots a_n^{M(a_n)}$ and all permutations of w represent M; the empty multiset is represented by λ, the empty string. The total multiplicity of elements of a multiset (this is also called the *weight* of the multiset) clearly corresponds to the length of a string representing it.

A few basic notions about multisets (union, inclusion, difference) are useful in membrane computing, but they are defined in a natural way; hence, we do not recall them here and refer to [1] for details.

2.3 Evolution Rules

The main way the chemicals present in the compartments of a cell evolve is by means of biochemical reactions which consume certain chemicals and produce other chemicals. In what follows, we consider the chemicals as unstructured, hence described by symbols from a given alphabet; we call these symbols *objects*. There are classes of P systems which also consider structured objects, especially described by strings, but we refer to the bibliography for details. In each compartment of a cell (of the membrane structure describing it), we have a multiset of objects, maybe the empty one. Corresponding to the biochemical reactions, we get *multiset rewriting rules*.

We write such a rule in the form $u \rightarrow v$, where u and v are multisets of objects (represented by strings over a given alphabet). The objects indicated by u are consumed and those indicated by v are produced. It is important to have in mind that both the objects and the rules are placed in the compartments of the membrane structure. The rules in a given compartment are applied only to objects in the same compartment. In order to make the compartments cooperate, we can move objects across membranes, and this can be achieved by adding *target indications* to the objects produced by a rule $u \rightarrow v$. The indications we use here are *here, in,* and *out,* with the meanings that an object associated with the indication *here* remains in the same region, one associated with the indication *in* goes immediately into an adjacent lower membrane, nondeterministically chosen, and *out* indicates that the object has to exit the membrane, thus becoming an element of the region surrounding it. For instance, we can have $aab \rightarrow (a, here)(b, out)(c, here)(c, in)$. Using this rule in a given region of a membrane structure means to consume two copies of a and one of b (they are removed from the multiset of that region), and one copy of a, one of b, and two of c are produced; the resulting copy of a remains in the same region, and the same happens with one copy of c (indication *here*), while the new copy of b exits the membrane, going to the surrounding region (indication *out*), and one of the new copies of c enters one of the children membranes, nondeterministically chosen. If no such child membrane exists, that is, the membrane with which the rule is associated is elementary, then the indication *in* cannot be followed, and the rule cannot be applied. In turn, if the rule is applied in the skin region, then b will exit into the environment of the system (and it is "lost" there; it can never come back, as there is no rule associated with the environment). In general, the indication *here* is not specified when giving a rule.

The evolution rules are classified according to the complexity of their left hand side. A rule with at least two objects in its left hand side is said to be *cooperative*; a particular case is that of *catalytic* rules, of the form $ca \rightarrow cv$, where c is an object (called catalyst) which assists the object a to evolve into the multiset v; rules of the form $a \rightarrow v$, where a is an object, are called *noncooperative*.

Biochemistry suggests various ways to extend the form of the rules, in particular ways to control their application. For instance, we can add *promoters* (objects which should be present in the compartment where the rule is applied) and *inhibitors* (objects which, if present, forbid the use of the rule). A *priority* relation can also be considered as a partial order relation on the set of rules present in a region; in each step, only rules with a maximal priority among the applicable rules can be used. A special ingredient is the membrane *dissolution* action: when applying a rule of the form $u \rightarrow v\delta$, besides the replacement of the multiset u by the multiset v, the membrane where the rule is applied is "dissolved"; its contents, object, and membranes alike, become part of the contents of the surrounding membrane, while its rules disappear with the membrane. The skin membrane is never dissolved.

There are several other types of rules for handling objects and also for handling membranes. The basic ones will be presented later. Now we pass to a crucial point in the definition of P systems: the ways of using the rules. Membranes, objects, and rules constitute the architecture of the computing model (its syntax); it is important to see how this model can function (its semantics).

2.4   Ways of Using the Rules

Having in mind the biochemical reality, the rules in a compartment of a membrane structure should be used in a *nondeterministic* (the objects to evolve and the rules by which they evolve are chosen in a nondeterministic manner) and *parallel* way. The parallelism adopted in membrane computing is the *maximal* one. More formally stated, we look to the *set* of rules and the multiset of objects from a given compartment and try to find a *multiset* of rules, by assigning multiplicities to rules, with two properties: (1) the multiset of rules is *applicable* to the multiset of objects available in the respective region, that is, there are enough objects to apply the rules a number of times as indicated by their multiplicities, and (2) the multiset is *maximal*, i.e., no further rule can be added to it (no multiplicity of a rule can be increased) such that the obtained multiset is still applicable.

Thus, an evolution step in a given region consists of finding a maximal applicable multiset of rules, removing from the region all objects specified in the left hand sides of the chosen rules (with multiplicities as indicated by the rules and by the number of times each rule is used), producing the objects from the right hand sides of the rules, and then distributing these objects as indicated by the targets associated with them.

Several alternatives are possible and were investigated in membrane computing: limited parallelism (only a given number of rules should be applied), sequential use of rules (only one at a time in each region), minimal parallelism (if a region can evolve, then at least one rule is used there), and asynchronous use of rules (no clock is assumed at the level of the system).

We continue here only with the maximal parallelism, and we define now *computations*, which are sequences of *transitions* between *configurations* of the P system, defined as above, by the maximally parallel nondeterministic use of rules in each configuration. Similar to Turing machines, we consider that a computation is successful if and only if it halts; it reaches a configuration where no rule can be applied to the existing objects. With a halting computation, we can associate a *result* in various ways. One possibility is to count the objects present in the halting configuration in a specified elementary membrane (*internal output*), or we can count the objects which leave the system during the computation (*external output*). In both cases the result is a number.

Starting from an initial configuration of the system and proceeding through transitions, because of the nondeterminism of the application of rules, we can get several halting computations, hence several results. In this way, a P system *computes* (or *generates*) a set of numbers. This corresponds to grammars in formal language theory. We can also use the system in the *accepting* mode, corresponding to automata: we start from the initial configuration, where some objects codify an input, and we accept that input if and only if the computation halts. A more general case is that of *computing* a function: we start with the argument introduced in the initial configuration, and we obtain the value of the function in the end of the computation.

2.5   A Formal Definition of a P System

Formally, a cell-like P system is a construct

$$\Pi = (O, C, \mu, w_1, \ldots, w_m, R_1, \ldots, R_m, i_{in}, i_{out}),$$

where O is the alphabet of objects, $C \subseteq O$ is the set of catalysts, μ is the membrane structure (with m membranes), given as an expression of labeled parentheses, $w_1, \ldots, w_m$ are strings over O representing multisets of objects present in the m regions of μ at the beginning of a computation, $R_1, \ldots, R_m$ are finite sets of evolution rules associated with the regions of μ, and i_{in}, i_{out} are the labels of input and output regions, respectively; i_{out} can be the environment, denoted *env*. If the system is used in the generative mode, then i_{in} is omitted, and if the system is used in the accepting mode, then i_{out} is omitted. The number m of membranes in μ is called the *degree* of Π.

In the generative case, the set of numbers computed by Π (in the maximally parallel nondeterministic mode) is denoted by $N(\Pi)$. The family of all sets $N(\Pi)$ computed by systems Π of degree at most $m \geq 1$ and using rules of α forms is denoted by $NOP_m(\alpha)$; if there is no bound on the degree of systems, then the subscript m is replaced with $*$. According to the previous classification, $\alpha \in \{ncoo, cat, coo\}$, with the obvious meaning.

2.6 A Simple Example

We illustrate the previous definitions with a simple example, of a computing P system, of a catalytic type, also using the membrane dissolution feature. The initial configuration of the system is given in a graphical representation in Fig. 2 and formally is the following one:

$\Pi = (O, C, \mu, w_1, w_2, R_1, R_2, i_{in}, i_{out})$, where:

$O = \{a, b_1, b_1', b_2, c, e\}$ (the set of objects),

$C = \{c\}$ (the set of catalysts),

$\mu = [_1 [_2]_2]_1$ (membrane structure),

$w_1 = c$ (initial objects in region 1),

$w_2 = \lambda$ (initial objects in region 2),

$R_1 = \{b_1 \rightarrow b_1, \ e \rightarrow e_{out}\}$ (rules in region 1),

$R_2 = \{a \rightarrow b_1 b_2, \ cb_1 \rightarrow cb_1', \ b_2 \rightarrow b_2 e, \ cb_1 \rightarrow cb_1'\delta\}$ (rules in region 2),

$i_{in} = 2$ (the input region),

$i_{out} = env$ (the output region = the environment).

This system computes the function $n \longrightarrow n^2$, for any natural number $n \geq 1$: any number n is introduced in the system in the form of n copies of the object a placed in region 2, and then the computation can start. It proceeds as follows.

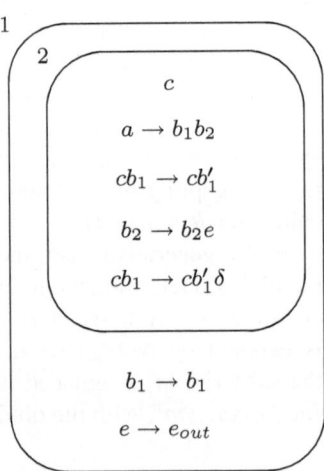

Fig. 2 A P system with catalysts and dissolution

The only rule to be applied is $a \rightarrow b_1 b_2$ in region 2. Because of the maximal parallelism, it has to be applied simultaneously to all copies of a. Thus, in one step, all n objects a are replaced by n copies of b_1 and n copies of b_2. From now on, the other rules from region 2 can be used. The catalytic rule $cb_1 \rightarrow cb_1'$ can be used only once in each step, because the catalyst is present in only one copy. This means that in each step one copy of b_1 gets primed. Simultaneously (because of the maximal parallelism), the rule $b_2 \rightarrow b_2 e$ should be applied as many times as possible, and this means n times, because we have n copies of b_2. In this way, in each step we change one b_1 to b_1' and we produce n copies of e (one for each copy of b_2). The computation should continue in region 2 (note that no rule can be applied in the skin region) as long as there are applicable rules. At any step, instead of $cb_1 \rightarrow cb_1'$, in region 2 we can use the rule $cb_1 \rightarrow cb_1'\delta$, which replaces one b_1 by b_1' but also dissolves membrane 2. All objects in region 2 are left free in region 1. If at least one object b_1 exists, then the computation will continue forever by means of the rule $b_1 \rightarrow b_1$ from region 1; hence, no result is obtained. Conversely, as long as the rule $cb_1 \rightarrow cb_1'\delta$ is not used, the rule $b_2 \rightarrow b_2 e$ in region 2 should be used; hence, again the computation is non-successful; it lasts forever if no copy of b_1 still exists. This means that the rule $cb_1 \rightarrow cb_1'\delta$ should be used, and this must be done in the moment when the last object b_1 is consumed, replaced by b_1'. Consequently, a catalytic rule is applied exactly n steps, simultaneously with using n times in parallel the rule $b_2 \rightarrow b_2 e$. In this way, n^2 copies of e are introduced in region 2 and left free in the skin region after dissolving membrane 2. The rule $e \rightarrow e_{out}$ will send immediately all these copies of e to the environment, and the computation halts. The result is the desired one, n^2.

The previous system can be easily transformed in a generative one: we only have to provide, nondeterministically, n copies of a to region 2, for all n. For instance, we can add a further membrane, with label 3, inside membrane 2, containing initially a copy of an object d and two rules:

$$d \rightarrow da, \ d \rightarrow a\delta.$$

After producing $m \geq 0$ copies of a by means of the rule $d \rightarrow da$, one introduces one further copy of a and one dissolves membrane 3 (rule $d \rightarrow a\delta$). From now on, the computation continues as above. For the modified system Π', we get $N(\Pi') = \{n^2 \mid n \geq 1\}$.

3   The Power of Catalytic P Systems

We have mentioned already that many classes of P systems are equivalent in power with Turing machines. This is not a surprise for cooperating P systems, but it is unexpected for catalytic P systems. Moreover, the number of catalysts sufficient for obtaining the computational completeness is also rather reduced, too.

Let us denote by $NP_m(cat_r)$ the family of sets of numbers $N(\Pi)$ computed (generated) by P systems with at most m membranes, using catalytic or noncooperative rules, containing at most r catalysts. When all the rules of a system are catalytic, we say that the system is *purely catalytic*, and the corresponding families of sets of numbers are denoted by $NP_m(pcat_r)$. When the number of membranes is not bounded by a specified m (it can be arbitrarily large), then the subscript m is replaced with $*$.

We denote by NRE the family of recursively enumerable (i.e., Turing computable) sets of natural numbers and by $NREG$ the family of semilinear sets of natural numbers (they are the length sets of Chomsky regular languages, hence the notation).

The following fundamental results are known:

Theorem 1   (i) $NP_2(cat_2) = NRE$, [7];
(ii) $NREG = NP_*(pcat_1) \subseteq NP_*(pcat_2) \subseteq NP_2(pcat_3) = NRE$, [10, 11].

Two resisting *open problems* appear here, related to the borderline between universality and non-universality: (1) are catalytic P systems with only one catalyst universal? (2) are purely catalytic P systems with two catalysts universal? The conjecture is that both these questions have a negative answer, but it is also felt that "one catalyst is almost universal": adding to P systems with one catalyst various features which, at the first sight, look weak enough, we already obtain the universality (see [8]); similar results were obtained also for purely catalytic P systems with two catalysts (see [5]).

Here we briefly recall (following [27]) the universality results for one catalyst P systems with additional ingredients:

- Introducing a *priority* relation among rules [22].
- Using *promoters* and *inhibitors* associated with the rules.
- Controlling the computation by means of controlling the *membrane permeability*, by actions δ (decreasing the permeability) and τ (increasing the permeability) [23].
- Besides catalytic and non-cooperating rules, also using rules for *membrane creation*, [19].
- Considering, instead of usual catalysts, *bi-stable catalysts*, [31], or *mobile catalysts*, [14].
- Imposing *target restrictions* on the used rules, [8]; the universality was obtained for P systems with 7 membranes, and it is an open problem whether or not the number of membranes can be diminished.
- Imposing to P systems the idea from *time-varying* grammars and splicing systems, [8]; the universality of time-varying P systems is obtained for one catalyst P systems with only one membrane, having the period equal to 6, and it is an open question whether the period can be decreased.
- Using in a transition only (labeled) rules with the same label—so-called *label restricted* P systems [15].

Several of these results were extended in [5] to purely catalytic P systems with two catalysts. It remains open to do this for all the previous results, as well as to look for further ingredients which, added to one catalyst P systems or to purely catalytic P systems with two catalysts, can lead to universality. It would be interesting to find such ingredients which work for one catalyst systems and not for purely catalytic systems with two catalysts, and conversely.

We end this section with a somewhat surprising issue: we know that $NP_2(cat_2) = NRE$, but no example of a P system with two catalysts which generates a nontrivial set of numbers (for instance, $\{2^n \mid n \geq 1\}, \{n^2 \mid n \geq 1\}$) is known. In fact, the problem is to find a system of this kind as simple as possible (otherwise, just repeating the construction in the proof from [7], starting from a register machine computing a set as above, we get an example, but of a large size). A first answer to this question is given in [33], where a catalytic P system with 54 rules is produced, but it is expected that this number could be reduced.

4 Efficiency: P Systems with Active Membranes

We proceed now to the second main question to investigate for any new computing model: the efficiency. We have mentioned that for P systems as considered above, the so-called Milano theorem was proved in [35]: such systems can be simulated in polynomial time by means of Turing machines; hence (assuming that $\mathbf{P} \neq \mathbf{NP}$, as one usually expects), these systems cannot solve $\mathbf{NP}$-complete problems in polynomial time. This is somewhat surprising because even noncooperative P systems can generate exponentially many objects in linear time: just consider a rule of the form $a \rightarrow aa$; because of the maximal parallelism, in n steps we get 2^n copies of object a. This exponential workspace does not help, and the intuition is that this happens because this space is not structured; the same rules are applied to all exponentially many objects. The situation changes if further membranes can be created, thus introducing a structure in the set of objects.

This also corresponds to the situation in biology, where also the membranes evolve: they can be divided, created, and destroyed, while also operations like exo-, endo-, and phagocytosis are met, etc. All of these kinds of operations were considered in membrane computing. Particularly interesting are the operations of membrane division and membrane creation, because they bring efficiency to P systems. We present here only the operation of membrane division, and the framework is that of *P systems with active membranes*, introduced in [24], which are constructs of the form

$$\Pi = (O, H, \mu, w_1, \ldots, w_m, R),$$

where $O, w_1, \ldots, w_m$ are as in a P system with multiset rewriting rules; H is a finite set of *labels* for membranes; μ is a membrane structure of degree $m \geq 1$, with polarizations associated with membranes, that is, "electrical charges" $\{+, -, 0\}$; and R is a finite set of rules, of the following forms:

(a) $[_h a \to v]_h^e$, for $h \in H, e \in \{+, -, 0\}, a \in O, v \in O^*$
    (object evolution rules)

(b) $a[_h \ ]_h^{e_1} \to [_h b]_h^{e_2}$, for $h \in H, e_1, e_2 \in \{+, -, 0\}, a, b \in O$
    (in communication rules)

(c) $[_h a \ ]_h^{e_1} \to [_h \ ]_h^{e_2} b$, for $h \in H, e_1, e_2 \in \{+, -, 0\}, a, b \in O$
    (out communication rules)

(d) $[_h a \ ]_h^e \to b$, for $h \in H, e \in \{+, -, 0\}, a, b \in O$
    (dissolving rules)

(e) $[_h a \ ]_h^{e_1} \to [_h b \ ]_h^{e_2} [_h c \ ]_h^{e_3}$, for $h \in H, e_1, e_2, e_3 \in \{+, -, 0\}, a, b, c \in O$
    (division rules for elementary membranes; in reaction with an object, the membrane is divided into two membranes with the same label, and possibly of different polarizations; the object specified in the rule is replaced in the two new membranes possibly by new objects; the remaining objects are duplicated and may evolve in the same step by rules of type (a)).

Note that each rule has specified the membrane where it is applied; the membrane is part of the rule; that is why we consider a global set of rules, R. The rules are applied in the maximally parallel manner, with the following details: a membrane can be subject of only one rule of types (b)–(e); inside each membrane, the rules of type (a) are applied in parallel; each copy of an object is used by only one rule of any type. The rules are used in a bottom-up manner: we use first the rules of type (a) and then the rules of other types; in this way, in the case of dividing membranes, the result of using first the rules of type (a) is duplicated in the newly obtained membranes. As usual, only halting computations give a result, in the form of the number of objects expelled into the environment during the computation.

Several generalizations are possible. For instance, a division rule can also change the labels of the involved membranes:

(e') $[_{h_1} a \ ]_{h_1}^{e_1} \to [_{h_2} b \ ]_{h_2}^{e_2} [_{h_3} c \ ]_{h_3}^{e_3}$,
    for $h_1, h_2, h_3 \in H, e_1, e_2, e_3 \in \{+, -, 0\}, a, b, c \in O$.

The change of labels can also be considered for rules of types (b) and (c). Also, we can consider the possibility of dividing membranes in more than two copies or even of dividing nonelementary membranes (in such a case, all inner membranes are duplicated in the new copies of the membrane).

P systems with active membranes can be used for computing numbers, in the usual way, but the main usefulness of them is in devising polynomial time solutions to computationally hard problems by a time-space trade-off. The space is created both by duplicating objects and by dividing membranes. An encoding of an instance of a decidability problem is introduced in the initial configuration of the P system (in the form of a multiset of objects), the computation proceeds, and if it halts and a special object yes is sent to the environment, then the respective instance has the positive answer.

A large research area starts at this point. The first important step is to formally define complexity classes for P systems with active membranes. This has been already done since several years—we refer to [32] and to its references. The respective classes refer to a parallel computing time, which also covers the steps for producing the exponential space necessary to the computation. Of course, very important are the ingredients used by the considered P systems: using or not using a specified type of rules (for instance, membrane dissolution rules)? how many polarizations, three (as in the initial definition) or a smaller number? how much time is allowed for constructing the system(s) which will solve a given decidability problem?

An intriguing question appears here. In classic computational complexity, a problem Q is solved in the *uniform* way: we have to start from Q (and its size parameters) when constructing the algorithm A_Q which solves Q and not from instances $Q(1), Q(2), \ldots$ of Q; A_Q depends on Q, while encodings of instances $Q(i)$ are introduced in A_Q and their answer is provided. Starting from an instance $Q(i)$ and building an algorithm $A_{Q(i)}$ for solving that instance do not look fair; we can solve $Q(i)$ during "programming" $A_{Q(i)}$ and then provide the answer in a time which is not the correct one.

Interesting enough, in many experiments in DNA computing, one however proceeds in this nonuniform way, constructing the "wet computer" starting from an instance of the problem, not from the problem itself. This can be accepted also in membrane computing and even in complexity theory, provided that the time for constructing $A_{Q(i)}$ is carefully limited. If this happens, then we cannot cheat "too much," working on solving the problem itself during the programming phase. We call *semi-uniform* a solution to a problem Q obtained in this way. (Again, formal definitions can be found in the literature, e.g., in [32].) The problem now is natural: which is the relation between uniform and semi-uniform complexity classes? (Surprisingly enough, the classic complexity theory seems to not have examined this natural question.) When a problem can be solved in a semi-uniform way, can it be also solved (in the same amount of time) in the uniform way? (In many cases, in membrane computing, uniform and semi-uniform families coincide—but not in all cases! Moreover, while initially semi-uniform solutions were reported to various problems, nowadays only uniform solutions are considered acceptable.)

Another "nonstandard" question concerns the possibility to use *precomputed resources*, and the suggestion comes again from biology. It is known that the brain contains a huge number of neurons, but in each moment only a small part of them seem to be active. The same is known for the liver, which in each moment uses only part of its cells, depending on the task it has to cope with. Can this idea be also used in computability? Roughly speaking, we can assume as given "for free" (we do not care which is the time for constructing it) a computing device which is initially "arbitrarily large," but it contains only a limited amount of information; a decidability problem is introduced in this initial configuration, the information spreads across the arbitrarily large workspace, and the answer is provided in a specified way. The strategy is rather natural; it can be useful in many circumstances where we have "enough" time before the computation itself (this can be the case, e.g., in cryptography); hence, we can prepare in advance an arbitrarily large "computer," without too much data inside, which is fed with the problem to solve in the moment when the problem appears. Still, this way of solving problems is not yet investigated in complexity theory. Actually, neither in membrane computing we do not have a formal framework for this approach, although this strategy has been used several times in solving computationally hard problems in a polynomial time (especially in cases when the exponential workspace cannot be produced during the computation).

Another technical detail, specific to membrane computing, is the fact that, in their general form, the P systems are nondeterministic computing devices. However, when solving a problem, we want to have a deterministic solution. This difficulty can be solved either by working only with deterministic P systems or, more adequately/realistic, by leaving the system to be nondeterministic, but asking to have a behavior which guarantees that the solution is the "real" one. This can be achieved if we can ensure that the system is *confluent*: the computation proceeds nondeterministically, but either eventually all computations reach the same configuration (strong confluence) and after that the computation continues in a deterministic way, or the computation is nondeterministic as a whole, but all computations halt and provide the same answer (weak confluence).

The complexity investigations are among the most active in membrane computing at this moment. Besides membrane division, membrane creation (by means of rules of the form $a \rightarrow [_h b]_h$, where a, b are objects and h is a label), string replication and other operations were used. The reader is referred to the complexity chapter from [30] and to the literature available at [36] for details—including many problems which are still open in this area.

5   Other Important Classes of P Systems

We briefly discuss now three important classes of P systems, with fundamental differences with respect to the P systems considered in the previous sections. For two of them (symport/antiport and spiking neural P systems) the motivation comes from biology; the third class (numerical P systems) has a motivation related to economics. Each of these variants of P systems gave birth to a strong branch of membrane computing, still with many open problems and research topics waiting to be addressed. (Numerical P systems also have a surprising area of applications—robot control. We strongly believe that these systems can find applications in many other areas where functions of several variables should be computed in an efficient way.)

We do not introduce here the tissue-like P systems, although their study is well developed, both in what concerns the theory (power and efficiency) and applications: the more general graph structure describing the arrangement of membranes/cells can cover more phenomena than the cell-like structure.

5.1 Symport/Antiport P Systems

In the functioning of the cell, one of the most interesting (and important) ways to pass chemicals across membranes is by means of protein channels, which can select in various ways the transported chemicals. We consider here the coupled passage of chemicals through protein channels, the operations called *symport* (two—or more— chemicals pass together across a membrane, in the same direction) and *antiport* (the case when the chemicals move in opposite directions).

We can formalize these operations by considering symport rules of the form (x, in) and (x, out) and antiport rules of the form $(z, out; w, in)$, where x, z, and w are multisets of arbitrary size; one says that the length of x, denoted $|x|$, is the *weight* of the symport rule and $\max(|z|, |w|)$ is the *weight* of the antiport rule. Such rules just move objects across membranes, but they can replace the multiset rewriting rules considered in the previous sections, and we get in this way a class of P systems which are again computationally universal, equivalent in power with Turing machines.

Formally, a *P system with symport/antiport rules* is a construct of the form

$$\Pi = (O, \mu, w_1, \ldots, w_m, E, R_1, \ldots, R_m, i_{in}, i_{out}),$$

where all components $O, \mu, w_1, \ldots, w_m, i_{in}, i_{out}$ are as in a P system with multiset rewriting rules, $E \subseteq O$, and $R_1, \ldots, R_m$ are finite sets of symport/antiport rules associated with the m membranes of μ. The objects of E are supposed to be present in the environment of the system with an arbitrary multiplicity.

We define transitions, computations, and halting computations in the usual way, making use of the nondeterministic maximally parallel mode of applying the rules. A system can be used in the generating, accepting, or computing mode.

The symport/antiport P systems were introduced in [21] and investigated in a large number of papers; the reader is referred to Chapter 5 of [30] for details and references. From this chapter we recall one example, illustrating the power of symport/antiport rules (used in the maximally parallel way), namely, Example 5.3. It is a P system working in the accepting mode: a number of objects is introduced in region 1, and this number is accepted if the computation halts. Here is the system:

$$\Pi = (\{a\}, [_1[_2[_3]_3]_2]_1, \lambda, \lambda, \lambda, \{a\}, R_1, R_2, R_3, 1),$$
$$R_1 = \{(aa, out; a, in)\},$$
$$R_2 = \{(a, in)\},$$
$$R_3 = \{(aa, out), (aa, in)\}.$$

The system is given in the graphical form in Fig. 3: the rules are written near the corresponding membranes; arbitrarily many copies of a are assumed to be present in the environment.

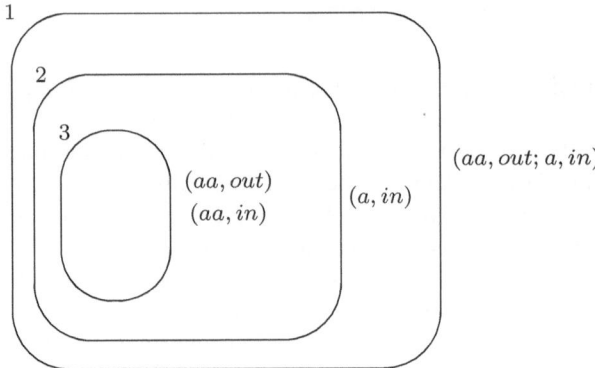

Fig. 3 A symport/antiport P system

We introduce a number n of copies of object a in region 1. By using the rule $(aa, out; a, in) \in R_1$ in a maximally parallel way, this number is repeatedly divided by 2, with a remainder in the case the reached number is odd. If a remainder exists, then the rule $(a, in) \in R_2$ must be used. We finish the halving of objects in region 1 with only one copy of a in region 2 if and only if n is of the form 2^k, for some $k \geq 0$; otherwise at least two copies of a arrive to region 2. In the latter case, the rules of R_3 can be used forever; hence, the computation never halts. Consequently, the set of numbers accepted by the system Π is $\{2^k \mid k \geq 0\}$.

Symport/antiport P systems (with reduced weights) are universal—we refer to [30] for details.

Not explored are the computational complexity properties of these systems; note that in the previous form, symport/antiport P systems do not have the possibility of dividing membranes (or other similar operations useful for generating an exponential workspace in a linear time), but such operations can be introduced and then "fypercomputations" are expected, in the sense of [26].

6 Spiking Neural P Systems

Spiking neural P systems (SN P systems) have a completely different architecture and functioning, as they start not from the cell but from the brain biology. Actually, we only consider here the neuron cooperation by means of spikes, a feature also much investigated in the neural computing (see, e.g., [18]). We do not define formally the SN P systems, but we only describe informally such a system, followed by an example.

In short, an SN P system consists of a set of *neurons* (represented by membranes) placed in the nodes of a directed graph (the arcs are called *synapses*) and containing *spikes*, denoted by the symbol a. Thus, the architecture is that of a tissue-like P system, with only one kind of objects present in the cells. The objects evolve by means of *spiking rules*, which are of the form $E/a^c \rightarrow a; d$, where E is a regular expression over $\{a\}$ and c, d are natural numbers, $c \geq 1, d \geq 0$. The meaning is that a neuron containing k spikes such that $a^k \in L(E), k \geq c$, can consume c spikes and produce one spike, after a delay of d steps. This spike is sent to all neurons to which a synapse exists outgoing from the neuron where the rule was applied. There also are *forgetting rules*, of the form $a^s \rightarrow \lambda$, with the meaning that $s \geq 1$ spikes are removed, provided that the neuron contains exactly s spikes. The system works in a synchronized manner, i.e., in each time unit, each neuron which can use a rule should do it, but the work of the system is sequential in each neuron: only (at most) one rule is used in each neuron. One of the neurons is considered to be the *output* one, and its spikes are also sent to the environment. The moments of time when a spike is emitted by the output neuron are marked with 1; the other moments are marked with 0. This binary sequence is called the *spike train* of the system—it might be infinite if the computation does not stop.

The result of a computation is encoded in the distance between the first two spikes sent into the environment by the (output neuron of the) system. Other ways to associate a result with a computation were considered; the spike train itself can be taken as the result of the computation, and in this way the system generates a binary sequence (a finite string, if the computation halts).

We recall an example from paper [12] where the SN P systems were introduced. It is given in Fig. 4, thus also suggesting the usual way of representing SN P systems. (The rules $E/a^c \rightarrow a; d$ with $L(E) = a^c$ are written in the simplified way $a^c \rightarrow a; d$.)

The neuron *out* spikes in step 1 by means of the rule $a^3 \rightarrow a; 0$. All its spikes are consumed. The spike emitted goes immediately to the environment and to neuron 1. The spike goes along the path $1, 2, \ldots, n - 2$ and gets doubled when passing from neuron $n - 2$ to neurons $n - 1$ and 0. Both these last neurons get fired. As long as

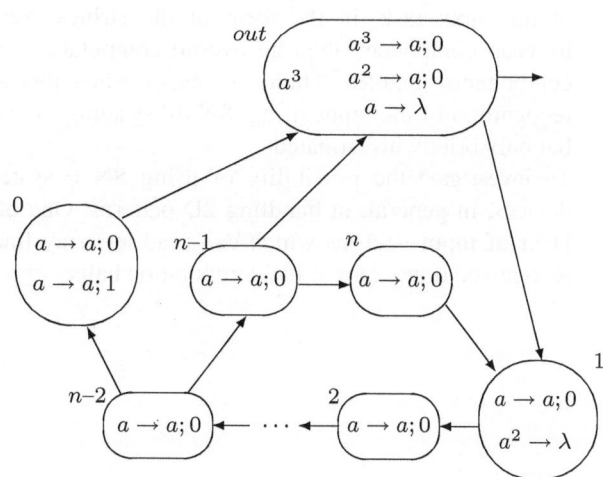

Fig. 4 An SN P system generating an arithmetical progression

neurons 0 and $n - 1$ spike in different moments (because neuron 0 can use either of its rules, hence also the one with delay), no further spike exits the system (neuron *out* gets only one spike and forgets it immediately), and one passes along the cycle of neurons $1, 2, \ldots, n - 1, n$ again and again. If neurons 0 and $n - 1$ spike at the same time (neuron 0 uses the rule $a \rightarrow a; 0$), then the system spikes again—hence in a moment of the form $ni, i \geq 1$. The spike of neuron *out* arrives at the same time in neuron 1 with the spike of neuron n, and this halts the computation, because of the rule $a^2 \rightarrow \lambda$, which consumes the spikes present in the system. Consequently, the system computes the arithmetical progression $\{ni \mid i \geq 1\}$.

There are several classes of SN P systems using various combinations of ingredients (rules of restricted forms, e.g., without a delay, without forgetting rules, or extended rules, e.g., producing more than one spike), as well as asynchronous SN P systems (no clock is considered; any neuron may use or not use a rule), with exhaustive use of rules (when enabled, a rule is used as many times as made possible by the spikes present in a neuron), with certain further conditions imposed to the halting configuration, etc. For most SN P systems with unbounded neurons (arbitrarily many spikes can be found in each of them), characterizations of Turing computable sets of natural numbers are obtained. When the neurons are bounded, characterizations of the family *NREG* are usually obtained. SN P systems can also be used in the accepting and the computing ways.

Many questions remain to be investigated in this area, and the membrane computing literature contains several collections of such research topics. We recall here only two, also mentioned recently in [27]:

- To further investigate the power and the properties of SN dP systems, that is, to combine the idea of distributed P systems introduced in [29] with that of spiking neural P systems. The dP systems consider "systems of P systems," each one with its own input, in the form of a string, communicating among them by means of antiport rules; the concatenation of the input strings is accepted if the whole system eventually halts. Note that we have here an explicit splitting of the input task, in the form of the strings "read" from the environment by each component, then an overall computation, with the cooperation of all components/modules. There are cases when this strategy can speed up the recognition of the input string. SN dP systems were already introduced in [13], but only briefly investigated.
- To investigate the possibility of using SN P systems as pattern recognition devices, in general, in handling 2D patterns. One of the ideas is to consider a layer of input neurons which can read an array line by line, and the array is recognized if and only if the computation halts.

7 Numerical P Systems

This class of P systems looks somewhat "exotic" in the framework of membrane computing. It takes only the membrane structure from cell-like P systems, but, instead of multisets of objects, uses numerical variables placed in compartments and evolving according to *production-repartition programs* inspired from economics.

Such a program has the form

$$F(x_{1,i}, \ldots, x_{k,i}) \rightarrow c_1|v_1 + c_2|v_2 + \cdots + c_n|v_n$$

where F is a function of k variables, $x_{1,i}, \ldots, x_{k,i}$ are (part of the) variables in region i, $c_1, \ldots, c_n$ are natural numbers, and $v_1, v_2, \ldots, v_n$ are variables from region i and from the parent and the children regions. The idea is that using the function F, one computes "the production" of region i at a given time, and this production is distributed to variables $v_1, v_2, \ldots, v_n$ proportionally with the coefficients $c_1, \ldots, c_n$. More formally, let

$$C = \sum_{s=1}^{n} c_s.$$

At a time instant $t \geq 0$, we compute $F(x_{1,i}(t), \ldots, x_{k,i}(t))$. The value $q = F(x_{1,i}(t), \ldots, x_{k,i}(t))/C$ represents the "unitary portion" to be distributed according to the repartition expression to variables $v_1, \ldots, v_n$. Thus, v_s will receive $q \cdot c_s, 1 \leq s \leq n$.

A production function may use only part of the variables from a region. Those variables "consume" their values when the production function is used (they become zero)—the other variables retain their values. To these values—zero in the case of variables contributing to the region production—one adds all "contributions" received from the neighboring regions.

Thus, a *numerical P system* is a construct of the form

$$\Pi = (\mu, (Var_1, Pr_1, Var_1(0)), \ldots, (Var_m, Pr_m, Var_m(0)), x_{j_0,i_0}),$$

where μ is a membrane structure with m membranes labeled injectively by $1, 2, \ldots, m$, Var_i is the set of variables from region i, Pr_i is the set of programs from region i (all sets Var_i, Pr_i are finite), $Var_i(0)$ is the vector of initial values for the variables in region i, and x_{j_0,i_0} is a distinguished variable (from a distinguished region i_0), which provides the result of a computation.

Such a system evolves in the way informally described before. Initially, the variables have the values specified by $Var_i(0), 1 \leq i \leq m$. A transition from a configuration at time t to a configuration at time $t + 1$ is made by (1) choosing nondeterministically one program from each region, (2) computing the value of the respective production function for the values of local variables at time t, and then (3) computing the values of variables at time $t + 1$ as directed by repartition protocols. A

Fig. 5 An example of a
numerical P system

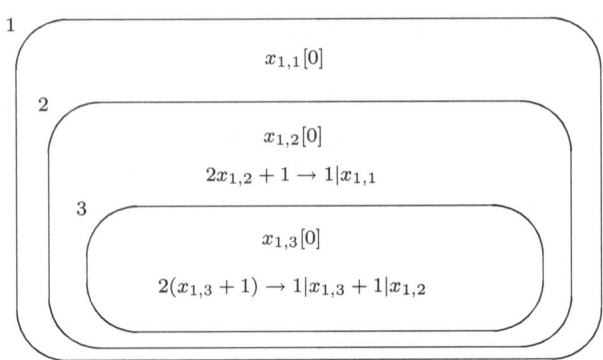

sequence of such transitions forms a computation, with which we associate a set of numbers, namely, the numbers which occur as positive values of the variable x_{j_0,i_0}; this set of numbers is denoted by $N^+(\Pi)$. If all numbers, positive or negative, are taken into consideration, then we write $N(\Pi)$.

Figure 5 gives an example of a deterministic numerical P system (only one program in each region), with the distinguished variable $x_{1,1}$. One can easily see that variable $x_{1,3}$ increases by 1 at each step, also transmitting its value to $x_{1,2}$. In turn, region 2 transmits the value $2x_{1,2} + 1$ to $x_{1,1}$, which is never consumed; hence, its value increases continuously. In the initial configuration all variables are set to 0. Thus, $x_{1,1}$ starts from 0 and continuously receives $2i + 1$, for $i = 0, 1, 2, 3, \ldots$, which implies that in $n \geq 1$ steps the value of $x_{1,1}$ becomes $\sum_{i=0}^{n-1}(2i + 1) = n^2$, and consequently $N(\Pi) = \{n^2 \mid n \geq 0\}$.

We denote by $NN^+ P_m(poly^n(r), div)$ the family of sets $N^+(\Pi)$ generated by numerical P systems with $m \geq 1$ membranes, using polynomials of degree $n \geq 0$ with at most $r \geq 0$ variables as production functions; div indicates the fact that we consider only systems whose programs have the property that the production at any moment, $F(x_{1,i}(t), \ldots, x_{k,i}(t))$, is divisible by the sum of repartition coefficients, C. Variants can be considered, e.g., with the remainder being lost, or carried to the next production in that region. If the system is deterministic, we add D in front of the notation. Any parameter which is not bounded is replaced with $*$.

Somewhat expected, numerical P systems are computationally complete [28]:

Theorem 2 $NRE = NN^+ P_8(pol^5(5), div) = NN^+ P_7(poly^5(6), div)$.

In the framework of robot control (see, e.g., [20, 34]), two important extensions were introduced. First, a catalyst-like control on using the programs was considered, introducing *enzymatic* programs, of the form

$$F(x_{1,i}, \ldots, x_{k,i})|_{e_{j,i}} \rightarrow c_1|v_1 + c_2|v_2 + \cdots + c_n|v_n,$$

where $e_{j,i}$ is a variable from Var_i different from $x_{1,i}, \ldots, x_{k,i}$ and from $v_1, \ldots, v_n$. Such a program is applicable at a time t only if $e_{j,i}(t) > \min(x_{1,i}(t), \ldots, x_{k,i}(t))$.

Using enzymes helps (in particular, the selection of positive values of the output variable can be done inside the system, not imposed as an external condition):

Theorem 3 $NRE = NNP_7(poly^5(5), enz)$.

Very important are the ways to define the transitions in a parallel way, using more programs at a time. Two possibilities were considered: (1) all programs in a region (which can be applied, under enzymes control) are applied, but each variable is used only once (this is called the *oneP* mode), and (2) as above, with each variable appearing as many times as necessary (denoted *allP*).

The following results were obtained in [16]:

Theorem 4 $NRE = NNP_1(poly^1(1), enz, allP) = NNP_1(poly^1(1), enz, oneP)$.

Natural problems appear in this context: What about the nonenzymatic systems? Improve the parameters in Theorem 2, for the sequential case.

Only recently numerical P systems were investigated from the computational complexity point of view [17], and the results are rather interesting. Here is the framework: a language $L \subseteq \{0, 1\}^*$ is decided by a numerical P system Π in polynomial time if Π contains two variables acc, rej, and after introducing the number $1x$ (for all $x \in L$) in a specified variable, Π halts in $O(|x|^k)$ time, and:

- If $x \in L$, then $acc = 1, rej = 0$.
- If $x \notin L$, then $acc = 0, rej = 1$.

We denote by **P-ENP**$(X), X \subseteq \{+, -, \times, \div\}$, the corresponding complexity class, when using enzymatic numerical P systems working in the *allP* mode; the set X indicates the operations used in the production functions. The following characterizations of **P** and **PSPACE** were obtained in [17]:

Theorem 5 (i) $P\text{-}ENP(+, -) = P$, (ii) $P\text{-}ENP(+, -, \times, \div) = PSPACE$.

Also in this case there are several questions which remain to be investigated: what about sequential systems, about systems working in the *oneP* mode, about nonenzymatic systems?

All these results (universality and efficiency—when using all four arithmetical operations) look very promising from the application point of view, while the robot control case is also encouraging. Numerical P systems deserve further research efforts.

8 Closing Remarks

This chapter has only presented some basic facts (notions and results) of membrane computing, at a rather informal level, always pointing also to research topics which wait to be investigated. Technical details and comprehensive lists of open problems can be found in the domain literature, in particular through the membrane computing website at [36]. From the point of view of application, we find particularly interesting the computational complexity results: P systems of various types can solve hard problems in a feasible time (due to the inherent massive parallelism, distribution, possibility of creating an exponential workspace in a linear time, by means of operations directly inspired from biology). It is important however to note that at this time there is no laboratory implementation of a P system. In exchange, there are many software products, helping to simulate P systems on usual computers, on grids and networks, on parallel hardware (such as GPU—Graphics Processing Units), and on other electronic supports. Based on such software and implementations, significant applications were reported, especially in biology, biomedicine, ecology, and approximate optimization. The domain is fast evolving, so that the reader interested in this research area is advised to watch progress through the mentioned website or to keep in touch with the membrane computing community.

References

1. C.S. Calude, Gh. Păun, G. Rozenberg, A. Salomaa (eds.) *Multiset Processing. Mathematical, Computer Science, and Molecular Computing Points of View*. LNCS, vol. 2235 (Springer, Berlin, 2001)
   The first international meeting devoted to membrane computing was organized already in the summer of 2000, in Curtea de Argeş, Romania, and it was concerned both with the developments in the emerging research area of membrane computing and with the mathematical and computer science investigations of multisets. This LNCS volume is the proceedings of the workshop, edited after the meeting.
2. G. Ciobanu, Gh. Păun, M.J. Pérez-Jiménez (eds.) *Applications of Membrane Computing* (Springer, Berlin, 2006)
   The volume presents several classes of applications (in biology and biomedicine, computer science, linguistics), as well as the software available at the time of editing the book, and a selective bibliography of membrane computing. Here are the sections of the chapter *Computer Science Applications*: Static sorting P systems; Membrane-based devices used in computer graphics; An analysis of a public key protocol with membranes; Membrane algorithms: approximate algorithms for NP-complete optimization problems; and Computationally hard problems addressed through P systems.
3. P. Frisco, M. Gheorghe, M.J. Pérez-Jiménez (eds.) *Applications of Membrane Computing in Systems and Synthetic Biology*. (Springer, Berlin, 2014)
   Different from volume [2], this time only applications in biology and biomedicine are concerned, at the level of year 2013, with a detailed biological motivation, in most cases reporting research done in interdisciplinary teams, including both biologists and computer scientists.
4. R. Freund, Particular results for variants of P systems with one catalyst in one membrane, in *Proc. Fourth Brainstorming Week on Membrane Computing*, vol. II (Fénix Editora, Sevilla, 2006), pp. 41–50
5. R. Freund, Purely catalytic P systems: Two catalysts can be sufficient for computational completeness, in *Proc. 14th Intern. Conf. on Membrane Computing* (Chişinău, Moldova, 2013), pp. 153–166
6. R. Freund, O.H. Ibarra, A. Păun, P. Sosík, H.-C. Yen, Catalytic P systems. Chapter 4 of [30]
7. R. Freund, L. Kari, M. Oswald, P. Sosík, Computationally universal P systems without priorities: two catalysts are sufficient. Theor. Comput. Sci. **330**, 251–266 (2005)

After a series of previous papers, the first one, by P. Sosík, where the universality of catalytic P systems was proved for systems with 8, then 6, catalysts, this paper established the best result in this respect: two catalysts suffice.

8. R. Freund, Gh. Păun, Universal P systems: One catalyst can be sufficient, in *Proc. 11th Brainstorming Week on Membrane Computing* (Fénix Editora, Sevilla, 2013), pp. 81–96

9. M. Gheorghe, Gh. Păun, M.J. Pérez-Jiménez, G. Rozenberg, Frontiers of membrane computing: Open problems and research topics. Int. J. Found. Comput. Sci. **24**(5), 547–623 (2013) (first version in *Proc. Tenth Brainstorming Week on Membrane Computing*, vol. I (Sevilla, 2012), pp. 171–249, January 30–February 3)
This paper circulated in the membrane computing community in the brainstorming version under the title of "mega-paper." In this form, it contains 26 sections, written by separate authors, covering most of the branches of this research area and presenting open problems and research topics of current interest. The titles of these 26 sections are worth recalling: A glimpse to membrane computing; Some general issues; The power of small numbers; Polymorphic P systems; P colonies and dP automata; Spiking neural P systems; Control words associated with P systems; Speeding up P automata; Space complexity and the power of elementary membrane division; The P-conjecture and hierarchies; Seeking sharper frontiers of efficiency in tissue P systems; Time-free solutions to hard computational problems; Fypercomputations; Numerical P systems; P systems formal verification and testing; Causality, semantics, behavior; Kernel P systems; Bridging P and R; P systems and evolutionary computing interactions; Metabolic P systems; Unraveling oscillating structures by means of P systems; Simulating cells using P systems; P systems for computational systems and synthetic biology; Biologically plausible applications of spiking neural P systems for an explanation of brain cognitive functions; Computer vision; and Open problems on simulation of membrane computing models.

10. O.H. Ibarra, Z. Dang, O. Egecioglu, Catalytic P systems, semilinear sets, and vector addition systems. Theor. Comput. Sci. **312**, 379–399 (2004)

11. O.H. Ibarra, Z. Dang, O. Egecioglu, G. Saxena, Characterizations of catalytic membrane computing systems, in *28th Intern. Symp. Math. Found. Computer Sci.*, ed. by B. Rovan, P. Vojtás. LNCS, vol. 2747 (Springer, 2003), pp. 480–489

12. M. Ionescu, Gh. Păun, T. Yokomori, Spiking neural P systems. Fundamenta Informaticae **71**(2–3), 279–308 (2006)
This is the paper where the spiking neural P systems were introduced, and two basic results of this area were proved: universality in the general case, and semilinearity of the computed sets of numbers in the bounded case. Similar results were later obtained for many classes of SN P systems.

13. M. Ionescu, Gh. Păun, M.J. Pérez-Jiménez, T. Yokomori, Spiking neural dP systems. Fundamenta Informaticae **11**(4), 423–436 (2011)

14. S.N. Krishna, A. Păun, Results on catalytic and evolution-communication P systems. New Generat. Comput. **22**, 377–394 (2004)

15. K. Krithivasan, Gh. Păun, A. Ramanujan, On controlled P systems. Fundamenta Informaticae **131**(3–4), 451–464 (2014)

16. A. Leporati, A.E. Porreca, C. Zandron, G. Mauri, Improving universality results on parallel enzymatic numerical P systems, in *Proc. 11th Brainstorming Week on Membrane Computing* (Fénix Editora, Sevilla, 2013), pp. 177–200

17. A. Leporati, A.E. Porreca, C. Zandron, G. Mauri, Enzymatic numerical P systems using elementary arithmetic operations, in *Proc. 14th Intern. Conf. on Membrane Computing* (Chişinău, Moldova, 2013), pp. 225–240

18. W. Maass, C. Bishop (eds.) *Pulsed Neural Networks* (MIT Press, Cambridge, 1999)

19. M. Mutyam, K. Krithivasan, P systems with membrane creation: Universality and efficiency, in *Proc. MCU 2001* ed. by M. Margenstern, Y. Rogozhin. LNCS, vol. 2055 (Springer, Berlin, 2001), pp. 276–287

20. A.B. Pavel, C.I. Vasile, I. Dumitrache, Robot localization implemented with enzymatic numerical P systems, in *Proc. Conf. Living Machines 2012*, LNCS, vol. 7375 (Springer, 2012), pp. 204–215

21. A. Păun, Gh. Păun, The power of communication: P systems with symport/antiport. New Generat. Comput. **20**, 295–305 (2002)

The symport/antiport P systems were introduced here, and their universality was proved for rules of various complexities/sizes. These results were improved in a large number of papers, until reaching universality for minimal symport and antiport rules.

22. Gh. Păun, Computing with membranes. J. Comput. Syst. Sci. **61**(1), 108–143 (2000) (and Turku Center for Computer Science-TUCS Report 208, November 1998, www.tucs.fi)

This is the paper where membrane computing was initiated. The cell-like P systems are introduced, both with symbol objects and string objects, and for both cases the universality was proved (using the characterization of Turing computable sets of numbers as the length sets of languages generated by context-free matrix grammars with appearance checking; later, more direct and simple proofs were obtained, starting from register machines). In the case of strings, both rewriting and splicing rules were investigated.

23. Gh. Păun, Computing with membranes—A variant. Int. J. Found. Comput. Sci. **11**(1), 167–182 (2000)

24. Gh. Păun, P systems with active membranes: attacking NP-complete problems. J. Autom. Lang. Combinat. **6**, 75–90 (2001)

Membrane division was introduced here, in the general framework of P systems with active membranes (the membranes are explicit parts of the object evolution rules), and a polynomial semi-uniform solution to SAT is provided. Later, uniform solutions were obtained (also for other **NP**-complete problems).

25. Gh. Păun, *Membrane Computing. An Introduction* (Springer, Berlin, 2002)

This is the first survey of membrane computing, systematizing the notions and the results at only a few years after the initiation of this research area. After an informal introduction ("Membrane computing—what it is and what it is not") and a chapter providing the biological and the computability prerequisites for the rest of the book, one presents the cell-like P systems with symbol objects and multiset rewriting rules, the systems with symport/antiport rules, the P systems with string objects, and then the tissue-like P systems; their computing power is investigated; then one passes to the computing efficiency ("Trading space for time"), considering P systems with membrane division, membrane creation, string replication, and precomputed resources. Two more chapters present "further technical results" and "(attempts to get) back to reality." The book ends with a list of open problems and of universality results.

26. Gh. Păun, Towards "fypercomputations" (in membrane computing), in *Languages Alive. Essays Dedicated to Jurgen Dassow on the Occasion of His 65 Birthday*ed. by H. Bordihn, M. Kutrib, B. Truthe. LNCS, vol. 7300 (Springer, Berlin, 2012), pp. 207–221

The term "fypercomputation" (coming from "fast computation" and reminding of "hypercomputation" = a computation going beyond the "Turing barrier") was coined to name situations when a computing device can solve **NP**-complete problems in polynomial time, hence when a significant efficiency speedup is obtained.

27. Gh. Păun, Some open problems about catalytic, numerical and spiking neural P systems, in *Proc. 14th Intern. Conf. on Membrane Computing* (Chişinău, Moldova, 2013), pp. 25–34

28. Gh. Păun, R. Păun, Membrane computing and economics: Numerical P systems. Fundamenta Informaticae **73**, 213–227 (2006)

29. Gh. Păun, M.J. Pérez-Jiménez, Solving problems in a distributed way in membrane computing: dP systems. Int. J. Comput. Commun. Cont. **5**(2), 238–252 (2010)

30. Gh. Păun, G. Rozenberg, A. Salomaa (eds.) *Handbook of Membrane Computing* (Oxford University Press, 2010)

The basics of membrane computing are given in the book [25] (translated in Chinese in 2013), but the domain has fast evolved beyond the contents of the volume; new classes of P systems were introduced; new results and applications were reported. This made both necessary and possible the editing of the present handbook, a comprehensive survey of membrane computing at the level of 2009. Its contents are a suggestive hint to the landscape of membrane computing: 1. An introduction to and an overview of membrane computing (Gh. Păun, G. Rozenberg); 2. Cell biology for membrane computing (D. Besozzi, I.I. Ardelean); 3. Computability elements for membrane computing (Gh. Păun, G. Rozenberg, A. Salomaa); 4. Catalytic P systems (R. Freund, O.H. Ibarra, A. Păun, P. Sosík, H.-C. Yen); 5. Communication P systems (R. Freund, A. Alhazov, Y. Rogozhin, S. Verlan); 6. P automata (E. Csuhaj-Varjú, M. Oswald, G. Vaszil); 7. P systems with string objects (C. Ferretti, G. Mauri, C. Zandron); 8. Splicing P systems (S. Verlan, P. Frisco); 9. Tissue and population P systems (F. Bernardini, M. Gheorghe); 10. Conformon P systems (P. Frisco); 11. Active membranes (Gh. Păun); 12. Complexity – Membrane division, membrane creation (M.J. Pérez-Jiménez, A. Riscos-Núñez, Á. Romero-Jiménez, D. Woods); 13. Spiking neural P systems (O.H. Ibarra, A. Leporati, A. Păun, S. Woodworth); 14. P systems with objects on membranes (M. Cavaliere, S.N. Krishna, A. Păun, Gh. Păun); 15. Petri nets and membrane computing (J. Kleijn, M. Koutny); 16. Semantics of P systems (G. Ciobanu); 17. Software for P systems (D. Díaz-Pernil, C. Graciani, M.A. Gutiérrez-Naranjo, I. Pérez-Hurtado, M.J. Pérez-Jiménez); 18. Probabilistic/stochastic models (P. Cazzaniga, M. Gheorghe, N. Krasnogor, G. Mauri, D. Pescini, F.J. Romero-Campero); 19. Fundamentals of metabolic P systems (V. Manca); 20. Metabolic P dynamics (V. Manca); 21. Membrane algorithms (T.Y. Nishida, T. Shiotani, Y. Takahashi); 22. Membrane computing and computer science (R. Ceterchi, D. Sburlan); 23. Other developments; 23.1. P Colonies (A. Kelemenová); 23.2. Time in membrane computing (M. Cavaliere, D. Sburlan); 23.3. Membrane computing and self-assembly (M. Gheorghe, N. Krasnogor); 23.4. Membrane computing and X-machines (P. Kefalas, I. Stamatopoulou, M. Gheorghe, G. Eleftherakis); 23.5. Q-UREM P systems (A. Leporati); 23.6. Membrane computing and economics (Gh. Păun, R.A. Păun); 23.7 Mobile membranes and mobile ambients (B. Aman, G. Ciobanu); 23.8. Other topics (Gh. Păun, G. Rozenberg)

31. Gh. Păun, S. Yu, On synchronization in P systems. Fundamenta Informaticae **38**(4), 397–410 (1999)
32. M.J. Pérez-Jiménez, A. Riscos-Núñez, A. Romero-Jiménez, Complexity— Membrane division and membrane creation. Chapter 12 of [30]
33. P. Sosík, A catalytic P system with two catalysts generating a non-semilinear set. *Romanian J. Inf. Sci. Technology* **16**(1), 3–9 (2013)
34. C.I. Vasile, A.B. Pavel, J. Kelemen, Implementing obstacle avoidance and follower behaviors on Koala robots using numerical P systems, in *Tenth Brainstorming Week on Membrane Computing*, vol. II (Sevilla, 2012), pp. 215–227
35. C. Zandron, C. Ferretti, G. Mauri, Solving NP-complete problems using P systems with active membranes, in *Proc. Unconventional Models of Computation* ed. by I. Antoniou et al. (Springer, 2000), pp. 289–301
   Among other results, one proves here the so-called "Milano theorem," saying that P systems without membrane division cannot solve **NP**-complete problems in polynomial time (unless if **P = NP**)
36. The P Systems Website, www.ppage.psystems.eu

A Panorama of Post-quantum Cryptography

Paulo S.L.M. Barreto, Felipe Piazza Biasi, Ricardo Dahab,
Julio César López-Hernández, Eduardo M. de Morais,
Ana D. Salina de Oliveira, Geovandro C.C.F. Pereira,
and Jefferson E. Ricardini

Abstract In 1994, Peter Shor published a quantum algorithm capable of factoring large integers and computing discrete logarithms in Abelian groups in polynomial time. Since these computational problems provide the security basis of conventional asymmetric cryptosystems (e.g., RSA, ECC), information encrypted under such schemes today may well become insecure in a future scenario where quantum computers are a technological reality. Fortunately, certain classical cryptosystems based on entirely different intractability assumptions appear to resist Shor's attack, as well as others similarly based on quantum computing. The security of these schemes, which are dubbed *post-quantum cryptosystems*, stems from hard problems on lattices, error-correcting codes, multivariate quadratic systems, and hash functions. Here we introduce the essential notions related to each of these schemes and explore the state of the art on practical aspects of their adoption and deployment, like key sizes and cryptogram/signature bandwidth overhead.

1 Introduction

In the 1990s, Peter Shor introduced new concerns to cryptography. He discovered a quantum algorithm able to factor large integers and compute discrete logarithms in finite fields in polynomial time, more precisely $O(\log^3 N)$ [84]. These concerns are due to the fact that the security of conventional techniques used in asymmetric cryptography is based precisely on these or related problems (e.g., RSA, ECC) [59, 81].

P.S.L.M. Barreto (✉) • F. Piazza Biasi • G.C.C.F. Pereira • J.E. Ricardini
Escola Politécnica, University of São Paulo, Av. Prof. Luciano Gualberto, trav 3, no. 158,
São Paulo (SP), Brazil,
e-mail: pbarreto@larc.usp.br; fbiasi@larc.usp.br; geovandro@larc.usp.br; jricardini@larc.usp.br

R. Dahab • J.C. López-Hernández • E.M. de Morais • A.D. Salina de Oliveira
Instituto de Computação, University of Campinas, Av. Albert Einstein, 1251, Campinas (SP),
Brazil,
e-mail: rdahab@ic.unicamp.br; jlopez@ic.unicamp.br; emorais@ic.unicamp.br;
anakarina@facom.ufms.br

© Springer International Publishing Switzerland 2014
Ç.K. Koç (ed.), *Open Problems in Mathematics and Computational Science*,
DOI 10.1007/978-3-319-10683-0_16

Another even more effective threat are the recent discoveries of classical algorithms for solving certain discrete logarithms used in asymmetric encryption [6].

Fortunately, there exist cryptographic schemes based on different computational problems that resist the known attacks with quantum computers. They became known as post-quantum cryptosystems: this is the case of cryptosystems based on lattices [42], error-correcting codes [55, 64], multivariate quadratic systems ($\mathcal{MQ}$) [28, 48], and hash functions [28, 29], excluding symmetric encryption buildings in general.

Given the new paradigm of Internet of Things, in which any object is able to connect to the Internet. A side effect of this interconnectivity is a possible vulnerability of these embedded systems. Attacks that have been primarily aimed at PCs can be launched against cars, cell phones, e-tickets, and RFIDs. In this scenario, the devices are typically characterized by shortages in energy supply (via battery) and limited processing power, storage, and often communication channels with low bandwidth (e.g., SMS).

Since embedded systems are typically deployed on a large scale, cost becomes designers' main concern. Therefore, security solutions for embedded systems must provide low cost, which can be achieved with tools that minimize overhead transmission, processing, and memory occupation. In this sense, symmetric encryption techniques already attend the required metrics, and asymmetric encryption is the bottleneck in the most cases.

Asymmetric cryptographic primitives for encryption and digital signatures are essential in a modern security framework. However, conventional techniques are not efficient enough in some aspects, which makes them unsuitable for embedded platforms, specially highly resource-constrained ones. In this context the absence of costly operations (operations with large integers, especially modular exponentiations) of post-quantum techniques makes them more attractive in such scenarios, as described previously.

The objective of this chapter is to introduce the basics of the main lines of post-quantum cryptography research (hash-based signatures, $\mathcal{MQ}$ systems, error-correcting codes, and lattices), as well as presenting the latest research focusing on improvements regarding key sizes and signature/cryptogram overheads of these schemes.

2 Hash-Based Digital Signature Schemes

Hash-based digital signature schemes became popular after Ralph Merkle's work [56] in 1979. The scheme proposed by Merkle *(MSS)* is inspired by Lamport and Diffie's one-time signature scheme [49]. The security of these signature schemes depends on the collision resistance and inversion resistance of the hash function used. The scheme *(MSS)* is considered practical, and although there is not a proof, it is believed to be resistant against quantum computers. The disadvantage of one-time signature schemes is that a key pair can only be used for one signature, although this signature can be verified an arbitrary number of times.

2.1 Hash Function

Cryptographic hash functions are used in security applications such as digital signatures, identification data and key derivation, among others. Formally, a hash function $h : \{0, 1\}^* \to \{0, 1\}^n$ takes as input an arbitrarily long string m and returns a fixed string r of size n, the *hash value* (i.e., $r = h(m)$).

Since the image $\{0, 1\}^n$ of h is a subset of $\{0, 1\}^*$, it is easy to see that more than one message is mapped to the same hash value (or digest). Some applications require that it be computationally infeasible for an attacker to find two random messages that generate the same digest; one such example is that of digital signatures, in which the hash of messages are signed, not the messages themselves.

2.2 Properties

The primary properties that a hash function $h : \{0, 1\}^* \longrightarrow \{0, 1\}^n$ should possess are pre image resistance, second pre image resistance, and collision resistance:

- *Pre image resistance:* Let r be a known digest. Then, it should be infeasible to find a value m such that $h(m) = r$.
- *Second pre image resistance:* Let m be a known message. Then, it should be infeasible to find m' such that $m' \neq m$ e $h(m') = h(m)$.
- *Collision resistance:* It should be infeasible to find a pair $m, m' \in \{0, 1\}^*$ such that $m' \neq m$ and $h(m') = h(m)$.

Another desirable property for practical applications is that the hash function be efficient (speed, memory, energy, etc.) when implemented on various computing platforms (hardware and/or software). It is easy to see that a function that is collision resistant is also second pre image resistant, but the reciprocal is not necessarily true.

2.3 Construction of Hash Functions

The design of hash functions has been based on various techniques such as block ciphers [54, 78, 93], the iterative Merkle-Damgård method [56], the sponge construction [16], and primitive arithmetic [25].

Standards based on these functions have evolved, mainly due to successive attacks advertised in the literature and specialized events. Recently, the National

Institute of Standards and Technology (NIST) selected the Keccak [15] algorithm as
the winner of a 5-year competition to create a new Secure Hash Algorithm standard.

2.4 Signature Schemes

A signature scheme *SIGN* is a triple of algorithms (*GEN*, *SIG*, *VER*) which satisfies
the following properties:

- The key generation algorithm *GEN* receives as input a security parameter 1^n and
  produces a key pair (X, Y), where X is the private key and Y is the public key.
- The signature generation algorithm *SIG* takes as input a message $M \in \{0, 1\}^*$
  and a private key X and produces a signature *Sig*, denoted by $Sig \leftarrow SIG_X(M)$.
- The signature verification algorithm *VER* takes as input a message M, a signature
  Sig of M, and a public key Y and produces a bit b, where $b = 1$ means that the
  signature is valid and $b = 0$ indicates that the signature is not valid.

2.5 One-Time Signature Schemes

One-time signature schemes first appeared in the work of Lamport [49] and
Rabin [27, Chapter "Digitalized signatures" by Michael O. Rabin]. Merkle [56]
proposed a technique to transform a one-time signature scheme into a scheme
with an arbitrary but fixed number of signatures. The following describes the
Lamport [49] and Winternitz [57] schemes.

2.5.1 Lamport-Diffie One-Time Signature Scheme

The Lamport-Diffie one-time signature scheme *(LD-OTS)* was proposed in [49]. Let
n be a positive integer, the security parameter of *LD-OTS*. *LD-OTS* uses a one-way
function

$$f : \{0, 1\}^n \to \{0, 1\}^n$$

and a cryptographic hash function

$$g : \{0, 1\}^* \to \{0, 1\}^n.$$

LD-OTS Key Pair Generation The signature key X consists of $2n$ bit strings of
length n chosen uniformly at random:

$$X = (x_0[0], x_0[1], \ldots, x_{n-1}[0], x_{n-1}[1]) \quad \in_R \quad \{0, 1\}^{(n, 2n)}.$$

The verification key Y is

$$Y = (y_0[0], y_0[1], \ldots, y_{n-1}[0], y_{n-1}[1]) \quad \in \quad \{0, 1\}^{(n, 2n)},$$

where $y_i[j] = f(x_i[j]), 0 \leq i \leq n - 1, j = 0, 1$.

LD-OTS Signature Generation The digest d of a message M is signed using the signature key X. Let $d = g(M) = (d_0, \ldots, d_{n-1})$. The signature of d is

$$Sig = (sig_0, \ldots, sig_{n-1}) = (x_0[d_0], \ldots, x_{n-1}[d_{n-1}]) \quad \in \quad \{0, 1\}^{(n, n)}.$$

The signature Sig is a sequence of n bits strings, each one of length n, and the signature elements are chosen from the key X according to the bits of the message digest. Then, for each bit d_i of d, it selects the corresponding string of length n from the private key X; that is, the algorithm selects $x_i[d_i]$. To verify a signature, the verifier first computes the message digest $d = g(M) = (d_0, \ldots, d_{n-1})$ and selects $y_i[d_i]$ from Y, the verification (public) key. Then, it checks whether

$$Sig = (f(sig_0), \ldots, f(sig_{n-1})) = (y_0[d_0], \ldots, y_{n-1}[d_{n-1}]).$$

If the signature is valid; otherwise it is refused. Observe that signature verification requires n evaluations of g while signing requires no evaluation of g.

Figure 1 illustrates the Lamport scheme. In this example, the one-way function used was $f(x) = x + 1 \mod 16$. For the public key generation, $2n$ evaluations of the

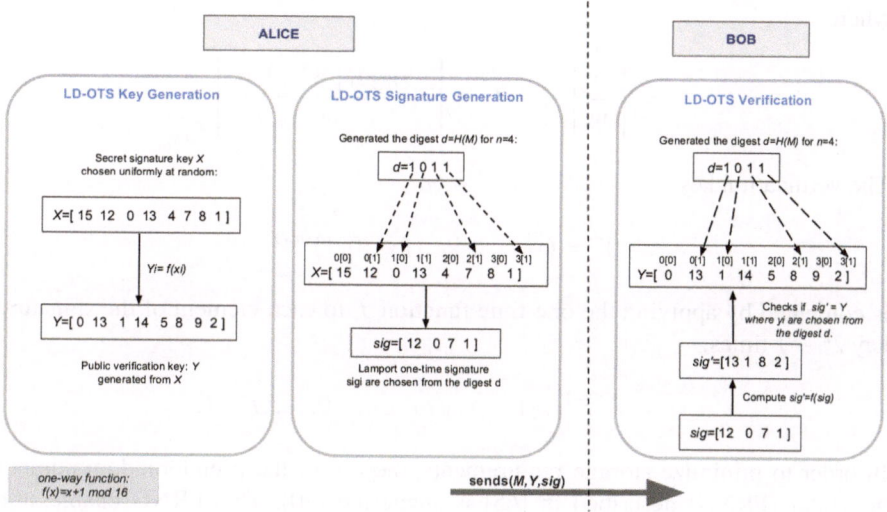

Fig. 1 Lamport one-time signature scheme example

one-way function are required, one for each element of X. For signature verification, n evaluations of the one-way function are required, one for each element of Sig.

2.5.2 Winternitz One-Time Signature Scheme

Winternitz proposed an improvement to Lamport's one-time signature scheme, reducing the size of the public and private keys. This scheme (W-OTS) was first mentioned in [57]. W-OTS uses a one-way function

$$f : \{0, 1\}^n \to \{0, 1\}^n$$

and a cryptographic hash function

$$g : \{0, 1\}^* \to \{0, 1\}^n,$$

where n is a positive integer. W-OTS uses a parameter w which is the number of bits to be signed simultaneously. Larger values for w result in smaller signature keys and longer times for signing and verification. A comparative analysis of the running times and key sizes in terms of parameter w is found in [29].

W-OTS Key Pair Generation Given parameter $w \in \mathbb{N}$, the private key is

$$X = (x_0, \ldots, x_{t-1}) \in_{\mathscr{R}} \{0, 1\}^{(n,t)},$$

where the x_i are chosen uniformly at random. The size t is computed as $t = t_1 + t_2$, where

$$t_1 = \left\lceil \frac{n}{w} \right\rceil, \qquad t_2 = \left\lceil \frac{\lfloor \log_2 t_1 \rfloor + 1 + w}{w} \right\rceil.$$

The verification key

$$Y = (y_0, \ldots, y_{t-1}) \in \{0, 1\}^{(n,t)}$$

is computed by applying the one-time function f to each element of the signature key $2^w - 1$ times:

$$y_i = f^{2^w - 1}(x_i), \qquad for \quad i = 0, \ldots, t - 1.$$

In order to minimize storage requirements, the use of the pseudorandom number generator (PRNG) described in [65] is suggested [20]. This PRNG enables the recovery of all signature keys based only on the initial seed SEED0, $SEED_{in} \to (RAND, SEED_{out})$.

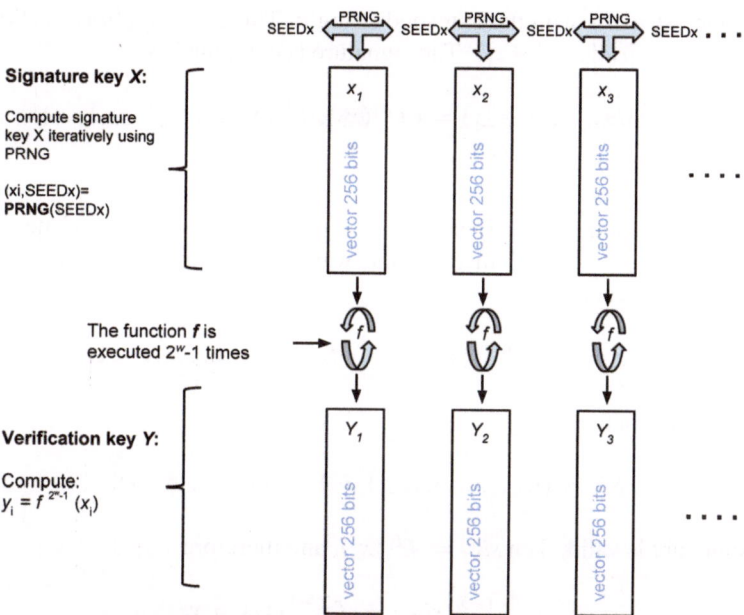

Fig. 2 Winternitz key pair generation

Figure 2 shows an example of key pair generation in the Winternitz signature scheme, using a PRNG and a one-way function. The PRNG computes $(SEED_x) \rightarrow (x_i, SEED_x)$. This scheme produces smaller signature keys than Lamport's, but it increases the number of one-way function evaluations from 1 to $2^w - 1$ for each element of the key signature.

W-OTS Signature Generation To generate the signature, first compute the message digest $d = g(M) = (d_0, \ldots, d_{n-1})$. If necessary, add zeros to the left of d, so as to make the bitlenght of d divisible by w. Then, d is split into t_1 binary blocks of size w, resulting in $d = (m_0 || \ldots || m_{t_1-1})$, where $||$ denotes concatenation. The m_i blocks are represented as integers in $\{0, 1, \ldots, 2^w - 1\}$. Now, a checksum c is computed as

$$c = \sum_{i=0}^{t_1-1} (2^w - m_i).$$

Since $c \leq t_1 2^w$, the length of the binary representation of c is less than $\lfloor \log_2 t_1 2^w \rfloor + 1 = \lfloor \log_2 t_1 \rfloor + w + 1$. If necessary, add zeros to the left of c in order to make the bitlength of string c divisible by w. Then, the extended string c can be divided into t_2 blocks $c = (c_0 || \ldots || c_{t_2-1})$ of length w. Let $b = d || c$ be the concatenation of the extended string d with the extended string c. Thus, $b = (b_0||b_1|| \dots ||b_{t-1}) = (m_0|| \dots ||m_{t_1-1}||c_0|| \dots ||c_{t_2-1})$. The signature is computed as

$$Sig = (sig_0, \dots, sig_{t-1}) = (f^{b_0}(x_0), f^{b_1}(x_1), \dots, f^{b_{t-1}}(x_{t-1})).$$

W-OTS Verification To verify signature $Sig = (sig_0, \dots, sig_{t-1})$ of message M, first calculate $(b_0, \dots, b_{t-1})$ in the same way it was computed during signature generation; then, compute

$$sig_i' = f^{2^w-1-b_i}(sig_i), \qquad for \quad i = 0, \dots, t-1.$$

Finally, check whether

$$Sig = (sig_0', \dots, sig_{t-1}') = Y = (y_0, \dots, y_{t-1}).$$

If the signature is valid, then $sig_i = f^{b_i}(x_i)$, and therefore,

$$f^{2^w-1-b_i}(sig_i) = f^{2^w-1}(x_i) = y_i$$

holds for $i = 0, 1, \dots, t-1$.

2.6 Merkle Digital Signature Scheme

In the Merkle digital signature scheme described below, the one-time signing key and the verification key are the leaves of the tree, and the public key is the root. A tree with height H and 2^H leaves will have 2^H one-time key pairs (public and private).

2.6.1 Merkle Key Generation

For the generation of the Merkle public key *(pub)*, which corresponds to the root of the Merkle tree, one must first generate 2^H one-time key pairs (public and private), for each leaf of the Merkle tree.

One-Time Key Pair Generation A one-time signature algorithm generates private keys $X[u]$ and public keys $Y[u]$, for each leaf of the Merkle tree, $u = 0, \dots, 2^H - 1$. Algorithm 2.1 describes the process of one-time key pair generation.

Merkle Public Key Generation (Pub) Algorithm 2.2 generates the Merkle tree public key *pub*. The input values are the initial leaf *leafIni* and tree height H. Each leaf node *node[u]* of the tree receives the corresponding verification key $Y[u]$.

Algorithm 2.1 Winternitz one-time key pair generation (*Leafcalc*) [56]

Require: Winternitz parameter t and w; seed $SEED_x$.
Ensure: verification key Y;
  for ($i = 0, i < t$, i++) **do**
    $(x[i], SEED_x) = PRNG(SEED_x)$;
    $y[i] = f^{2^w - 1}(x[i])$;
  end for
  $Y = g(y[0] || \ldots || y[t - 1])$;
  return Y;

Algorithm 2.2 Merkle public key generation (CalcRoot) [56]

Require: Leaf *leafIni*; tree height H; seed $SEED_{in}$.
Ensure: The root of the tree *pub*.
  Create a stack *stackNode*.
  $SEED[0] = SEED_{in}$
  for ($u = leafIni, u < 2^H$, u++) **do**
    $(SEED_x, SEED[u + 1]) = PRNG(SEED[u])$;
    $node[u].digest = $ Leafcalc$(t, SEED_x)$
    Push $node[u]$ in the stack *stackNode*
    while The nodes at the top of the *pilhaNo* has the same height **do**
      Pop $node[right]$
      Pop $node[left]$
      Compute $node[parent].digest = g(node[left].digest || node[right].digest)$
      if $node[parent].height = H$ **then**
        return $(node[parent])$
      else
        Push $node[parent]$ into *stackNode*
      end if
    end while
  end for

The inner nodes of the Merkle tree *node[parent]* contain the hash value of the concatenation of their left and right children, *node[left]* and *node[right]*, respectively. Each time a leaf u is calculated and stacked in *stackNode*, the algorithm checks if the nodes at the top of the *stackNode* have the same height. If the nodes have the same height, the two nodes will be unstacked and the hash value of their concatenation will be pushed into *stackNode*. The algorithm terminates when the root of the tree is found.

Figure 3 shows the order in which the nodes are stacked on the tree according to Algorithm 2.2. The nodes in gray represent the nodes that have already been generated. For example, the $4th$ node generated (leaf $u = 2$) received $Y[2]$. The $3rd$ node is the hash result of the concatenation of the nodes 1 and 2.

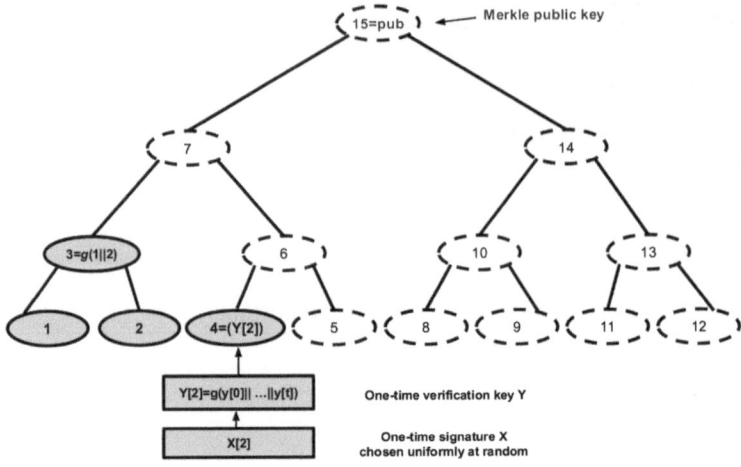

Fig. 3 Merkle public key generation (pub)

2.6.2 MSS Signature Generation

Scheme *MSS* allows the generation of 2^H signatures for a tree of height H. Suppose we want to sign $M[u]$ messages, for $u = 0, .., 2^H - 1$. Each message $M[u]$ is signed with the one-time signature key $X[u]$ resulting in a signature $Sig[u]$.

An authentication path *Aut* is used to store the nodes in the path needed to authenticate leaf $Y[u]$, eliminating the need for sending the whole tree to the receiver.

The Merkle signature *SIG* consists of one-time signature $Sig[u]$ for leaf u, the corresponding verification key $Y[u]$, the index u (index leaf), and its authentication path, $Aut = (Aut[0], .., Aut[H - 1])$. Therefore, the signature is

$$SIG = (u, Sig[u], Y[u], (Aut[0], \ldots, Aut[H - 1])).$$

The Classic Authentication Path Algorithm The classic authentication path algorithm *(Path Regeneration Algorithm)* [56] computes node authentication path *Aut* for each tree leaf, needed to authenticate public key *pub* of the Merkle tree. This algorithm uses two stack variables, *Aut* and *Aux*. Stack *Aut* contains the path of current authentication and stack *Aux* saves the next authentication nodes that will be needed. *Aut* is formed by right siblings at each level of the authentication path connecting the leaf to the root of the Merkle tree.

We now describe the computation of authentication paths. The first authentication path is generated during the execution of Algorithm 2.2. The next authentication path is generated if a new signature is required. In Fig. 4, the nodes in gray show the first authentication path *Aut* for leaf $u = 0$.

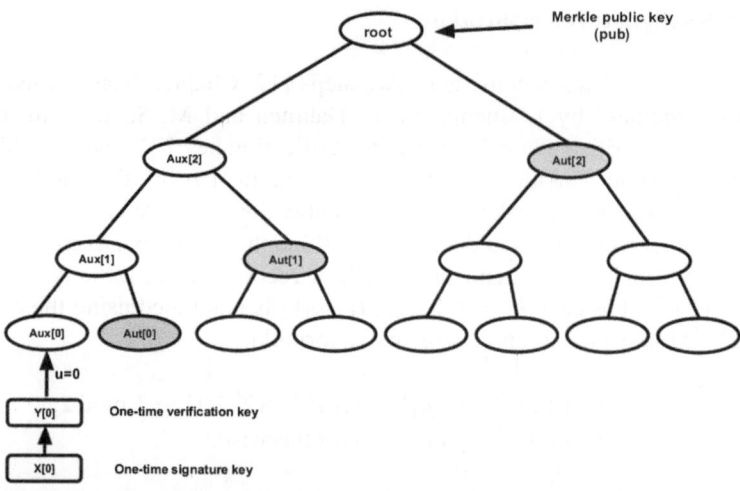

Fig. 4 Implementation of Algorithm 2.2 with the first authentication path

---

Algorithm 2.3 The Path Regeneration Algorithm [56]

Require: Tree height H; seed $SEED_{in}$.
Ensure: One authentication path Aut.
  for $(u = 0, u < 2^H, u + +)$ **do**
    for $(h = 0, h < H, h + +)$ **do return** Aut of leaf u
      A signature with leaf u is done
      for $(h = 0, h < H, h + +)$ **do**
        if $(u + 1)/(2^h) = 0$ **then**
          Update $Aut[h] = Aux[h]$
          $node_{Ini} = (u + 1 + 2^h) \oplus 2^h$.
          $Aux[h] = \text{CalcRoot}(node_{Ini}, h, SEED_{in})$.
        end if
      end for
    end for
  end for

Output and Update Phases Algorithm 2.3 shows the steps for producing the authentication path for the next leaf u in the tree. The algorithm starts by signing leaf $u = 0$; then, the leaf is updated in one unit, and the next authentication path is computed efficiently since only the nodes that change in the path will be updated.

Algorithm 2.3 updates authentication nodes by executing function

$$CalcRoot(node_{Ini}, h, SEED_{in}).$$

Function *CalcRoot* executes Algorithm 2.2 for node $node_{Ini}$. After 2^h rounds, the value of the selected node will be computed.

2.6.3   MSS Signature Verification

The signature verification consists of two steps [12, Chapter "Hash-based Digital Signature Schemes" by J. Buchmann, E. Dahmen and M. Szydlo]: in the first, signature Sig is verified using the one-time verification key $Y[i]$ and the underlying one time algorithm; in the second step, the public key of the Merkle tree is validated, and the receiver calculates its authentication path, reconstructing the path $(p[0], \ldots, p[h])$ from leaf i to the root, for all heights h. Index i is used to decide the order in which the authentication path is reconstructed. Initially, for leaf i, $p[0] = g(Y[i])$. For each $h = 1, 2, \ldots, H$, $p[h]$ is computed using the condition (if $\lfloor i/(2^{h-1}) \rfloor \equiv 1 \bmod 2$) and the recursive formula

$$p[h] = \begin{cases} g(Aut[h-1] \| p[h-1]) & \text{if } \lfloor i/(2^{h-1}) \rfloor \equiv 1 \bmod 2; \\ g(p[h-1] \| Aut[h-1]) & \text{otherwise.} \end{cases}$$

Finally, if value $p[H]$ is equal to the public key pub, the signature is valid.

2.7   CMSS: An Improved Merkle Signature Scheme

The $CMSS$ scheme [20] is a variation of the MSS scheme which allows the increase of the number of signatures from 2^{20} to 2^{40}. In addition, $CMSS$ reduces key pair generation time, signature generation time, and private key size. In [20] it was demonstrated that $CMSS$ is competitive in practice, by presenting a highly efficient implementation within the Java Cryptographic Service Provider FlexiProvider and showing that the implementation can be used to sign messages within Microsoft Outlook.

In the $CMSS$ scheme, two MSS authentication trees are used, a subtree and a main tree, each one with 2^h leaves, where $h = H/2$. Thus, we increase the number of signatures in relation to MSS. Note that MSS becomes impractical for $H > 25$ since private keys are too large and the key pair generation takes too much time. For example, to generate 2^{20} signature keys, two trees with 2^{10} leaves are generated with $CMSS$, while with MSS, a single tree with 2^{20} leaves is constructed. Therefore, key generation time is reduced.

In order to improve signature generation time, $CMSS$ uses Szydlo's algorithm [89], which is more efficient for constructing authentication paths. This algorithm was implemented in [24], in which the purpose is to balance the number of calculated leaves in each authentication path.

As for reducing the private key size, a pseudo-number random generator $PRNG$ [65] is used, where only the seed of the $PRNG$ is stored. By using a hash function of n bits and the Winternitz parameter t, the signature key will have $(t \cdot n)$ bits. Thus, one needs only to store a seed of n bits.

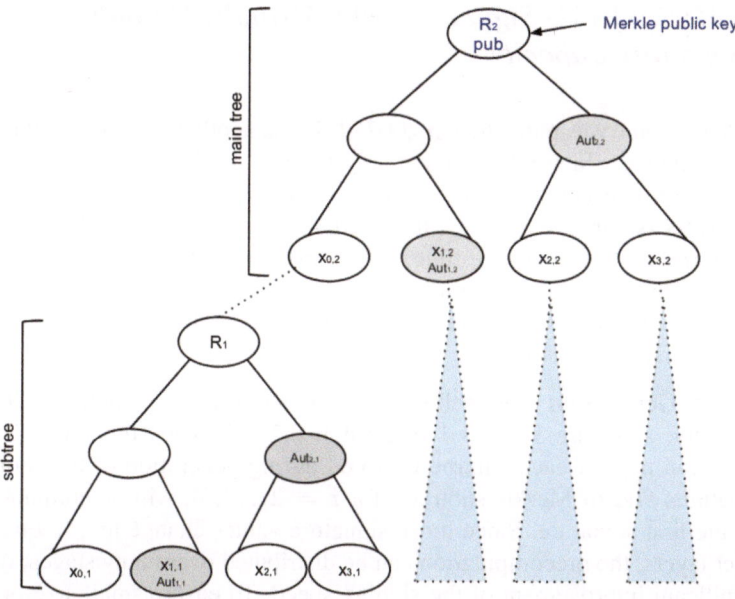

Fig. 5 CMSS signature scheme

The *CMSS* public key is the root of the main tree. The messages are signed using the leaves of the subtree. After the first 2^h signatures have been generated, a new subtree is constructed and used to generate the next 2^h signatures.

CMSS Key Generation For key pair generation, the *MSS* key pair generation is called twice. The subtree and its first authentication path are generated. Then, the main tree and its first authentication path are computed.

The *CMSS* public key is the root of the main tree. *CMSS* uses the Winternitz one time signature scheme. Figure 5 shows the *CMSS* scheme.

CMSS Signature Generation *CMSS* signature generation is carried out in various parts. First, the one-time signature of the message is computed using the leaf of subtree. After that, the one-time signature of the root of the subtree is computed using the leaf of the main tree. This signature will be recalculated in the next signature only if all the leaves of the current subtree have already been used. Then, the authentication path of both trees (main and subtree) is appended in the signature and the next authentication paths are computed. Thus, the next subtree is partially constructed, and the *CMSS* private key is updated.

CMSS Verification To verify the *CMSS* signature, it is required the checking of the roots of both subtrees and both one-time signatures.

2.8   GMSS: Merkle Signatures with Virtually Unlimited Signature Capacity

The *GMSS* scheme was published in 2007 [23]. It is another variation of the Merkle signature scheme, which allows a virtually unlimited 2^{80} number of messages to be signed with one key pair. The basic construction of *GMSS* consists of a tree with T layers (subtrees). Subtrees in different layers may have different heights. To reduce the cost of signature time, *GMSS* distributes the cost of one signature generation across previous signatures and key generation. Thus, this scheme allows for the choice of different parameters w of Winternitz in different subtrees, in order to produce smaller signatures.

GMSS Key Generation   For each subtree, the one-time key generation algorithm computes the signature keys and Algorithm 2.2 calculates the roots. The first authentication path of each subtree is stored during generation of the root. Then, the signatures Sig_τ of Merkle subtrees, for $\tau = 2, \ldots, T$, will be computed to be used in the first signature. Since those signature values change less frequently for the upper layers, the precomputation can be distributed over many stages, resulting in a significant improvement of the signing speed. To ensure small size of private keys, only the seed of the *PRNG* needs to be stored.

GMSS Signature Generation   The root of a subtree is signed with the one-time signature key corresponding to the parent tree. $Root_\tau$ denotes the root of the tree τ. Sig_τ denotes the one-time signature of $Root_\tau$, which is generated using the leaf l of parent τ. The message digest d is signed using the leaves on the deepest layer T.

The number of messages that can be signed with a *GMSS* key is $S = 2^{h_1 + \cdots + h_T}$, where $h_1, \ldots h_T$ are the heights of the subtrees. The *GMSS* signature consists of:

- the index leaf s;
- the one-time signatures Sig_d and $Sig_{\tau_{i,j_i}}$ for $i = 2, \ldots, T$, $j = 0, \ldots, 2^{h_1 + \cdots + h_{i-1}} - 1$.
- authentication paths $Aut[\tau_{i,j_i}, l_i]$ of leaves l_i, for $i = 1, \ldots, T$, $j = 0, \ldots, 2^{h_1 + \cdots + h_{i-1}} - 1$.

During the signature generation roots $Root_{\tau_{i,1}}$ are also calculated, as are the authentication paths $Aut[\tau_{i,1}, 0]$ of trees $\tau_{i,1}$, for $i = 2, \ldots, T$. The signature generation is split into two parts. The first, online part, computes Sig_d. The second, offline part, precomputes the authentication paths and one-time signatures of the roots required for upcoming signatures.

GMSS Verification   The GMSS signature verification is essentially the same as that of schemes *MSS* and *CMSS*: the verifier checks the one-time signatures Sig_d and $Sig_{\tau_{i,j_i}}$ for $i = 2, \ldots, T$ and $j = 0, \ldots, 2^{h_1 + \cdots + h_{i-1}} - 1$. Therefore, she verifies the roots $Root_\tau$ for $\tau = 2, \ldots, T$, and the public key using the corresponding authentication path.

2.9 XMSS: eXtended Merkle Signature Scheme

The hash-based signature scheme XMSS [22] is a variation of MSS, and it was the first practical forward secure signature with minimal security requirements. This scheme uses a function family F and a hash function family G. *XMSS* is efficient, provided that G and F are efficient. The parameters of *XMSS* are $n \in \mathbb{N}$, the security parameter; $w \in \mathbb{N}(w > 1)$, the Winternitz parameter; $m \in \mathbb{N}$, the message length; $H \in \mathbb{N}$, the tree height; the one-time signature keys $x \in \{0, 1\}^n$, chosen randomly with uniform distribution; a function family

$$F_n = \{f_K : \{0, 1\}^n \to \{0, 1\}^n | K \in \{0, 1\}^n\};$$

and a hash function g_K, chosen randomly with uniform distribution from the family

$$G_n = \{g_K : \{0, 1\}^{2n} \to \{0, 1\}^n | K \in \{0, 1\}^n.\}$$

The one-time signature key x is used to construct the one-time verification y, by applying the function family F_n. In [22] the family function used was $f_K(x) = g(Pad(K)||Pad(x))$, for a key $K \in \{0, 1\}^n$, $x \in \{0, 1\}^n$. $Pad(z) = (z||10^{b-|z|-1})$, for $|z| < b$, where b is the size of the hash function block.

The *XMSS* scheme uses a slightly modified version of the WOTS proposed in [21]. This modification makes collision resistance unnecessary: the iterated evaluations of a hash function is replaced by a random walk through the function family F_n, as follows: for $K, x \in \{0, 1\}^n$, $e \in \mathbb{N}$, and $f_K \in F_n$, the function $f_K^e(x)$ is $f_K^0(x) = K$. For $e > 0$, the function is $f_K^e(x) = f_{K'}(x)$, where $K' = f_K^{e-1}(x)$.

Modified WOTS Key Pair Generation First compute the Winternitz parameters

$$l_1 = \left\lceil \frac{m}{\log_2(w)} \right\rceil, \qquad l_2 = \left\lfloor \frac{\log_2(l_1(w-1))}{\log_2(w)} \right\rfloor + 1, \qquad l = l_1 + l_2.$$

The public verification key is

$$Y = (y_1, \dots, y_l) = (f_{sk_1}^{w-1}(x), \dots, f_{sk_l}^{w-1}(x)),$$

where sk_i is the private signature key chosen uniformly at random and f^{w-1} as defined above.

Modified WOTS Signature Generation This scheme signs messages of binary length m. The message bits are processed in base w representation. The message is $M = (m_1, \dots, m_{l_1})$, $m_i \in \{0, \dots, w - 1\}$. The checksum $C = \sum_{i=1}^{l_1}(w - 1 - m_i)$ in base w representation, and the length l_2 are appended to M, resulting in $b = (b_1, \dots, b_l)$. The signature is

$$Sig = (sig_1, \dots, sig_l) = (f_{sk_1}^{b_1}(x), \dots, f_{sk_l}^{b_l}(x)).$$

Modified WOTS Verification To check the signature, the verifier constructs the values $b = (b_1, \ldots, b_l)$ as in the signature generation and then checks the equality

$$(f_{\text{sig}_1}^{w-1-b_1}(x), \ldots, f_{\text{sig}_l}^{w-1-b_l}(x)) = (y_1, \ldots, y_l).$$

XMSS Public Key Generation *XMSS* is a modification of the Merkle tree. A tree of height H has $H + 1$ levels. *XMSS* uses the hash function g_K and bitmasks (bitmaskTree) $(b_{l,j}\|b_{r,j}) \in \{0,1\}^{2n}$, chosen uniformly at random, where $b_{l,j}$ is the left *bitmask* and $b_{r,j}$ is the right *bitmask*. The nodes on level j, $0 \leq j \leq H$, are written $NODE_{i,j}$, $0 \leq i < 2^{H-j}$, and $0 < j \leq H$. The nodes are computed as

$$NODE_{i,j} = g_K((NODE_{2i,j-1} \oplus b_{l,j})\|(NODE_{2i+1,j-1} \oplus b_{r,j})).$$

The bitmasks are the main difference to the other Merkle tree constructions, since they allow one to replace the collision resistant hash function family. Observe, in Fig. 6, how the tree nodes $NODE_{i,j}$ in the *XMSS* scheme are constructed at each level j, to generate the public key of the tree.

To generate a leaf of the *XMSS* tree, an *L*-tree is used. The one-time public verification keys $(y_1, \ldots, y_l)$ are the first l leaves of an *L*-tree. If l is not a power of 2, then there are not sufficiently many leaves. A node that has no right sibling is lifted to a higher level of the *L*-tree until it becomes the right sibling of another node. The hash function uses new bitmasks (bitmaskLtree). The bitmaskLtree are the same for each of those trees. The *XMSS* public key contains the bitmaskTree, bitmaskLtree, and the root of the *XMSS* tree.

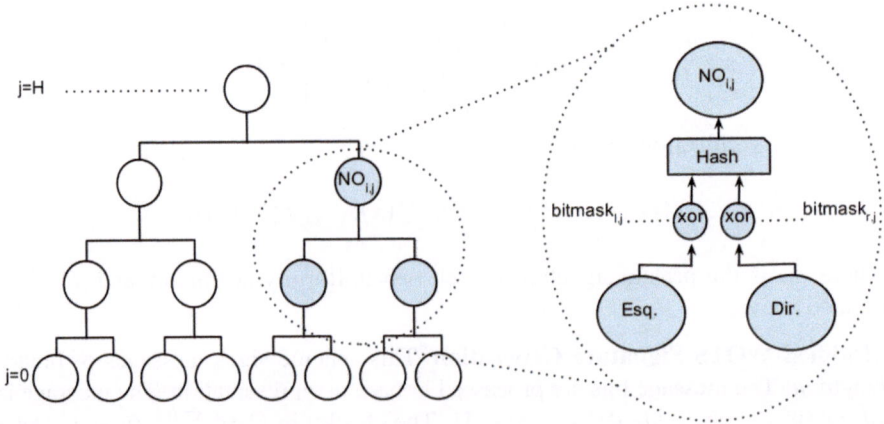

Fig. 6 XMSS signature scheme [21]

2.10 Security of the Hash-Based Digital Signature Schemes

In this section we present the main results known about the security of hash-based digital signature schemes.

In [12, Chapter "Hash-based Digital Signature Schemes" by J. Buchmann, E. Dahmen and M. Szydlo], it was proved that the Lamport-Diffie one-time signature scheme has existential unforgeability under an adaptive chosen message attack ($CMAsecure$), assuming that the underlying one-way function is pre image resistant. In the same work, it was also proved that the Merkle signature scheme has existential unforgeability under the assumption that the hash function is collision resistant and the underlying one-time signature scheme has existential unforgeability.

On the security of $XMSS$, in [21], it was proved the following result: if H_n is a second preimage-resistant hash function family and F_n a pseudorandom function family, then $XMSS$ is existentially unforgeable under chosen message attacks. In addition, in the same paper, it was shown that $XMSS$ is forward secure under some modifications on the key generation process.

Hülsing [44] showed that W-OTS is existentially unforgeable under adaptive chosen message attacks. In the same work it was also shown that scheme $XMSS^{MT}$ is secure; more specifically, it is proved the following result: if H_n is a second-preimage-resistant hash function family and F_n is a pseudorandom function family, then $XMSS^{MT}$ is a forward secure signature scheme.

2.11 Implementation Results

In this section we present a summary of recent works on the implementation of variants of the Merkle signature scheme.

We use the following notation: time to generate keys (t_{key}), time to generate a signature (t_{sig}), and time to verify a signature (t_{ver}). Table 1 shows timings which were obtained in the following works:

- $CMSS$ scheme [20] software implementation on a Pentium M 1.73 GHz, 1 GB of RAM running Microsoft Windows XP for 2^{40} signatures and $w = 3$;
- $GMSS$ scheme [23] software implementation on a Pentium computer dual core 1.8 GHz for 2^{40} signatures ($w_1 = 9$ and $w_2 = 3$ were 390 min, 10.7 ms and 10.7 ms);
- $XMSS$ scheme [22] software implementation on an Intel(R) Core (TM) $i5$ M 540, 2.53GHz computer with Infineon technology;
- $CMSS$ scheme [85] hardware implementation on a novel architecture on an FPGA Platform (Virtex-5);
- $XMSS$ scheme [66] software implementation on an Intel Core $i7-2670$ QMCPU, 2.20 GHz with 6 GB of RAM.

Table 1 Implementation results

Schemes	Hash	H	w	t_{key}	t_{sig} (ms)	t_{ver} (ms)
CMSS [06]	SHA2	40	(3,3)	120.7 min	40.9	3.7
GMSS [07]	SHA1	40	(9,3)	390 min	10.7	10.7
XMSS [11]	SHA2	20	4	408.6 s	6.3	0.51
CMSS [11]	SHA2	30	4	820 ms	2.7	1.7
XMSS [13]	SHA2	20	4	553 s	2.7	0.31

In Table 1 the size of all public keys is 32 bytes, except for the *XMSS* scheme, that also has to store the bitmasks. The private key and signature are smaller in the *XMSS* scheme, since in the other schemes it is necessary to store information of more than one tree. The *XMSS* scheme presented the best timings for signing and verification on a software implementation, given that only one authentication path needs to be updated and checked for each signature. However, the *XMSS* is only recommended for applications requiring up to 2^{20} signature keys, since the generation of more keys is too time consuming. A Multi Tree *XMSS* ($XMSS^{MT}$) [79] based on algorithms *CMSS* and *GMSS* is recommended for applications that require a large numbers of signatures.

3 Multivariate Schemes

Multivariate public key cryptosystems (MPKC) constitute one of the main public key families considered potentially resistant against the powerful quantum computers. The security of MPKC schemes is based upon the difficulty of solving nonlinear system of equations over finite fields. In particular, in most cases, such schemes are based upon multivariate systems of *quadratic* equations because of computational advantages. This last problem is known as multivariate quadratic problem or $\mathcal{MQ}$ Problem, and it was shown to be NP-complete by Patarin [69]. MPKC has been developed more intensively in the last two decades. It was shown that, in general, encryption schemes were not as secure as it was believed to be, while signatures constructions can be considered viable.

The idea behind MPKC is to define a trapdoor one-way function whose image is a nonlinear system of multivariate equations over a finite field. The public key is given by a set of polynomials:

$$\mathcal{P} = \{p_1(x_1, \ldots, x_n), \cdots, p_m(x_1, \ldots, x_n)\}$$

where each p_k is a nonlinear polynomial (usually quadratic) in the variables $\mathbf{x} = (x_1, \cdots, x_n)$:

$$p_k(x_1, \ldots, x_n) := \sum_{1 \leq i \leq j \leq n} P_{ij}^{(k)} x_i x_j + \sum_{1 \leq i \leq n} L_i^{(k)} x_i + c^{(k)}, 1 \leq k \leq m \qquad (1)$$

Fig. 7 Pure quadratic map or transform

$$x\,P^{(k)}\,x^T = h_k \quad (k = 1, \dots, m)$$

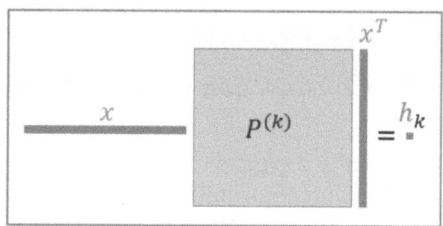

and all the coefficients and variables are in $\mathbb{F}_q$. In order to make the previous definition simpler, we will adopt vector notation, which is closer to practical implementations:

$$p_k(\mathbf{x}) := \mathbf{x}P^{(k)}\mathbf{x}^T + L^{(k)}\mathbf{x}^T + c^{(k)}, 1 \le k \le m \tag{2}$$

where $P^{(k)} \in \mathbb{F}_q^{n \times n}$ is a $n \times n$ matrix, whose entries are the quadratic terms coefficients of $p_k(x_1, \dots, x_n)$, $L^{(k)} \in \mathbb{F}_q^n$ is a vector whose entries are the linear terms coefficients of $p_k(x_1, \dots, x_n)$ and $c^{(k)}$ denotes the constant term of $p_k(x_1, \dots, x_n)$. Finally, $\mathbf{x}$ is the row vector of variables $[x_1, \dots, x_n]$. Figure 7 illustrates the pure quadratic transformation (or map) $\mathbf{x}P^{(k)}\mathbf{x}^T$ (whose evaluation provides a certain element of the finite field denoted by $h_k \in \mathbb{F}_q$).

A formal definition for the $\mathcal{MQ}$ Problem is given as follows.

Definition 1 ($\mathcal{MQ}$ Problem) Solve the random system $p_1(\mathbf{x}) = p_2(\mathbf{x}) = \cdots = p_m(\mathbf{x}) = 0$, where each p_i is quadratic in variables $\mathbf{x} = (x_1, \dots, x_n)$. All coefficients and variables are in $K = \mathbb{F}_q$, the field with q elements.

In other words, the target of the $\mathcal{MQ}$ Problem is to find a solution $\mathbf{x}$ for a given map $\mathscr{P}$. In 1979, Garey and Johnson proved [33, page 251] that the decision variant of the $\mathcal{MQ}$ Problem over binary finite fields is NP-complete.

On the other hand, the proposed $\mathcal{MQ}$ signature schemes in literature do not rely their security only on the original $\mathcal{MQ}$ Problem. In order to invert the trapdoor one-way function, which means finding the original private system (or an equivalent), it is necessary to solve a related problem called the Isomorphism of Polynomials Problem or *IP* Problem, proposed by Patarin [70].

Definition 2 (Isomorphism of Polynomials Problem) Let $m, n \in \mathbb{N}$ be arbitrarily fixed. Further denote $\mathscr{P}, \mathscr{Q} : \mathbb{F}_q^n \to \mathbb{F}_q^m$ two multivariate quadratic maps and $\mathbb{T} \in \mathbb{F}_q^{m \times m}$, $\mathbb{S} \in \mathbb{F}_q^{n \times n}$ two bijective linear maps, such that $\mathscr{P} = \mathbb{T} \circ \mathscr{Q} \circ \mathbb{S}$. Given $\mathscr{P}$ and $\mathscr{Q}$, find $\mathbb{T}$ and $\mathbb{S}$.

In other words, the *IP* Problem goal is to find $\mathbb{T}$ and $\mathbb{S}$ for a given pair $(\mathscr{P}, \mathscr{Q})$. Note that, originally, $\mathbb{S}$ was defined as an affine instead of linear transformation [71].

But, Braeken et al. [18, Sec. 3.1] noticed that the constant part is not important for the security of certain $\mathcal{M}\mathcal{Q}$ schemes and thus can be omitted.

3.1 Construction of $\mathcal{M}\mathcal{Q}$ Keys

Generically, a typical $\mathcal{M}\mathcal{Q}$ private key consists of two linear transformations $\mathbb{T}$ and $\mathbb{S}$ along with a quadratic transformation $\mathcal{Q}$. Note that $\mathcal{Q}$ presents certain particular trapdoor structure. We will present two distinct trapdoor structures in Sect. 3.2 for the UOV and Rainbow signature schemes. The trapdoor structure will allow the signer to easily solve the public $\mathcal{M}\mathcal{Q}$ system in order to generate valid signatures. The public key is simply given by the composition $\mathcal{P} = \mathbb{T} \circ \mathcal{Q} \circ \mathbb{S}$. For some signature schemes it is not necessary to explicitly use the map $\mathbb{T}$, since it is reduced to the identity [12, Chapter 6].

The main difference among distinct $\mathcal{M}\mathcal{Q}$ signature schemes falls in the trapdoor structure of $\mathcal{Q}$. Since public keys have the same structure in most schemes, verifying a signature follows the same procedure, in other words checking if a given signature $\mathbf{x}$ is a solution of a public quadratic system $p_k(\mathbf{x}) = h_k, 1 \le k \le m$. For other trapdoor constructions, the reader can see, for example, [94].

It is worth to mention an obvious optimization in the public matrices $P^{(k)}$ defined over odd characteristic fields that provides a reduction by a factor about two in the space representation. From the definition of the summation of the quadratic part of $p_k(x_1, \ldots, x_n)$ (Eq. 1), the coefficient of the term $x_i x_j$ is $P_{ij}^{(k)} + P_{ji}^{(k)}$; thus, one can update the coefficient $P_{ij}^{(k)}$ of the $P^{(k)}$ with the value $P_{ij}^{(k)} + P_{ji}^{(k)}$ and the coefficient $P_{ji}^{(k)}$ with zero for $i \le j \le n$, which makes the matrix $P^{(k)}$ upper triangular. After applying this representation one is able to define a unique public matrix called the *public matrix of coefficients*, denoted M_P. Each row of M_P is given by the linearization of the coefficients of each upper triangular matrix $P^{(k)}$. Figure 8 illustrates this construction.

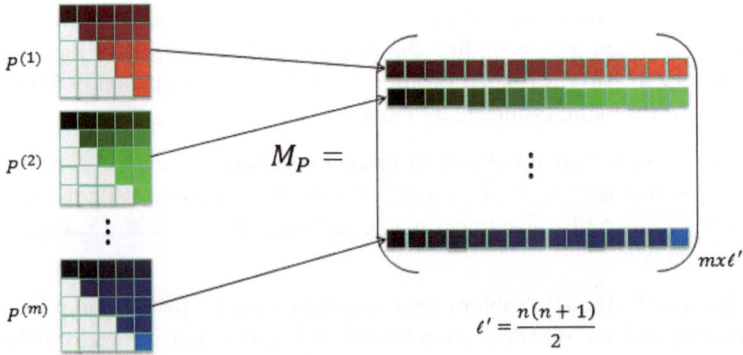

Fig. 8 Public matrix of coefficients

3.2   UOV and Rainbow $\mathcal{MQ}$ Signatures

One of the main still secure $\mathcal{MQ}$ signature families is the Unbalanced Oil and Vinegar (UOV) construction which was proposed by Patarin [48]. The name Oil and Vinegar came from the fact that variables $(x_1, \cdots, x_n)$ of a certain quadratic private system are separated in two subsets $O = (x_1, \cdots, x_m)$ and $V = (x_{m+1}, \cdots, x_n)$, in such a way that variables of the first set are never mixed in a term of the quadratic system.

Formally, the trapdoor consists of a purely quadratic map, called the central map, $\mathcal{Q} : \mathbb{F}^n \to \mathbb{F}^m$ with

$$\mathcal{Q} = \{f_1(u_1, \ldots, u_n), \ldots, f_m(u_1, \ldots, u_n)\}$$

and

$$f_k(u_1, \ldots, u_n) := \sum_{1 \le i \le j \le n} Q_{ij}^{(k)} u_i u_j \equiv u Q^{(k)} u^{\mathsf{T}} \tag{3}$$

The central map has an additional restriction in its polynomials $f_k(u_1, \ldots, u_n)$. It is imposed that a certain part of its coefficients be zeros. The set of variables u is divided in two subsets: the one of vinegar variables u_i with $i \in V = \{1, \cdots, v\}$ and the one of oil variables u_i with $i \in O = \{v+1, \cdots, n\}$ of $m = n - v$ elements. The restriction on the polynomials f_k is that they have no term combining any two oil variables. That assures that we do not have quadratic (or crossed) terms in oils. Thus, we only have terms combining the following sort of variables $v \times v$ and $o \times v$. Patarin showed that given this construction one can fix arbitrary values for the vinegars and then get a linear system in the oils. This remaining linear system will have a solution with high probability, i.e., $1 - 1/q$, and can be solved using Gaussian elimination with complexity $\mathcal{O}(n^3)$. The structure of the private polynomials is the following:

$$f_k(u_1, \cdots, u_n) := \sum_{i,j \in V, i \le j} Q_{ij}^{(k)} u_i u_j + \sum_{i \in V, j \in O} Q_{ij}^{(k)} u_i u_j \tag{4}$$

In order to generate a signature $\mathbf{x} \in \mathbb{F}_q^n$ of a given message, particularly of its hash $h \in \mathbb{F}_q^m$, the signer have to invert the map $P(\mathbf{x}) = Q(S(\mathbf{x})) = h$. Defining $\mathbf{x}' = \mathbf{x} \cdot S$, one first solves the multivariate system, $\mathbf{x}' Q^{(k)} \mathbf{x}'^{\mathsf{T}} = h_k, 1 \le k \le m$, finding $\mathbf{x}'$. Finally, the signature $\mathbf{x} = \mathbf{x}' S^{-1}$ is computed.

As explained before, the structure of the matrices $Q^{(k)}$ allows to efficiently solve the $\mathcal{MQ}$ system, by choosing v vinegar variables at random and then solving the resulting system for the remaining m oil variables. If the linear system has no solution, repeat the process by choosing new vinegar variables until it has a valid solution.

A signature $\mathbf{x}$ for h is valid, if and only if, all polynomials p_k constituting the public key have their evaluation satisfied, i.e., $p_k(x_1, \cdots, x_n) = \mathbf{x}P^{(k)}\mathbf{x}^T = h_k$, $1 \le k \le m$. The consistency of the verification $P(\mathbf{x}) \overset{?}{=} h$ is shown next:

$$p(\mathbf{x}) = \mathbf{x}P\mathbf{x}^\mathsf{T}$$
$$= \mathbf{x}(Q \circ S)\mathbf{x}^\mathsf{T}$$
$$= \mathbf{x}(SQS^\mathsf{T})\mathbf{x}^\mathsf{T}$$
$$= (\mathbf{x}'S^{-1})(SQS^\mathsf{T})(\mathbf{x}'S^{-1})^\mathsf{T}$$
$$= \mathbf{x}'(S^{-1}S)Q(S^\mathsf{T}(S^{-1})^\mathsf{T})\mathbf{x}'^\mathsf{T}$$
$$= \mathbf{x}'IQI\mathbf{x}'^\mathsf{T}$$
$$= \mathbf{x}'Q\mathbf{x}'^\mathsf{T}$$
$$= h.$$

Historically, UOV signatures came from *Oil and Vinegar* or OV construction [68], where the number of vinegars and oils are the same (balanced oil and vinegar), but that construction was shown insecure [47]. Next, it was redesigned a way to make it secure by unbalancing the amount of each subset ($v > m$), what originated the Unbalanced Oil and Vinegar (UOV) signature [48]. Figure 9 illustrates the structure of each UOV private polynomial.

In order to hide the trapdoor structure at polynomials f_k, an invertible linear transformation $\mathbb{S} \in \mathbb{F}_q^{n \times n}$ is applied to the right of $\mathcal{Q}$. So the resulting public map is $\mathcal{P} = \mathcal{Q} \circ \mathbb{S}$. The private key is given by the pair $sk := (\mathcal{Q}, \mathbb{S})$ and the public key is composed by polynomials $\mathcal{P} := p(x_1, \cdots, x_n) = \{p_1(x_1, \cdots, x_n), \cdots, p_m(x_1, \cdots, x_n)\}$. So, it becomes clear that the security of the system is not directly based on the $\mathcal{MQ}$ Problem and indeed recovering the private

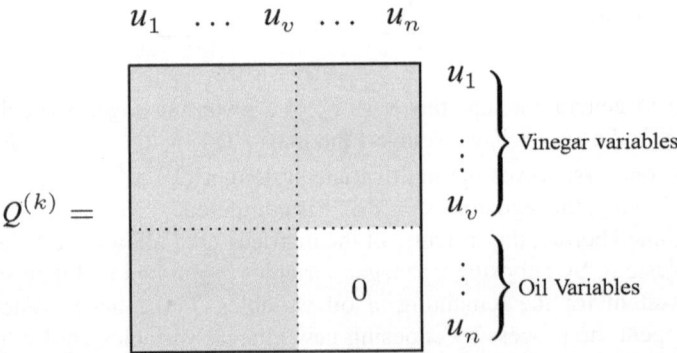

Fig. 9 UOV central map

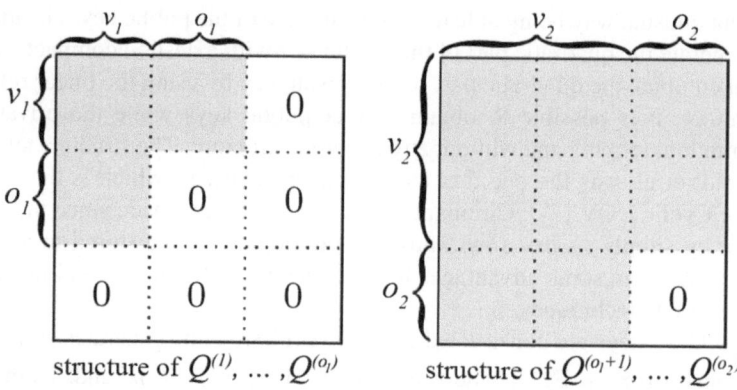

Fig. 10 Rainbow central map $\mathcal{Q}$ with two bands

key is related to the difficult to decompose $\mathcal{P}$ in $\mathcal{Q}$ and $\mathbb{S}$, in other words, to solve the *IP* Problem.

An important variant of the UOV scheme is the Rainbow [28] signature. It was proposed by Ding and Schmidt, whose main advantage is the shorter signature footprints attained compared to UOV [91, Section 3].

The basic idea of the Rainbow signature is to separate the m private UOV equations into smaller bands and partitioning the variables accordingly; in other words, each band has its own *oils* and *vinegars*. After a band is processed, all of its variables become the vinegars for the next band and so on until the last band is processed.

Typically the central map is divided into only two bands, since this configuration has been shown the most suitable in the sense that it avoids certain structural attacks and keeps the signatures reasonably short [91].

Rainbow central map $\mathcal{Q}$ with two bands, for example, is divided in two layers as shown in Fig. 10 where v_1 and o_1 are the number of vinegars and oils of the first layer and v_2 and o_2 are the number of vinegars and oils of the second layer. Note that $v_2 = o_1 + v_1$.

The signature procedure is similar to UOV one, choosing vinegars at random for the first band in order to be able to compute its oils, as it is done in UOV. Then, these computed variables (vinegars plus oils) are used as vinegars for the next band.

3.3 The Cyclic UOV Signature

An interesting step towards the reduction of UOV/Rainbow key sizes was made by means of the Cyclic UOV/Rainbow constructions [74, 77]. Petzoldt et al. noticed the existence of a linear relation between part of the public quadratic map and the private quadratic map. That relation was exploited in order to construct key pairs in a different unusual way being able to reduce the size of the public key. The idea is to firstly generate the quadratic part of the public key with a desired compact structure and then compute the quadratic part of the private key by using the linear relation.

Therefore, it is possible to obtain shorter public keys while the private ones remain random looking and without any apparent structure. The structure suggested by Petzoldt et al. was the one that uses circulant matrices, which is the origin of the name Cyclic UOV [77]. Circulant matrices are very compact, since they can be represent by simply its first row. Thus, the public key can be stored in a efficient manner, apart from some advantages in processing like Karatsuba and fast Fourier transform (FFT) techniques.

Cyclic UOV keys are constructed as follows. Firstly, one generates an invertible linear transform $\mathbb{S} \in \mathbb{F}_q^{n \times n}$, where $S_{ij} \xleftarrow{\$} \mathbb{F}_q, 1 \le i, j \le n$, and, from $\mathbb{S}$, one computes the aforementioned linear relation and denoted by $A_{\mathrm{UOV}} := \alpha_{ij}^{rs}$:

$$\alpha_{ij}^{rs} = \begin{cases} S_{ri} \cdot S_{si}, & \text{i=j} \\ S_{ri} \cdot S_{sj} + S_{rj} \cdot S_{si}, & \text{otherwise.} \end{cases}$$

In order to illustrate how the public and private matrices of coefficients, M_P and M_F, are related, we have initially Figs. 11 and 12 that separate the proper parts of these matrices.

Blocks B of M_P and F of M_F obey the relation $B := F \cdot A_{\mathrm{UOV}}(S)$. Thus, for the key generation, one may first generate matrix M_P with B with circulant structure and then computing $F := B \cdot A_{\mathrm{UOV}}^{-1}(S)$. That methodology was able to reduce UOV public key size in about 6 times for the security level of 80 bits.

As mentioned above, $\mathcal{M2}$ signatures have been developed more intensively in the last two decades. Many constructions were purposed toward key size reduction which is the main disadvantage today. Table 2 shows some of them.

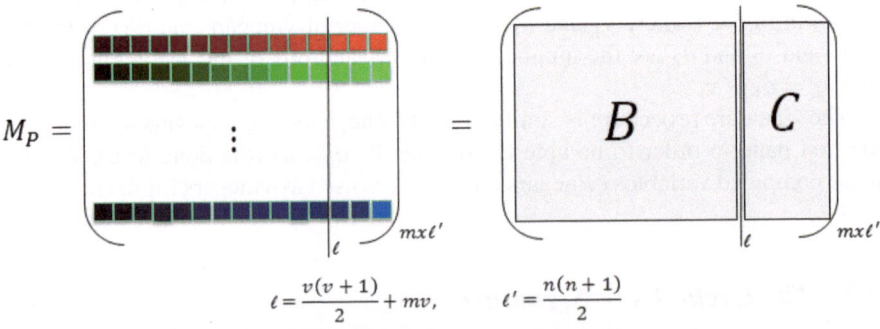

$$\ell = \frac{v(v+1)}{2} + mv, \qquad \ell' = \frac{n(n+1)}{2}$$

Fig. 11 Cyclic UOV: public matrix of coefficients

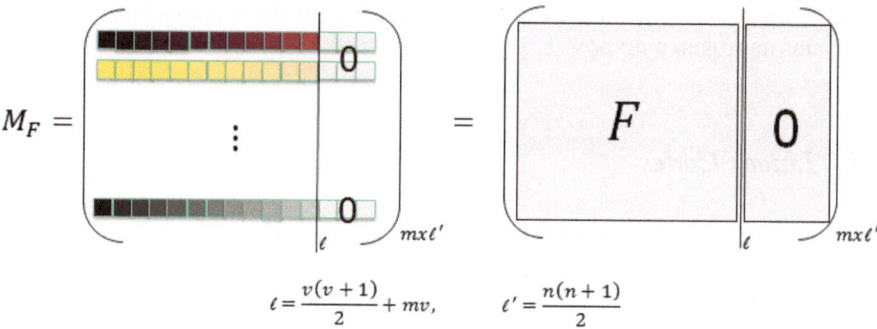

$$\ell = \frac{v(v+1)}{2} + mv, \qquad \ell' = \frac{n(n+1)}{2}$$

Fig. 12 Cyclic UOV: private matrix of coefficients

Table 2 $\mathcal{M} \mathcal{Q}$ signatures evolution

Construction	$|sk|$	$|pk|$	$|hash|$	$|sig|$	Ref.
Rainbow($\mathbb{F}_{2^4}$, 30, 29, 29)	75.8 KiB	113.4 KiB	232	352	[75]
Rainbow($\mathbb{F}_{31}$, 25, 24, 24)	59.0 KiB	77.7 KiB	232	392	[75]
CyclicUOV($\mathbb{F}_{2^8}$, 26, 52)	14.5 KiB	76.1 KiB	208	624	[74]
NC-Rainbow($\mathbb{F}_{2^8}$, 17, 13, 13)	25.5 KiB	66.7 KiB	384	672	[95]
Rainbow($\mathbb{F}_{2^8}$, 29, 20, 20)	42.0 KiB	58.2 KiB	272	456	[75]
CyclicLRS($\mathbb{F}_{2^8}$, 26, 52)	71.3 KiB	13.6 KiB	208	624	[76]
UOVLRS($\mathbb{F}_{2^8}$, 26, 52, 26)	71.3 KiB	11.0 KiB	208	624	[76]
CyclicRainbow($\mathbb{F}_{2^8}$, 17, 13, 13)	19.1 KiB	10.2 KiB	208	344	[74]

4 Code-Based Schemes

In this section we will discuss the theory and practice of cryptosystems based on error-correcting codes.

Coding theory aims at ensuring that when transmitting a collection of data over a channel subject to noise (i.e., the perturbations in the data), the recipient of this transaction can recover the original message. For this, one must find efficient ways to add redundant information to the original message such that if the message reaches the recipient containing errors (existing inversion in certain bits in case of binary messages), the receiver can decode it.

In the cryptographic context, the primitive adds errors in a word of an error-correcting code and compute a syndrome relative to the parity check matrix of this code.

The first construction was a public key encryption scheme proposed by Robert J. McEliece in 1978 [55]. The private key is a random, binary, and irreducible Goppa code (which will be reviewed in Sect. 4.1.1), and the public key is a random generator matrix with a permuted version of this code. The ciphertext is a codeword in which some errors were introduced, and only the owner of the private key can correct these errors (and thus decrypt the message). A few years later some parameter modifications were necessary to keep the security level high, but it remains unbroken until now.

4.1 Linear Codes

For a better technical understanding of this section, we first explain some basic concepts used within the code-based cryptography.

Matrix and vector indices will be numbered from 0 throughout this context, unless otherwise stated. Let p be a prime, and let $q = p^m$ for some integer $m > 0$. $\mathbb{F}_q$ denotes the finite field with q elements. The degree of a polynomial $g \in \mathbb{F}_q[x]$ is denoted by $\deg(g)$. It is also defined the notion of *Hamming weight* and *Hamming distance*:

Definition 3 The *Hamming weight* of a vector $u \in \mathscr{C} \subseteq \mathbb{F}_q^n$ is the number of nonzero coordinates on it, i.e., $\mathrm{wt}(u) = \#\{i, \ 0 \leq i < n \mid u_i \neq 0\}$. The *Hamming distance* between two vectors $u, v \in \mathscr{C} \subseteq \mathbb{F}_q^n$ is the number $\mathrm{dist}(u, v)$ of coordinates that these vectors differ from each other, i.e. $\mathrm{dist}(u, v) := \mathrm{wt}(u - v)$.

Now we will introduce some useful concepts to the task of encoding messages. The first refers to the *linear code*, which can be defined as

Definition 4 A (binary) linear $[n, k]$ error-correcting code $\mathscr{C}$ is a subspace of $\mathbb{F}_2^n$ of dimension k.

A vector $u \in \mathscr{C}$ is also called codeword (or, briefly, a word) of $\mathscr{C}$.

As a vector space, $\mathscr{C}$ is represented by a base, which can be written as a *generator matrix*:

- A generator matrix G of $\mathscr{C}$ is a matrix over $\mathbb{F}_q$ such that $\mathscr{C} = \langle G \rangle$, where $\langle G \rangle$ indicates the vector space generated by the rows of G. Normally the rows of G are independent and the matrix has dimension $k \times n$; in other words, $\exists G \in \mathbb{F}_q^{k \times n}$: $\mathscr{C} = \{uG \in \mathbb{F}_q^n \mid u \in \mathbb{F}_q^k\}$.
- We say that a generator matrix G is in the systematic form if its first k columns form the identity matrix.
- The so-called dual code $\mathscr{C}^\perp$ is the orthogonal code of $\mathscr{C}$ to the scalar product over $\mathbb{F}_q$ and is a linear code of dimension $n \times (n - k)$ over $\mathbb{F}_q$.

Alternatively, $\mathscr{C}$ is fully featured as the core of a linear transformation specified by a *parity check matrix* (or abbreviated *parity matrix*):

- A parity matrix H over $\mathscr{C}$ is a generator matrix of $\mathscr{C}^\perp$. In other words, $\exists H \in \mathbb{F}_q^{r \times n}$: $\mathscr{C} = \{v \in \mathbb{F}_q^n \mid Hv^\mathsf{T} = 0^r \in \mathbb{F}_q^r\}$, where $r = n - k$ is the codimension of $\mathscr{C}$ (i.e., the dimension of the orthogonal space $\mathscr{C}^\perp$).

It is easy to see that G and H, although not uniquely defined (because there is no one single basis for $\mathscr{C}$ or to $\mathscr{C}^\perp$), are related by $HG^\mathsf{T} = 0 \in \mathbb{F}_q^{r \times k}$.

The linear transformation defined by a parity matrix is called *syndrome function* of the code. The value of this transformation over any vector $u \in \mathbb{F}_q^n$ is called *syndrome* of this vector. Clearly, the syndrome of any codeword is always null.

Definition 5 The *distance* (or *minimum distance*) of a $\mathscr{C} \subseteq \mathbb{F}_q^n$ code is the minimum Hamming distance between words of $\mathscr{C}$, i.e., dist($\mathscr{C}$) = min{wt(u) | $u \in \mathscr{C}$}.

We write $[n, k, d]$ for a code $[n, k]$ whose minimum distance is (at least) d. If $d \geqslant 2t + 1$, it is said that the code is capable of correcting at least t errors, in the sense that there is no more than one codeword with a Hamming distance no more than t from any vector of $\mathbb{F}_q^n$.

Several computational problems involving codes are intractable, starting with the actual determination of the minimum distance of a code. The following problems are important for code-based cryptography:

Definition 6 (General Decoding) Let $\mathbb{F}_q$ be a finite field, and let (G, w, c) be a triple consisting of a matrix $G \in \mathbb{F}_q^{k \times n}$, an integer $w < n$, and a vector $c \in \mathbb{F}_q^n$. The *general decoding problem (GDP)* is the question if there is a vector $m \in \mathbb{F}_q^k$ such that $e = c - mG$ has Hamming weight wt(e) $\leqslant w$.

The search problem associated with the GDP is to calculate the vector m given the word with errors c.

Definition 7 (Syndrome Decoding) Let $\mathbb{F}_q$ be a finite field, and let (H, w, s) be a triple consisting of an $H \in \mathbb{F}_q^{r \times n}$, an integer $w < n$, and a vector $s \in \mathbb{F}_q^r$. The *syndrome decoding problem (SDP)* is whether there is a vector $e \in \mathbb{F}_q^n$ with Hamming weight of wt(e) $\leqslant w$ such that $He^\mathsf{T} = s^\mathsf{T}$.

The problem associated with the SDP consists in computing the error pattern e given its syndrome $s_e := eH^\mathsf{T}$.

Both the general decoding problem and the problem of syndrome decoding for linear codes are NP-complete [9].

In contrast to the overall results, the knowledge of the structure of certain codes makes the GDP and SDP soluble in polynomial time. A basic strategy to define code-based cryptosystems is therefore keep secret the information about the structure of the code and publish a code associated without any apparent structure (hence, by hypothesis hard to decode).

4.1.1 Goppa Codes

One of the most important families of linear error-correcting codes for cryptographic purposes is the Goppa codes:

Definition 8 Given a prime number p, $q = p^m$ for some $m > 0$, a sequence $L = (L_0, \ldots, L_{n-1}) \in \mathbb{F}_q^n$ of distinct elements, and a monic polynomial $g(x) \in \mathbb{F}_q[x]$ of degree t (called generator polynomial) such that $g(L_i) \neq 0$ to $0 \leqslant i < n$, the

Goppa code $\Gamma(L, g)$ is the code $\mathbb{F}_p$-alternate corresponding to $GRS_t(L, D)$ over $\mathbb{F}_q$, where $D = (g(L_0)^{-1}, \ldots, g(L_{n-1})^{-1})$.

The distance of an irreducible *binary* Goppa code is at least $2t + 1$ [43], and therefore a Goppa code can correct up to t errors (using, e.g., Patterson's algorithm [72]), sometimes a little more [11]. Appropriate decoding algorithms can still decode t errors when the generator $g(x)$ is not irreducible but free of squares. For example, one can see equivalently a binary Goppa code as an alternate code defined by the generator polynomial $g^2(x)$, in which case any alternate is able to decode t errors. Codes called *wild* extend this result under certain circumstances [14]. For all other cases, there is no known decoding method capable of correcting more than $t/2$ errors.

Equivalently we can define Goppa codes in terms of its syndrome function:

Definition 9 Let $L = (L_0, \ldots, L_{n-1}) \in \mathbb{F}_q^n$ be a sequence (called *support*) of $n \leq q$ distinct elements, and let $g \in \mathbb{F}_q[x]$ be a monic irreducible polynomial of degree t such that $g(L_i) \neq 0$ for all i. For any word $e \in \mathbb{F}_p^n$ is defined a polynomial *Goppa syndrome* $s_e \in \mathbb{F}_q[x]$ as

$$s_e(x) = \sum_{i=0}^{n-1} \frac{e_i}{x - L_i} \mod g(x). \tag{5}$$

The syndrome is a linear function of e. We also present an alternative definition for Goppa codes:

Definition 10 The *Goppa code* $[n, n - mt]$ over $\mathbb{F}_p$ supported L and generator polynomial g is the core function syndrome (Eq. 5), i.e., the set of $\Gamma(L, g) := \{e \in \mathbb{F}_p^n \mid s_e \equiv 0 \mod g\}$.

Writing $s_e(x) := \sum_i s_i x^i$ for some $s \in \mathbb{F}_q^n$, we can show that $s^\mathsf{T} = He^\mathsf{T}$ with

$$\begin{aligned} H &= \text{toep}(g_1, \ldots, g_t) \\ &\quad \cdot \text{vdm}_t(L_0, \ldots L_{n-1}) \\ &\quad \cdot \text{diag}(g(L_0)^{-1}, \ldots, g(L_{n-1})^{-1}) \end{aligned} \tag{6}$$

Thus, $H = TVD$ where T is a Toeplitz matrix $t \times t$, V is a Vandermonde matrix $t \times n$, and D is a diagonal matrix $n \times n$ according to the following definitions:

Definition 11 Given a sequence $(g_1, \ldots, g_t) \in \mathbb{F}_q^t$ for some $t > 0$, the *Toeplitz matrix* $\text{toep}(g_1, \ldots, g_t)$ is the matrix $t \times t$ with components $T_{ij} := g_{t-i+j}$ for $j \leq i$ and $T_{ij} := 0$ in other cases, namely,

$$\text{toep}(g_1, \ldots, g_t) = \begin{bmatrix} g_t & 0 & \ldots & 0 \\ g_{t-1} & g_t & \ldots & 0 \\ \vdots & \vdots & \ddots & \vdots \\ g_1 & g_2 & \ldots & g_t \end{bmatrix}.$$

Definition 12 Given $t > 0$ and a sequence $L = (L_0, \ldots, L_{n-1}) \in \mathbb{F}_q^n$ for some $n > 0$, the *Vandermonde matrix* $\mathrm{vdm}(t, L)$ is the matrix $t \times n$ with components $V_{ij} = L_j^i$, i.e.,

$$
\mathrm{vdm}(t, L) = \begin{bmatrix}
1 & \cdots & 1 \\
L_0 & \cdots & L_{n-1} \\
L_0^2 & \cdots & L_{n-1}^2 \\
\vdots & \ddots & \vdots \\
L_0^{t-1} & \cdots & L_{n-1}^{t-1}
\end{bmatrix}.
$$

Definition 13 Given a sequence $(d_0, \ldots, d_{n-1}) \in \mathbb{F}_q^n$ for some $n > 0$, we denote by $\mathrm{diag}(d_0, \ldots, d_{n-1})$ a *diagonal matrix* with components $D_{jj} := d_j, 0 \leqslant j < n$, and $D_{ij} := 0$ in other cases, namely,

$$
\mathrm{diag}(d_0, \ldots, d_{n-1}) = \begin{bmatrix}
d_0 & 0 & \cdots & 0 \\
0 & d_1 & \cdots & 0 \\
\vdots & \vdots & \ddots & \vdots \\
0 & 0 & \cdots & d_{n-1}
\end{bmatrix}.
$$

4.2 Decodability

All codes $[n, k]$ with distance d satisfy the *Singleton limit*, which states that $d \leqslant n - k + 1$. The existence of a binary linear code $[n, k]$ with distance d is guaranteed since:

$$
\sum_{j=0}^{d-2} \binom{n-1}{j} < 2^{n-k}.
$$

This is called the *Gilbert-Varshamov (GV) boundary*. Random binary codes achieve the GV bound, in the sense that the above inequality is very close to equality [53]. There is no known family of binary codes, however, that can be decoded in subexponential time until the GV limit nor known subexponential algorithm for decoding general codes to the GV limit.

Consider a code $\mathbb{F}_p$-alternant with length n and able to decode t errors, derived from a code GRS over $\mathbb{F}_{p^m}$. The syndrome space have size p^{mt}. However, decodable syndromes are only those that match the error vector with weight not exceeding t. In other words, only $\sum_{w=1}^{t} \binom{n}{w}(p-1)^w$ nonzero syndromes are uniquely decodable, and thus its density is

$$
\delta = \frac{1}{p^{mt}} \sum_{w=1}^{t} \binom{n}{w}(p-1)^w.
$$

If the code length is a fraction $1/p^c$ of the maximum length for any $c \geq 0$, i.e., $n = p^{m-c}$, the density can be approximated by

$$\delta \approx \frac{1}{p^{mt}} \binom{n^t}{t!} (p-1)^t = \frac{(p^{m-c})^t (p-1)^t}{p^{mt} t!} = \left(\frac{p-1}{p^c}\right)^t \frac{1}{t!}.$$

A particularly good case is therefore $\delta \geq 1/t!$, which occurs when $p^c/(p-1) \leq 1$, i.e., $c \leq \log_p(p-1)$ or $n \geq p^m/(p-1)$. Unfortunately this also means that, for binary codes, the highest densities are reached only by codes of maximum length or nearly maximum; otherwise the density is reduced by a factor 2^{ct}. For codes of maximum length ($n = p^m$ and hence $c = 0$), the density simplifies to $\delta \approx (p-1)^t/t!$ that achieves the relative minimum $\delta \approx 1/t!$ for binary codes.

We will also be interested in the particular case of error patterns if a particular magnitude prevails over the others and more especially when all the error magnitudes are equal. In this case, the density of decodable syndromes is $\delta \approx (p-1)/t!$ which again reaches the minimum $\delta \approx 1/t!$ in binary codes.

4.3 Code-Based Cryptosystems

The original McEliece encryption schemes [55] and Niederreiter [64], despite the historic name, but inaccurate and undue, as *cryptosystems*, are best described as *trapdoor one-way functions* than as full encryption methods themselves. Functions of this nature can be transformed in various ways in cryptosystems, for example, Fujisaki-Okamoto transform.

Interestingly, McEliece and Niederreiter commonly show a substantial speed advantage over traditional processing schemes. For example, a code of length n presents time complexity $O(n^2)$, while Diffie-Hellman/DSA systems, as well as the operations of the RSA private exponent system, have time complexity $O(n^3)$ and keys with n bits.

For simplicity, the descriptions of McEliece and Niederreiter schemes below assume that patterns of correctable errors are binary vectors of weight t, but variants with broader patterns of error are possible, as the ability to decode the underlying code. Simple and effective criteria for choosing parameters are provided in Sect. 4.3.3. Each encryption scheme consists of three algorithms: MakeKeyPair, Encrypt, and Decrypt.

4.3.1 McEliece

- MakeKeyPair. Given the desired level of security λ, choose a prime p (commonly $p = 2$), a finite field $\mathbb{F}_q$ with $q = p^m$ for some $m > 0$, and a Goppa code $\Gamma(L, g)$ with support $L = (L_0, \ldots, L_{n-1}) \in (\mathbb{F}_q)^n$ (with distinct elements) and generator polynomial $g \in \mathbb{F}_q[x]$ of degree t and free of squares satisfying

$g(L_j) \neq 0, 0 \leqslant j < n$. Let $k = n - mt$. The choice is guided so that the cost of decoding a code $[n, k, 2t + 1]$ is at least 2^λ steps. Compute a systematic generator matrix $G \in \mathbb{F}_p^{k \times n}$ for $\Gamma(L, g)$, i.e., $G = [I_k \mid -M^T]$ for any matrix $M \in \mathbb{F}_p^{mt \times k}$ and I_k an identity matrix of order k. The private key is $sk := (L, g)$ and the public key is $pk := (M, t)$.

- Encrypt. To encrypt a plaintext $d \in \mathbb{F}_p^k$, we choose a vector $e \xleftarrow{\$} \{0, 1\}^n \subseteq \mathbb{F}_p^n$ with weight $\mathrm{wt}(e) \leqslant t$, and compute the encrypted text $c \leftarrow dG + e \in \mathbb{F}_p^n$.
- Decrypt. To decrypt an encrypted text $c \in \mathbb{F}_p^n$ with the knowledge of L and g, we compute the decodable syndrome c, apply in it a decoder to determine the error vector e, and recover the plain text d from the first k columns of $c - e$.

4.3.2 Niederreiter

- MakeKeyPair. Given the desired level of security λ, choose a prime p (commonly $p = 2$), a finite field $\mathbb{F}_q$ with $q = p^m$ for some $m > 0$, and a Goppa code $\Gamma(L, g)$ with support $L = (L_0, \ldots, L_{n-1}) \in (\mathbb{F}_q)^n$ (of distinct elements) and a generator polynomial $g \in \mathbb{F}_q[x]$ of degree t and free of squares satisfying $g(L_j) \neq 0, 0 \leqslant j < n$. Let $k = n - mt$. The choice is guided so that the cost of decoding a code $[n, k, 2t + 1]$ is at least 2^λ steps. Compute a systematic parity matrix $H \in \mathbb{F}_p^{mt \times n}$ for $\Gamma(L, g)$, i.e., $M = [M \mid I_{mt}]$ for some matrix $M \in \mathbb{F}_p^{mt \times k}$ and I_{mt} the identity matrix of order mt. Finally, choose as public parameter a function of rank permutation $\phi : \mathscr{B}(n, t) \rightarrow \mathbb{Z}/\binom{n}{t}\mathbb{Z}$. The private key is $sk := (L, g)$ and the public key is $pk := (M, t, \phi)$.
- Encrypt. To encrypt a plaintext, $d \in \mathbb{Z}/\binom{n}{t}\mathbb{Z}$ is represented d as a error pattern $e \leftarrow \phi^{-1}(d) \in \{0, 1\}^n \subseteq \mathbb{F}_p^n$ of weight $\mathrm{wt}(e) = t$, and compute as a ciphertext syndrome $s \leftarrow eH^\mathsf{T} \in \mathbb{F}_p^{mt}$.
- Decrypt. To decrypt an encrypted text $s \in \mathbb{F}_p^{mt}$ with the knowledge of L and g, this syndrome becomes another one decodable, applies to income results a decoder to determine the error vector e, and recovers from this the plaintext $d \leftarrow \phi(e)$.

4.3.3 Parameters for Code-Based Cryptosystems

The classical schemes of McEliece and Niederreiter, implemented on the class of Goppa codes, remain safe until the present date, in contrast to implementations on many other families of codes proposed [41, 67]. Indeed, Goppa codes have weathered well the intense attempts of cryptanalysis, and despite considerable progress in the area [10] (see also [12] for review), they remain essentially intact for cryptographic purposes that have been suggested in the pioneering work of McEliece [55].

Table 3 suggests parameters for the underlying codes of cryptosystems such as McEliece or Niederreiter and size $|pk|$ in bits of the resulting public key. Only *generic* Goppa codes are considered irreducible.

We notice that in this generic Goppa codes scenario, these schemes are adversely affected by very large keys compared to conventional counterparts. That is the importance of seeking ways to reduce the key sizes, maintaining intact the level of security associated.

The first steps toward the goal of reducing the size of the keys without reducing the level of security in post-quantum cryptosystems were given by Monico et al. through codes with low-density parity check (matrix) (*LDPC codes*) [61], after that by Gaborit with quasi-cyclic codes [31], and Baldi and Chiaraluce through a combination of both [4].

4.4   LDPC and QC-LDPC Codes

LDPC codes were invented by Robert Gallager [32] and are linear codes obtained from sparse bipartite graphs. Suppose that $\mathbb{G}$ is a graph with n nodes on the left side (called message nodes) and r nodes on the right side (called verification nodes), as can be seen in Fig. 13 below. The graph gives rise to a linear code of size n and block size of at least $n - r$ as follows: the n coordinates of the code words are associated

Table 3 Parameters for McEliece/Niederreiter using generic binary Goppa codes

m	n	k	t	lg WF	$|pk|$
11	1893	1431	42	80.025	661122
12	2887	2191	58	112.002	1524936
12	3307	2515	66	128.007	1991880
13	5397	4136	97	192.003	5215496
13	7150	5447	131	256.002	9276241

Message nodes   Verification nodes

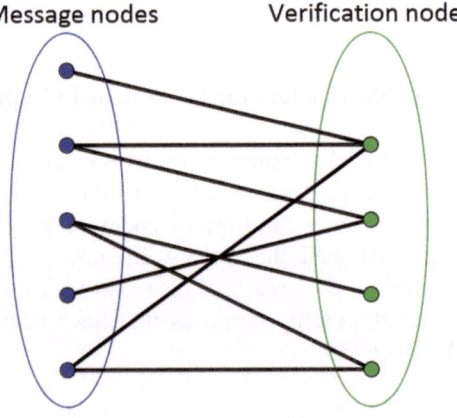

Fig. 13 Bipartite graph

with n message nodes. The code words are the vectors $(c_1, \ldots, c_n)$ such that, for all verification nodes, the sum of positions between neighboring message nodes is zero.

The graph representation is analogous to the matrix representation looking at the adjacency matrix of the graph: H is a binary matrix $r \times n$ whose entry (i, j) is 1 if and only if the ith check node is connected to the jth message node in the graph. Then the LDPC code defined by the graph is the set of vectors $c = (c_1, \ldots, c_n)$ such that $H \cdot c^{\mathsf{T}} = 0$. The matrix M is called *parity matrix* for the code. Conversely, any binary matrix $r \times n$ gives rise to a bipartite graph between n message nodes and r verification nodes, and the code defined to null space of M is precisely the code associated with that graph. Therefore, any linear code has a representation such as a code associated with a bipartite graph (note that this graph is not defined solely by the code). However, not every binary linear code has a representation as a *sparse* bipartite graph. If there is, then the code is called low-density paritycheck code.

An important subclass of LDPC codes which have advantages over other codes in the same class of codes is the quasi-cyclic low-density parity check (QC-LDPC) [90]. In general, a $[n, k]$ QC-LDPC code satisfies $n = n_0 b$ and $k = k_0 b$ (and thus also $r = r_0 b$) for some b, n_0, k_0 (and r_0) and admits a parity matrix consisting of $n_0 \times r_0$ blocks of circulating sparse $b \times b$ submatrices. A particularly important case is when $b = r$ (and $r_0 = 1$ and $k_0 = n_0 - 1$), since a systematic parity matrix for this code is fully defined by the first line of each $r \times r$ block. It is said that the parity matrix is in the *circulant form*.

However, it was shown that all these proposals contain vulnerabilities that make them unsuitable for cryptographic purposes [67]. Indeed, in these methods, the trapdoor was essentially protected by any other mechanism including a private permutation of the underlying code. The attack strategy in this scenario is to obtain a soluble system of linear equations that the components of the permutation matrix must satisfy and was set up successfully because of the overly restrictive nature of the secret permutation (since it needs to preserve the quasi-cyclic structure of the result) and the fact that the secret code is a subcode of a very particular public code.

An attempt to fix the proposal of Baldi and Chiaraluce was presented [5]. More recently, Berger et al. [8] showed how to avoid the problems of the original Gaborit scheme and removed vulnerabilities previously known through two techniques:

1. Extract public keys shortened by blocks of very long private codes, exploring a theorem due to Wieschebrink about NP-completeness to distinguish shortened codes [92];
2. Working with the subfield subcodes of an intermediate field between the original field and the extension field of the original GRS code adopted by construction.

These techniques have been applied with some success to quasi-cyclic codes. However, almost all of this family of codes was subsequently broken due to structural failure of security, more precisely a relationship between the secret structure and certain multivariate quadratic equation systems [30].

Historical and experiential wisdom suggests, therefore, to restrict the search for more efficient parameters of code-based cryptosystems to the class of Goppa codes.

4.5 MDPC and QC-MDPC Codes

An interesting subclass of the LDPC codes consists of moderate density parity check codes (MDPC) and their quasi-cyclic variant (QC-MDPC) [60].

These codes, introduced by Misoczki et al., have densities low enough to enable decoding by simple (and arguably more efficient) methods of belief propagation and Gallager's *bit flipping*. Yet densities are high enough to prevent attacks based on the presence of very sparse words in the dual code as seen in the Stern attack [87] and variants, without ruining the error correction capability, as well as keeping decoding attacks based on information set [10, 13] also unfeasible.

Moreover, to prevent structural attacks as proposed by Faugère et al. [30] and Leander and Gauthier [36], oriented encryption codes should be maintained as much as possible without structure except for the secret trapdoor that allows private decryption, and in the case of quasi-cyclic codes, external symmetries allow an efficient implementation. Finally, the circulant symmetry can introduce security weaknesses as pointed out by Sendrier [83], but with respect to attacking performance, it induces only a polynomial gain (specifically linear), and a small adjustment in the parameters completely eliminates this problem. Typical densities in this case are in the range from 0.4 to 0.9 % of the code size, an order of magnitude above LDPC codes, but much better than previously mentioned MDPC, and certainly appropriate for Gallager codes. The construction is also as random as possible, maintaining only the desired density and circulant geometry. Furthermore, the code size is much higher than typical values for MDPC.

4.6 Method for Gallager's Hard Decision Decoding (Bit Flipping)

In this section we describe *Gallager's hard decision decoding algorithm*, or more simply *bit flipping*, following the concise and clear description of Huffman and Pless [46]. This algorithm is necessary to recover the original message from the encrypted codeword with errors.

We assume that the codeword is encrypted with a binary LDPC code $\mathbb{C}$ for transmission and the vector c is received. To calculate the syndrome $s = cH^T$, each bit received from c affects at most d_v components of this syndrome. If only the jth bit of c contains an error, then the corresponding d_v with component s_i of s is equal to 1, indicating the parity check equations that are not satisfied. Even if you have a few other bits with error among those who contributed to the calculation of s_i, it is expected that several of d_v components of s are equal to 1. This is the basis of the decoding algorithm of Gallager, both *hard decision decoding* and *bit flipping*:

1. Compute cH^T and determine the unsatisfied parity checks (namely, the parity checks where the components of cH^T equal 1).

2. For each of the n bits, compute the number of unsatisfied parity checks involving that bit.
3. Flip the bits of c that are involved in a number of unsatisfied parity check equations overcoming some threshold.
4. Repeat steps 1, 2, and 3 until either $cH^\top = 0$, in which case c has been successfully decoded, or until a certain bound in the number of iterations is reached, in which case decoding of the received vector has failed.

The *bit-flipping* algorithm is not the best method for decoding LDPC codes; in fact, the belief propagation technique [32, 46] is known for its ability to exceed correction errors. However, belief propagation decoders involve a computation with a *probability* increasingly refined for each bit of the received word c containing an error, incurring floating-point arithmetic and high-precision approaches that suit the process and computationally expensive algorithms. In a scenario where the number of errors is fixed and known in advance, as is the case for cryptographic applications, parameters can be adjusted so that complex and expensive decoding methods, such as belief propagation, are no longer needed.

4.7 Digital Signatures with Error Correcting Codes

After unsuccessful attempts to create a digital signature scheme based on error-correcting codes [2, 88], in 2001, Courtois, Finiasz, and Sendrier proposed a promising scheme [26].

4.7.1 CFS

The *CFS* has been proposed as a System of Digital Signatures based on McEliece Cryptographic System. By definition, a system of digital signature must provide a way to sign any document in such a way that uniquely identifies its author, and which has an efficient public signature verification algorithm. For these tasks, a linear code must be chosen, illustrated below as $\mathscr{C}$. So, CFS uses a public hash function h to compress the document m by computing the vector $h(m)$. Decoding this hash with the chosen error correction code algorithm, we obtain a vector c', corresponding to the signature of the message m. For signature verification, simply encrypt c', received with the message m, and verify that if it corresponds to the calculation of the hash of the message m, as follows:

- Make Key Pair:

  1. Choose a Goppa code $G(L, g(X))$;
  2. Compute a corresponding $(n-k) \times n$ parity check matrix H;
  3. Compute $V = SHP$, where S is a random binary invertible matrix $(n-k) \times (n-k)$ and P is a random permutation matrix $n \times n$.

  The private key is G, and the public key is (V, t).

- Signature:

  1. Find the short $i \in \mathbb{N}$ such that, for $c = h(m,i)$ and $c' = S^{-1}c$, c' is a decodable syndrome of G.
  2. Using the decoding algorithm of G, compute the error vector e', whose syndrome is c', i.e., $c' = H(e')^t$.
  3. Compute $e^t = P^{-1}(e')^t$.

  Therefore, the signature is the pair (e, i).
- Signature Verification:

  1. Compute $c = Ve^t$.
  2. Accept iff $c = h(m,i)$.

Although CFS is a still safe signature scheme after going through many cryptanalysis, it is not suitable for standard applications commonly used today, since besides the size of public keys to sign the cost is too large for a set of reliable parameters.

5 Lattice-Based Schemes

From the mathematical point of view, historically lattices have been studied since the 18th century by mathematicians such as Lagrange and Gauss. However, the interest in cryptography starts more recently with Ajtai's work, that proves the existence of one-way functions based on the hardness of the shortest vector problem (SVP). The versatility and flexibility of lattice based cryptography, in terms of possible cryptographic features and simplicity of the basic operations, make it one of the most promising lines of research in cryptography. Moreover, some lattice schemes are supported by security demonstrations that rely on the worst-case hardness of certain problems.

Lattice-based cryptography can be divided in two categories: (i) those with a security proof, as, for example, is the case of Ajtai's construction or cryptosystems based on the LWE problem, whose encryption and decryption are quadratic or even cubic algorithms involving the manipulation of a matrix A, associated with the public key, which is not efficient when compared to conventional cryptography; and (ii) those without a security proof, but with efficient implementations, for example, the NTRU cryptosystem. A recent result [86] reduces the security of NTRU-based cryptosystems to the worst-case problem over ideal lattices. Although hard problems over lattices may not be hard over ideal lattices, no polynomial algorithm is known to solve them, even when considering a polynomial approximation factor or the utilization of quantum computation.

5.1 Basic Definitions

Definition 14 Let $\mathbb{R}^m$ be a m-Dimensional Euclidean Vector Space, and $B = \{b_1, \ldots, b_n\}$ be a set of n linearly independent vectors, the lattice $\mathscr{L}$ in $\mathbb{R}^m$ is the additive subgroup, that consists of all linear combinations of B with integer coefficients, in other words:

$$\mathscr{L}(b_1, \ldots, b_n) = \left\{ \sum_{i=1}^{n} x_i b_i : x_i \in \mathbb{Z} \right\},$$

where the vectors $b_1, \ldots, b_n$ are the called basis vector of $\mathscr{L}$ and the set B is called *lattice basis*.

Alternatively, it is possible to define lattices using matrix notation. Let $B \in \mathbb{R}^{m \times n}$ be a matrix, the lattice generated by B is defined as $\mathscr{L} = \{Bx \mid x \in \mathbb{Z}^n\}$, such that the determinant $\det(B)$ is independent from the basis choice and corresponds geometrically to the inverse of the lattice point density in $\mathbb{Z}^m$.

Definition 15 Given the lattice $\mathscr{L}(B)$, the basis vectors can be seen as edges of a dimension n parallelepiped. Thus, we define $\mathscr{P}(B) = \{Bx \mid x \in [0, 1)^n\}$, denominated the *fundamental parallelepiped* of B. We can also define a symmetric fundamental parallelepiped as $\mathscr{P}_{1/2}(B) = \{Bx \mid x \in [-1/2, 1/2)^n\}$, the *centralized fundamental parallelepiped* of B. Figures 14 and 15 show examples of fundamental parallelepipeds on dimension 2.

Theorem 1 *Let $\mathscr{L}$ be a lattice and let $\mathscr{P}(B)$ be the fundamental parallelepiped of $\mathscr{L}$. Then, given an element $w \in \mathscr{L}$, we can write w as $w = v + t$, with $v \in \mathscr{L}$ and $t \in \mathscr{P}(B)$, such that t is uniquely determined. This operation is equivalent to a modular reduction, where the vector t is interpreted as $w \pmod{\mathscr{P}(B)}$ (Fig. 16).*

The volume of the fundamental parallelepiped is related with the determinant of B, and given by $\mathrm{Vol}(\mathscr{P}(B)) = |\det(B)|$. Given two basis $B = \{b_1, \ldots, b_n\}$ and $B' = \{b'_1, \ldots, b'_n\}$ for lattice $\mathscr{L}$, we have that $\det(B) = \pm \det(B')$.

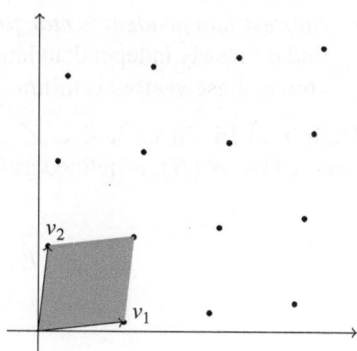

Fig. 14 $\mathscr{P}(B)$

Fig. 15 $\mathscr{P}_{1/2}(B)$

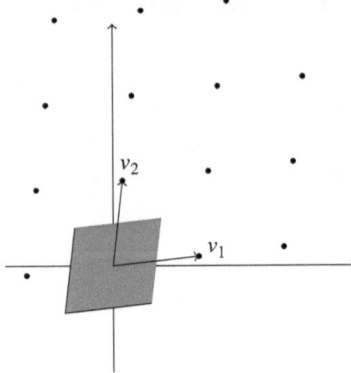

Fig. 16 Reduction modulo $\mathscr{P}(B)$

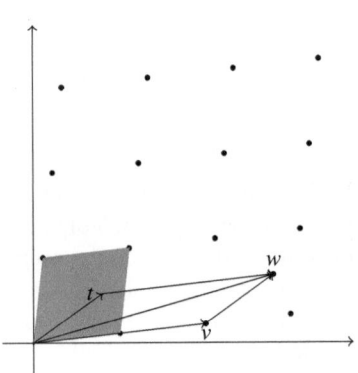

The most important computational problem in lattices is the *shortest vector problem* (SVP), it is defined as follows: given the lattice $\mathscr{L}(B)$, one has to find a nonzero vector with minimum norm. In practice, it is used an approximation factor $\gamma(n)$ such that we look for a vector whose norm is less than the minimum multiplied by $\gamma(n)$.

The following problems are also important for cryptographic purposes:

- *closest vector problem* (CVP). Given lattice $\mathscr{L}(B)$ and a vector $t \in \mathbb{R}^m$, the goal is to find the vector $v \in \mathscr{L}(B)$ closest to t;
- *shortest independent vector problem* (SIVP). Given basis $B \in \mathbb{Z}^{m \times n}$, we must find n linearly independent lattice vectors $(v_1, \ldots, v_n)$, such that maximum norm among these vectors is minimum.

Definition 16 Given lattice $\mathscr{L}$ and basis $B = (v_1, \ldots v_n)$, the *Hadamard ratio*, denoted by $\mathscr{H}(B)$, is defined as follows:

$$\mathscr{H}(B) = \left(\frac{|\det \mathscr{L}|}{\prod_{1 \leq i \leq n} ||v_i||} \right)^{1/n}.$$

It is easy to show that, for any basis B, we have that $0 \leq \mathcal{H}(B) \leq 1$. Furthermore, the closer this ratio is to 1, the "more orthogonal" is the basis.

A particularly important class in lattices is that of *q-ary lattices class*, denoted by Λ_q. Given an integer q, the vector coordinates are restricted to be elements in $\mathbb{Z}_q$. Given the matrix $A \in \mathbb{Z}_q^{n \times m}$, the q-ary lattice is determined by the rows of A, instead of the columns. That is, it is formed by vectors $y = A^T s \pmod{q}$, for $s \in \mathbb{Z}^n$. The orthogonal q-ary lattice, $\Lambda_q^\perp$, corresponding to matrix A, is given by vectors y such that $Ay = 0 \pmod{q}$. Given the lattice $\mathcal{L}$, the *dual lattice*, $\mathcal{L}^*$, is formed by vectors y, such that $\langle x, y \rangle \in \mathbb{Z}$, for $x \in \mathcal{L}$. In particular, the q-ary orthogonal lattice, $\Lambda_q^\perp(A)$, is the same as $q\Lambda_q(A)^*$.

5.1.1 LLL Algorithm

The LLL Algorithm is important in lattice because the practical security analysis in general are based this algorithm. In fact, the LLL can be used to tackle the SVP and related problems, as we will see later. In this section we will describe the LLL algorithm. Given a lattice and a basis for it, LLL computes a new basis, with Hadamard ratio closer to 1. In other words, the LLL algorithm performs a basis reduction, because the computed basis has lower norm and greater orthogonality than the original one.

In a vector space with a basis $(v_1, \cdots, v_n)$, an orthonormal basis can easily be obtained by using the Gram-Schmidt algorithm. In lattices we can apply a similar approach using Gauss reduction. The idea used in Gauss reduction is the same as in Gram-Schmidt algorithm, where we have $\mu_{ij} = v_i v_j^* / ||v_j^*||^2$, but the values μ_{ij} are not necessarily integers. Thus, Gauss reduction considers the closest integers $\lfloor \mu_{ij} \rceil$. The algorithm ends when this closest integers are zero, a condition that only in dimension 2 is sufficient to prove that the shortest vector was found.

Definition 17 Let $B = (v_1, \ldots, v_n)$ be a basis for lattice $\mathcal{L}$ and let $B^* = (v_1^*, \ldots, v_n^*)$ be the Gram-Schmidt orthogonal basis. The basis B is called *LLL-reduced* if the following conditions are satisfied:

Algorithm 5.1 Gauss reduction

Require: A basis (v_1, v_2).
Ensure: Returns a basis with shortest vector (v_1^*) and with a vector v_2^* that cannot be reduced by subtracting v_1.
    Let $v_1^* = v_1$ and $v_2^* = v_2$.
    while true **do**
        if $||v_2^*|| < ||v_1^*||$ **then**
            Swap v_1^* and v_2^*.
        end if
        Compute $m = \lfloor v_1^* . v_2^* / ||v_1^*||^2 \rceil$.
        if $m = 0$ **then return** (v_1^*, v_2^*).
        end if
        Swap v_2^* and $v_2^* - mv_1^*$.
    end while

Algorithm 5.2 LLL

Require: A basis $(v_1, \ldots, v_n)$.
Ensure: Returns a basis with the shortest vector (v_1^*) and a vector v_2^* that cannot be reduced by
subtracting v_1.
  $k = 2$.
  $v_1^* = v_1$.
  while $k \leq n$ **do**
    for $j = 1$ till $j = k - 1$ **do**
      $v_k = v_k - \lfloor \mu_{k,j} \rceil v_j^*$.
    end for
    if $\|v_k^*\|^2 \geq (\frac{3}{4} - \mu_{k,k-1}^2)\|v_{k-1}^*\|$ **then**
      $k = k + 1$.
    else
      Swap v_{k-1} and v_k.
    end if
  end while
    return $(v_1, \ldots, v_n)$.

Norm condition: $|\mu_{i,j}| = \frac{v_i.v_j^*}{\|v_j^*\|^2} \leq \frac{1}{2}$ for all $1 \leq j < i \leq n$.

Lovász condition: $\|v_i^*\|^2 \geq (\frac{3}{4} - \mu_{i,i-1}^2)\|v_{i-1}^*\|^2$ for all $1 < i \leq n$.

Theorem 2 *Let B be an LLL-reduced basis for lattice $\mathscr{L}$; then, B solves the SVP
problem with approximation factor $2^{(n-1)/2}$.*

It is important to justify the choice of value $3/4$. If this value were replaced by 1,
we would have a Gauss reduction. However, there is no proof that the algorithm
would end in polynomial time. In fact, any value strictly less than 1 would be
enough. Thus, cryptosystems based on SVP and CVP must have their parameters
well chosen in order to avoid attacks based on the LLL algorithm.

In general, given a basis $(v_1, \ldots, v_n)$, it is possible to obtain a new basis satisfying
the norm condition, just by subtracting multiples of $v_1, \ldots, v_{k-1}$ from v_k, in order
to reduce the absolute value of v_k. If the norm condition is satisfied, we verify if
Lovász condition is also satisfied; if not, the vectors are reordered and the procedure
is repeated, executing the norm reduction again.

5.2 Lattice Based Hash

The first lattice-based cryptosystem was proposed by Ajtai [1]. This work is very
important because it presented a worst-case reduction, in the sense that an attack
to the cryptosystem leads to solutions of hard instances of problems on lattices. In
particular, inverting the hash function has, in average, the same complexity as the
SVP problem on dual lattices in the worst case.

Specifically, given integers n, m, d, q, we build a cryptographic hash family, f_A :
$\{0, \ldots, d - 1\}^m \to \mathbb{Z}_q^n$, indexed by matrix $A \in \mathbb{Z}_q^{n \times m}$. Given a vector y, we have

---

Algorithm 5.3 Ajtai's hash

Require: Integers $n, m, q, d \geq 1$. A matrix A chosen uniformly in $\mathbb{Z}_q^{n \times m}$. A vector $y \in \{0, \ldots, d-1\}^m$.
Ensure: A vector $f(y) \in \mathbb{Z}_q^n$.
    return $f(y) = A.y \pmod{q}$.

---

that $f_A(y) = Ay \pmod{q}$. Algorithm 5.3 describes these operations. A possible parameter choice is $d = 2, q = n^2, m \approx 2n \log q / \log d$, such that we obtain a compression factor of 2.

The scheme's security follows from the fact that if one is able to find a collision $f_A(y) = f_A(y')$, then it is possible to compute a short vector, $y - y'$ in $\mathcal{L}_q^*(A)$.

This proposal is really simple and can be efficiently implemented; however in practice, hash functions are designed in an ad-hoc way, without theoretical guarantees provided by a security proof, so that they are faster than Ajtai's construction. Moreover, if an attacker has access to sufficiently many hash values, then it is possible to recover the fundamental domain of $\mathcal{L}_q^*(A)$, allowing an attacker to compute collisions easily.

In 2011, Stehle and Steinfeld [86] proposed a collision-resistant hash function family with better algorithms, whose construction will be important to digital signature schemes, as we are going show in Sect. 5.4.

5.3 Lattice-Based Encryption

5.3.1 GGH

The GGH cryptosystem [42] allows us to easily understand the use of lattices in public key cryptography. This cryptosystem uses the orthonormality of the basis in the key pair definition. The private key is defined as a basis B_{priv}, formed by almost orthonormal vectors, namely, vectors with Hadamard ratio close to 1.

In general, the cryptosystem works as follows:

- The encryption algorithm adds noise $r \in \mathbb{R}^n$ to the plaintext $m \in \mathcal{L}$, obtaining the ciphertext $c = m + r$;
- The decryption algorithm must be able to remove the inserted noise. Alternatively, it is necessary to solve an instance of the CVP problem.

Figure 17 shows a two-dimensional lattice, with basis given by vectors v_1 and v_2, almost orthogonal. Figure 18 shows a different basis to the same lattice, composed by vectors whose Hadamard ratio is close to zero.

In lattices with high dimension, if basis orthonormality is closer to zero, then the CVP problem becomes harder. Thus, we can define the public key as a basis B_{pub}, such that $\mathcal{H}(B_{\text{pub}})$ is close to zero. Furthermore, if we know the private key

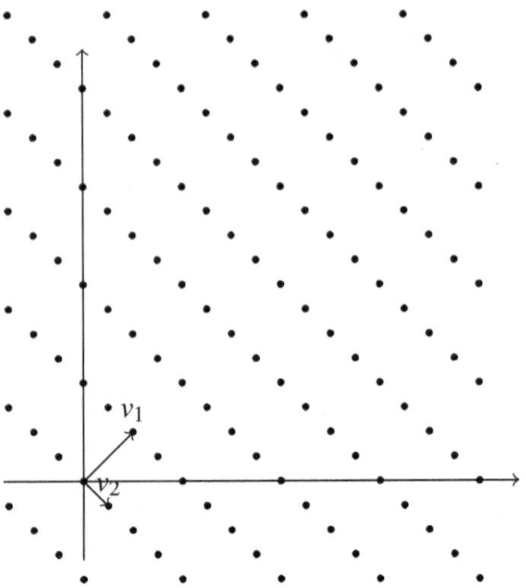

Fig. 17 Good basis

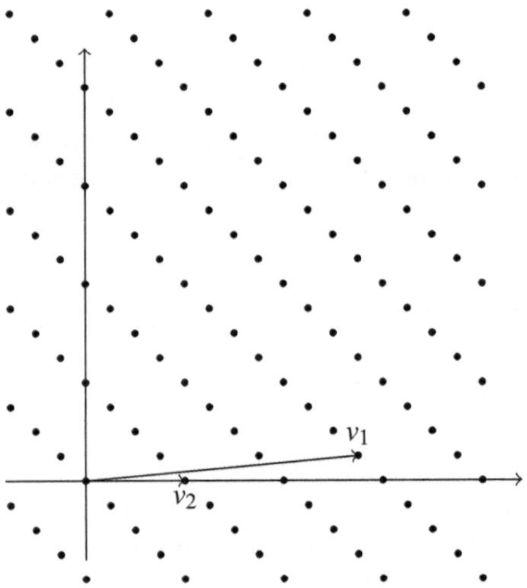

Fig. 18 Bad basis

B_{priv}, then it is possible to use Babai's algorithm [3], defined below, to recover the plaintext.

The general idea of Babai's algorithm is to represent the vector c using private basis B_{priv}, solving the linear system in n equations. As $c \in \mathbb{R}^{n \times n}$, to obtain a lattice point $\mathcal{L}$, each coefficient $t_i \in \mathbb{R}^n$ must be approximated to the nearest integer a_i, where this operation is denoted by $a_i \leftarrow \lfloor t_i \rceil$. This procedure is simple and works very well, since the basis B_{priv} is sufficiently orthonormal, reducing rounding errors.

One way to attack the cryptosystem is trying to reduce the basis B_{pub}, in order to obtain shorter vectors, with Hadamard ratio close to 1. In dimension two the problem can be easily solved using Gauss reduction (algorithm 5.1). For higher dimensions the problem is considered hard, although in 1982 there was a great advance, with

---

Algorithm 5.4 Babai's algorithm

Require: Dimension n lattice $\mathscr{L}$; a vector $c_{B_{\text{pub}}} = (c_1, \ldots, c_n)$, where $c_i \in \mathbb{R}$; and a basis $B_{\text{priv}} = (s_1, \ldots, s_n)$, sufficiently orthonormal.
Ensure: The vector $m \in \mathscr{L}$ that solves the CVP problem with respect to c and $\mathscr{L}$.
Solve the linear system $c = t_1 s_1 + \ldots + t_n s_n$, on variables t_i, for $1 \leq i \leq n$.
for $i = 0$ till $i = n$ **do**
    $a_i \leftarrow \lfloor t_i \rceil$
end for
      return $m \leftarrow a_1 s_1 + \ldots a_n s_n$

---

the invention of LLL algorithm [50]. Thus, the cryptosystem parameters must be designed to resist LLL basis reduction.

5.3.2 NTRU

The NTRU cryptosystem [45] is originally constructed over polynomial rings but can also be defined over lattices, because the underlying problem can be interpreted as being SVP and CVP. Hence, the solution of these problems would mean an attack to the cryptosystem if parameters are not carefully chosen.

The cryptosystem uses the following polynomial rings: $R = \mathbb{Z}[x]/(x^N - 1)$, $R_p = (\mathbb{Z}/p\mathbb{Z})[x]/(x^N - 1)$ and $R_q = (\mathbb{Z}/q\mathbb{Z})[x]/(x^N - 1)$, where N, p, q are positive integers.

Definition 18 Given positive integers d_1 and d_2, we define $\mathscr{T}(d_1, d_2)$ as the class of polynomials that have d_1 coefficients equal to 1, d_2 coefficients equal to -1 and the remaining coefficients equal to zero. These polynomials are called *ternary polynomials*.

The parameters are given by (N, p, q, d), where N and p are prime numbers, $(p, q) = (N, q) = 1$ and $q > (6d + 1)p$. The private key corresponds to the polynomials $f(x) \in \mathscr{T}(d + 1, d)$ and $g(x) \in \mathscr{T}(d, d)$. Public key is given by polynomial $h(x) \equiv F_q(x).g(x)$, where $F_q(x)$ is the multiplicative inverse of $f(x)$ in R_q.

Given message $m(x) \in R$, with coefficients in the interval $[-p/2, p/2]$, $r(x)$ is randomly chosen and the ciphertext is computed by $e(x) \equiv ph(x).r(x) + m(x)$ (mod q).

To decrypt, we first compute the function $a(x) \equiv f(x).e(x)$ (mod q), such that its coefficients are in the interval $[-q/2, q/2]$. The message $m(x)$ is obtained by computing $m(x) \equiv F_p(x).a(x)$ (mod p):

- **KeyGen.** Choose $f \in \mathscr{T}(d + 1, d)$ such that f has inverse in R_q and R_p. Choose also $g \in \mathscr{T}(d, d)$. Compute F_q as the inverse of f in R_q and, analogously, F_p is the inverse of f in R_p. The public key is given by $h = F_q.g$.
- **Encrypt.** Given plaintext $m \in R_p$, choose randomly $r \in \mathscr{T}(d, d)$ and compute $e \equiv pr.h + m$ (mod q), where h is a public key.

- **Decrypt.** Compute $a = \lfloor f.e \rceil_q \equiv \lfloor pg.r + f.m \rceil_q$. Finally, the message can be obtained computing $m \equiv F_p.a \pmod{p}$.

5.3.3   LWE (Learning with Errors)-Based Encryption

In this section, we are going to present a cryptosystem based on the LWE problem, that is, an efficient proposal with security proof based on worst-case problems over lattices [80]. This proof was a quantum reduction; in other words, it shows that a cryptosystem vulnerability implies the existence of a quantum algorithm to solve hard problems over lattices. In 2009, Peikert gave a classical reduction for the security proof [73].

Definition 19 The *LWE problem* consists in finding the vector $s \in \mathbb{Z}_q^n$, given the equations $\langle s, a_i \rangle + e_i = b_i \pmod{q}$, for $1 \leq i \leq n$. The values e_i are small errors that were inserted according to the distribution $\mathscr{D}$, generally taken as a Gaussian distribution.

In 2010, Lyubaskevsky, Peikert, and Regev used polynomial rings to define the scheme RLWE [52]. Let $f(x) = x^d + 1$, where d is a power of 2. Given the integer q and an element $s \in R_q = \mathbb{Z}_q[x]/f(x)$, the *ring-LWE problem* over R_q, with respect to the distribution $\mathscr{D}$, is defined as that of finding s satisfying equations $s.a_i + e_i = b_i \pmod{R_q}$, for $1 \leq i \leq n$, such that a_i and b_i are elements of R_q, and modular reduction on R_q is the same as reducing by the polynomial modulo $f(x)$ and its coefficients modulo q. The LWE based cryptosystem problem can be constructed as follows:

- **KeyGen.** Choose randomly $a \in R_q$ and generate s and e in R using distribution $\mathscr{D}$. The private key is given by s, while the public key is given by $(a, b = a.s+e)$.
- **Encrypt.** To encrypt d bits, it is possible to interpret these bits as R polynomial coefficients. The encryption algorithm then chooses $r, e_1, e_2 \in R$, using the same distribution $\mathscr{D}$ and computes (u, v) in the following way:

$$u = a.r + e_1 \pmod{q},$$

$$v = b.r + e_2 + \lfloor q/2 \rfloor.z \pmod{q}.$$

- **Decrypt.** To decrypt, the algorithm computes

$$v - u.s = (r.e - s.e_1 + e_2) + \lfloor q/2 \rfloor.z \pmod{q}.$$

According to the parameters choice, we have that $(r.e - s.e_1 + e_2)$ has maximum size $q/4$, such that each plaintext bit can be computed verifying each coefficient from the obtained result. If the coefficient is closer to 0 than to $q/2$, then the corresponding bit is 0; otherwise it is 1.

Some concepts in this section, for example, the cyclotomic polynomial ring and the Gaussian distribution $\mathscr{D}$, were recently incorporated to the NTRU scheme, allowing us to construct a semantically secure scheme efficient for lattice-based encryption [86], whose public and private keys and encryption and decryption algorithms have complexity $\tilde{O}(\lambda)$, such that λ is the security parameter.

5.3.4 Homomorphic Encryption

In 2009, Gentry proposed the construction of a fully homomorphic encryption scheme [37], solving a problem open since 1978, when Rivest, Adleman, and Dertouzos conjectured the existence of privacy homomorphisms [62], such that the encryption function is also an algebraic homomorphism. In other words, it is possible to add and multiply encrypted texts, so that when decrypted we obtain the corresponding result with respect to the same operation, executed using corresponding plaintexts.

If the plaintext space is given by $\{0, 1\}$, then bit addition is equivalent to logic exclusive disjunction, while multiplication is equivalent to logic conjunction. Hence, it is possible to compute any Boolean circuit over encrypted data, which implies that we can evaluate any algorithm homomorphically with encrypted arguments, obtaining an encrypted output.

Using homomorphic encryption it is possible to delegate algorithm computation to a server, maintaining input confidentiality. This is interesting for cloud computing, because it allows the construction of applications such as encrypted databases, encrypted disks, encrypted search engines, etc.

The computational complexity of performing homomorphic encryption is, nevertheless, still a hindrance for its practical utilization. Recently, Brakerski proposed the use of the LWE problem to construct fully homomorphic encryption [19], reducing the algorithms' complexities and achieving polylogarithmic complexity per operation. Brakerski used a new way to manage noise growth, which allowed us to execute a greater number of multiplications. In particular, he proposed a modulo reduction algorithm, which implicitly reduced the noise growth rate. An algorithm called dimension reduction allows us to replace the bootstrapping procedure by a new method (similar in many aspects), which leads to better parameters. Nevertheless, even considering recent optimizations, homomorphic encryption remains not practical.

5.4 Digital Signatures

GGH and NTRU cryptosystems can be converted into digital signature schemes [12]. However, such proposals do not have a security proof and, in fact, there are attacks that allow us to recover the private key given a sufficiently large number of signatures [63].

In 2007, Gentry, Peikert, and Vaikuntanathan [39] created a new kind of trapdoor function f with an extra property: an efficient algorithm that, using the trapdoor, samples elements from the preimage of f. A composition of Gaussians is used to obtain a point close to a lattice vector. This distribution has standard deviation greater than the basis vector within maximum norm, such that reduction by fundamental parallelepiped has indistinguishable distribution from uniform. Furthermore, this construction does not reveal the lattice geometry, because the normal distribution is spherical. Given message m and a hash function H that maps plaintexts to the preimage of f, we compute the point $y = H(m)$. The signature is given by $\delta = f^{-1}(y)$. To verify the signature, we compute $f(\delta) = H(m)$. This kind of construction was proposed by Bellare and Rogaway [7], using trapdoor permutations and modeling H as a random oracle. Therefore, a digital signature scheme is obtained, with existential unforgeability under adaptive chosen plaintext attack. We can use a Gaussian to generate the noise e, such that $f(e) = y$ and $y = v + e$, for a point v chosen uniformly in the lattice. Thus, the construction has a security proof based on lattices worst-case problems.

This construction can be viewed with respect to two functions: $f_A(x) = Ax$ (mod q), Ajtai's construction, and $g_A(s, e) = A^T s + e$, LWE problem, such that the first function is surjective and the second is injective. In 2012, Micciancio and Peikert [58] showed a simple, secure, and efficient way to invert g_A and sample from the preimage of f_A, allowing the construction of an efficient digital signature scheme. In this proposal, the Gaussian composition allowed parallelism (in later work [39], and subsequent proposals [86], it was inherently sequential), leading to a concrete improvement. The optimizations described above can be used in applications based on function g_A or sampling from preimage of f_A; hence, it is not only important for digital signatures, but also to secure encryption construction in the adaptive chosen ciphertext attack.

5.5   Other Applications

Lattice-based cryptography is interesting not only because it resists to quantum attacks but also because it is a flexible alternative to the construction of cryptosystems. In particular, the ring-LWE problem has become more and more important, as it allows us to construct stronger trapdoor functions, with better parameters for both security and performance [58].

Gentry [38] analyzed how flexible a cryptosystem can be, considering not just fully homomorphic encryption, but also with respect to access control. Thus, lattice-based cryptography seems to be, according to Gentry, a feasible alternative to explore the limits of possible applications with cryptography. Among other applications, we emphasize the following:

- **multilinear maps**. Bilinear pairings can be used in different contexts, as, for example, in identity-based encryption. The generalization of this concept, called multilinear maps, is very useful and, although no proposal appeared for a while, many applications were suggested. Using the noise concept, also present in homomorphic encryption, Garg, Gentry, and Halevi achieved the construction of multilinear maps [34];

- **identity-based encryption**. For some time, identity-based encryption was only achievable by using bilinear pairings. Using lattices, many proposals were put forward [17, 39], built upon the dual scheme $\mathscr{E}$, composed by algorithms {DualKeyGen, DualEnc, DualDec}, as pointed out in Sect. 5.3.3. Specifically, DualKeyGen computes the private key as the error e, chosen using the Gaussian distribution, while the public key is given by $u = f_A(e)$. To encrypt a bit b, the algorithm DualEnc randomly chooses s, chooses x and e' according to the Gaussian, and computes $c_1 = g_A(s, x)$ e $c_2 = u^T s + e' + b.\lfloor q/2 \rfloor$. The ciphertext is $\langle c_1, c_2 \rangle$. Finally, DualDec computes $b = c_2 - e^T c_1$. Then, given the hash function H, modeled as a random oracle mapping identities to public keys of the dual cryptosystem, the identity-based encryption scheme was constructed as follows:

  - **Setup**. Choose the public key $A \in \mathbb{Z}_q^{n \times m}$ and the master key as the trapdoor s, according to the description in Sect. 5.4;
  - **Extraction**. Given the identity id, we compute $u = H(\text{id})$ and the decryption key $e = f^{-1}(u)$, using the trapdoor preimage sampling algorithm with trapdoor s;
  - **Encrypt**. Given bit b, return $\langle c_1, c_2 \rangle = \text{DualEnc}(u, b)$;
  - **Decrypt**. Return DualDec($e, \langle c_1, c_2 \rangle$).

- **functional encryption**. Functional encryption is a new primitive in cryptography, that opens new horizons [51]. In this system, a public function $f(x, y)$ determines what the user that knows the key y can infer from a ciphertext, denoted by c_x, according to parameter x. Within this model, the one who encrypts a message m can previously choose what kind of information is obtained after decryption. Moreover, a trusted party is responsible for key s_y generation, which can be used to decrypt c_x, returned as output for $f(x, y)$, without necessarily revealing information about m. With this approach it is possible to define an identity-based encryption scheme as a functional encryption special case, such that $x = (m, \text{id})$ and $f(x, y) = m$ if and only if $y = \text{id}$. A recent result [35] proposes the construction of a functional encryption scheme based on lattices, being able to deal with any polynomial-size Boolean circuit;

- **attribute-based encryption**. This is a functional encryption special case, such that $x = (m, \phi)$ and $f(x, y) = m$ if and only if $\phi(y) = 1$. Namely, the decryption works since y, the decryptor's attribute, satisfies the predicate ϕ, such that the encryptor can determine a access control policy (predicate ϕ) for the cryptosystem. There are proposals to achieve this kind of operations based on the LWE problem [82], and the multilinear map construction mentioned above has been used by Sahai and Waters [40] to propose an attributed-based scheme for any Boolean circuit, showing one more time the versatility of lattice-based cryptography;

- **obfuscation**. There is a negative result proving that obfuscation is impossible in a certain security model. However, it was recently proposed the construction of *indistinguishability obfuscation* which, in a different security model, is proved to be the best possible approach. The LWE problem was used to construct this kind of primitive [35], being part of a functional encryption construction. Such schemes, therefore, although versatile, are relevant mostly for their theoretical importance rather than for their practical applications.

Concluding Remarks

As we have seen, not all.is lost for the deployment of efficient and flexible cryptosystems in a scenario where large quantum computers are a technological reality. Many proposals have already attained a fairly good level of maturity, and one can even discern some patterns in schemes based on different underlying security assumptions, in the sense of there existing strikingly similar schemes based on codes, lattices, $\mathcal{MQ}$ systems and sometimes even hash functions. Determining how far the analogies can go (and why) is an interesting line for future investigation.

At the same time, practical considerations are ever more often being addressed in the literature, as they are as important as theoretical ones in a truly post-quantum scenario where conventional systems would have to be replaced.

The fact that post-quantum schemes can also provide functionalities not available elsewhere has already been, and is likely to continue to be, a strong additional motivation for further research in the area.

Acknowledgements Paulo S. L. M. Barreto, Ricardo Dahab and Julio López acknowledge support by the Brazilian National Council for Scientific and Technological Development (CNPq) research productivity grants 306935/2012-0, 311530/2011-7, and 309258/2011-1, respectively.

References

1. M. Ajtai, Generating hard instances of lattice problems (extended abstract), in *Proceedings of the Twenty-Eighth Annual ACM Symposium on Theory of Computing*, STOC '96 (ACM, New York, 1996), pp. 99–108
2. M. Alabbadi, S.B. Wicker, A digital signature scheme based on linear error-correcting block codes, in *Advances in Cryptology – Asiacrypt '94*, vol. 917 of *Lecture Notes in Computer Science* (Springer, New York, 1994), pp. 238–348
3. L Babai, On lovsz lattice reduction and the nearest lattice point problem. Combinatorica **6**(1), 1–13 (1986)
4. M. Baldi, F. Chiaraluce, Cryptanalysis of a new instance of McEliece cryptosystem based on QC-LDPC code, in *IEEE International Symposium on Information Theory – ISIT 2007* (IEEE, Nice, 2007), pp. 2591–2595
5. M. Baldi, F. Chiaraluce, M. Bodrato, A new analysis of the McEliece cryptosystem based on QC-LDPC codes, in *Security and Cryptography for Networks – SCN 2008*, vol. 5229 of *Lecture Notes in Computer Science* (Springer, Amalfi, 2008), pp. 246–262

6. R. Barbulescu, P. Gaudry, A. Joux, E. Thomé, A quasi-polynomial algorithm for discrete logarithm in finite fields of small characteristic. HAL-INRIA technical report, http://hal.inria.fr/hal-00835446/ (2013)

7. M. Bellare, P. Rogaway, Random oracles are practical: A paradigm for designing efficient protocols, in *Proceedings of the 1st ACM conference on Computer and communications security* (ACM, 1993), pp. 62–73

8. T.P. Berger, P.-L. Cayrel, P. Gaborit, A. Otmani, Reducing key length of the McEliece cryptosystem, in *Progress in Cryptology – Africacrypt 2009*, Lecture Notes in Computer Science (Springer, Gammarth, 2009), pp. 77–97

9. E. Berlekamp, R. McEliece, H. van Tilborg, On the inherent intractability of certain coding problems. IEEE Trans. Inf. Theory **24**(3), 384–386 (1978)

10. D. Bernstein, T. Lange, C. Peters, Smaller decoding exponents: ball-collision decoding, in *Advances in Cryptology – Crypto 2011*, vol. 6841 of *Lecture Notes in Computer Science* (Springer, Santa Barbara, 2011), pp. 743–760

11. D.J. Bernstein, List decoding for binary Goppa codes, in *Coding and Cryptology—Third International Workshop, IWCC 2011*, Lecture Notes in Computer Science (Springer, Qingdao, 2011), pp. 62–80

12. D.J. Bernstein, J. Buchmann, E. Dahmen, *Post-Quantum Cryptography* (Springer, Heidelberg, 2008)

13. D.J. Bernstein, T. Lange, C. Peters, Attacking and defending the McEliece cryptosystem, in *Post-Quantum Cryptography – PQCrypto 2008*, vol. 5299 of *Lecture Notes in Computer Science* (Springer, New York, 2008), pp. 31–46. http://www.springerlink.com/content/68v69185x478p53g

14. D.J. Bernstein, T. Lange, C. Peters, Wild McEliece, in *Selected Areas in Cryptography – SAC 2010*, vol. 6544 of *Lecture Notes in Computer Science* (Springer, Waterloo, 2010), pp. 143–158

15. G. Bertoni, J. Daemen, M. Peeters, G. Van Assche, Keccak specifications. Submission to NIST (2010). http://keccak.noekeon.org/Keccak-specifications.pdf

16. G. Bertoni, J. Daemen, M. Peeters, G. Van Assche, Sponge functions. ECRYPT Hash Workshop 2007 (2007). Also available as public comment to NIST from http://www.csrc.nist.gov/pki/HashWorkshop/Public_Comments/2007_May.html

17. D. Boneh, C. Gentry, M. Hamburg, Space-efficient identity based encryption without pairings, in *FOCS*, pp. 647–657 (2007)

18. A. Braeken, C. Wolf, B. Preneel, A study of the security of unbalanced oil and vinegar signature schemes, in *Topics in Cryptology – CT-RSA 2005*, vol. 3376 of *Lecture Notes in Computer Science* (Springer, New York, 2005), pp. 29–43

19. Z. Brakerski, V. Vaikuntanathan, Efficient fully homomorphic encryption from (standard) lwe. Electron. Colloq. Comput. Complex. **18**, 109 (2011)

20. J. Buchmann, C. Coronado, E. Dahmen, M. Dring, E. Klintsevich, CMSS – an improved merkle signature scheme, in *Progress in Cryptology INDOCRYPT 2006*, vol. 4329 of *Lecture Notes in Computer Science* (Springer, New York, 2006), pp. 349–363

21. J. Buchmann, E. Dahmen, S. Ereth, A. Hlsing, M. Rckert, On the security of the Winternitz one-time signature scheme, in *Progress in Cryptology – AFRICACRYPT 2011*, vol. 6737 of *Lecture Notes in Computer Science* (Springer, New York, 2011), pp. 363–378

22. J. Buchmann, E. Dahmen, A. Hlsing, XMSS-a practical secure signature scheme based on minimal security assumptions, in *Cryptology ePrint Archive - Report 2011/484*. ePrint (2011)

23. J. Buchmann, E. Dahmen, E. Klintsevich, K. Okeya, C. Vuillaume, Merkle signatures with virtually unlimited signature capacity, in *Applied Cryptography and Network Security – ACNS 2007*, vol. 4521 of *Lecture Notes in Computer Science* (Springer, New York, 2007), pp. 31–45

24. J. Buchmann, E. Dahmen, M. Schneider, Merkle tree traversal revisited, in *Post-Quantum Cryptography – PQCrypto 2008*, vol. 5299 of *Lecture Notes in Computer Science* (Springer, New York, 2008), pp. 63–78

25. S. Contini, A.K. Lenstra, R. Steinfeld, VSH, an Efficient and Provable Collision Resistant Hash Function. Cryptology ePrint Archive, Report 2005/193 (2005). http://eprint.iacr.org/

26. N. Courtois, M. Finiasz, N. Sendrier, How to achieve a McEliece-based digital signature scheme, in *Advances in Cryptology – Asiacrypt 2001*, vol. 2248 of *Lecture Notes in Computer Science* (Springer, Gold Coast, 2001), pp. 157–174
27. R.A. DeMillo, D.P. Dobkin, A.K. Jones, R.J. Lipton, *Foundations of Secure Computation* (Academic Press, New York, 1978)
28. J. Ding, D. Schmidt, Rainbow, a new multivariable polynomial signature scheme, in *International Conference on Applied Cryptography and Network Security – ACNS 2005*, vol. 3531 of *Lecture Notes in Computer Science* (Springer, New York, 2005), pp. 164–175
29. C. Dods, N. Smart, M. Stam, Hash based digital signature schemes, in *Cryptography and Coding*, vol. 3796 of *Lecture Notes in Computer Science* (Springer, New York, 2005), pp. 96–115
30. J.-C. Faugère, A. Otmani, L. Perret, J.-P. Tilllich, Algebraic cryptanalysis of McEliece variants with compact keys, in *Advances in Cryptology – Eurocrypt 2010*, vol. 6110 of *Lecture Notes in Computer Science* (Springer, Nice, 2010), pp. 279–298
31. P. Gaborit, Shorter keys for code based cryptography, in *International Workshop on Coding and Cryptography – WCC 2005* (ACM Press, Bergen, 2005), pp. 81–91
32. R.G. Gallager, Low-density parity-check codes. Information Theory, IRE Transactions on **8**(1), 21–28 (1962)
33. M.R. Garey, D.S. Johnson, *Computers and Intractability – A Guide to the Theory of NP-Completeness* (W. H. Freeman and Company, New York, 1979)
34. S. Garg, C. Gentry, S. Halevi, Candidate multilinear maps from ideal lattices, in *Advances in Cryptology – EUROCRYPT 2013*, pp. 1–17 (2013)
35. S. Garg, C. Gentry, S. Halevi, M. Raykova, A. Sahai, B. Waters, Candidate indistinguishability obfuscation and functional encryption for all circuits, IACR Cryptology ePrint Archive **2013**, 451 (2013)
36. V. Gauthier, G. Leander, Practical key recovery attacks on two McEliece variants, in *International Conference on Symbolic Computation and Cryptography – SCC 2010* (Springer, Egham, 2010)
37. C. Gentry, *A fully homomorphic encryption scheme*. PhD thesis, Stanford University, 2009. crypto.stanford.edu/craig
38. C. Gentry, Encrypted messages from the heights of cryptomania, in *TCC*, pp. 120–121 (2013)
39. C. Gentry, C. Peikert, V. Vaikuntanathan, Trapdoors for hard lattices and new cryptographic constructions, in *Proceedings of the 40th Annual ACM Symposium on Theory of Computing, STOC '08* (ACM, New York, 2008), pp. 197–206
40. C. Gentry, A. Sahai, B. Waters, Homomorphic encryption from learning with errors: Conceptually-simpler, asymptotically-faster, attribute-based, in *Advances in Cryptology – CRYPTO '89*, vol. 8042 of *Lecture Notes in Computer Science* (Springer, New York, 2013), pp. 75–92
41. J.K. Gibson, The security of the Gabidulin public key cryptosystem, in *Advances in Cryptology – Eurocrypt '96*, vol. 1070 of *Lecture Notes in Computer Science* (Springer, Zaragoza, 1996), pp. 212–223
42. O. Goldreich, S. Goldwasser, S. Halevi, Public-key cryptosystems from lattice reduction problems, in *Advances in Cryptology – CRYPTO '97*, vol. 1294 of *Lecture Notes in Computer Science* (Springer, New York, 1997), pp. 112–131
43. V.D. Goppa, A new class of linear error correcting codes. Problemy Peredachi Informatsii **6**, 24–30 (1970)
44. A. Hülsing, *Practical forward secure signatures using minimal security assumptions*. PhD thesis, TU Darmstadt, 2013
45. J. Hoffstein, J. Pipher, J.H. Silverman, Ntru: A ring-based public key cryptosystem, in *Lecture Notes in Computer Science* (Springer, New York, 1998), pp. 267–288
46. W.C. Huffman, V. Pless, *Fundamentals of Error-Correcting Codes* (Cambridge University Press, Cambridge, 2003)
47. A. Kipnis, A. Shamir, Cryptanalysis of the oil and vinegar signature scheme, in ed. by H. Krawczyk. *Advances in Cryptology – Crypto 1998*, vol. 1462 of *Lecture Notes in Computer Science* (Springer, New York, 1998), pp. 257–266

48. A. Kipnis, J. Patarin, L. Goubin, Unbalanced oil and vinegar signature schemes, in ed. by J. Stern. *Advances in Cryptology – EUROCRYPT '99*, vol. 1592 of *Lecture Notes in Computer Science* (Springer, New York, 1999), pp. 206–222

49. L. Lamport, Constructing digital signatures from a one way function, in *SRI International*. CSL-98 (1979)

50. A.K. Lenstra, H.W. Lenstra, L. Lovsz, Factoring polynomials with rational coefficients. Math. Ann. **261**(4), 515–534 (1982)

51. A. Lewko, T. Okamoto, A. Sahai, K. Takashima, B. Waters, Fully secure functional encryption: Attribute-based encryption and (hierarchical) inner product encryption, in H. Gilbert. *Advances in Cryptology – EUROCRYPT 2010*, vol. 6110 of *Lecture Notes in Computer Science* (Springer, Berlin/Heidelberg, 2010), pp. 62–91

52. V. Lyubashevsky, C. Peikert, O. Regev, On ideal lattices and learning with errors over rings. Adv. Cryptology EUROCRYPT 2010 **6110/2010**(015848), 1–23 (2010)

53. F.J. MacWilliams, N.J.A. Sloane, *The Theory of Error-Correcting Codes*, vol. 16 (North-Holland Mathematical Library, Amsterdam, 1977)

54. S.M. Matyas, C.H. Meyer, J. Oseas, Generating strong one-way functions with cryptographic algorithm, IBM Techn. Disclosure Bull., 1985

55. R. McEliece, A public-key cryptosystem based on algebraic coding theory. The Deep Space Network Progress Report, DSN PR 42–44, 1978. http://ipnpr.jpl.nasa.gov/progressreport2/42-44/44N.PDF. Acesso em:.

56. R.C. Merkle, Secrecy, Authentication, and Public Key Systems. Stanford Ph.D. thesis, 1979

57. R.C. Merkle, A digital signature based on a conventional encryption function, in *Advances in Cryptology – CRYPTO '87*, vol. 435 of *Lecture Notes in Computer Science* (Springer, New York, 1987), pp. 369–378

58. D. Micciancio, C. Peikert, Trapdoors for lattices: Simpler, tighter, faster, smaller, in ed. by D. Pointcheval, T. Johansson. *Advances in Cryptology EUROCRYPT 2012*, vol. 7237 of *Lecture Notes in Computer Science* (Springer, Berlin/Heidelberg, 2012), pp. 700–718

59. V.S. Miller, Use of elliptic curves in cryptography, in *Advances in Cryptology — Crypto '85* (Springer, New York, 1986), pp. 417–426

60. R. Misoczki, N. Sendrier, J.-P. Tilllich, P.S.L.M. Barreto, MDPC-McEliece: New McEliece variants from moderate density parity-check codes. Cryptology ePrint Archive, Report 2012/409, 2012. http://eprint.iacr.org/2012/409

61. C. Monico, J. Rosenthal, A. Shokrollahi, Using low density parity check codes in the McEliece cryptosystem, in *IEEE International Symposium on Information Theory – ISIT 2000* (IEEE, Sorrento, 2000), p. 215

62. E.M. Morais, R. Dahab, Encriptao homomrfica, in *XII Simpsio Brasileiro em Segurana da Informao e de Sistemas Computacionais: Minicursos*, SBSeg (2012)

63. P. Nguyen, O. Regev, Learning a parallelepiped: Cryptanalysis of ggh and ntru signatures, in S. Vaudenay. *Advances in Cryptology - EUROCRYPT 2006*, vol. 4004 of *Lecture Notes in Computer Science* (Springer, Berlin/Heidelberg, 2006), pp. 271–288

64. H. Niederreiter, Knapsack-type cryptosystems and algebraic coding theory. Prob. Control Inf. Theory **15**(2), 159–166 (1986)

65. NIST, *Federal Information Processing Standard FIPS 186-3 – Digital Signature Standard (DSS) – 6. The Elliptic Curve Digital Signature Algorithm (ECDSA)* (National Institute of Standards and Technology (NIST), Gaithersburg, 2012). http://csrc.nist.gov/publications/fips/fips186-3/fips_186-3.pdf

66. A. K. D. S. Oliveira, J. López. Implementação em software do Esquema de Assinatura Digital de Merkle e suas variantes, in *Brazilian Symposium on Information and Computer Systems Security – SBSeg 2013* (SBC, 2013)

67. A. Otmani, J.-P. Tillich, L. Dallot, Cryptanalysis of two McEliece cryptosystems based on quasi-cyclic codes. Math. Comput. Sci. **3**(2), 129–140 (2010)

68. J. Patarin, The oil and vinegar signature scheme, in *Dagstuhl Workshop on Cryptography* (1997). Transparencies

69. J. Patarin, L. Goubin, Trapdoor one-way permutations and multivariate polynomials, in *ICICS'97*, vol. 1334 of *Lecture Notes in Computer Science* (Springer, New York, 1997), pp. 356–368

70. J. Patarin, Hidden fields equations (hfe) and isomorphisms of polynomials (ip): Two new families of asymmetric algorithms, in ed. by U. Maurer. *Advances in Cryptology – EURO-CRYPT '96*, vol. 1070 of *Lecture Notes in Computer Science* (Springer, Berlin/Heidelberg, 1996), pp. 33–48

71. J. Patarin, L. Goubin, N. Courtois, Improved algorithms for isomorphisms of polynomials, in *Advances in Cryptology – EUROCRYPT '98* (Springer, New York, 1998), pp. 184–200

72. N.J. Patterson, The algebraic decoding of Goppa codes. IEEE Trans. Inf. Theory **21**(2), 203–207 (1975)

73. C. Peikert, Public-key cryptosystems from the worst-case shortest vector problem: extended abstract, in *Proceedings of the 41st Annual ACM Symposium on Theory of Computing, STOC '09* (ACM, New York, 2009), pp. 333–342

74. A. Petzoldt, S. Bulygin, J. Buchmann, CyclicRainbow – a multivariate signature scheme with a partially cyclic public key, in ed. by G. Gong, K. Gupta. *Progress in Cryptology – Indocrypt 2010*, vol. 6498 of *Lecture Notes in Computer Science* (Springer, Berlin/Heidelberg, 2010), pp. 33–48

75. A. Petzoldt, S. Bulygin, J. Buchmann, Selecting parameters for the Rainbow signature scheme, in ed. by N. Sendrier *Post-Quantum Cryptography – PQCrypto 2010*, vol. 6061 of *Lecture Notes in Computer Science* (Springer, Berlin/Heidelberg, 2010), pp. 218–240. Extended Version: http://eprint.iacr.org/2010/437

76. A. Petzoldt, S. Bulygin, J. Buchmann, Linear recurring sequences for the UOV key generation, in *International Conference on Practice and Theory in Public Key Cryptography – PKC 2011*, vol. 6571 of *Lecture Notes in Computer Science* (Springer, Berlin/Heidelberg, 2011), pp. 335–350

77. A. Petzoldt, S. Bulygin, J. Buchmann, Cyclicrainbow - a multivariate signature scheme with a partially cyclic public key, in ed. by G. Gong, K.C. Gupta. *INDOCRYPT*, volume 6498 of *Lecture Notes in Computer Science* (Springer, New York, 2010), pp. 33–48

78. B. Preneel, Analysis and design of cryptographic hash functions. PhD thesis, Katholieke Universiteit Leuven, 1983

79. L. Rausch, A. Hlsing, J. Buchmann, Optimal parameters for $xmss^{MT}$, in *CD-ARES 2013*, vol. 8128 of *Lecture Notes in Computer Science* (Springer, New York, 2013), pp. 194–208

80. O. Regev, The learning with errors problem (invited survey), in *IEEE Conference on Computational Complexity* (IEEE Computer Society, Washington, DC, 2010), pp. 191–204

81. R.L. Rivest, A. Shamir, L. Adleman, A method for obtaining digital signatures and public-key cryptosystems. Commun. ACM **21**, 120–126 (1978)

82. A. Sahai, B. Waters, Attribute-based encryption for circuits from multilinear maps. CoRR, abs/1210.5287 (2012)

83. N. Sendrier, Decoding one out of many, in ed. by B-Y. Yang. *Post-Quantum Cryptography – PQCrypto 2011*, vol. 7071 of *Lecture Notes in Computer Science* (Springer, Berlin/Heidelberg, 2011), pp. 51–67. 10.1007/978-3-642-25405-5-4

84. P.W. Shor, Polynomial-time algorithms for prime factorization and discrete logarithms on a quantum computer. SIAM J. Comput. **26**, 1484–1509 (1997)

85. A. Shoufan, N. Huber, H. Molter, A novel cryptoprocessor architecture for chained merkle signature scheme, in *Microprocessors and Microsystems* (Elsevier, Amsterdam, 2011), pp. 34–47

86. D. Stehlé, R. Steinfeld, Making ntru as secure as worst-case problems over ideal lattices, in *Proceedings of the 30th Annual International Conference on Theory and Applications of Cryptographic Techniques: Advances in Cryptology*, EUROCRYPT'11 (Springer, Berlin, Heidelberg, 2011), pp. 27–47

87. J. Stern, A method for finding codewords of small weight. Coding Theory Appl. **388**, 106–133 (1989)

88. J. Stern, Can one design a signature scheme based on error-correcting codes? in *Advances in Cryptology – ASIACRYPT'94*, vol. 917 of *Lecture Notes in Computer Science* (Springer, New York, 1994), pp. 426–428

89. M. Szydlo, Merkle tree traversal in log space and time, in *Advances in Cryptology – Eurocrypt 2004*, vol. 3027 of *Lecture Notes in Computer Science* (Springer, New York, 2004), pp. 541–554

90. R.M. Tanner, Spectral graphs for quasi-cyclic LDPC codes, in *IEEE International Symposium on Information Theory – ISIT 2001* (IEEE, Washington, DC, 2001), p. 226

91. E. Thomae, A generalization of the Rainbow band separation attack and its applications to multivariate schemes. Cryptology ePrint Archive, Report 2012/223, 2012. http://eprint.iacr.org/2012/223.

92. C. Wieschebrink, Two NP-complete problems in coding theory with an application in code based cryptography, in *IEEE International Symposium on Information Theory – ISIT 2006* (IEEE, Seattle, 2006), pp. 1733–1737

93. R.S. Winternitz, Producing a one-way hash function from DES, in *Advances in Cryptology – CRYPTO '83* (Springer, New York, 1983), pp. 203–207

94. C. Wolf, B. Preneel, Taxonomy of public key schemes based on the problem of multivariate quadratic equations. IACR Cryptology ePrint Archive **2005**, 77 (2005)

95. T. Yasuda, K Sakurai, T. Takagi, Reducing the key size of Rainbow using non-commutative rings, in *Topics in Cryptology – CT-RSA 2012*, vol. 7178 of *Lecture Notes in Computer Science* (Springer, New York, 2012), pp. 68–83